The Carcinogenicity of Metals
Human Risk through Occupational and Environmental Exposure

Issues in Toxicology

Series Editors:

Titles in the Series:

1: Hair in Toxicology: An Important Bio-Monitor
2: Male-mediated Developmental Toxicity
3: Cytochrome P450: Role in the Metabolism and Toxicity of Drugs and other Xenobiotics
4: Bile Acids: Toxicology and Bioactivity
5: The Comet Assay in Toxicology
6: Silver in Healthcare
7: *In Silico* Toxicology: Principles and Applications
8: Environmental Cardiology
9: Biomarkers and Human Biomonitoring, Volume 1: Ongoing Programs and Exposures
10: Biomarkers and Human Biomonitoring, Volume 2: Selected Biomarkers of Current Interest
11: Hormone-Disruptive Chemical Contaminants in Food
12: Mammalian Toxicology of Insecticides
13: The Cellular Response to the Genotoxic Insult: The Question of Threshold for Genotoxic Carcinogens
14: Toxicological Effects of Veterinary Medicinal Products in Humans: Volume 1
15: Toxicological Effects of Veterinary Medicinal Products in Humans: Volume 2
16: Aging and Vulnerability to Environmental Chemicals: Age-related Disorders and their Origins in Environmental Exposures
17: Chemical Toxicity Prediction: Category Formation and Read-Across
18: The Carcinogenicity of Metals: Human Risk through Occupational and Environmental Exposure

How to obtain future titles on publication:
A standing order plan is available for this series. A standing order will bring delivery of each new volume immediately on publication.

For further information please contact:
Book Sales Department, Royal Society of Chemistry, Thomas Graham House, Science Park, Milton Road, Cambridge, CB4 0WF, UK
Telephone: +44 (0)1223 420066, Fax: +44 (0)1223 420247
Email: booksales@rsc.org
Visit our website at www.rsc.org/books

Preface

Metals have played a decisive role in the development of human civilisations from earliest times. They have shaped developments in engineering, science and medicine and in the past century many Nobel Laureates were founded on research using metals. Marie Curie focussed her early research on the magnetic properties of steel but was later to perform fundamental research into the radioactivity of metals and the properties of uranium. Paul Ehrlich made notable contributions to the understanding of cancer and introduced the arsenic-related therapy Salvarsan as one of the first effective cures for syphilis and other infections prevalent at the time. Countless other memorable contributions could be included, but whilst we accept the value of metals and metalloid elements in industry and medicine, since the 1950s at least, clinicians, environmentalists and toxicologists have become aware that few substances in daily use or to which humans are exposed in daily life or in occupational environments and in medicine are entirely safe, and that a modicum of risk arises through excessive exposure, abuse or accident.

The present review re-evaluates epidemiological and occupational health studies, experimental studies in animals and *in vitro* experiments relating to the toxicity of metal and metalloid elements for which evidence of carcinogenicity has been presented. Human carcinogenic risk is substantiated in relation to arsenic, beryllium, thorium, chromium, radioactive elements, probably lead, and some nickel and cobalt compounds, and respirable silica particles, but the carcinogenicity of iron, aluminium, titanium, tungsten, antimony, bismuth, mercury, *cis*-platin, precious metals, and certain related compounds in humans is unresolved. The toxicity and carcinogenicity of each element is specific but correlates poorly with its position in the Periodic Table. Carcinogenicity differs according to the valency of the ion and its ability to interact with and penetrate membranes in target cells and to bind, denature or induce mutations by genotoxic or epigenetic mechanisms. The influence of lifestyle, environmental

Issues in Toxicology No. 18
The Carcinogenicity of Metals: Human Risk through Occupational and Environmental Exposure
By Alan B. G. Lansdown

Published by the Royal Society of Chemistry, www.rsc.org

contaminants and human factors in the interpretation of epidemiological studies is discussed. Further studies are indicated to investigate the interaction between xenobiotic elements and genotype as an explanation for regional variations in population response. The relevance of experimental studies in isolation in predicting human risk through metal exposures is questioned. *In vitro* studies in mammalian cell lines and bacterial reversion tests provide evidence that certain metals and metalloid elements are capable of inducing mutagenic and clastogenic changes, but they provide limited information on target organ susceptibility, inherent protective mechanisms within the intact body or immunomodulation.

Alan B. G. Lansdown
Imperial College London

Foreword

The ancient Egyptians had knowledge of the dangers of antimony, copper, crude arsenic and lead, and, in a number of cultures in the last three millennia, minerals have been used as poisons. A more clearly defined role for the widespread effects of these elements on human health and in disease states has been identified and characterised during the last two centuries. Initially, the need for a particular element was often discovered by observations of deficiency states in particular locations (iodine for goitre, as an example) but as a wider interest in comparative epidemiology developed, conditions induced by large local excesses of particular minerals were also identified and could be attributed to the effects of this excess (arsenic in drinking water in central Europe, say). Observations relating to the effects of therapeutic interventions had also shown that mercury and its salts, gold and silver, all used in manner that was hoped to confer benefit, may all cause evident human toxicity.

That we need many minerals in varying amounts is clear for reasons set out in early chapters of this book. As with vitamins, it is also clear that an excessive intake can be harmful – and that acute and chronic toxicity may result from exposure to excessive intake. These exposures may occur for a number of reasons and from varied sources, many of which have been recognised in comparatively recent times.

Occupational disease (wrist drop in painters) was recognised as a marker of toxic exposure and helped to define the cause of the toxic effects seen. In a clear historical example, mining of uranium-bearing ore in Schneeberg (Germany) and Jachimov (Czechoslovakia) both for metals and the manufacture of uranium dyes had been carried out for centuries and was known to be associated with lung disease – both pulmonary fibrosis and carcinoma of the lung, although this distinction was not evident to contemporary observers when the link was published in 1879. The development of industry and of industrial processes together with the gradual development of health care relating to those

Issues in Toxicology No. 18
The Carcinogenicity of Metals: Human Risk through Occupational and Environmental Exposure
By Alan B. G. Lansdown

Published by the Royal Society of Chemistry, www.rsc.org

working in industry and better record keeping relating to workers and the local environment (an often undervalued element in identifying causality in disease processes) made clear that industrial development has produced well-defined problems such as the presence of organic mercury compounds in effluvia (Minamata Bay).

Associations of a different kind, relating to long-term exposures and to exposures to levels of toxin that did not produce acute illness or evident direct toxicity, were harder to identify. Although potentially carcinogenic actions of minerals were often investigated after singular associations between occupations and uncommon tumours had been identified, modern methods of diagnosis and record keeping were needed to provide the means to question potential causality between exposures and common tumours. These associations would often promote a search for pathogenetic mechanisms by experimentation.

Epidemiological research is a valuable weapon in identifying apparently causative factors in disease. Although causes may be defined in a number of ways, in pragmatic terms it is clear that if the elimination of a causal factor results in a change in disease incidence, its relevance to public health is evident. This is what epidemiological research has sought to achieve in the field of exposure to minerals examined in this book. But the epidemiological approach has its dangers and before constructing a hypothesis, it should be remembered that the strength of any association, consistency of results in different studies and consistent experimental evidence are the most powerful discriminants in examining links. Experimental work must be constructed around a hypothesis of action that is clearly defined for it to be capable of translation between species, say.

This book is a comprehensive survey of a major health concern (carcinogenesis) relating to the use of minerals. It considers all those elements about which human health concerns have been thought to exist, having defined a view of carcinogenicity that is internationally adopted in regulatory circles and which is clearly set out in initial chapters.

Sir Colin Berry
Emeritus Professor of Pathology,
Queen Mary's College, London, UK

The Carcinogenicity of Metals
Human Risk through Occupational and Environmental Exposure

Alan B. G. Lansdown
Imperial College London, UK
Email: a.lansdown@imperial.ac.uk

RSC Publishing

Issues in Toxicology No. 18

ISBN: 978-1-84973-718-0
ISSN: 1757-7179

A catalogue record for this book is available from the British Library

Published by The Royal Society of Chemistry,
Thomas Graham House, Science Park, Milton Road,
Cambridge CB4 0WF, UK

Registered Charity Number 207890

For further information see our web site at www.rsc.org

I should like to dedicate this publication to my grandchildren, Emma, Rosie, Molly, Caroline and Christopher. Whatever aspirations they may develop in the future, I hope that their endeavours will bring personal rewards and satisfaction as have come my way in the study of metals over more than 40 years.

This volume is also dedicated to my wife, Veronica, who has given me loving support and friendship throughout my endeavours.

I acknowledge with sincere gratitude the fruitful conversations and constructive criticism provided by many friends and acquaintances at the former Charing Cross and Westminster Medical School, Birkbeck College and the British Industrial Biological Research Association.

Contents

Issues in Toxicology No. 18
The Carcinogenicity of Metals: Human Risk through Occupational and Environmental Exposure
By Alan B. G. Lansdown

Published by the Royal Society of Chemistry, www.rsc.org

CHAPTER 1
Introduction

1.1 Introduction

Metals and metalloid elements are ubiquitous in the human environment (Figure 1.1). They are present to a varying extent in the rocks and soils throughout the world and exist in the air we breathe and in our food or drinking water. Natural deposits in some parts of the world are extensive and in the case of lead and arsenic are prominent sources of local health problems.[1,2] Inland waterways, estuaries and open sea contain the largest natural sources of metals and their compounds. In addition, these waters accumulate metal residues eluted from inland sources, pesticides and agrochemicals, factory wastes and sludges, deposits from landfill sites and even domestic waste. Metal residues enter local streams, lakes and rivers to be disseminated into open water through tides, offshore currents and adverse weather conditions. This is well illustrated by discharges of silver residues into the San Fransisco Bay area in California (the so-called Great Silver Estuary) where sediments in one year were as high as 8800 kg.[3] Other notable examples include the Minamata Bay catastrophe in Japan in 1953 where an estimated 27 tons of mercury compounds were discharged into sea waters,[4] and local disasters following release of cadmium residues into rivers by mining companies (Figure 1.2).[5,6] In such cases, cadmium is concentrated in local food sources such that fish in the rivers start to die and rice irrigated with river water fails to grow. Cadmium poisoning is related to the human disease Itai-Itai, which causes softening of the bones and kidney failure.[7,8] Cadmium and cadmium compounds are now listed as human carcinogens.[9,10] Sea water possibly contains all stable and some radioactive metal and metalloid elements listed in the Periodic Table, albeit some being present in minute quantities.

Ecologists, environmentalists and regulatory toxicologists throughout the world are justifiably concerned that high concentrations of toxic metals discharged into sea water concentrate in marine deposits, fish and marine life and

Issues in Toxicology No. 18
The Carcinogenicity of Metals: Human Risk through Occupational and Environmental Exposure
By Alan B. G. Lansdown

Published by the Royal Society of Chemistry, www.rsc.org

Figure 1.1 Metals and metalloid elements in the human environment. Soils in the Vetagrande region of Zacateras in Mexico are rich in lead, mercury, silver, gold, zinc and copper.
(With kind permission of Dr JOSÉ RODRÍGUEZ, of the Fundación Universitaria Iberoamercana, Mexico.)

Figure 1.2 Human contamination through food chains. Firefighters pour polyaluminium chloride into a pool to dilute the cadmium-polluted water in the Longjiang river in China.
(Photo: AFP/Getty Images.)

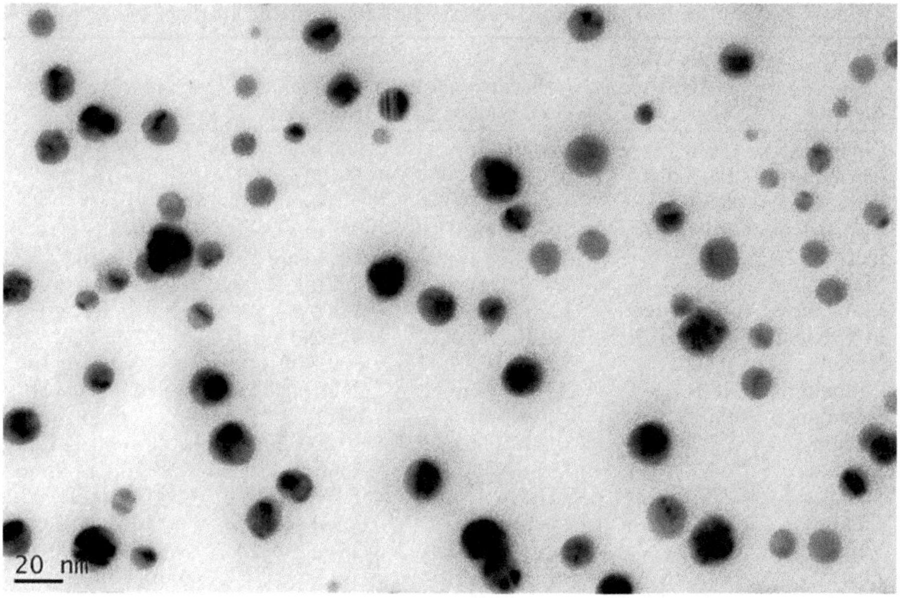

Figure 1.3 Nanoparticles of metallic silver <20 nm diameter. The high surface to volume ratio increases their capacity for ionisation and changes physico-chemical properties: the grain boundary phenomenon.
(By courtesy of Dr S. Misra, Natural History Museum, London.)

enter human food chains. Other major ecological and human health concerns relate to the discharge of metal particles into the air by volcanic action, natural erosion of rocks and shales, emissions and effluents from mining, extraction and refining from metal industries, and incineration of commercial and domestic waste. Plants and food animals in contaminated pastures accumulate lead, mercury, cadmium and other xenobiotic elements. Other concerns relate to the increasing use of nanotechnology and the production of minute metal particles of 20 nm or less for commercial purposes.[11–13] Nanoparticles in the air present special problems. They are considered to have different surface properties, and the physico-chemical properties of their grain boundaries may be more injurious to health.[14] Nanoparticles of silver are probably more than 100-fold more soluble than silver foil or filings.[15] Special health problems of pulmonary fibrosis, pneumoconiosis, chronic respiratory disease and even cancer are recognised following inhalation of industrial dust and nanoparticles of respirable size of gold, silver, chromium, silica and nickel in industrial environments (Figure 1.3).

1.2 Metals as Nutrients

The human body has evolved over many millennia to depend upon certain metals and metalloid elements as constituents of cellular structure or intercellular matrices, electrolytes, or as components or co-factors of key enzyme

Table 1.1 Metal and metalloid ions as nutrients in the human body.

Element	*Total body concentration (g)*	*Concentration in circulation*	*Recommended daily intake*	*Daily excretion*
Calcium	1500	8.8–10.5 mg dL^{-1}	800 mg	100–350 mg
Chromium	0.05–0.10	0.5–5.0 μg L^{-1}	5–100 μg	5–10 μg L^{-1}
Cobalt	0.0015	<1 μg/L	1–5 μg vit. B_{12}	<1 μg/L
Copper	0.1–15	80–153 mg dL^{-1}	2–5 mg	10–30 μg
Iron	4–4.5	4000 mg	10–14 mg	<100 μg
Magnesium	25	1.4–2.4 mg dL^{-1}	350 mg	75–150 μg
Manganese	10–20 μg	1–200 μg L^{-1}	1–8 mg	0.1–1.2 μg
Molybdenum	0.009	30–700 nmol L^{-1}	0.5–2.0 mg	22–173 μg
Nickel	0.10	3–8 μg L^{-1}	300–900 μg	10–100 μg L^{-1}
Potassium	180	3.5–5.0 mmol L^{-1}	2500 mg	66–85 mmol[a]
Selenium	0.013	1.3–4.3 μmol L^{-1}	60 μg	17 nmol
Silicon	0.024	1000–5000 μg L^{-1}	10 000–50 000 μg	15 000 μg L^{-1}
Sodium	64	135–145 mmol L^{-1}	2500 mg	129–181 mmol[a]
Strontium	3.5–4 mmol	12 μ L^{-1}	1–3 mg	144 μg L^{-1}
Tin	0.017	140 μg L^{-1}	3.5–17 mg	10–20 μg L^{-1}
Vanadium	0.018	10 μg L^{-1}	25 μg	<10 μg L^{-1}
Zinc	1–2	80–110 μg dL^{-1}	7–17 mg	0.4–0.6 mg

[a]Urinary levels vary according to sex and levels of salt intake.

systems or biosynthetic pathways (Table 1.1).[16–19] Metalloenzymes containing calcium, magnesium, zinc, iron and copper are important at critical stages of the cell cycle and may have a role in carcinogenic transformation. Patterns of uptake, metabolism, metal-binding proteins, cellular metabolism and excretion are well defined for all nutrients, although optimal levels for good health for minor trace metals such as molybdenum, vanadium, chromium and nickel are still debatable. The roles of blood concentrations, hormones or other factors regulating uptake, levels in the systemic circulation, tissue accumulation and excretion are imperfectly understood.

Macro- and trace nutrients are defined broadly as substances required at appropriate concentrations for optimal health and wellbeing. Demands for different nutrients vary according to age, sex and physiological state (especially pregnancy and lactation). The body displays characteristic signs of metal ion deficiency through malnutrition, dietary imbalances and malabsorption syndromes, through genetic or acquired disease processes. These conditions regress when deficiencies are corrected, as illustrated by iron deficiency anaemia (IDA), hypozincaemia, hypocalcaemia and cobalt deficiency (manifest through suboptimal Vitamin B_{12} levels).[20–24] The role of tin and strontium as trace metal nutrients is still unclear.

Metal ions interact in the body and ionic balances determined by carrier proteins, are critical in regulating the programmed sequence of proliferation in stem cells, maintenance of cellular architecture, cell-to-cell adhesion and functional differentiation.[25–28] Calcium, for example, interacts with zinc, magnesium, copper and iron and imbalances in metal-to-metal ionic ratios can be detrimental at specific phases in the cell cycle and in the post-mitotic functional differentiation in tissues such as skin, bone, bone marrow and gastrointestinal mucosae with high stem cell populations.[29] Calcium is a particularly important macro-nutrient and more than 70 calcium-binding proteins are present in the body, notably the so-called "EF-hand proteins", cahederins, calmodulin and S-100 proteins. Most display binding sites for other metal ions, notably strontium, lead, aluminium and mercury.[30,31] Strontium mimics calcium and can substitute for it in biological systems, particularly musculoskeletal tissues.

Elements such as silver, arsenic, aluminium, bismuth, platinum and lead have no nutritional function but are present occasionally in the body at low levels (Table 1.2). Several bind to proteins such as metallothioneins, ferritin, calmodulin, *etc.* and can impair the availability of essential nutrients if present to excess (Table 1.3). Arsenic accumulates in bone and displaces calcium from hydroxyapatite binding; clinical studies in Bangladesh and elsewhere have shown that arsenic in drinking water is a cause of retarded body growth and brittle bones.[32] Other xenobiotic elements such as lead, cadmium, mercury and antimony are also cumulative poisons which deposit in liver, neurological tissues, kidney and bone with potential toxicological effects.

The human body exhibits a variety of inherent protective mechanisms against the toxic effects of excesses and imbalances in nutrient metal or

Table 1.2 Xenobiotic metal and metalloid elements present in the human body.

Aluminium	Antimony
Barium	Bismuth
Gold	Mercury
Silver	Tungsten (Wolfram)
Zirconium	Arsenic
Beryllium	Cadmium
Lead	Platinum[a]
Titanium	

[a] Platinum-group metals include palladium, rhodium, ruthenium, iridium, and osmium which commonly occur together in nature.

Table 1.3 Metallothionein and metal-binding in human tissue.

Metallothionein	*Expression*	*Metal induction and binding*
I	Epithelia	Zn, Cu, Cd, Hg, Ag, Au, As, Pt
II	Epithelia	Zn, Cu, Cd, Hg, Ag, Au, As, Pt
III	Brain	Zn, Cu, Cd
IV	Squamous epithelia, tongue	Zn, Cu, Cd, Co

metalloid ions, as well as uptake of xenobiotic ions by ingestion, inhalation or percutaneous absorption. The main protective mechanisms seen include:

- Gastrointestinal physiology and factors that modulate metal ion absorption
- Intestinal commensal bacteria that detoxify, oxidise or reduce metal or metalloid ions
- Dietary factors such as phytate, plant fibres and organic matter that bind metal ions
- Epidermal cytokeratins that strongly bind metal cations, thereby controlling percutaneous absorption
- Intra- and inter-cellular metal binding proteins that chelate or otherwise bind xenobiotic ions or modulate their uptake and metabolism
- Pulmonary alveolar macrophages that phagocytose and "mop-up" inhaled particles
- Selective uptake and competitive receptor binding on cell membranes
- Metal-binding proteins.

Metal-binding proteins including calmodulin, calbindin, caeruloplasmin and the cysteine-rich metallothioneins (MT) serve critical functions as cytoprotective agents. The MT are induced by and play an instrumental role in the metabolism of key nutrients such as zinc, copper and selenium, but they strongly bind ions including arsenic, bismuth, cadmium, gold, silver and mercury. Transferrin is a key iron-binding protein, but this multivalent molecule also binds bismuth, aluminium, indium, vanadium and gallium, any of which, if present to excess, disturb iron metabolism.

The majority of metal and metalloid elements are toxic to some extent in humans. At least 12 are carcinogenic under some circumstances. Toxicity and carcinogenicity can occur under a variety of conditions but, mining, refining, heavy metal industries and exposure through contaminated drinking water are major sources of exposure. Whereas haematite ore is relatively harmless, mining of the ore in many parts of the world presents risks of lung and other cancers through inhalation of the radioactive gas radon.[33] A second example is seen with gold mining. Gold is not carcinogenic but miners exposed to arsenic are exposed to lung cancer.[34]

1.3 Diagnosis of Carcinogenicity

The US National Toxicology Programme (NTP), US Environmental Protection Agency (EPA), US Department of Health and Human Services (DHHS), International Agency for Research on Cancer (IARC) and World Health Organization (WHO) have reviewed published work over the past 100 years and, on the basis of collated observations from epidemiological studies, case reports and experimental studies in laboratory animals, have classified known carcinogenic materials in five main categories (Table 1.4).[9,35–39] Authoritative guidance on the carcinogenicity of metals and other environmental

Table 1.4 IARC Classification of Carcinogens (IARC, 2006).

Category	*Designation*	*Classification*
Group 1	Carcinogenic to humans	Sufficient evidence of carcinogenicity in humans. Exceptionally, where there is insufficient evidence for carcinogenicity in humans but there is sufficient evidence in experimental animals and strong evidence the agent acts through a relevant mechanism of carcinogenicity
Group 2	Category contains agents for which the degree of evidence for carcinogenicity in humans is almost sufficient, or where there are no human data but for which there is evidence of carcinogenicity in animals	
Group 2A	Probably carcinogenic to humans	Limited evidence of carcinogenicity in humans and sufficient evidence of carcinogenicity in experimental animals. Alternatively, where inadequate evidence of carcinogenicity in humans but sufficient evidence of carcinogenicity in experimental animals and strong evidence that carcinogenesis is mediated by a mechanism that operates in humans
Group 2B	Possibly carcinogenic to humans	Agents, mixtures and exposure circumstances for which there is limited evidence of carcinogenicity in humans and less than sufficient evidence of carcinogenicity in experimental animals
Group 3	Not classified as carcinogenic to humans	Inadequate evidence for carcinogenicity in humans and inadequate or limited evidence of carcinogenicity in experimental animals. Agents showing strong evidence that mechanisms of carcinogenicity in experimental animals do not operate in humans. Not a determination of non-carcinogenicity but may indicate that more research is needed
Group 4	Probably not carcinogenic in humans	Evidence suggesting a lack of carcinogenicity in humans and experimental animals.

contaminants is contained within the 12 Reports on Carcinogenicity (RoC), monographs of IARC working parties and numerous authoritative independent reviews.[106]

A "cancer hazard" is defined by the IARC in their Preamble to the *Monographs*,[106] as:

a. An "agent" capable of causing malignant neoplasms in one or more organ systems under some circumstances
b. An agent or related compound capable of "increasing the incidence of malignant neoplasms, reducing their latency, or increasing their severity or multiplicity".[10,36]

A "cancer risk" is an estimate of the carcinogenic effects expected through occupational or environmental exposure to a carcinogenic agent. Where an agent is shown to induce an increased incidence of benign neoplasms, this may be taken into account in judgements of carcinogenicity. The terms "neoplasm" and "tumour" are used interchangeably. The IARC Expert Working Parties classify the term "agents" broadly to include individual elements and related compounds, complex mixtures, occupational exposures, lifestyle factors and other potentially carcinogenic exposures. The classification of carcinogenic agents is updated regularly as newer information comes to hand.

Scientific judgement as to whether exposure to an element, chemical compound, mining or extraction process or finished product constitutes a proven or anticipated human carcinogenic risk depends upon a balanced, scientific and statistically valid assessment of:

- Occupational and environmental health reports, human case and forensic studies
- Regulatory style experimental studies in animals
- Short-term laboratory *in vitro* tests to demonstrate: mutagenicity, DNA damage, cell transformation, clastogenicity, genotoxicity and molecular toxicity.[39–47]

The RoC have documented certain agents as "reasonably carcinogenic to humans" on the basis of their being "structurally related to a class of substances whose members are listed as carcinogens or are reasonably anticipated to be human carcinogens". In all, conclusions are based on a consideration of all relevant information. "This is not limited to dose response, route of exposure, chemical structure, metabolism, pharmacokinetics, sensitive sub-populations, genetic effects or other data relating to mechanisms of action or factors that may be unique for a given substance."[9]

1.4 Mechanisms of Carcinogenicity as Applied to Metals and Metalloid Elements

The scientific community has moved on far beyond the initial concepts of chemical carcinogenesis, founded on the studies of Isaac Berenblum and Phillip Shubic in the 1940s, that chemical carcinogenesis involves at least two stages – induction and promotion.[48] Recent advances in molecular genetics dictate that, these days, greater emphasis should be placed on mechanism-based carcinogenesis and the action of xenobiotics on cellular growth, mitotic homeostasis and the activation and expression of oncogenes.[9,42,49–52]

Early concepts of multi-step chemical carcinogenesis envisaged an initial (induction) phase involving DNA damage, chromosomal change, impairment of DNA replication and repair followed by one or more promotional phases in which this pre-neoplastic state is promoted to frank tumour formation.[53–61] Promoters such as croton oil may have marginal or no carcinogenic activity but serve as mitogens, motivating transformed cells to colony formation and metastasis. Phorbolester A_1, isolated from croton oil, was shown to invoke increased permeability in nuclear membranes preceding excitation of DNA synthesis and nuclear enlargement. Other non-carcinogenic promoters include non-specific stress factors such as noise, disturbances in diurnal rhythms, dietary factors, infection, immuno-suppression and oxidative stress. Complete carcinogens are defined as substances capable of inducing irreversible mutagenic changes in target cells with or without metabolic transformation, followed by transformation/promotion of stem cells to tumour formation.[53]

Biochemical and molecular evidence emphasises that elements such as arsenic, cadmium, chromium (VI), cobalt and nickel compounds can evoke carcinogenic changes through mechanisms other than direct genotoxicity, DNA binding or chromosomal aberrations.[62] These so-called "*epigenetic changes*", cumulatively leading to altered signal transduction, regulation in gene expression and carcinogenesis, include chronic inflammation, immunosuppression, oxidative change and induction of reactive oxygen species, changes in DNA-methylation patterns and activation of hormonal receptors.[63,64] Growth factors, cytokines and other subcellular or intracellular factors are probably involved. Epigenetics is a new and challenging aspect of carcinogenesis and is well illustrated by studies in molecular genetics of unequivocal carcinogens such as arsenic.[51] Plausible studies now suggest that epigenetics should be defined as a "*study of heritable changes in gene function that occur without any direct changes in DNA sequence*". Epigenetic effects influence gene expression and regulatory mechanisms controlling tissue-specific cellular receptors, signal transducers and effector molecules.[65]

Alterations in DNA-methylation patterns probably constitute a significant part of the carcinogenic process and involve transcriptional inactivation or activation of cancer-related genes.[51] Molecular studies with arsenic emphasise that carcinogenesis is principally a "*disease of stem cells*" which express a range of cell surface markers responsive to stem-cell maintenance-related genes.[66] They may also involve covalent modifications in the amino acid residues in histones around which DNA is wrapped.[67–69] Changes in the methylation status of cytosine bases in cytosine–phosphate–guanine dinucleotides (*i.e.* CpG islands) within the DNA molecule act in a form of gene "silencing". Grønbaek viewed cancer developing when "*cells acquire specific growth advantages through a stepwise accumulation of heritable changes in gene function*" modulated by tumour suppressor genes that inhibit cell growth and oncogenes that promote cell growth and survival.[69]

1.5 Epidemiological Evidence

Numerous epidemiological studies are published claiming to demonstrate that exposure to metal or metalloid elements in industrial environments or through contamination of food, drink or air is a cause of human cancer. Few are scientifically sound and many fail to demonstrate a clear correlation between exposure to metal/metalloid and evidence of tumour induction/promotion.[70–75]

Observer bias is evident in some older studies but long lag phases of 20 years or more years between presumed exposure and evidence of tumours (*e.g.* arsenic, lead and cadmium), failure adequately to allow for human lifestyle factors, and incomplete reporting complicate the true evaluation of risk in many epidemiological studies.[76–78] Few industrial environments, mining, smelting and refining operations contain a single toxic element, and in the case of electroplating, steel production, the electronics industry, and waste metal recycling, workers are exposed to several toxic and potentially carcinogenic materials capable of inducing, promoting or otherwise modifying chemically induced or idiopathic cancers. The World Health Organization reported, in 2008, 12.7 million new cases and 7.6 million deaths, and a total of 107 agents, mixtures and exposure situations as carcinogenic to humans.[79] They noted that environmental causes of cancer include factors in the environment such as air pollution, ultraviolet (UV) radiation and indoor radon exposure but that "...every tenth lung cancer is closely related to risks in the workplace". These include complications due to environmental contaminants such as microcrystalline silica. Microcrystalline silica of respirable size is an acknowledged carcinogen, and exposure to quartz dusts in industry is a cause of chronic respiratory distress and increased incidence of lung cancer.[9,10] Radon is a colourless, odourless and tasteless natural radioactive gas released as a degeneration product of uranium that occurs naturally in all rocks, soils and deep in the Earth's core. Radioactive emissions are experienced at very low levels in homes and dispersed in the general environment but higher concentrations are experienced in metal mining, smelting and refining; the emissions are harmful and are recognised environmental factors impacting upon the incidence of environmental carcinogenesis.[80–84] The US EPA estimated that as many as 20,000 lung cancer deaths are caused each year by radon exposure and, in financial terms, an annual cost of more than $2 billion in direct and indirect health care costs. The values for action levels for environmental radon show a wide range, but concentrations between 100 and 400 Bq m^{-3} are used.[82] Radon exposure is now known to be largely responsible for lung tumours reported in miners of gold, tin and haematite.[9,83]

1.6 Lifestyle Factors

The so-called "lifestyle" factors, which are specific to certain races, geographical areas and socio-economic groups, present considerable difficulty in the evaluation of environmental and industrial carcinogens. The principal lifestyle factors encountered repeatedly in this review include:

- Cigarette smoking
- Alcohol consumption
- Locality of exposure (urban *vs.* rural communities, geographical areas, geophysical disasters)
- Demographic factors (age, sex, race, geographical area, *etc.*)
- Diet.[84,85]

Cigarette smoking amongst workers is a major lifestyle factor influencing the incidence of occupationally related lung cancers.[86–88] Cigarette smoking and exposure to mainstream smoke, passive inhalation of cigarette smoke in bars, casinos and other public places and exposure to cigarette-related products as a cause of lung cancer is a massive topic in itself, and outside the scope of the present review. However, it is important to recognise that:

- Most brands of cigarette contain significant residues of toxic metals, including cadmium, beryllium, mercury, arsenic and lead, together with at least 40 non-metallic mutagens, including benzene, benz[a]anthracene, napthalenes, dibenzo[c,g]carbazole, tobacco-specific nitrosamines and benzo[α]pyrene, all of which have been identified and classified as carcinogens through experimental studies in rodents[89,90]
- Wide variations exist in cigarette tobacco according to the areas in which it is grown and materials used in curing, processing, production and packaging[91,92]
- Local regulations relating to quality, refinement and production vary according to cigarette type
- The radionuclides polonium (^{210}Po) and uranium (^{235}U and ^{238}U) may occur in tobacco smoke.

All radioactive isotopes, as emitters of ionising radiation, are potentially carcinogenic, but levels present in most cigarettes are probably insufficient to have an impact upon the incidence of smoking-related lung cancers. Professor Stephen Hecht, an international expert in smoking-related cancer, drew up a list of harmful and potentially carcinogenic constituents of tobacco smoke or smokeless tobacco, filters and additives using criteria accepted by the IARC, EPA and NTP in the USA. He discussed, in detail, putative carcinogenic mechanisms for 20 hydrocarbons.[86,89]

1.7 Laboratory Models and Diagnosis

Animal models have been used in diagnostic and investigative medicine since the time of the ancient Greeks, and such notable names as Hippocrates, Galen of Pergammon and William Harvey used monkeys and dogs to study the circulation of the blood and the vascular network of arteries and veins.[93] With greatly improved refinement, animal models still have an important contribution these days in investigative medicine and there is abundant evidence that regulatory style animal studies provide a useful guide to the pathogenicity and carcinogenicity of metals/metalloids and related compounds.[10,44,66,94–97] They cannot provide a reliable surrogate for humans in predicting human risk from environmental carcinogens but there is good evidence illustrated throughout this volume to show that animal tests are complementary to human epidemiological studies and provide a useful guide to the pathogenicity where human studies do not exist or where observations are equivocal. Extrapolation of experimental results from animals will always be complex, subjective and equivocal and no non-human species can realistically reproduce the wide racial, cultural, geographical, behavioural and genetic variability seen in the human race.

A large battery of tests in bacteria and in human and mammalian cell lines in culture have been developed over the last 50 years. They are designated for evaluation of the capacity of metals and metalloid ions and vast numbers of xenobiotic materials to evoke mutagenicity, DNA damage and other sub-cellular changes predisposing to malignancy.[43,98,99] In the absence of epidemiological or experimental animal studies, these *in vitro* genetic toxicology tests are helpful to regulators in making decisions as to which substances should be regarded as potentially carcinogenic to humans. They provide a data-base for use in structure–activity analyses.[98–100] The Ames *Salmonella* reverse mutation test was the first genotoxicity test recommended and experience shows that it will produce positive results for about 60% of genotoxic carcinogens (sensitivity ~60%).[43,101,102] The late Professor Leon Golberg stressed to the American Medical Association, in 1979, that the Ames test for mutagens is able to tell us "little about the effects of chemicals in man and should be regarded only as a Litmus test; it ignores the fact that many suspected carcinogens are actually pro-carcinogens that require a specific triggering mechanism".[103] This triggering mechanism can be expected to produce derivatives capable of DNA interaction, mutagenic change, chromosomal damage, *etc.* as an integral stage in the multi-step process leading chemical carcinogenesis.[104,105]

The present volume aims to present a critical re-evaluation of clinical and experimental data upon which judgements on the carcinogenic properties of metals and metalloid elements are made. The toxicity of elements emitting ionising radiation is a large and specialised subject and not within the scope of the present book, but all radioactive isotopes are known human carcinogens as documented in IARC *Monographs.*[106] To illustrate certain clinical problems, reference is made as appropriate to certain radio-isotopes, including thorium, gold (^{195}Au and ^{198}Au) and other metals used in diagnostic and experimental medicine. As improved investigation, and wider chemical and toxicological knowledge comes to hand, so improved environmental and industrial regulation achieves improved human health and safety.

In the past 20 years, specific health problems with implications of carcinogenicity have led to considerable research and controversy. For completeness, the present discussion addresses the basis and possible resolution of public concerns relating to risk associated with such problems as:

- Metal-on-metal prostheses and metal implants
- Aluminium-related breast cancer
- Semiconductor exposure
- Silicone breast implants
- Antimony oxide as a flame retardant in cot mattresses
- Nanotechnology and its impact on human health.

The EPA and National Institutes of Health in the USA and the Scientific Committee on Emerging & Newly Identified Health Risks (SCENIHR) in

Europe meet regularly to discuss current issues and publications are available for consultation on these and other issues of public interest and concern. The report of the SCENIHR in March 2007 relating to preliminary opinion on "*The appropriateness of the risk assessment methodology in accordance with the Technical Guidance Documents for new and existing substances for assessing the risks of nanomaterials*" is one example.[107]

References

1. I. Thornton, in *Metals in the Global Environment*, International Council on Metals and the Environment (ICME), Ottawa, Canada, 1995.
2. J. O. Nriagu and J. M. Pacyna, *Nature*, 1988, **333**, 134.
3. A. R. Flegal and G. J. Smith, *Estuaries*, 1993, **16**, 547.
4. M. Harada, *Crit. Rev. Toxicol.*, 1995, **25**, 1.
5. H. Muntau and R. Baudo, *IARC Sci. Publ.*, 1992, **118**, 133.
6. Occupational Safety and Health Administration (OSHA) US Dept. Labor, *Cadmium*, OSHA, Washington DC, 2004, p. 3136.
7. M. R. Fox, R. M. Jacobs, A. O. Jones and B. E. Fry, *Environ. Health Perspect.*, 1979, **28**, 107.
8. J. Godt, F. Scheidig, C. Grosse-Siestrup, V. Esche, P. Brandenburg, A. Reich and D. A. Groneberg, *J. Occup. Med. Toxicol.*, 2006, **1**, 22.
9. US Department of Health and Human Services (DHHS), *Report on Carcinogens*, National Toxicology Program (NTP), Research Triangle Park, NC, 12th edn, 2011.
10. IARC, *Monographs on the Evaluation of Carcinogenic Risks to Humans*, IARC, Lyons, 2012, p. 100C.
11. Centers for Disease Control and Prevention (CDCP), *Nanotechnology*, CDCP, Atlanta, 2012.
12. European Commission, Director General for Health and Consumers, Scientific Committee on Emerging and Newly Identified Health Risks, *Risk Assessment of Products Nanotechnologies*, Brussels, 2009.
13. J. C. Bonner, *Proc. Am. Thorac. Soc.*, 2010, **7**, 138.
14. P. A. Borm, D. Robbins, S. Haubold, T. Kuhlbusch, H. Fissan, K. Donaldson, R. Schins, V. Stone, W. Kreyling, J. Lademann, J. Krutmann, D. Warheit and E. Oberdorster, *Part. Fibre Toxicol.*, 2006, **3**, 11.
15. R. E. Burrell, *Ostomy. Wound Manag.*, 2003, **49**(Suppl), 19.
16. J. E. Underwood, *Trace Elements in Human and Animal Nutrition*, Academic Press, Inc., London, 1977.
17. W. Mertz, *Proc. Nutr. Soc.*, 1974, **33**, 307.
18. N. S. Scrimshaw and V. R. Young, *Sci. Am.*, 1976, **235**, 51.
19. D. M. Czujka-Narins, in *Food, Nutrition and Diet Therapy*, ed. L. K. Mehan and M. V. Kraus, Saunders, Philadelphia, 1979, vol. 2, p. 114.
20. A. B. G. Lansdown, *Wound Rep. Regen.*, 2002, **10**, 271.
21. A. B. G. Lansdown, U. Mirastschijski, N. Stubbs, E. Scanlon and M. S. Agren, *Wound Rep. Regen.*, 2007, **15**, 2.

22. A. S. Prasad, *Ann. Rev Nutr.*, 1985, **5**, 341.
23. K. Liu and A. J. Kaffes, *Eur. J. Hepatol.*, 2012, **24**, 109.
24. I. Chanarin, *J. Clin. Pathol.*, 1987, **40**, 978.
25. M. K. Heng, M. K. Song and M. C. Heng, *Br. J. Dermatol.*, 1993, **129**, 280.
26. S. MacNeil, R. A. Dawson, G. Crocker, W. F. G. Tucker, B. Bittiner, J. G. Singleton, T. Hunter and D. F. Tierney, *Br. J. Dermatol.*, 1993, **128**, 143.
27. C. W. Heizmann, G. Fritz and B. W. Schäfere, *Front. Biosci.*, 2002, **7**, 1356.
28. P. Navarro, M. Gomez, A. Pizarro, C. Gamallo, M. Quintilla and A. Cano, *J. Cell. Biol.*, 1991, **115**, 517.
29. J. F. Whitfied, R. P. Bird, B. R. Chakravarty, J. Isaacs and P. Moley, *J. Cell. Biochem.*, 1995, **59**, 74.
30. A. Lewit-Bentley and S. Réty, *Curr. Opin. Struct. Biol.*, 2000, **10**, 637.
31. H. J. Vogel, *Calcium Binding Protein Protocols, Reviews and Case Studies*, Humana Press, New Jersey, USA, 2002, p. 172.
32. M. Vahter, *Ann. Rev. Nutr.*, 2009, **29**, 381.
33. S. Y. Chen, R. B. Hayes, S. R. Liang, Q. G. Li, P. A. Stewart and A. Blair, *Br. J. Ind. Med.*, 1990, **47**, 175.
34. B. K. Armstrong, J. C. McNaulty, L. J. Levitt, K. A. Williams and M. S. Hobbs, *Br. J. Ind. Med.*, 1979, **36**, 199.
35. L. Tomatis, *Environ. Health Perspect.*, 1993, **101**(Suppl. 5), 149.
36. International Agency for Research on Cancer (IARC), *IARC Scientific Publications*, IARC, Lyon, 2012, p. 100.
37. A. Berlin, M. Draper, K. Hemminki and H. Vainio, *IARC Scientific Publications*, IARC, Lyon, 1984, No. 595.
38. R. B. Hayes, *Cancer Caus. Contr.*, 1997, **8**, 371.
39. D. Beyersmann, *Toxicol. Lett.*, 2002, **127**, 63.
40. D. Beyersmann and A. Hartwig, *Arch. Toxicol.*, 2008, **82**, 493.
41. E. J. Tokar, L. Benbrahim-Tallaa and M. P. Waalkes, *Met. Ions Life Sci.*, 2011, **8**, 375.
42. T. R. Durham and E. T. Snow, *Cancer, Cell Structures, Carcinogens and Genomic Instability*, ed. L. P. Bignold, Birkhäuser Verlag, Switzerland, 2006, p. 97.
43. B. N. Ames, J. McCann and E. Yamasaki, *Mutat. Res.*, 1975, **31**, 347.
44. IARC, *IARC Monographs on the Evaluation of the Carcinogenic Risk of Chemicals to Humans,* World Health Organization (WHO), IARC, Lyon, 1980, Suppl. 2.
45. W. Bal, A. M. Protas and K. S. Kasprzak, *Met. Ions Life Sci.*, 2011, **8**, 319.
46. A. Yamamoto, Y. Kohyama and T. Hanawa, *J. Biomed. Mater. Res.*, 2002, **59**, 176.
47. J. E. Klaunig, L. M. Kamendulis and B. A. Hocevar, *Toxicol. Pathol.*, 2010, **38**, 96.

48. I. Berenblum and P. Shubik, *Br. J. Cancer*, 1947, **1**, 379.
49. S. Venitt, *Clin. Chem.*, 1994, **40**, 1421.
50. A. Léonard and A. Bernard, *Envir. Health Perspect. Suppl.*, 1993, **101**, 127.
51. C. Michailidi, A. G. Papavassiliou and C. Troungos, *Curr. Mol. Med.*, 2012, **12**, 237.
52. P. Koedrith and Y. R. Seo, *Int. J. Mol. Sci.*, 2011, **12**, 9576.
53. E. Faber, *Cancer Res.*, 1964, **44**, 4217.
54. I. Berenblum, *Carcinogens: Identification and Mechanisms of Action*, ed. A. C. Griffin and C. R. Shaw, Raven Press, New York, 1979, p. 25.
55. T. S. Argyris, *Am. J. Pathol.*, 1980, **98**, 639.
56. F. Marks and G. Fürstenberger, *Br. J. Cancer*, 1986, **115**(Suppl. 31), 1.
57. T. S. Argyris, *Am. J. Pathol.*, 1980, **98**, 639.
58. R. Chibber and M. Ord, *Bil. Met.*, 1990, **3**, 213.
59. T. J. Slaga, *Environ. Health Perspect.*, 1983, **50**, 3.
60. H. Bach and K. Goerttler, *Virchows Arch. Abt. B Zellpath.*, 1971, **8**, 196.
61. E. Hecker, *Naturwissenschaften*, 1967, **54**(282), 112.
62. A. Arita and M. Costa, *Metallomics*, 2009, **1**, 222.
63. K. Salnikow, W. Su, M. V. Blagosklonny and M. Costa, *Cancer Res.*, 2000, **60**, 3375.
64. G. K. Harris and X. Shi, *Mutat. Res.*, 2003, **533**, 183.
65. A. Bird, *Genes Gene Devt.*, 2002, **16**, 6.
66. E. J. Tokar, W. Qu and M. P. Waalkes, *Toxicol. Sci.*, 2010, **120**, S192.
67. Y. Kondo, L. Shen, S. Ahmed, Y. Boumber, C. Charo, T. Yamochi, T. Urano, K. Furukawa, B. Kwabi-Addo, D. L. Gold, Y. Sekido, T. H. Huang and J. P. Issa, *Nature Genet.*, 2008, **40**, 741.
68. S. Mishra, S. P. Dwivedi and R. B. Singh, *Open Pharmac. J.*, 2010, **3**, 188.
69. K. Grønbaek, C. Hother and P. A. Jones, *Acta Path. Microbiol. Immunol. Scand.*, 2007, **115**, 1039.
70. F. W. Sunderman, *Dis. Chest*, 1968, **54**, 527.
71. F. W. Sunderman, *Fd. Cosmet. Toxicol.*, 1971, **9**, 105.
72. G. Kazantzis, *Environ. Health Perspect.*, 1981, **40**, 143.
73. T. Sorahan, D. C. Burges, L. Hamilton and J. M. Harrington, *Occup. Environ. Med.*, 1998, **55**, 236.
74. E. Molderheim, B. Andersen, N. E. Freystein and E. Sletten, *Inorg. Chemic. Acta*, 1998, **273**, 41.
75. C. L. Berry, *Toxicology*, 2010, **267**, 7.
76. A. Luch, *Cancer*, 2005, **5**, 113.
77. A. J. Arlberg, J. G. Ford and J. M. Samet, *Chest*, 2007, **132**, 29S.
78. T. Cotter, D. Perez, A. Dessaix, D. Baker, M. Murphy, J. Crawford, J. Denney and J. F. Bishop, *Cancer and Lifestyle Factors*, Cancer Institute NSW, Sydney, 2007.
79. World Health Organization, *Environmental and Occupational Cancers*, WHO, Geneva, Fact Sheet No. 350, p. 201.

80. National Research Council Committee on Health Risks of Exposure to Radon: BEIR VI, *Health Effects of Exposure to Radon*, National Academy Press, Washington, DC, 1999.
81. S. Darby, D. Hill, H. Deo, A. Auvinen, M. J. Barros-Dios, H. Baysson, F. Bochicchio, R. Falk, S. Farchi, A. Figueiras, M. Hakama, I. Heid, N. Hunter, L. Kreienbrock, M. Kreuzer, F. Lagarde, I. Mäkeläinen, C. Muirhead, W. Oberaigner, G. Pershagen, E. Ruosteenoja, A. S. Rosario, M. Tirmarche, L. Tomásek, E. Whitley, H. E. Wichmann and R. Doll, *Scand. J. Work, Environ. Health*, 2006, **32**(Suppl. 1), 1; erratum in *Scand. J. Work, Environ. Health*, 2007, **33**(1), 80.
82. WHO, International Radon Project, *Survey on Radon Guidelines, Programs and Activities*, WHO, Geneva, 2007.
83. IARC, *Summaries and Evaluation: Haematite and ferric oxide, ferric oxide underground haematite mining with exposure to radon*, IARC, Geneva, 2007.
84. H. Bartsch and E. Hietanen, *Environ. Health Perspect.*, 1996, **104**, 569.
85. G. N. Wogan, S. S. Hecht, J. S. Felton, A. H. Conny and L. A. Loeb, *Semn. Cancer Biol.*, 2004, **14**, 473.
86. S. S. Hecht, *J. Natl. Cancer Inst.*, 1999, **91**, 1194.
87. Z. Feng, W. Hu, W. N. Rom, M. Costa and M.-S. Tang, *Carcinogenesis*, 2003, **24**, 771.
88. J. C. Stavrides, *Free Radical Biol. Med.*, 2006, **41**, 1017.
89. S. S. Hecht, *Nat. Rev. Cancer*, 2003, **3**, 733.
90. H. Oliveira, E. A. N. Fernandes, M. A. Bacchi, G. A. Sarriés and F. S. Tagliaferro, *J. Radioanal. Nucl. Chem.*, 2000, **244**, 299.
91. E. A. Menden, V. J. Elias, L. W. Michael and H. G. Petering, *Environ. Sci. Technol.*, 1972, **6**, 830.
92. W. S. Rickert and M. J. Kalserman, *Environ. Sci. Toxicol.*, 1994, **28**, 924.
93. W. C. Aird, *J. Thromb. Haemostat.*, 2011, **9**(Suppl. 1), 1118.
94. US DHHS, *Guidance for Industry, Carcinogenicity Study for Protocol Submissions*, FDA, Center for Drug Evaluation and Research (CEDR), Washington, DC, 2002.
95. A. Knight, J. Bailey and J. Balcombe, *Alter. Lab. Anim.*, 2006, **34**, 19–27.
96. A. Knight, J. Bailey and J. Balcombe, *Altern. Lab. Anim.*, 2006, **34**, 29–38.
97. A. Knight, *Japan. Soc. Altern. Anim. Exp.*, 2008, **14**(Special Issue), 125.
98. R. D. Storer, *Toxicol. Lett.*, 2000, **112–113**, 55767.
99. K. Mortelmans and E. Zeiger, *Mutation Res.*, 2000, **455**, 29.
100. E. Zeiger, B. Anderson, S. Haworth, T. Lawlor and K. Mortelmans, *Environ. Mol. Mutagen.*, 1992, **19**(Suppl. 21), 2.
101. J. Ashby, R. W. Tennant, E. Zeiger and S. Stasiewicz, *Mutat. Res.*, 1989, **223**, 73.
102. J. Ashby and R. W. Tennant, *Mutat. Res.*, 1991, **257**, 229.
103. L. Golberg, *Drug. Cosmet. Ind.*, 1979, **124**, 38; 77.

104. E. Farber, *Cancer Res.*, 1984, **44**, 4217.
105. E. Farber, *Cancer Res.*, 1984, **44**, 5463.
106. IARC, *IARC Monographs,* Lyon, 2012, **100D;** *Lancet Oncology*, **10**, 751.
107. European Commission Scientific Committees, *Scientific Committee on Newly Identified Health Risks (SCENIHR)*, 2007.

Part 1
Elements of Importance as Nutrients

CHAPTER 2

Iron

2.1 Introduction

Iron has been known to man for longer than most other elements. Artefacts have been recovered from human civilisations of the "Iron Age" 3000 years ago, but evidence of early mining dates from prehistoric times. Iron is a transitional metal and the fourth most abundant element in the Earth's crust. It exhibits two main oxidataion states, Fe(II) and Fe(III), and compounds of both are found naturally. Iron is also relatively abundant in the universe and asteroids and much of the Earth's core is composed of iron. Large quantities of iron have been recovered from meteoritc debris. Iron deposits are commonly seen as black sands along beaches and stream banks.

Iron is mined mainly as haematite (Fe_2O_3) in Europe, North and South America and China (Figure 2.1), but other ores including magnetite ($Fe_2O_3 + Fe_3O_4$), iron pyrites (FeS_2), goethite, limonite or siderite, and taconite (a low-grade siliceous iron) are more prevalent naturally in certain areas. Banded iron formations (BIF) of taconite found in North America are fine-grained metamorphosed sedimentary rocks composed predominantly of magnetite and silica (as quartz). Iron ore deposits are frequently associated with vanadium, aluminium and titanium but health concerns amongst miners have increasingly been associated with exposure to radon and radon daughters, silaceous dusts and arsenic poisoning. Lead poisoning is an additional hazard. Nowadays, iron and ferrous metals (cast iron, carbon steel, high speed steel, *etc.*) have a vast range of industrial uses in engineering, automotive manufacture, structural components in building, and steels with specific properties of hardness, durability, tensile strength, resistance to corrosion and malleability.

Many thousands of workers are exposed to iron in industrial dusts, aggregates and fumes in mining pits, ore crushing and grinding, extraction processes, furnace working, smelting, refining, steel works and other iron-related industries. Siderosis, pneumoconiosis, silicotuberculosis and other respiratory

Issues in Toxicology No. 18
The Carcinogenicity of Metals: Human Risk through Occupational and Environmental Exposure
By Alan B. G. Lansdown

Published by the Royal Society of Chemistry, www.rsc.org

Figure 2.1 Haematite (from the West Cumberland Minefield).

diseases are prevalent in workers employed in haematite mines where blasting and pneumatic drilling are commonly used to release iron ore from hard rock (see Figure 2.2, below).[1–3] Pulmonary and tracheal cancer is an industrial hazard in iron (haematite) mining but levels of risk vary greatly according to geographical area, the type and duration of working in the pits, safety procedures in operation (including effective dust control systems, face masks and high-pressure mist systems), and the nature of environmental contaminants such as radon, carbon monoxide and methane gas.[4,5] Considerable efforts have been made in recent years to control cancer risks in iron mining and related industries, and local and national authorities publish stringent guidelines. Ferrous metals are magnetic and contain variable amounts of carbon and other metals including nickel, chromium, cadmium, vanadium and tungsten to improve tensile strength, resilience, resistance to corrosion and other physical properties. Many of these metal additives are toxic and potentially carcinogenic in humans and other animals. Stainless steel, for example, comprises 18% chromium and 8% nickel to provide a characteristic resistance to corrosion. The most commonly used ferrous metals include cast iron, mild steel, high speed steel, stainless steel and high tensile steel.

Iron is an essential nutrient for most living organisms, with key roles in cell physiology, proliferation and functional differentiation, haemoglobin synthesis and blood quality, oxygen transport, resistance to stress and disease, immunomodulation, synthesis of myoglobin and oxygen supply to muscles, and prevention of fatigue.[6–8] The human body contains 3.5–4 g of iron, most being bound in haemoglobin in circulating erythrocytes, but some is stored in the liver, muscle and reticuloendothelial tissues. The human body has no means of excreting iron, and regulation is principally through limiting intestinal absorption or through iron loss in perspiration, desquamation of epidermal keratinocytes, nail and hair growth. Human requirements for iron are provided by dietary intake (red meat, vegetables, cereal, nuts, seafood, *etc.*), although in clinical cases of iron deficiency anaemia parenteral injection of iron supplements including iron dextran, iron sucrose and preparations of ferrous sulfate, ferrous fumarate, ferrous gluconate and polysaccharide–iron complexes is available. These are beneficial for those patients unable to tolerate oral iron supplements, but, as discussed below, controversy exists over their capacity to invoke injection site tumours and other complications.[9–11]

A third and worrying aspect of iron toxicity concerns the iron-overload conditions such as the hereditary and idiopathic haemochromatoses, which have been associated with hepatocellular carcinoma and possibly breast cancers.[12,13] Dietary iron overload was first described by Strachan in 1929 in peoples of sub-Saharan Africa. It seems that these people were accustomed to drinking home-brewed beer with a high iron content.

Iron interacts with other metals in the body, largely as a result of ion binding in the intestinal mucosa, circulation and elsewhere. Ferritin is a major iron-binding protein that will bind aluminium, zinc, beryllium, cadmium and copper in a competitive fashion.[14–16] Iron interacts with cobalt and nickel in intestinal uptake, which is modulated by vitamin C.[17,18] The role of iron in haem synthesis is impaired by the action of lead and possibly nickel in inhibiting the essential enzyme delta-aminolaevulinate dehydratase in bone marrow, and thus erythroid cell differentiation.[19–22] This is a reliable and quantitative indicator of lead exposure in humans.

2.2 Iron in Human Nutrition

McCance and Widdowson first demonstrated the essentiality of body iron-balance and metabolisable iron for normal health and tissue oxygenation through regulation of haemoglobin synthesis erythropoietin and serum proteins.[23] Iron plays a fundamental role in the growth and functional maturation of all tissues in the human body, principally through its role in haem synthesis and oxygen-carrying function in the blood, but also in DNA synthesis, immune responsiveness and enzymes regulating energy metabolism.[6–8,24–27] Iron (Fe) plays a critical role in cell proliferation, and Fe deficiency results in G_1/S-phase arrest of the cell cycle and apoptosis. However, the precise role of Fe in cell-cycle control remains unclear.[27]

Iron is absorbed intestinally by ferrous ion Fe(II)-binding cell surface receptors, endocytic vacuolation and intracellular ferritin binding. Ferric iron Fe(III) is reduced to ferrous ion in the acidic medium of the stomach and in the presence of ferri-reductase. Iron absorption is intimately linked to a specific transporter protein, Nramp-2/DCT1, which is upregulated in states of iron insufficiency and chronic hypoxia.[28] A further basolateral intestinal membrane transporter, ferro-oxidase (hephaestin), has been identified. This is a caeruloplasmin-like protein involved in the efflux of iron from intestinal epithelial cells. Subsequent metabolism of Fe(II) involves a sequence of intracellular co-ordination complexes involving low molecular weight transferrin-carrier proteins and incorporation into mitochondrial and intracellular enzymes. One or more of these co-ordination complexes may involve linkage to water molecules or molecular oxygen as required in the oxygen-transporting function of haemoglobin. Intestinal iron absorption modulated through the action of vitamin C and iron-binding proteins, such as transferrin and lactoferrin, is probably regulated by body demands for iron or through hormonal action.[8,17] In humans, most of the body iron is bound within erythrocytes, with lower concentrations in bone and soft tissues. Homeostatic mechanisms controlling iron metabolism in the human body are unclear. Intestinal iron absorption is impaired by chelators including phytate, phenolic compounds and plant fibres in the diet, or through competitive inhibition by lead, cadmium, nickel and cobalt binding at intestinal receptor proteins.[26,29] Ethylenediamine tetraacetic acid (EDTA) has good iron-binding capacity. Iron-containing dietary supplements and parenteral iron are readily available to treat iron deficiency anaemia and related conditions.[30] As discussed in more detail later, iron supplements such as iron–dextran are a cause of concern on account of their ability to evoke anaphylactic reactions, and injection site sarcomas in rats.[31]

Much is still unclear regarding the function of carrier proteins in iron homeostasis, but recent research suggests that extracellular transferrin functions as a primary iron carrier protein, regulating its availability and mobilisation in ferro-enzyme synthesis, energy metabolism, synthesis and transcription of RNA and DNA, and production of nucleotides, hormones and neurotransmitters.[6] Ferritin modulates intracellular iron storage and serves in binding and protecting tissues from excess iron. Ferritin synthesis is stimulated through increased demand for iron in inflammatory conditions, stress, mitosis, cell differentiation and repair processes.[9,32] Increased ferritin levels have been reported in some cancer cells without corresponding increases in iron-binding capacity. Recent research suggests that abnormalities in molecular upregulation of ferritin synthesis in cancer cells can deregulate iron homeostasis.[33,34]

In vitro studies in hybridoma cells show that increased cell membrane permeability to Fe(II) raises lipid peroxidise synthesis and iron storage, and leads to oxidative stress and DNA base modifications.[35] In non-neoplastic cells, intracellular iron is critical in controlling the mitotic cycle, notably completion of the S-phase of DNA replication, respiration and oxidative phosphorylation through the action of mitochondrial ferro-enzymes and electron-transport proteins.[9] Intra- and extracellular iron balance activates transferrin in the differentiation and maturation of lympho-myeloid cells in culture. In children with iron deficiency anaemia, consistently low apoptotic responses in neutrophils and monocytes are normalised by iron therapy.[36] Inhibition of iron transport and iron deficiency syndrome resulting from desferrioxamine binding or other chelators leads to decreased mitosis in erythropoietic tissues.[37]

Lead (Pb)(II) is a well known ferro-enzyme inhibitor, and depletion of δ-aminolaevulinic acid, dehydratatase ferrocheletase and other enzyme-dependent events results in disturbed iron homeostasis and anaemia.[19–22] Depletion of ferro-enzymes serves as a biomarker for lead exposure.

High iron intake can cause intestinal damage and impair copper and zinc uptake but importantly the association between chronic high iron levels in diet and colorectal cancer is a cause for concern. Whilst the full extent of the problem is unresolved, it is predicted that long-term induction of reactive oxygen species via the Fenton reaction may lead to cell damage as a consequence of lipid peroxidation and oxidative DNA and protein damage, and ultimately genomic instability and cancer.[38,39]

2.3 Occupational Exposures and Carcinogenic Risk in Haematite Mining

Although products made of iron and ferrous metals are encountered in everyday life, the greatest toxic risks of cancer are encountered in the mining and iron extraction, smelting and foundry industries (Figure 2.2). Workers are exposed to haematite dust in deep mines where blasting and pneumatic drills have been in regular use since the 1920s. The risks of cancer through iron exposures in surface workers and those employed in building and construction industries, automobile and aircraft manufacture, tool and equipment production and pigments for use in paints, plastics, printing, ceramics and textiles are similar to national averages. However, epidemiological and statistical evidence from studies in Europe, Asia and North and South America show conclusively that occupational exposure to haematite dust and mine contaminants increases the risk of lung cancer.[1,2,5] The cancer risk is significantly higher amongst underground workers but not surface workers, and over the past 50 years at least, there has been considerable debate over the principle causative factor(s). The discussion focuses upon whether increased lung cancer is attributable to inhalation of:

- Ferric oxide (haematite) dust
- Carcinogenic metal/metalloid contaminants
- Silica dust
- Radon and radioactive emissions
- Cigarette smoking
- Or to a combination of more than one cause.

An early statistical evaluation of risks encountered in haematite mining in four Cumberland pits in the UK over the period 1948–1967 showed that, of 5811 miners over the age of 15 dying in the 20-year period, 42 deaths were attributable to lung cancer, 74 to other cancers and 174 to respiratory diseases.[40] Risks of lung cancer varied widely according to profession but the number of lung cancer deaths observed was 50% higher than expected among all iron miners (underground and surface workers); cancers in other tissues and presumably

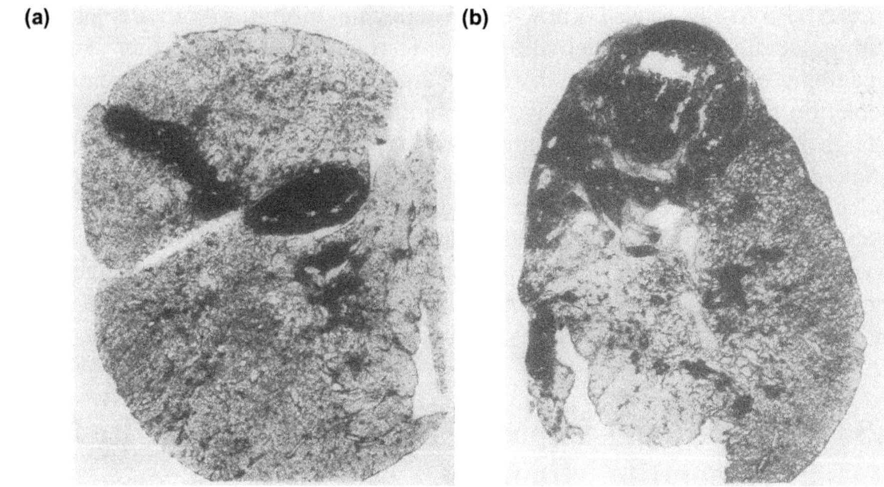

Figure 2.2 Pneumoconiosis or siderosis: lung pathology in Durham iron miners. The pictures show a histological profile of iron particles conglomerated in the lung alveoli at low and high power magnification: (a) progressive massive fibrosis involving the upper lobe and upper part of the lower lobe (P.M. 5016); (b) advanced stage of progressive massive fibrosis (P.M. 2939). (Reproduced from Faulds, *J. Clin Path*, 1957, **10**, 187.)

Table 2.1 Comparison of observed and expected mortality among iron mine employees in Cumberland mines 1947–1967. (Numbers expected from national experience in parentheses.)

Profession	*Clinical observation*		
	Lung cancer	*Other cancers*	*Respiratory disease*
Underground workers	36 (21)	65 (59)	159 (60)
Surface workers	6 (7)	9 (17)	13 (16)
All iron miners	42 (29)	74 (75)	174 (76)

attributable to other causes were similar to national averages (Table 2.1). The statistics also showed that miners working underground were exposed to a 75% higher risk of lung cancer mortality. Post-mortem reports of lung cancer fatalities in the Cumberland pits showed that most of the cancers seen were of the "oat-cell" type, a tumour type prevalent in uranium miners exposed to radioactive emissions.[41] Coal mining in Cumberland and elsewhere involves exposure to silica dust, haematite dust and radon.[42] There is no doubt that cigarette smoking amongst miners was a contributory factor in lung cancer incidence, but this was not shown on the *post-mortem* reports analysed.

Inhalation of haematite iron oxide *per se* is not considered by Boyd *et al.* to be a major cause of lung cancer in miners,[40,43] although it does pose a serious risk of pneumoconiosis and pulmonary fibrosis, with silica dust.[1] Siderosis caused by the retention of haematite dust in pulmonary and tracheal epithelia and alveolar macrophages in miners and above-ground workers employed in grinding, cracking and foundry work is expected to contribute to chronic respiratory disease and increase fatality.[44,45] Experimental studies in rodents have shown that haematite dust activates the carcinogenic effects of tar droplets instilled intra-tracheally in Syrian hamsters.[46]

Haematite as mined commonly contains unspecified amounts of impurities including Ti, Al and Mg, none of which is an established carcinogen. Conflicting reports of carcinogenicity in miners exposed to iron oxide dusts have been seen at permitted exposure limits (10 mg m^{-3}, measured as total particulate iron oxide dust and fume) (OHSA) but, presently, views are that lung and other cancers associated with iron/haematite mining are primarily attributable to the radioactive gas radon and not to iron or related ores. In a study of 171 workers employed in iron turneries and grinding professions, pulmonary changes manifest as siderosis were identified as "a ground glass effect" or reticulation on radiographic analysis.[47] The incidence was unrelated to age but to duration of iron oxide exposure. In 10 professions, chronic iron oxide inhalation was not linked to pulmonary or other cancers, and respiratory distress was noted in few workers. On the other hand, inhalation of foundry fumes in an iron and steel factory was associated with a 1.4 times increased risk of lung cancer, but in this situation the workers were exposed to atmospheres containing iron oxide, microcrystalline silica, organic binders and other additives in steel manufacture, many of which are carcinogens in chronic rodent studies and *in vitro* mutagenicity screens.[2] Rats implanted intra-bronchially with iron oxide pellets for 2 years failed to show a carcinogenic response. Additionally, workers exposed to factory and foundry fumes were probably exposed to carcinogenic polycyclic aromatic hydrocarbons in the fumes.

Risks posed by chronic silica dust exposures are well appreciated in haematite and other metal mining operations. Inhalation of microcrystalline silica of respirable size is listed as a profound human carcinogen by IARC and is discussed in detail in Chapter 8.[31,48] Chen *et al.* recorded that, of 490 fatalities in underground haematite miners, 42% died of silicosis or silicotuberculosis and 20% died of cancers (5.9% lung cancers).[3] Haematite mining is associated with an excess risk of non-malignant respiratory disease and lung cancer, with fatality rates greatly in excess of national averages. The IARC list foundry workers, furnace and ladle repair workers and fitters as trades most at risk of silica-related lung disease in the iron industry, and levels of respirable dust and quartz dust of 1.2 mg m^{-3} and 0.052 mg m^{-3} were given for at-risk furnace workers.[5]

There is overwhelming evidence now that exposure to radon gas in mining is a major cause of cancer. Radon is a colourless, odourless gas, with a half-life of 3.8 days that is present in mining as a degeneration product of uranium and radium. Like other actinide elements, radon is a known human carcinogen through emissions of α-particles.[3,31,49] A survey showed that the radon exposure in three or four Cumberland mines ranged from 30 to >300 pCi l^{-1}, and that none of the 22 measurements was below the permissible level recommended by the International Commission on Radiological Protection at the time (30 μCi l^{-1}). The link between radon exposure and lung cancer was first noted in silver, copper, nickel, cobalt, bismuth, arsenic radium and uranium mines and it was estimated that 40–75% of miners regularly employed died of lung cancer.[50] Since then, radon-related cancer has been recorded in coal, gold, arsenic, lead, tin and many other mining sites.

On the basis of available information from occupational health surveys and experimental evidence there is little to link exposure to haematite with lung or other cancers. There is good evidence to associate cancer risks in mining with exposure to microcrystalline dust in mines, foundry areas and in grinding industries, and with exposures to carcinogenic elements lead, arsenic, cadmium, *etc.* and emissions of carcinogenic metal/metalloid contaminants in smoke and polycyclic aromatic hydrocarbons in combustibles. It is unclear whether or to what extent these agents act independently as cancer inducers or co-operatively in the fashion of multi-step inducer–promoter carcinogenic mechanisms.

Stainless steel devices used in orthodontic surgery contain iron, nickel and chromium, the latter elements being associated with DNA damage in cultured cells. In an attempt to understand the risks associated with steel alloys used in dentistry, micronucleus tests and comet assays to detect DNA damage were conducted on buccal cells isolated from patients.[51] Both screens showed that the devices were capable of inducing DNA damage and showed genotoxic properties in orthodontic patients. However, steel in orthodontic devices does not seem to be a cause of oral cancers, which result more from the effects of smoking, betel quid/tobacco chewing, alcohol, carcinogen-contaminated diets and viral infections. The oral cancer prevalent in many parts of the world relates much to oral hygiene, overall health and dietary balances rather than the use of iron in dental devices.

2.4 Iron Overload and Hereditary Haemochromatosis

The Scientific Advisory Committee on Nutrition emphasised that inherited or acquired iron overload conditions are potentially harmful.[8] Consumption of excessive iron in red meat leading to a surge in the release of free radicals is a known cause of cardiovascular disease and possibly colorectal cancer.[8,38,39] Progression of colorectal cancers is possibly associated with increased expression of intestinal iron "import" proteins and a block in iron export due to decreased expression and aberrant location of the proteins hephaestin and ferroportin. These proteins promote cell proliferation and reduce cell adhesion. Hephaestin is a ferroxidase and homologous to the plasma cupro-protein caeruloplasmin. All residues involved in copper binding and disulfide bond formation in caeruloplasmin are conserved in hephaestin.

The most convincing evidence for the carcinogenic effects of the iron overload state is illustrated by hereditary or idiopathic haemochromatosis. Haemochromatosis mainly affects white Caucasian people of Northern Europe and North America, and affected patients are reportedly at a 20–200-fold risk of developing cardiomyopathies, cirrhosis and intrahepatic cancer.[52–55] Men are at greater risk than women. Other human cancers associated with iron overload conditions include breast cancer, leukaemias and colorectal cancers. Hereditary haemochromatosis is an autosomal recessive disorder caused by mutations in the human leucocyte antigen (HLA)-linked *HFE* gene, which is expressed on mucosal cell membranes and regulates the amount of iron absorbed from food.[56] Abnormally high levels of iron accumulate in parenchymal tissues, notably the liver, pancreas and skin, in patients exhibiting the mutations Cys282>Tyr and His63->Asp who have been linked to adult haemochromatosis. Diabetes is an effect of pancreatic iron overload, and iron overload also leads to an overproduction of toxic oxygen reactive species. Free oxygen radical damage increases the risk of liver, breast and colorectal cancers and childhood leukaemia.[12,57–60] Genetically modulated changes in iron transport and intracellular mobility also may underlie increased cell proliferation and reduced adhesion in colorectal carcinogenesis.[39]

Iron overload may lead to hepatocellular carcinoma through cirrhosis or in a direct fashion,[61] and the condition may be exacerbated by certain viral infections including hepatitis B and hepatitis C.[12,55] Iron overload conditions are more common in persons with appropriate HFE-genotypes,[62] but can occur under a range of clinical conditions including chronic alcoholism, cystic fibrosis, viral infections and adverse reactions to drugs. Additionally, Wilson's disease is an inherited disorder that causes excessive amounts of copper to accumulate in the liver and results in hardening of hepatocyte cellular walls and cirrhosis. The liver cannot metabolise or reduce the copper burden and the cirrhosis intensifies, creating a life-threatening crisis. Release of reactive oxygen radicals and increased oxidative activity is held to be responsible for both iron-induced cirrhosis and progression to hepatocellular cancers.[63,64]

2.5 Iron–Dextran

Iron–dextran complex was first developed in 1953 and the new complex soon became a widely used parenteral therapy for iron deficiency anaemia.[65] It is an iron complex containing ferric hydroxide and dextran with 5% of iron, with recommendations that it be administered by intramuscular injection into the gluteal muscle or by slow intravenous injection or intravenous infusion. Dose is determined by the body weight of the patient and level of iron-deficit (*British National Formulary*). The actual formula has been controversial but the presumed formula of $[3FeO_2OH–4(C_6H_{10}O_5).2H_2O]_n$ (Figure 2.3) is cleaved, particularly in Kupffer cells of the liver, to release metabolisable iron which binds ferritin and an innocuous saccharide moiety.[66,67] The relative molecular weight of the molecule is presumed to be between 150 000 and 300 000 Daltons. Iron–dextran given intravenously can be toxic and indications are that it should be given as parenteral iron therapy only to those patients with an inability either to absorb or to tolerate the oral intake of iron.[66]

Early observations that repeated injections of iron–dextran induce highly aggressive subcutaneous sarcoma in rats suggested that the complex can act as

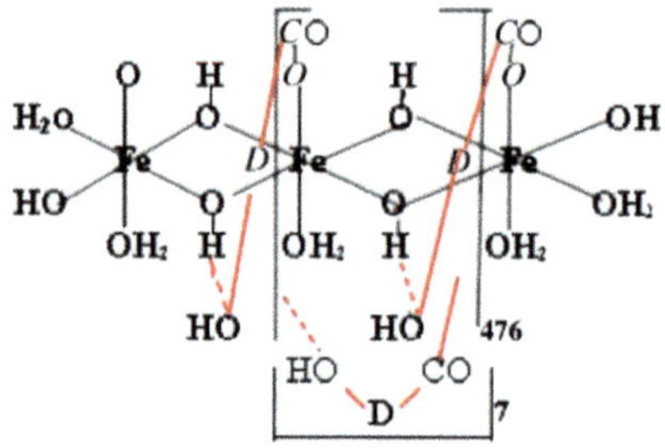

Figure 2.3 Proposed iron dextran core molecule.
(From London, *J. Pharma. Sci.*, 2004, **93**, 1838).

both inducer and promoter and is thus a complete carcinogen. Iron dextran therapy is listed in the 12th RoC as "reasonably anticipated to be carcinogenic in humans".[31] Richmond and others observed that repeated intramuscular injections of iron–dextran complex in rats led to iron-laden histiocytic and highly aggressive pleomorphic spindle cell tumours and fibrosarcomas at injection sites after 11–16 months, whereas the lower molecular weight dextran moiety was without effect.[68–70] In contrast, Langvad observed that mice injected with iron–dextran developed lymphoreticular tumours at sites distant from the injection site after a prolonged latent period, and that these tumours were unrelated to iron deposits.[71]

In clinical experience, Fielding reported that, in the 22 years following the introduction of iron–dextran therapy, only nine cases of malignancy had been recorded and only one patient showed a case history sufficient to implicate parenteral iron supplementation as a possible carcinogen.[72] Injection site tumours in anaemic patients purportedly due to iron–dextran injections include pleomorphic fibrosarcomas, squamous cell tumours, reticulum cell sarcoma, myxoid liposarcoma and rhabdomyosarcoma.[73–75] These variations in response to iron–dextran therapy may reflect individual variations in cellular sensitivities and inflammatory responses to local iron overload, or immune states.[76] Further variables relate to the actual amount of iron administered and the duration of the latent period between the final injection and appearance of sarcoma, which in some cases was in excess of 2 years.[77] The validity of subcutaneous sarcoma in rats in predicting human risk associated with iron–dextran therapy is seriously questioned.[78,79]

2.6 Experimental Studies

The capacity of iron and iron overload conditions to elicit carcinogenic changes is supported by experimental evidence that shows:

1. The ability of iron to promote formation of free radicals and oxidative changes in key tissues is a cause of mitotic abnormalities and mutagenic change.

2. That supplementary iron promotes carcinogenic changes initiated by 7,12-dimethylbenz[α]anthracene (DMBA), urethane and other carcinogens.[39,80–83]
3. Repeated subcutaneous injections of iron dextran and other parenteral iron supplements are a cause of sarcoma formation.[31,73]

Iron and some inorganic iron compounds express weak and inconsistent mutagenicity in *Salmonella typhimurium* reversion tests, lymphoma cell lines and hamster embryo cells but most studies emphasise that excess iron is a cause of cytotoxicity through generation of reactive oxygen species and related oxidative stress.[84–86] Fe(II) sulfate was shown to be cytotoxic and to induce polyploidy in human lymphoma cells treated during the G_1 phase, and it inhibited DNA synthesis. Delay in passage of cells through the cell cycle was attributed to an accumulation of cells in S and G_2 to M phases.[87] Mixed evidence is provided from *in vitro* and *in vivo* studies to show that reactive oxygen radicals and oxidative stress attributable to iron overload lead to carcinogenic change.[88] Iron-catalysed oxidative DNA damage is not random and probably involves upregulation of "target genes" regulating synthesis of iron transporter or regulatory proteins.[89]

Genetic and hormonal changes leading to increased iron accumulation, and related biochemical and cytopathic changes, predispose cells to neoplastic transformation. Experimental work has demonstrated that iron overload is a cause of DNA damage and biochemical changes associated with hepatocellular carcinoma, mammary carcinoma or renal carcinoma in rodents.[90–93] Experimental work has also shown that excess iron initiates mutagenic and genotoxic effects in human cells in culture. Similar observations were seen in mice given an excess of iron in their diet; genotoxic and mutagenic changes in bone marrow and peripheral lymphocytes were reported.[94] In workers exposed to fumes in iron and steel foundries, the relative risk of lung cancer was increased by a factor of 1.4, but this has not been substantiated in rodent bioassays in the absence of other contaminants.[1,2]

Experimental evidence has shown that iron can promote development of tumours in rats and mice induced by 7,12-dimethylbenz[a]anthracene (DMBA) or other known carcinogens.[80–81,95] Thus, mice fed a low iron diet (7 ppm), insufficient to cause liver damage, and given intraperitoneal urethane showed an 86% increase in the number of pulmonary adenomas.[83] At higher concentrations iron supplementation was toxic to mice; it increased liver weight and the iron content of hepatocytes, Kupffer cells and splenic macrophages and showed a dose-related pattern of liver pathology.[83] In two-step skin carcinogenesis studies with benzoyl peroxidase-induced dermal tumours in mice, iron overload led to an increased tumour incidence, correlating with iron concentrations in tissue.[80,81] Iron-induced oxidative stress and free radical formation were associated with raised xanthine oxidase and ornithine dicarboxylase activity and ^{3}H-thymidine uptake by cutaneous DNA. Further studies using DMBA or 12-*O*-tetradecanoylphorbol-13-acetate showed that iron-induced peroxidative damage in mouse skin enhanced tumour induction and reduced lag phases, presumably by a similar mechanism. In a transgenic strain of mice (Ah-responsive C57BL/10ScSn), oxidative stress induced by a single dose of iron significantly potentiated the induction and growth of liver tumours evoked by polychlorinated hydrocarbons (PCB).[94] The PCB mixture induced a state of porphyria through inhibition of uroporphyrinogen dicarboxylase in hepatic haem synthesis. Where body iron load and supplementary iron promote experimental tumour induction, the mechanism possibly involves a complex interaction between cell membrane permeability and the oxidative effects of reactive oxygen species on cellular physiology.[90,94]

In a rodent model, iron levels and haemocrit were inter-related in determining body weight gain and mammary tumour incidence in rats treated with 1-methyl-1-nitrosourea.[95] Animals maintained on a low iron diet (2 ppm) became iron deficient and showed a reduced carcinogenic effect, haemocrit and weight gain. When iron levels and haemacrit normalised on a normal ion diet, so the protective effect diminished. Higher dietary iron levels promoted tumour formation.

References

1. J. Faulds, *J. Clin. Path.*, 1957, **10**, 187.
2. C. D. Humfrey, L. S. Levy and S. P. Faux, *Food Chem. Toxicol.*, 1996, **34**, 1103.
3. S. Y. Chen, R. B. Hayes, S. R. Liang, Q. G. Li, P. A. Stewart and A. Blair, *Brit. J. Ind. Med.*, 1990, **47**, 175.
4. IARC, *Monographs of the Evaluation of Carcinogenic Risk to Humans*, IARC, Lyon, 1984, **34**, p. 1.
5. IARC, *Monographs of the Evaluation of Carcinogenic Risk to Humans*, Suppl., IARC, Lyon, 1987, **7**, p. 1.
6. J. H. Brock, *Iron in Immunity, Cancer and Inflammation*, ed. M. de Souza and J. H. Brock, J. Wiley & Sons, New York, 1989, p. 35.
7. H. Sussman, *Iron in Immunity, Cancer & Inflammation*, ed. M. De Sousa and J. H. Brock, John Wiley & Sons, New York, 1989, p. 261.
8. Scientific Advisory Committee on Nutrition (SACN), *Iron and Health*, SACN, London, 2010.
9. V. J. Kumpf and E. G. Holland, *Ann. Pharmacol.*, 1990, **24**, 162.
10. D. L. Burns and J. J. Pomposelli, *Kidney Int.*, 1999, **55**, S119.
11. P. I. Folb, *The Safety of Iron Dextran and a Comparison with Iron Sucrose for Intravenous Use: A short report to the World Health Organization Advisory Committee on the Safety of Medicines*, WHO, Geneva, 2004.
12. J. F. Blanc, P. Bioulac-Sage and C. Balabaud, *Bull. Acad. Nat. Med.*, 2000, **184**, 355.
13. J. G. Gao, W.-S. Yang, W. Zhang, S. Gao, J. Wang and Y.-B. Xiang, *Asian Pacific J. Cancer Prevent.*, 2012, **13**, 743.
14. D. J. Price and J. G. Joshi, *J. Biol. Chem.*, 1983, **25**, 10873.
15. C. T. Settlemire and G. Matrione, *J. Nutr.*, 1967, **92**, 153.

16. J. Flemming and J. G. Joshi, *Proc. Nat. Acad. Sci.*, 1987, **84**, 7866.
17. L. Hallberg, M. Brune and L. Rossander, *Hum. Nutr: Appl. Nutr.*, 1986, **40**, 97–113.
18. W. M. Ringsdorf and E. Cheraskin, *Oral Surg. Oral Med. Oral Pathol.*, 1982, **53**, 231.
19. S. S. Bottomly and U. Muller-Eberhard, *Semin. Hematol.*, 1988, **25**, 282.
20. W. W. Ku, D. Slowiejko, L. L. Bestervelt, M. R. Buroker and W. N. Piper, *Toxicol. in Vitro*, 1990, **4**, 763.
21. M. J. Warren, J. B. Cooper, S. P. Wood and P. M. Shoolingin-Jordan, *Trans Biochem. Sci.*, 1998, **23**, 217.
22. T. Sakai, *Ind. Health*, 2000, **38**, 127.
23. R. A. McCance and E. M. Widdowson, *Lancet*, 1937, **ii**, 680–684.
24. R. A. McCance and E. M. Widdowson, *Ann. Rev. Biochem.*, 1944, **13**, 315.
25. A. B. G. Lansdown, *Int. J. Cosmet. Sci.*, 2001, **23**, 129–137.
26. S. Miret, J. J. Simpson and A. T. McKie, *Ann. Rev. Nutr.*, 2003, **23**, 283.
27. D. Fu and D. D. Richardson, *Blood*, 2007, **110**, 752.
28. T. J. Peters, *Royal College of Pathologists (UK) Bulletin*, College Symposia, 2000, **109**, 1.
29. G. O. Latunde-Dada, S. Shirali, A. T. McKie, R. J. Simpson and T. J. Peters, *Eur. J. Clin. Invest.*, 2004, **34**, 626.
30. C. E. Carpenter and A. W. Mahoney, *Crit. Rev. Food. Sci. Nutr.*, 1992, **31**, 333.
31. US Department of Health and Human Resources, *12th Report on Carcinogens*, US DHHS, Research Triangle Park, NC, 2011.
32. E. Fibach, A. M. Konijn and E. A. Rachmilewicz, *Am. J. Hematol.*, 1985, **18**, 143–151.
33. J. T. Hazard and J. W. Drysdale, *Nature*, 1977, **265**, 755.
34. S. I. Shpyleva, V. P. Trondyak, O. Kovalchuk, A. Starlard-Davenport, V. P. Chekhun, P. A. Beland and I. P. Pogribny, *Breast Cancer Res. Treat.*, 2011, **126**, 63.
35. T. H. Zastawney, S. A. Altman, L. Randers-Eichhorn, J. A. Lumpkin, M. Dizdaroglu and G. Rao, *Free Radic. Biol. Med.*, 1995, **18**, 1013.
36. S. G. Berrak, M. Angaji and C. Turkkan, *Cell Prolif.*, 2007, **40**, 741.
37. H. G. Van Eijk and G. de Jong, *Biol. Trace Elem. Res.*, 1992, **35**, 13.
38. M. J. Brookes, S. Hughes, F. E. Turner, G. Reynolds, N. Sharma, T. Ismail, G. Berx, A. T. McKie, N. Hotchin, G. J. Anderson, T. Iqbal and C. Tselepis, *Gut*, 2006, **55**, 1449.
39. A. C. Chua, A. B. Klopcic, I. C. Lawrance, J. K. Olynyk and D. Trinder, *World J. Gastroenterol.*, 2010, **16**, 663.
40. J. T. Boyd, R. Doll, J. S. Faulds and J. Leiper, *Br. J. Ind.*, 1970, **27**, 97.
41. G. Saccomanno, V. E. Archer, R. P. Saunders, L. A. James and P. A. Beckler, *Health Phys.*, 1964, **10**, 1195.
42. M. J. Duggan, P. J. Soilleux, J. C. Strong and D. M. Howell, *Br. J. Industr. Med.*, 1970, **27**, 106.
43. C. G. Billings and P. Howard, *Monaldi Arch. Chest. Dis.*, 1993, **48**, 304.
44. A. I. G. MacLaughlin and H. E. Harding, *Arch. Ind. Health*, 1956, **14**, 350.

45. E. L. Kennaway and N. M. Kennaway, *Br. J. Cancer*, 1947, **1**, 260.
46. U. Saffiotti, F. Cefis, L. H. Kolb and P. Shubik, *J. Air Poll. Control Ass.*, 1965, **15**, 23.
47. M. Buchell, J. Gerrard, M. H. Jupe, A. I. G. MacLaughlin and K. M. A. Perry, *Monaldi Arch. Chest. Dis.*, 1993, **48**, 304.
48. IARC, *Monograph on the Evaluation of Chemical Risks to Humans*, IARC, Lyon, 1997, p. 68.
49. C. Edling and O. Axelson, *Br. J. Ind. Med.*, 1983, **40**, 182.
50. R. Doll, *Br. J. Ind. Med.*, 1959, **16**, 181.
51. G. H. Westphalen, L. M. Menezes, D. Prá, G. C. Garcia, V. M. Schmitt, J. A. Henriques and R. Medina-Silva, *Genet. Mol. Res.*, 2008, **7**, 1259.
52. C. Niederau, R. Fuscher, A. Sonneberg, W. Stremmel, H. J. Trampisch and G. Strohmeyer, *N. Engl. J. Med.*, 1985, **313**, 1256.
53. M. Elmberg, R. Hultkrantz, A. Ekbom, L. Brandt, S. Ilsson, R. Olsson, S. Lindgren, L. Wööf, P. Stål, S. Walllerstedt, S. Almer, H. Sandberg-Gertzén and J. Askling, *Gastroenterology*, 2003, **125**, 1733.
54. K. V. Kowdley, *Gastroenterology*, 2004, **127**, S79.
55. M. C. Kew, *Cancer Lett.*, 2009, **286**, 38.
56. T. Cox, *Bull. Royal Coll. Pathol. (UK)*, 2000, **109**, 3.
57. J. B. Porter, *Haemoglobin*, 2009, **33**(Suppl.), S37.
58. X. Huang, *Mutat. Res.*, 2003, **533**, 153.
59. B. K. Abraham, C. Justenhoven, B. Pesch, V. Harth, G. Weirich, C. Baisch, S. Rabstein, Y. D. Ko, T. Brüning, H. P. Fischer, S. Haas, S. Brod, C. Oberkanins, U. Hamann and H. Brauch, *Cancer Epidemiol. Biomarkers Prev.*, 2005, **14**, 1102.
60. M. T. Dorak, R. K. Mackay, C. L. Relton, M. Worwood, L. Parker and H. G. Hall, *Pediatr. Blood Cancer*, 2009, **53**, 1242.
61. R. S. Blumberg, S. Chopra, R. Ibrahim, J. Crawford, F. A. Farraye, J. B. Zeldis and M. D. Berman, *Gastroenterology*, 1988, **95**, 1399.
62. D. J. Brandhagen, W. Alvarez, T. M. Therneau, K. E. Kruckeberg, S. N. Thibodau, J. Ludwig and M. K. Porayko, *Hepatology*, 2000, **31**, 456.
63. M. T. Kuo and N. Savaraj, *Mol. Carcinogen.*, 2006, **45**, 701.
64. X. Quan, S.-O. Lim and G. Jung, *FEBS Lett.*, 2011, **585**, 3436.
65. E. London, *J. Pharm. Sci.*, 2004, **93**, 1838.
66. V. J. Kumpf and E. G. Holland, *DICP Ann. Pharmacother.*, 1990, **24**, 162.
67. B. B. Danielson, *J. Am. Soc. Nephrol.*, 2004, **15**, S93–S98.
68. H. G. Richmond, *Br. Med. J.*, 1959, **1**, 947.
69. F. J. C. Roe and R. L. Carter, *Int. J. Cancer*, 1967, **2**, 370.
70. R. L. Carter, *Br. J. Cancer*, 1969, **23**, 559.
71. E. Langvad, *Int. J. Cancer*, 1968, **3**, 415.
72. J. Fielding, *Scand. J. Hematol.*, 1977, **32**(Suppl.), 100–104.
73. C. E. G. Robinson, D. N. Bell and J. H. Sturdy, *Br. Med. J.*, 1960, **2**, 648.
74. J. D. Crowley and W. J. Still, *Br. Med. J.*, 1960, **1**, 1411.
75. A. E. MacKinnon and J. Bancewicz, *Br. Med. J.*, 1973, **2**, 277.
76. D. L. Burns, E. A. Mascoli and B. R. Bistrian, *Nutrition*, 1995, **11**, 163.
77. G. Greenberg, *Br. Med. J.*, 1976, **2**, 1508–1509.

78. P. Grasso, *Br. Med. J.*, 1973, **2**, 667.
79. H. K. Weinbren, R. Salm and G. Greenberg, *Br. Med. J.*, 1978, **1**, 683–685.
80. H. Rezazedeh and M. Athar, *Cancer Lett.*, 1997, **126**, 135.
81. H. Rezazeheh, P. K. Julka and M. Athar, *Skin Pharmacol. Appl. Skin Pharmacol.*, 1998, **11**, 98.
82. B. A. Diwan, K. S. Kasprzak and L. M. Anderson, *Carcinogenesis*, 1997, **18**, 1757.
83. F. L. Omara and B. R. Blakley, *Can. J. Vet. Res.*, 1993, **57**, 209.
84. V. C. Dunkel, R. H. San, H. E. Seifried and P. Whittaker, *Environ. Mol. Mutagen.*, 1999, **33**, 28.
85. Z. Elias, O. Poirot, O. Schneider, A. M. Marande, M. C. Danière, F. Terzetti, S. Pezerat, J. Fournier and R. Zalma, *Cancer Detect. Prev.*, 1995, **19**, 405–414.
86. M. Auffan, W. Achouak, J. Rose, M. A. Roncata, C. Chanéac, D. T. Waite, A. Masion, J. C. Woicik, M. R. Weisner and J. Y. Bottero, *Environ. Sci. Technol.*, 2008, **42**, 6730.
87. P. D. Lima, M. C. Vasconcellos, R. A. Montenegro, C. M. Sombra, M. O. Bahia, L. V. Costa-Lotufo, C. O. Pessoa, M. O. Moraes and R. R. Burbano, *Toxicol. in Vitro*, 2008, **22**, 723.
88. S. Okada, *Int. J. Clin. Oncol.*, 1998, **3**, 191.
89. S. Toyokuni, *Cancer Sci.*, 2009, **100**, 9.
90. A. Fukuda, T. Osawa, H. Oda, S. Tokokuni, K. Satoh and K. Uchida, *Arch. Biochem. Biophys.*, 1996, **329**, 39.
91. G. A. Asare, K. S. Mossanda, M. C. Kew, A. C. Paterson, C. P. Kahler-Venter and K. Siziba, *Toxicology*, 2006, **219**, 41.
92. D. Prá, S. I. Franke, R. Guilian, M. L. Yoheama, J. F. Dias, B. Erdtmann and J. A. Henriques, *Biometals*, 2008, **21**, 289.
93. G. C. Kabat and T. E. Rohan, *Cancer Causes Control*, 2007, **18**, 1047.
94. A. G. Smith, P. Carthew, B. Clothier, D. Constantin, J. E. Francis and S. Madra, *Toxicol. Lett.*, 1995, **82–83**, 945.
95. H. J. Thompson, K. Kennedy, M. Witt and J. Juzefyk, *Carcinogenesis*, 1991, **12**, 111–114.

CHAPTER 3
Zinc

3.1 Introduction

Zinc is a bivalent metal with chemical properties similar to those of cadmium and mercury in Group 12 (formerly 2B) of the Periodic Table. Metallic zinc is not found freely in natural deposits but zinc ores are widely distributed in the Earth's crust. China, Australia, Peru, Europe and Canada are the biggest zinc mining countries where metal sulfides abound in volcanic deposits, limestone, dolomite and the so-called Sedex deposits of shale, siltstone and sandstone. In all, 555 zinc-containing minerals have been identified, but ores of greatest commercial value are shown in Table 3.1. Sphalerolite or zinc blende (ZnS) is the principle ore but deposits often contain silver, gold, lead, arsenic, silica, bismuth, tin and lead. Other commonly found zinc minerals include marmatite (a zinc–iron sulfide complex), zincite (zinc oxide) and calamine (smithsonite), found in North America. Zinc deposits close to the Earth's surface are often converted to oxides and carbonates. Hemimorphite is a complex blue crystalline form of zinc with the formula $Zn_4Si_2O_7(OH)_2 \cdot (H_2O)$, found in Central America and in the veins and beds in stratified calcareous rocks in Siberia. These lustrous crystals are of gemstone value. They are suspected of having healing properties, although this may be folklore.

Zinc is a blue–white lustrous metal with a low melting point. It has 38 isotopes, some with half-lives ranging from a few milliseconds to several years, but five exist in a stable state. The physical and chemical properties of zinc make it valuable for a wide range of products. As a modest electrical conductor and malleable metal, it can be formed into sheets or wires for use in electrical components, semiconductors, batteries and fluxes in soldering and welding. Since zinc does not deteriorate because of weather conditions, it has found extensive use in metal plating and anticorrosion galvanised coating for iron and steel in building industries. Zinc silicates have many industrial uses. Large quantities of zinc are utilised to produce die castings used extensively by the

Issues in Toxicology No. 18
The Carcinogenicity of Metals: Human Risk through Occupational and Environmental Exposure
By Alan B. G. Lansdown

Published by the Royal Society of Chemistry, www.rsc.org

Table 3.1 Zinc ores of greatest commercial value.

Name of ore[a]	*Chemical composition*	*Concentration of extractable zinc (%)*	*Where found*
Sphalerite	ZnS	67	China, Canada, USA, Peru, India, Mexico, Australia, Central Europe
Hemimorphite	$Zn_4SiO_2O_7(OH)_2.H_2O$	54.2	Central Europe
Smithsonite	$ZnCO_3$	52	North America
Hydrozincite	$Zn_5(OH)_6.(CO_3)_2$	56	Central Europe
Zincite	ZnO	80.3	USA
Willemite	Zn_2SiO_4	58.5	USA
Franklinite	$(Zn,Fe^{2+})(Fe^{3+})_2O_4$ or $(Zn,Fe^{2+},Mn^{2+})(Fe^{3+},Mn^{3+})_2O_4$	15–20	New Jersey (USA)

[a]Many zinc ores naturally contain other metals including gold, copper silver arsenic, cadmium, lead, manganese and iron. There are many crystalline forms of differing crystal colour and configuration.

automotive and hardware industries. Brass is the best known of the zinc-containing alloys. It contains zinc (15–40%) and copper as its main constituents. The earliest brass, called calamine brass, dates to Neolithic times, when it was probably made by reduction of mixtures of zinc and copper ores. In ancient documents, including the Bible, the term "brass" is often used to denote bronze, an alloy of copper with tin. An alloy called Prestal(R), containing 78% zinc and 22% aluminium, is reported to be almost as strong as steel and as easy to mould as plastic. This alloy said to be so mouldable that it can be used in the production of inexpensive ceramics or cement die casts. The brilliant colours of zinc compounds such as zinc chromate (yellow) and zinc arsenate (orange) make them an obvious choice for use in pigments and preservatives in paints, analytical reagents in production of textiles, antifungal agents and fireproofing. Zinc silicate (willemite, Zn_2SiO_4), mined in New Jersey, can take the form of green crystals which are fluorescent under shortwave ultraviolet light. Manufacture and use of products containing zinc chromate or zinc arsenate present occupational risks in view of the toxicity and carcinogenicity of the arsenate and chromate anionic moieties.[1–4]

In recent years, considerable commercial use has been made of zinc oxide preparations in the wound-care industry, in zinc oxide paste, zincated bandages and related products. These are completely safe in use apart from occasional reports of allergy due to non-zinc components, and have proven of commendable value in tissue repair and regeneration, anti-infective protection and skin care.[5–7] The beneficial effects of zinc therapy have been recognised for many years, and clinical and experimental research has revealed the value of Zn(II) ion in wound closure, burn wound therapy, pilonodal sinus repair and treatment of infectious lesions. Zinc metalloenzymes are essential in cell proliferation and in the biosynthesis of matrix metalloproteins. Recent work has demonstrated that zinc-finger proteins promote cell proliferation, DNA synthesis, RNA packaging and transcriptional activation.[8,9] Zinc pyrithione (zinc omadine, ZPT, bis(1-hydroxy-2(1H)-pyridineselonoato-O,S)zinc), has been developed and approved by the US Food and Drug Administration (FDA) as an efficacious anti-dandruff preparation.[10,11]

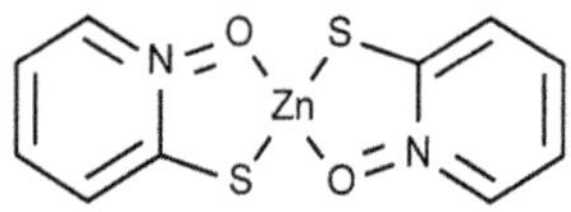

Zinc pyrithione

Zinc is well documented as an essential trace element in the human body and the biochemical basis of its role in the physiology of growth and tissue repair is well understood.[5,7,12,13] It is second to iron in importance as a nutrient and since the pioneer work of Elsie Widdowson and Robert McCance in the Dunn Laboratory in Cambridge, the nutritional role of zinc has been widely researched in animals and humans, and its importance in developing countries is well documented. Widdowson and McCance emphasised the significance of adequate zinc in the growth and development of children. The literature on this subject is enormous and much research has focused upon the clinical problems associated with zinc deficiency syndromes that abound in certain parts of the world, and the importance of inherited and acquired zinc deficiency syndromes, of which acrodermatitis enteropathica is best known.

Acrodermatitis is a rare lethal autosomal recessive disorder that affects young children. It is characterised clinically by a triad of conditions: alopecia, diarrhoea and dermatitis. Affected children show profound growth retardation, rough skin, alopecia, hypogonadism and infertility, reduced immune status, and failure to thrive, unless treated early.[14,15] Moynahan recommended small supplements of 35 mg of zinc sulfate daily as a effective treatment; the optimal daily intake of zinc is believed to be 150 mg in divided doses. Recent gene mapping and profiling relates acrodermatitis to a defect on chromosome 8, mapped to 8q24, with the defective gene being identified as SLC39A4.[16] This encodes a transporter protein, Zip 4, regulating the absorption and metabolism of zinc in the intestinal mucosa.[13,17]

People fed well balanced diets show a low incidence of zinc deficiency syndrome, but low blood zinc levels and symptoms of zinc deficiency are recorded in people subject to alcoholic cirrhosis, certain types of liver disease, active tuberculosis and Down syndrome.[5,18] Zinc and iron deficiencies prevalent in developing countries may affect nearly 2 billion individuals. Other major causes of dietary zinc deficiency states relate to the diets high in plant fibres and organic matter (phytates) consumed in Egypt, Iran and other middle eastern countries.[12,19,20] Villagers from Iran and Egypt eating diets comprised largely of bread and beans with a high content of phytate show deficiency in both iron and zinc. Plant fibres and organic matter are well known to bind or chelate free zinc ion, thereby reducing concentrations available for absorption.[12,19,20] The condition is particularly severe in young children and is a cause of retarded growth and infertility similar to that seen in patients with the acrodermatitis syndrome. Zinc deficiency in pregnancy is a cause of fetal growth retardation and problems with lactation. Zinc deficiency in children is a cause of immune dysfunction, aberrant B and T cell maturation and function, and increased susceptibility to intercurrent infectious diseases.[21] Zinc is possibly involved in the activation and suppression of immunoregulatory genes operating on lymphopoietic tissues including the thymus and lymph nodes. The thymus is highly sensitive to zinc deficiency syndrome and the thymic peptide thymulin, known for its immunomodulatory role, is zinc dependent.

Zinc is ubiquitous in the human environment and whilst all nutritional requirements are provided by food and drinking water, zinc can be absorbed into the body through inhalation of particles in the air and a small amount by percutaneous absorption. Hyperzincaemia can be corrected clinically by parenteral or dietary administration of non-proprietary zinc sulfate solution. Zinc oxide has been beneficial in treating skin wounds associated with local zinc (and copper) loss.[22] Percutaneous absorption of zinc is normally very low, even in the industrial environment where workers are in contact with metallic zinc and zinc alloys for long periods. Metallic zinc oxidises to a low level in the atmosphere and is insoluble in water, but whereas Zn(II) ion is released in the presence of acids, most binds strongly to epidermal keratins.[23] Some absorption does take place when zinc oxide is applied to damaged skin in amphiphilic formulations in wound care products and in sunscreens.[22,24,25] Zinc is absorbed through inhalation of zinc particles (ZnO) in the atmosphere and in industrial areas where zinc oxide dust and zinc emissions are released from combustion sources and industrial emissions. As discussed later, these emissions are a potential cause of "zinc fume fever" and severe respiratory distress, but not cancer.[26,27]

3.2 Zinc in Nutrition

The Zn(II) ion is an essential component of all living cells, and as long ago as 1869 Jules Raulin, working with Louis Pasteur, noted that mitotic failure occurred when the mould *Aspergillus niger* was deprived of zinc.[28] Subsequently, the phenomenon was observed in a range of bacteria, flagellates and other microflora and unicellular organisms. Vallee and Falchuk summarised "...that zinc is the only pre-, post- and transition element to be non-toxic whilst being indispensible for all forms of life and critical to transmission of the genetic message, development, growth and differentiation".[13] The largest part of the body zinc burden is provided by absorption of Zn(II) ion from food and drinking water, a small amount is absorbed percutaneously (especially from lipophilic vehicles), and hyperzincaemia has been observed following inhalation of zinc and zinc oxide dusts in occupational situations.[27,28] A 70 kg man can be expected to contain 2–3 g zinc.[5,6] Daily requirements for adults have been estimated at about 15 mg per day, and normozincaemia is in the range on 70–130 μg 100 ml^{-1}. Net zinc absorption is normally 20–30% of that ingested daily. Hyperzincaemic conditions are observed in human volunteers consuming or administered parenteral zinc sulfate.

Zinc absorbed into the body binds readily to albumins and macroglobulins. In the circulation; at least 50% of zinc ion is bound to albumins, the remainder to α_2-macro-globulins and transferrin (Figure 3.1).[29] In the blood, 75% of zinc is contained within erythrocytes, 22% in plasma and 3% in white blood cells; mean human serum (or plasma) zinc is within the narrow range of 70–130 μg 100 ml^{-1}.[5] As protein complexes, zinc is distributed widely throughout the body with highest concentrations located in brain, bone, liver, prostate, pancreas, thymus, muscle and basal stem cell populations of dermal and gastrointestinal epithelia (Table 3.2). The plasma pool is minor, 2–8%, but zinc reserves are labile and immunologically important. According to Rink and Gabriel, there is no specialised zinc storage in the human body and daily intake is essential in maintaining cellular homeostasis.[30] In the skin, levels of zinc are highest in proliferating in basal stem cell populations but progressively decline as cells undergo functional differentiation and maturation. Zinc gradients

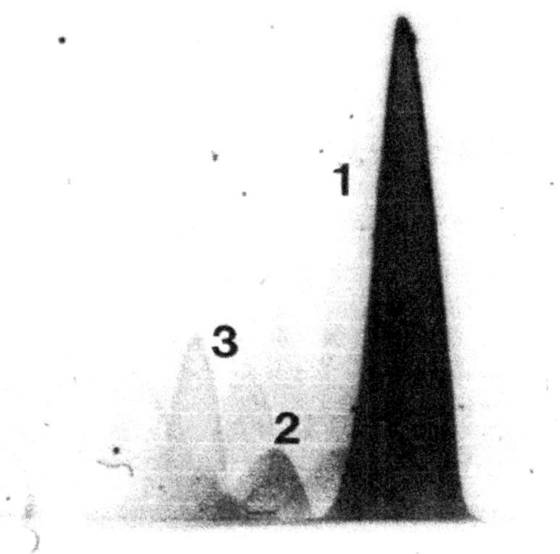

Figure 3.1 Autoradiograph of ^{65}Zn bound to serum proteins in a human serum sample: 1. albumin, 2. α_2-macroglobulin, 3. transferrin.[29]

Table 3.2 Zinc content of human organs.[30]

Tissue	*Zinc content* *Mg g^{-1} tissue weight*	*% Whole body zinc burden*
Hair	150	0.1
Bone	100	29
Liver	58	5.0
Kidney	55	0.7
Skin	32	6.0
Heart	23	0.4
Brain	11	1.5
Plasma	1	0.1

across the normal epidermis and in wound sites are critical in determining the functional maturation of the skin in health and injury.[5–7] Zinc concentrations in epidermal tissues and wound sites following injury are six-fold higher than in dermal tissues.[31] The skin and hair are sensitive indicators of zinc deficiency arising through nutritional deficiency, alcoholic stress and the lethal genome associated with acrodermatitis enteropathica.[13,14,32–35] Tissue injury, inflammation or systemic hypozincaemia trigger increased absorption of zinc to satisfy physiological needs, but excess zinc is voided *via* the skin, hair and nails as is seen with many other trace and xenobiotic metal ions.[35] Regulatory mechanisms for gastrointestinal zinc absorption are not known but are thought to involve a signalling process regulated by carrier proteins such as metallothioneins (MT), zinc transporters, hormones and cytokines (?interleukin 2).[12,13] Zinc is absorbed following topical application as zinc oxide or a soluble salt, but Zn(II) ion binds readily to serum proteins and to metallothioneins.[38,39]

3.3 Zinc in Cell Biology

Zinc absorbed intracellularly through the action of transporter proteins locates to all segments of the cell with at least 50% bound in nucleoproteins; the remainder is in cell membranes, mitochondria and cytosolic fractions.[13] Zinc is a recognised constituent of at least 300 key enzyme systems of which RNA/DNA polymerases, carbonic anhydrase, carboxypeptidases, alcohol dehydrogenase, alkaline phosphatase and zinc-finger proteins are of particular importance.[7,13,34,36,37,40–43] Zinc-finger proteins are amongst the most abundant proteins in eukarotic organisms where they function in DNA binding and post-transcriptional gene expression.[8,13] Matrix metalloproteinases constitute a large group of zinc-containing enzymes with a central role in wound repair and regeneration.[44,45] Vallee and Falchuk identified three main classes of zinc-related enzymes: oxidoreductases, transferases and hydrolases.[13] Zinc binds in these enzymes in a catalytic, coactive or structural manner. Zinc proteins/enzymes participate either directly in chemical catalysis or otherwise become involved in maintaining protein structure and stability.[46] Zinc metalloenzymes function at all synthetic phases of the cell cycle, reflecting its role in proliferation and intra-ionic balances; Zn : Cu, Zn : Ca and Ca : Mg are now known to be critical for the normal maturation and migration of post-mitotic cells and in maintenance of homeostatic rhythms in the skin and gastrointestinal tract, development of immune function and in endocrine secretion.

Zinc metalloenzymes and Zn–MT complexes perform an important role in intracellular activities and regulatory function, disturbances in which may underlie the role of the element in carcinogenic change or promotion of tumour formation (Figure 3.2). Evidence is presented to show that high zinc can be used to induce apoptosis in tumour cells.[47,48] In cultured hepatocytes, Zn–MT locates in the intermembrane spaces of mitochondria where it regulates respiration.[49] Mitochondrial zinc homeostasis is achieved by the action of thermodynamic and/or kinetic forces and through the action of certain cuproenzymes. The cyclo-oxygenase Cox-17, acting as a specific copper donor

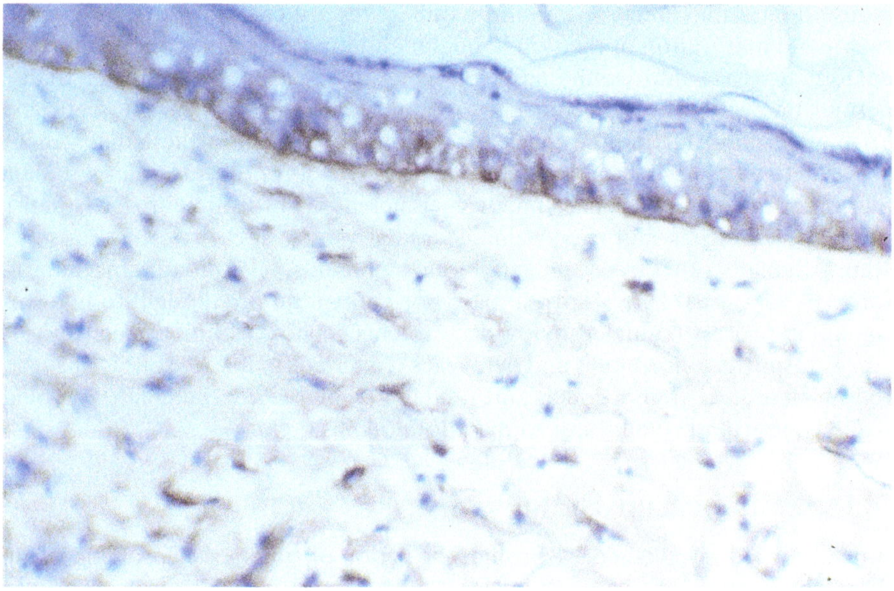

Figure 3.2 Metallothionein demonstrated in the basal layer of the epidermis following topical zinc oxide application.

or chaperone, is the only protein known to control homology of MT complexes and intracellular metabolism of zinc. Zn–MT binding occurs in a competitive manner with other metal ions including copper, silver, cadmium, mercury, selenium and arsenic.[38,39,49–51] Induction of MT synthesis is probably mediated through transcriptional control of the genes encoding MT, with the numbers of their specific mRNA transcripts increasing in response to zinc or other metals.[13] Copper and zinc interact in nutrition and intracellular biology, and high dietary intake of zinc inhibits copper absorption. Cadmium and lead ions also displace zinc from Zn–MT complexes and induce toxic change.[52] Immuno-histochemical demonstration of MTI or MTII serves as a convenient cellular marker of zinc in tissue and its location in wound healing. Intracellular zinc concentrations increase in epidermal cells and epidermal constructs cultured in the presence of zinc pyrithione (ZPT).[53] The mechanism is not clear but is thought to involve an upregulation of stress response genes encoding MT, heat shock proteins, antioxidants and the cell cycle inhibitor p21. ZPT is marginally toxic to epidermal cells in tissue culture, possibly through the action of the biocidal pyrithione moiety which depress adenosine triphosphate (ATP) synthesis in bacterial cells.

Zinc is located in all cellular compartments as enzyme cofactors or structural components. Its functions are tightly controlled through genetic expression and by metal ion balances. Approximately 50% of cellular zinc is within the nucleus, bound to RNA and DNA.[43,54] Normal human blood analyses show that erythrocytes contain 75%, plasma 22% and leucocytes 3% of whole blood zinc, with minor differences between males and females.[55,56] Individual leucocytes contain about 25 times the amount of zinc found in erythrocytes (Table 3.3).

Table 3.3 Normal blood zinc concentrations.[55,56]

Tissue	*Zinc concentration*
Whole blood	8.8 ± 2.0 μg ml^{-1}
Plasma	3.0 ± 1.6 μg ml^{-1}
Erythrocyte	0.014 ± 0.002 μg per 1 million cells
Leucocyte	0.032 ± 0.013 μg per 1 million cells

Cells deprived of zinc show mitotic arrest and failed functional differentiation. The cell cycle can be blocked at any stage, depending upon cell type and the level and duration of zinc deprivation. Zinc-deficient brain cells are blocked in G_0/G_1 transition phase, whereas hepatocytes are more sensitive during the G_1 period.[57,58] Human malignant lymphoblastic cells cultured in zinc-deficient medium showed reduced cell growth with an accumulation of cells in S phase; they failed to progress to subsequent stages in mitosis.[59] In this model, zinc acted in the form of a T-cell specific growth factor in the G_1 phase of the cell cycle. In a squamous cancer cell line, zinc deprivation led to suppression of zinc-finger protein expression and failure in growth and protein synthesis in the G_0–G_1 phase, leading to apoptosis.[60] In each case, we can presume that zinc sufficiency is responsible for upregulation and enhancing the expression of genes controlling specific phases of the cell cycle. Other evidence points to the role of zinc metalloenzymes in maintaining cell membrane stability and integrity, chromatin restructuring and immunosufficiency.[30] In summary, zinc performs a central role in regulating mitosis and apoptosis, protein synthesis and folding, and lipid binding, and in synthesis of zinc-dependent matrix metalloproteinases and other key enzymes involved in wound healing.[61] Research in molecular genetics in the past 50 years has shown that zinc metalloenzymes and zinc–protein complexes comprise a complex interactive network maintaining intracellular homeostasis as well as responses to signalling in synthetic and catalytic enzymes involved in tissue repair and regeneration.

Zinc-related intracellular mechanisms have evolved to permit cells to cope with zinc-deficiency stresses and adverse changes in their microenvironment. Experimental studies indicate that low tissue zinc and hypozincaemia seem to trigger transporter genes or import genes of the Zip1-6 type involved in Zn(II) ion absorption. In lactating mammary tissue, increased zinc uptake is modulated through transporter proteins located on cell surfaces.[62] Confocal microscopy has revealed that the transporter Zip-3 protein located on the cell membrane is transiently relocalised to the intracellular compartment in lactating cells actively absorbing Zn(II) under the influence of prolactin. The regulation of this absorptive process is not fully understood but research into acrodermatitis enteropathica suggests that it takes the form of a negative feedback, possibly involving hormones, growth factors, cytokines, *etc.*, which closes down when the normozincaemic state is achieved. Reduced zinc uptake, in mammary tissue at least, was signalled by the action of a Zip-3 "silencing" protein acting on cell membranes, continued action of which led to apoptosis and cell death.

3.4 Toxicity and Carcinogenicity

Zinc is a relatively non-toxic metal and its uptake and inter- and intracellular metabolism are regulated by a complex network of proteins including metallothioneins, which also serve as cytoprotectants against metal overload.[7,38,39,52] Cellular uptake of zinc and other metal ions is tightly regulated but hyperzincaemic situations arise as a result of:

- Idiopathic increased zinc-binding capacity of serum proteins
- Increase in intestinal absorption due to reduced metallothionein synthesis in mucosal cells[6]
- Ingestion of high concentrations of zinc from food
- Ingestion of acid foods and drinks stored in galvanised containers
- Parenteral administration of zinc supplements
- Inhalation of zinc fumes and dusts.

Increased blood and tissue zinc can be expected to upregulate expression of zinc-dependant genes and transporter proteins. Clinical and biochemical changes resulting from hyperzincaemia in dietary or occupational circumstances are reversible and without lasting changes in tissues and organs such as the liver, brain, haemopoietic tissue and pancreas which are sensitive to zinc deficiency states. Rare cases are recorded of hyperzincaemia in patients on renal dialysis or those consuming high levels of elemental zinc for idiopathic zinc deficiency syndrome or psychological reasons, in whom blood zinc may be many times higher than normal.[63,64] Characteristically, symptoms of hyperzincaemia include nausea, vomiting, fevers, abdominal pains, muscular cramps and blood disorders. One good example relates to a 20-year-old woman with psychiatric problems and a zinc-dependency syndrome. She consumed an average of 2300 mg of elemental zinc daily for four months and her blood zinc levels peaked at 7760 μg dl^{-1}. Her serum copper was greatly reduced 8 μg dl^{-1}, and the copper-binding protein caeruloplasmin was almost undetectable. The patient became severely anaemic and showed markedly elevated alkaline phosphatase levels, but these normalised when she was given oral copper therapy with vitamin C. Dietary intake of zinc at 3–20 times the recommended daily intake has been shown to impair serum lipid and cholesterol balance, with a detrimental influence on pancreatic function resulting in increased release of alkaline phosphatase, amylase and lipase.[65,66] Alkaline phosphatase is used as a marker of zinc metalloenzyme changes.[37] Zinc supplements have been shown in some studies to decrease Zn/Cu-superoxide dismutase activity owing to the antagonistic relationship between high zinc uptake and copper absorption.

Recent evidence has shown that increased cellular zinc can be toxic through its action as a mild antioxidant and its ability to invoke release of oxygen reactive species and oxygen stress in tissues.[63–70] This aspect of zinc toxicology is almost entirely restricted to *in vitro* studies at the moment, many of which are specifically directed to the study of the cytotoxicity of nanoparticulate zinc. Small size particles show higher cytotoxicity than the larger particles and induce greater mitototic inhibition, membrane damage and production of reactive oxygen species. Cytotoxicity in human lung cells was shown to be directly related to particle size and surface charge. Other cytotoxic changes attributable to zinc oxide-induced reactive oxygen species include increased lipid peroxidation, oxygen stress-related responses in epidermal constructs, excitation of inflammatory cytokines, apoptosis and cell death.[48,70–72] Prolonged oxidative stress, disturbances in lipid metabolism and elaboration of inflammatory cytokines may have a part in certain chronic conditions including atherosclerosis and cardiovascular disease, neurological disorders, immunological disorders and some malignancies.[73] Whilst there is no unequivocal evidence that high and prolonged zinc administration is a cause of human cancer, mutagenicity or changes in DNA, further research is clearly warranted.

3.4.1 Experimental Studies in Animals

Few satisfactory experimental studies have been conducted in animal models with a view to revealing the carcinogenic properties of zinc compounds. Concentrations of the solutions administered by the oral route (in drinking water) or parenteral injection have been high, and unrealistic in terms of human exposure. Many of the studies seen have focused upon the ability of zinc to induce neoplasia in the prostate, which shows a higher zinc level than most other tissues. In these studies, benign hyperplasia and prostatic cancer resulted.[74] Design of the experiments and statistical validation of the results are inadequate in predicting human disease. Ethical considerations in the older studies are unsound.

Early experimental studies showed that injection of soluble zinc compounds such as sulfate, nitrate or chloride directly into the testes of cockerels could induce teratoma or mixed tumours of the genital glands.[75–77] In his initial study in 1928, Michalowski induced teratoma by injection of small quantities of 5% zinc chloride directly into the testis.[75] Falin confirmed that the tumours were hormonally dependent and developed only when birds were injected during the January to March period.[76] The tumours were described as being of an embryonal character and composed of derivatives of all three germ layers. The stroma consisted of connective tissue "which differentiates in diverse directions". Other tissues identified in the teratomas included ossifying cartilage, smooth muscle fibres, neurological and optic tissues and mucus-secreting glandular structures. Considerable inter-strain variation was reported in the light and heavy strains of birds used in the studies, but the highest incidence of tumours occurred in White Leghorn chickens injected at 18 months of age.[78]

Teratomas have been induced also in mice or Syrian hamsters by intra-testicular or intra-scrotal injection of copper sulfate or zinc chloride with androgen stimulation.[79–81] The natural incidence of tumours of any type in hamsters is low, but Guthrie and Guthrie induced a low incidence of embryonal carcinoma in hamsters by injecting 0.05 ml of 4% zinc chloride during the season of high testicular growth.[82] Tumours were seen in only two animals after 10 weeks, but 25% of the tissue was subject to severe coagulative necrosis. The tumours were similar in structure and comprised dense cellular growth with prominent cells containing large nuclei, prominent nucleoli and numerous mitotic figures. Clearly the local concentrations of zinc injected were highly toxic and exceeded inherent protective mechanisms against metal overload (MT, protein binding, *etc.*). In six other animals, large areas of the testes were necrotic and no neoplastic tissue was visible. This study is statistically and ethically unsound and provides little predictive information for the carcinogenic risk of zinc salts in humans. Tumours induced by injecting zinc into hamsters were histologically different from those reported in chickens.

Mice administered massive doses of 5000 ppm of soluble zinc oleate (10% Zn) in the diet for 2 years developed a profound anaemia.[83,84] A lower dose of 1250 ppm zinc oleate failed to increase the incidence of any tumour type (notably lymphoma, lung tumours) significantly above control levels. However, the authors drew attention to a marginally increased dose-related incidence of hepatoma. This observation has not been confirmed by subsequent experiments. When high levels of zinc sulfate (13.3% Zn) (1000 or 5000 ppm) were administered in drinking water, there was no increase in tumour incidence of any type in a strain of mice. Experiments performed in presumed tumour-resistant and tumour-susceptible strains of mice dosed with 0–200 mg zinc chloride daily for 3 years in a multigeneration study also failed to provide statistically valid data on the carcinogenicity of zinc.[85] The experiments were statistically unsound even though 24 of 74 A/Sn and 33 of 76 C3H animals developed tumours. Female animals of a Porton strain of mice given 121.7 mg Zn m^{-3} zinc oxide/hexachlorophene dust daily for 5 days weeks for 20 weeks (100 exposures) showed a statistically significant increase in alveologenic carcinoma within 13 months following exposure, but no tumours were observed in guinea pigs or rats exposed in the dose range 1–121.7 mg Zn/m^{-3} dust for an equivalent period.[86] Hexachlorophene is toxic and potentially carcinogenic in mice.

When rats were injected intra-testicularly with zinc chloride (47% Zn), tumour incidence was low but one seminoma, nine interstitial cell tumours and one embryonal carcinoma developed.[87] In a second study, attempts to induce seminoma in albino rats by injection of 0.15 ml of 5% zinc chloride or 10% zinc sulfate failed, although 1 animal of 37 showed a massive retroperitoneal fibrosarcoma near the pancreas and another developed leukaemia with enlarged spleen and mesenteric lymph nodes.[88] A 2-year study in rats given 10, 25 or 50 ppm zinc pyrithione in their diet failed to provide evidence of carcinogenicity.[89]

Many reports show that zinc ion can modify carcinogenic responses to carcinogenic metals such as nickel and cadmium, and responses to hydrocarbons.[90–93] Low dietary zinc had negligible effect on the induction or growth of most tumours, whereas in animals with oesophageal cancer resulting from *N*-nitrosomethylbenzylamine, tumour growth regressed. In a 20-week experiment where oesophageal cancer was induced in rats by methylbenzylnitrosamine, excess zinc in the diet offered no protection, and where diets were changed in the course of the experiment, zinc was implicated in the induction and promotion of neoplasia. The intracellular mechanism is not entirely clear but would seem to implicate zinc-related enzymes in cell proliferation. Considerable experimental evidence exists to show that oral or injected zinc salts mitigate cadmium-induced testicular and prostatic tumours markedly, possibly through induction of MT.[90] Zinc-induced MT will bind cadmium, mercury, silver and other xenobiotic elements as a cytoprotective measure.[38,52,94]

Zinc chromate, arsenate and dithiocarbamate are carcinogenic in animals through the genotoxic and mutagenic action of the anion moieties; zinc chromate induced centrosome amplification and aneuploidy in cultured human lung fibroblasts.[95] Testicular tumours have been reported in mice, Syrian hamsters, and quail cocks following direct intra-testicular injection of zinc chloride, sulfate and nitrate.[96]

3.4.2 Human Epidemiological and Case studies

Epidemiological and case studies have demonstrated that workers and volunteers exposed to zinc experience transitory biochemical, haematological and physiological changes resulting from the antioxidant effects of raised blood zinc levels. Inhalation of zinc and zinc oxide dusts and fumes in occupational and factory environments is a cause of respiratory distress and the so-called "zinc fume fever", but evidence that zinc or zinc compounds are a cause of human cancer is very limited. Symptoms of metal fume fever may occur up to 24 hours post-exposure with cough, dyspnoea, sore throat, chest tightness, headache, fever, rigors, myalgia, arthralgia and sometimes a metallic taste, nausea, vomiting and blurred vision. Chest X-ray may show transient ill-defined opacities but there are typically no delayed sequalae. In a clinical trial involving 12 healthy men aimed at comparing the influence of inhaling 500 μg m^{-3} ultrafine zinc oxide, the same volume of fine zinc oxide particles, and filtered air, no obvious differences were seen among the three groups after 2 hours.[25] If the results of an experiment in rats are true for humans, inhaled zinc oxide dissolves in broncho-alveolar fluids to be cleared from the lungs within hours. It is notable that zinc oxide instilled into rat lungs at concentrations of 20–1000 μg per animal strongly induced MT within two days and even at the lowest installates increased β-glucuronidase activity and the protein content of pulmonary lavage.

Numerous studies are published where hyperzincaemic states have been induced in volunteers by administration of supplementary zinc in the diet. Individuals given 100–300 mg Zn per day, many times the recommended dietary allowance of 15 mg Zn per day, developed nausea, vomiting, epigastric pain, abdominal cramp, lethargy and fatigue within a few days. Copper metabolism and the copper-binding protein caeruloplasmin were depressed, but in most cases these symptoms regressed and the body chemistry normalised as body zinc levels normalised. Excess zinc and copper in some people with high hepatic zinc deposits has been associated with progressive cholestasis, notably in Cree and Ojibwa-Cree children in Canada.[97]

Higher concentrations of airborne zinc particles and zinc oxide dust are expected in mining, refining, soldering, galvanising, alloy production and manufacturing industries. Workers in these professions will be exposed to environments containing dust and fumes from zinc ores together with other contaminants; some of which, such as arsenic and cadmium, are known carcinogens. In two epidemiological studies workers did not show an increased incidence of lung or other cancers associated with occupational exposure to zinc in zinc and copper refining industries.[98,99] Zinc oxide dust is regarded primarily as a nuisance dust, but exposures to high concentrations are a potential cause of pulmonary inflammation and respiratory distress, but not lung cancer. Epidemiologic studies of zinc refinery workers reviewed in 1991, found no correlation between industrial zinc exposures and lung or other types of cancer.[100–102] However, in view of the improved environmental monitoring systems and health records required now in mines and other industries, causation of cancer-related mortalities in zinc industries requires revision. Strong evidence exists for a relationship between exposure to such compounds as zinc chromate in pigments and preservatives, and zinc arsenate in pesticides and carcinogenic risk, but the association exists between the anionic moieties *i.e.* arsenate As(III) and chromate Cr(VI) and cancer, and not the Zn(II) ion.[2]

3.5 Conclusions

Zinc should be regarded as one of the least toxic elements in the human body. It serves vital functions in maintaining body growth and maturation, tissue homeostasis, and repair following injury, immunity against infection, and overall health and vitality.[5–7,24,73] Considerable clinical attention has been given to the importance of zinc in nutrition, and intrinsic and dietary factors influencing its absorption and metabolism. Moynahan and others have contributed much to the understanding of potentially lethal zinc deficiency syndromes such as acrodermatitis enteropathica, and the value of injectable zinc sulfate therapy as an efficacious treatment.[14,15] There is no evidence that zinc is carcinogenic in humans following long-term inhalation of metallic zinc or zinc oxide particles in industrial settings. Metallic zinc and most of the inorganic and organic compounds are not recognised as being carcinogenic in humans by the US Department of Health and Human Safety (USDHHS) or the International Agency for Research on Cancer (IARC), and there is very limited evidence that administration of supra-optimal levels of zinc compounds will induce testicular cancers in experimental animals. Zinc arsenate and zinc chromate are carcinogenic in view of the properties of the arsenate and chromate anions released in body fluids.

The Zn(II) compounds, notably zinc oxide, used in wound care products, anti-infective preparations and in semiconductor materials, have failed to provide positive results in *Salmonella typhimurium* reverse mutation assays, and there is little evidence for mutagenic change or clastogenic properties in cultured cell systems.[103] Clinical and experimental evidence shows that supplementary zinc promotes mitosis in sensitive cells through up-regulation of RNA/DNA polymerases, integrins and zinc-dependent enzymes.[104] Zinc is a mild oxidising agent and induces release of oxygen reactive radicals and oxidative stress, but this is not linked to genotoxic change. Zinc supplementation can mitigate the carcinogenic action of certain other metals through induction of cytoprotective metal binding proteins (MBP) such as MTI or MTII, or through the zinc-finger proteins.[105] Carcinogens including nickel, chromium, arsenic and cadmium interact with and displace zinc from zinc-finger proteins as an integral part of generating free radicals to induce DNA damage.[67,106]

References

1. Environmental Protection Agency (EPA), *Toxicological Review of Zinc Compounds (CAS No. 7440-66-6)*, Washington, DC, 2005.
2. Occupational Safety and Health Administration, *Occupational Safety and Health Guideline for Zinc Chromate*, OSHA, Washington, DC, 1993.
3. S. Kapaj, H. Peterson, K. Liber and P. Bhattachayra, *J. Environ. Sci. Health: A Toxic Hazard Subs. Environ. Eng.*, 2006, **41**, 2399.
4. K. Jomova, Z. Jenisova, M. Feszterova, S. Baros, J. Liska, D. Hudecova, C. J. Rhodes and M. Valko, *J. Appl. Toxicol.*, 2011, **31**, 95.
5. J. A. Halstead and J. C. Smith, *Lancet*, 1970, **i**, 322.
6. R. D. Lindeman and B. J. Mills, *Mineral Electrol. Metab.*, 1980, **3**, 223.
7. A. B. G. Lansdown, U. Mirastschijski, N. Stubbs, E. Scanlon and M. S. Ågren, *Wound Repair Regen.*, 2007, **15**, 2.
8. J. H. Laity, B. M. Lee and P. E. Wright, *Curr. Opinion Struct. Biol.*, 2001, **11**, 39.
9. M. Papworth, P. Kolasinska and M. Minczuk, *Gene*, 2006, **366**, 27.
10. W. T. Gibson, M. Chamberlain, J. F. Parsons, J. E. Brunskill, D. Leftwich, S. Lock and R. J. Safford, *Food Chem. Toxicol.*, 1985, **23**, 93.
11. W. T. Gibson, W. S. Hardy and M. H. Groom, *Food Chem. Toxicol.*, 1985, **23**, 103.
12. A. S. Prasad, *Spec. Topics Endocrinol. Metab.*, 1985, **7**, 45.
13. B. L. Vallee and K. H. Falchuk, *Physiol. Rev.*, 1993, **73**, 79.
14. E. J. Moynahan, *Lancet*, 1974, **ii**, 399.
15. P. M. Barnes and E. J. Moynahan, *Proc. Roy. Soc. Med.*, 1973, **66**, 327.
16. E. Maverakis, M. A. Fung, P. J. Lynch, M. Draznin, D. J. Michael, B. Ruben and N. Fazel, *J. Am. Acad. Dermatol.*, 2007, **556**, 116.
17. M. L. Ackland and A. Michalczyk, *Genes Nutr.*, 2006, **1**, 41.
18. M. J. Eckhardt, T. C. Harford, C. T. Kaelber, E. S. Parker, L. S. Rosenthal, R. S. Ryback, G. C. Salmoiraghi, E. Vanderveen and K. R. Warren, *J. Am. Med. Assoc.*, 1981, **246**, 648.
19. W. J. Pories, J. H. Henzel, C. G. Rob and W. H. Strain, *Lancet*, 1967, **i**, 121.
20. A. S. Prasad, *Br. Med. J.*, 2003, **326**, 409.
21. C. L. Keen and M. E. Gershwin, *Ann. Rev. Nutr.*, 1990, **10**, 415.
22. J. E. Derry, W. M. McLean and J. B. Freeman, *J. Parenteral Enteral Nutr.*, 1983, **7**, 131.
23. A. B. G. Lansdown, *Food Chem. Toxicol.*, 1991, **29**, 57.

24. M. S. Ågren, *Dermatologica*, 1990, **180**, 36.
25. B. Gulson, M. McCall, M. Korsch, L. Gomez, P. Casey, Y. Oytam, A. Taylor, M. McCulloch, J. Trotter and L. Kinsley, *Toxicol. Sci.*, 2010, **118**, 149.
26. R. G. Cooper, *Indian J. Occup. Environ. Med.*, 2008, **12**, 10.
27. J. L. Malo, J. Malo, A. Cartier and J. Dolovich, *Eur. Resp. J.*, 1990, **3**, 111.
28. L. L. Raulin, *Ann. Sci. Nat.*, 1869, **11**, 93.
29. B. J. Scott and A. R. Bradwell, *Clin. Chem.*, 1983, **29**, 629.
30. L. Rink and P. Gabriel, *Proc. Nutr. Soc.*, 2000, **59**, 541.
31. M. M. Molokhia and B. Portnoy, *Br. J. Dermatol.*, 1969, **81**, 759.
32. R. I. Ecker and A. L. Schroeter, *Arch. Dermatol.*, 1978, **114**, 937.
33. D. Gaveau, F. Piette, A. Cortot, V. Dumur and H. Burgoend, *Arch. Dermatol. Venereol.*, 1987, **114**, 39.
34. A. S. Prasad, *Nutrition*, 1995, **11**, 93.
35. M. Wilhelm, D. Hafner, I. Lombck and F. K. Ohnesorge, *Sci. Total Environ.*, 1991, **103**, 199.
36. J. R. Schwartz, R. G. Marsh and Z. D. Draeios, *Dermat. Surg.*, 2005, **31**, 837.
37. M. Yamaguchi and R. Yamaguchi, *Biochem. Pharmacol.*, 1986, **35**, 773.
38. I. Bremner, *Progr. Food Nutr.*, 1987, **11**, 1.
39. A. B. G. Lansdown, *Wound Rep. Regen*, 2001, **82**, 35.
40. G. J. P. Filion, S. Zhenilo, S. Salozhin, D. Yamada, E. Prokhortchouk and P.-A. Defossez, *Mol. Cell. Biol.*, 2006, **26**, 169.
41. E. Nagata, F. F. Liew, C. Uwatoko, S. Kiyonaka, Y. Mori, Y. Katsuda, M. Endo, H. Sugiyama and T. Mori, *Angew. Chem. Int. Ed.*, 2012, **51**, 2421.
42. A. S. Prasad and D. Oberleas, *Lab. Clin. Med.*, 1973, **82**, 461.
43. R. E. Thiers and B. L. Vallee, *J. Biol. Chem.*, 1957, **226**, 911.
44. M. S. Ågren, *Arch. Dermatol. Res.*, 1999, **291**, 583.
45. W. C. Parkes, C. L. Wilson and Y. S. López-Boado, *Nature Rev. Immunol.*, 2004, **4**, 617.
46. K. A. McCall, C.-C. Huang and C. A. Fierke, *J. Nutr.*, 2000, **130**, S1437.
47. R. B. Franklin and L. C. Costello, *J. Cell. Biochem.*, 2009, **106**, 750.
48. M. J. Akhtar, M. Ahamed, S. Kumar, M. M. Khan, J. Ahmad and S. A. Alrokayan, *Ind. J. Nanomed.*, 2012, **7**, 845.
49. B. Ye, W. Maret and B. L. Vallee, *Proc. Nat. Acad. Sci.*, 2001, **98**, 2317.
50. U. Rana, R. Kothinti, J. Meeusen, N. M. Tabatabai, S. Kresoski and D. H. Peterling, *J. Inorgan. Biochem.*, 2008, **102**, 489.
51. D. M. Durnan and R. D. Palmiter, *J. Biol. Chem.*, 1981, **256**, 5712.
52. A. B. G. Lansdown, B. Sampson and A. M. Rowe, *Int. J. Exp. Pathol.*, 2001, **82**, 35.
53. S. D. Lamore and G. T. Wondrak, *Biometals*, 2011, **24**, 875.
54. T. Fujii, *Nature*, 1954, **174**, 1108.
55. B. L. Vallee and J. G. Gibson, *J. Biol. Chem.*, 1948, **176**, 435.
56. B. L. Vallee and J. G. Gibson, *Blood*, 1949, **4**, 455.

57. M. S. Clegg, J. M. Rogers, R. M. Tucker, L. S. Hurley and C. L. Keen, *Fed. Proc.*, 1986, **45**, 1086.
58. J. K. Chesters, in *Zinc in Human Biology*, ed. C. F. Mills, Springer-Verlag, London, 1989, p. 109.
59. A. S. Prasad, F. W. Beck, L. Endre, W. Handschu, M. Kukurunga and G. Kumar, *J. Lab. Clin. Med.*, 1996, **128**, 51.
60. W. Guo, Y. B. Zou, Y. G. Jiang, R. W. Wang, Y. P. Zhao and Z. Ma, *Tumour Biol.*, 2011, **32**, 810.
61. M. S. Ågren, U. Ostenfeld, F. Kallehave, Y. Gong, K. Raffn, M. E. Crawford, K. Kiss, A. Friis-Moller, C. Gluud and L. N. Jorgensen, *Wound Repair Regen.*, 2006, **14**, 526.
62. S. L. Kelleher and B. Lönnerdahl, *Am. J. Physiol.*, 2005, **258**, C1042.
63. E. D. Gallery, J. Blomfield and S. F. Dixon, *Br. Med. J.*, 1972, **4**, 331.
64. C. C. Pfeiffer, R. Papaioannou and A. Sohler, *Orthomolec. Psychiatr.*, 1980, **9**, 79.
65. S. Hughes and S. Samman, *J. Am. Coll. Nutr.*, 2006, **25**, 285.
66. J. Nriagu, *Zinc Deficiency in Human Health*, Elsevier, 2007.
67. B. Sarkar, *Nutr.*, 1995, **11**(Suppl.), 646.
68. T. Xia, M. Kovochich, M. Liong, L. Mädler, B. Gilbert, H. Shi, J. I. Yeh, J. I. Zink and A. E. Nel, *ACS Nano.*, 2008, **2**, 2121.
69. M. Baek, M. K. Kim, H. J. Cho, J. A. Lee, J. Yu, H. E. Chung and S. J. Choi, *J. Phys.: Conf. Ser.*, 2011, **304**, 012044 doi.
70. V. Sharma, D. Anderson and A. Dhawan, *Apoptosis*, 2012, **17**, 852.
71. A. Lipovsky, Z. Tzitrinovich, H. Friedmann, G. Applerot, A. Gedanken and R. Lubart, *J. Phys. Chem. C*, 2009, **113**, 1599772.
72. R. K. Dutta, B. P. Nenavathu, M. K. Gangishetty and A. V. Reddy, *Colloids Surf. B Interfaces*, 2012, **94**, 143.
73. A. S. Prasad, *Curr. Opinion Clin. Nutr. Metab. Care*, 2009, **12**, 646.
74. V. Y. Zaichick, T. V. Sviridova and S. V. Zaichick, *Int. Urol. Nephrol.*, 1997, **29**, 656.
75. I. Michalowski, *Virchows Arch. F. Path Anat.*, 1928, **267**, 27.
76. L. I. Falin, *Am. J. Canc.*, 1940, **18**, 199.
77. J. Guthrie, *Br. J. Canc.*, 1964, **18**, 130.
78. R. L. Carleton, N. B. Friedman and E. Bomze, *Cancer*, 1953, **6**, 464.
79. V. M. Bresler, *Probl. Oncol.*, 1959, **5**, 24.
80. V. M. Bresler, *VIIth International Cancer Congress Abstracts*, Moscow (Medgiz), 1962, 202.
81. C. L. Stevens and C. C. Little, *Proc. Nat. Acad. Sci., Wash.*, 1954, **40**, 1080.
82. J. Guthrie and O. A. Guthrie, *Canc. Res.*, 1974, **34**, 2612.
83. M. Walters and F. J. C. Roe, *Food Cosmet. Toxicol.*, 1965, **3**, 271.
84. E. Boyland and F. J. C. Roe, *Food Cosmet. Toxicol.*, 1963, **1**, 161.
85. E. Halme, *Vitalstoffe*, 1961, **6**, 59.
86. T. C. Marrs, H. F. Colgrave, J. A. Edgington, R. F. Brown and N. L. Cross, *Arch. Toxicol.*, 1988, **62**, 123.

87. M. R. Riviere, I. Chouroulinkov and M. Guerin, *Bull. Assoc. Fr. Etudes Canc.*, 1960, **47**, 55.
88. J. Guthrie, *Br. J. Cancer*, 1956, **10**, 134.
89. Arch Chemicals, Inc., *Materials Safety Data*- Zinc Omadine™ powder, EPA, Reg. No. 1258-1223, 2003.
90. M. P. Waalkes, S. Rehm, C. W. Riggs, R. M. Bare, D. E. Devor, L. A. Poirier, M. L. Wenk and J. R. Henneman, *Canc. Res.*, 1989, **49**, 4282.
91. L. Y. Y. Fong, T. Nguyen and J. L. Farber, *J. Nat. Canc. Inst.*, 2001, **93**, 1525.
92. R. Lee, W. Woo, A. Kummar, H. Duminy and Z. Xu, *Exper. Biol. Med.*, 2003, **228**, 689.
93. M. Newberne, T. F. Schrager and S. Broitman, *Pathobiology*, 1997, **65**, 39.
94. A. T. Mile, G. M. Hawkesworth, J. H. Beattie and V. B. Rodilla, *Crit. Rev. Biochem.*, 2000, **35**, 35.
95. A. L. Holmes, S. S. Wise, A. E.-M. Aboueissa, W. Lingle, J. Salisbury, J. Gallagher and J. P. White, *Chem. Res. Toxicol.*, 2010, **23**, 386.
96. M. Costa, *Metal Carcinogenesis Testing: Principles and In Vitro Methods*, Springer, 1980.
97. M. J. Phillips, C. A. Ackerley, R. A. Superina, E. A. Roberts, R. M. Filler and G. A. Levy, *Lancet*, 1996, **347**, 866.
98. J. N. Logue, M. D. Koontz and M. A. Hattwick, *J. Occup. Med.*, 1982, **24**, 398.
99. J. S. Neuberger and J. G. Hollowell, *Sci. Total Environ.*, 1982, **25**, 287.
100. W. S. Beckett, D. A. Chalupa, A. Pauly-Brown, D. M. Speers, J. C. Stewart, M. W. Frampton, M. J. Utell, L.-S. Huang, C. Cox, W. Zareba and G. Oberdörster, *Am. J. Resp. Crit. Care Med.*, 2005, **171**, 1129.
101. American Conference on Governmental Industrial Hygienists (ACGIH), *Documentation of the Threshold Limit Values and Biological Exposure Indices*, ACGIH, Cincinnati, OH, 6th edn, 1991.
102. US Department of Health and Human Services, National Toxicology Program, *Report on Carcinogens*, NTP Research Triangle Park, NC, 12th edn, 2011.
103. A. Léonard, G. B. Gerber and F. Léonard, *Mutat. Res.*, 1986, **165**, 343.
104. I. Tennaud, I. Sainte-Marie, O. Jumbou, P. Litoux and B. Dréno, *Br. J. Dermatol.*, 1999, **140**, 26.
105. Y. Qin, D. Thomas, C. P. Fontaine and R. A. Colvin, *Neurosci. Lett.*, 2009, **450**, 206.
106. A. Hartwig, *Toxicol. Lett.*, 1998, **102–103**, 235.

CHAPTER 4

Chromium and Chromates

4.1 Introduction

Chromium is a hard, shiny, steel-grey transitional metal of Group VI of the Periodic Table. It may be the fourth most abundant metal in the Earth's crust and is widely distributed in the soils and seas around the world, particularly in South Africa, Russia, southeast and northern Europe and the Philippines where it is mined. It is a reactive element with four main oxidation states Cr(II), Cr(III), Cr(IV) and Cr(VI), of which only Cr(III) and Cr(VI) are stable (Table 4.1). Metallic chromium is readily oxidised and does not occur naturally, the commonest chrome ores being chromite ($FeCr_2O_4$), crocoites (Siberian red lead) and many brightly coloured silicates. These ores are often associated with iron, aluminium and magnesium in major mining sites. Cr(II) compounds readily oxidise to Cr(III) analogues which have micro-trace metal value in mammalian nutrition. Cr(VI) compounds are presently classified as known human carcinogens on the sufficiency of human epidemiological evidence and case studies supported by experimental evidence of mutation and genotoxic effects in bacterial tests and human cells in culture.[1,2] In the USA, Cr(VI) accounts for at least 35% of all chromium found in the human environment.[3] Despite the value of Cr(III) as a micro-trace element, Cr(VI) compounds, notably the chromates and dichromates, are amongst the most toxic metals in the human environment; persons exposed to them occupationally experience a high incidence of chrome ulcers, chromium-related dermatitis and contact allergy, respiratory diseases through inhalation of gases and fumes in welding and foundry work, and cancers of the lung and respiratory tract, skin and intestinal tract.

Humans are exposed to chromium and its compounds in everyday life through dermal contact, ingestion in drinking water and food, inhalation of chromium dusts and vapours in industry, and through implantation of orthopaedic prostheses composed of chromium alloy. Chromium particles of

Issues in Toxicology No. 18
The Carcinogenicity of Metals: Human Risk through Occupational and Environmental Exposure
By Alan B. G. Lansdown

Published by the Royal Society of Chemistry, www.rsc.org

Table 4.1 Common trivalent, Cr(III) and hexavalent, Cr(VI) compounds of chromium.

Cr(III)	*Cr(VI)*
Chromic acetate	Chromium (IV) oxide
Chromic (III) nitrate	Ammonium dichromate
Chromic (III) chloride	Ammonium chromate
Chromic chloride stearate complex	Barium chromate
Chromic (III) chloride	Calcium chromate
Ferrochromite	Chrome orange pigments ($CaCrO_4$)
Chromic (III) oxide (Cr_2O_3)	Chromic acid (H_2CrO_4)
Chromic (III) phosphate	Chromium chloride ($CrCl_6$)
Chromic (III) sulphate	Chromium trioxide (CrO_3)
Sodium chromite	Lead chromate
Potassium chromic sulfate (chrome alum)	Potassium chromate
	Potassium dichromate
	Sodium chromate
	Sodium dichromate
	Strontium chromate
	Zinc chromate

about 1 μm diameter are commonly present in the air but settle to the ground or waterways after a short time, aided by rain or snow. Airborne Cr(VI) is reduced environmentally to Cr(III) by iron and vanadium compounds and by sulfides and sulfuric acid. Cr(III) and Cr(VI) residues in drinking water are controlled, and in the USA the drinking water standard is 0.1 mg Cr l^{-1};[4] in the European Union it is 0.05 mg mg Cr l^{-1}. Unintentional contact is highest in areas close to mining and refining sites, and in the vicinity of industrial and domestic waste sites. A case is reported where 5000 tonnes of chromium slag from a local mining area were deposited in the soil in Hunnan in southwest China, thereby endangering the health of the local communities for generations.[5] Incineration of waste releases chromium in gases and fumes into the atmosphere. Certain chromium compounds such as phosphate, as used in safety matches, are explosive. Lit matches release hazardous chromate into the atmosphere.

Metallic chromium has a vast range of industrial applications, among which the steel industry is of greatest importance. Chromium alloys with iron (ferrichrome), nickel, cobalt, copper, manganese, molybdenum and silicone are exceedingly hard, durable and resistant to corrosion and entirely suitable for use in engineering, automobile manufacture and in construction industries (Table 4.2). Chromium and chromium–nickel coating is a highly efficient means of preserving steel and providing a tough, shiny and corrosion-proof finish. The vivid reds, greens and yellow colours of chromium compounds (especially chromates) are valuable as colorants in paints, paper, plastics, textiles, ceramics, glass, tattoos and even cosmetics (eye shadows and soaps). Other major uses of chromium compounds include leather tanning, wood preservatives and chrome plating (with nickel), all of which are associated with toxicological problems. Chromated copper arsenate (CCA) has been used as a preservative for timber treatment since the mid-1930s.

Table 4.2 Common chromium-containing alloys.

Allegheny metal/Enduro metal
Stainless steel
Chromel
Stellite
Duralumin
Illium
Nobilium
Vitallium

A major problem faced by workers in chromate-related and pigment industries concerns the profound contact allergy and delayed hypersensitivity of the element.[6] (Nickel is similarly a profound contact allergen.) It is not within the context of the present volume to discuss this aspect of chromium-related toxicity other than to note the extensive literature on dermatitis, chrome ulcers and burns attributable to chromate exposure in builders and construction workers in regular contact with cement, concrete and building aggregates which contain Cr(VI) and Cr(III) residues. Cr(VI) is a major health risk but on rare occasions Cr(III) may be a cause of sensitisation. The risks associated with Cr(VI) in industry are closely associated with the solubility of chromates and dichromates and their capacity to irritate and penetrate the skin and cell membranes. Metallic chromium is not commonly regarded as being allergenic but cases are recorded where patients with chromium-alloy prostheses have shown symptoms of allergy, necessitating removal of the devices.

In recent years, extensive use has been made of chromium alloys with nickel, cobalt, molybdenum, titanium and other metals in the so-called metal-on-metal (MoM) orthopaedic prostheses.[7] Hip arthroplasty devices, including MoM hip systems, have been available for use within the United States and Europe for over 50 years. They are frequently used to relieve pain and restore joint function in patients with chronic hip pain or disease which is not responsive to more conservative therapy. This specialised subject has recently been recognised as a considerable healthcare concern since excessive release of metal ions and metal fragments into the circulation and surrounding tissues over several years has led to adverse local reactions and a marked accumulation of chromium, nickel and cobalt deposits in liver, spleen, kidney and other tissues. Progressive deterioration and failure of the devices has led to many prostheses being removed and litigation cases in the USA and Europe. Risks of carcinogenicity arising through excessive release of chromium and cobalt are real and subject to extensive research and case studies.[8]

4.2 Chromium as a Trace Metal

Chromium as Cr(III) is now recognised as an essential micro-trace element in human and animal nutrition, largely as a result of pioneering work conducted by the late Walter Mertz in the early 1950s.[9] He demonstrated that liver necrosis in rats could be prevented by addition of selenium and chromium to the diet. From this work chromium became identified as a "glucose tolerance factor" with key roles in the potentiation of insulin and insulin-dependent functions of lipid and carbohydrate metabolism and cell membrane function.[10,11] Chromium may promote insulin-related functions in the presence of reduced insulin secretion. Henry Schroeder was the first to conduct experiments in the so-called "controlled trace element environments" where he demonstrated that chromium-deficient rats were subject to severe diabetes and premature death.[12,13] Chromium deficiency, leading to a depression in fatty acid and cholesterol synthesis, has been identified as a factor in atherosclerosis as well as nerve and brain disorders.[14,15]

Human nutritional requirements for Cr(III) are normally low but controversy exists as to what levels are tolerable in the diet without evidence of toxic change. Baker's yeast is a major dietary source of chromium but beef, dairy products, apples and bananas contain high levels. The US Office of Dietary Supplements has issued guidelines on dietary requirements for chromium Cr(III) in men and women (including during pregnancy and lactation) according to age, but an upper tolerable level of dietary intake was not known in 2001.[16] Current estimates for a safe and adequate daily dietary intake of Cr(III) are 50–200 μg per day for children >7 years to adults (younger children require less).[17] Anderson and Kozlovsky found that normal dietary intakes in adult women (25 μg) and men (33 μg) were below the recommended minimum levels. Higher dietary chromium possibly with injections of chromium picolinate supplement (Figure 4.1) were recommended for patients with Type II diabetes mellitus.[18,27,28]

Studies in diabetic patients have substantiated that serum chromium levels are lower than normal and that supplementary chromium is therapeutically beneficial in correcting insulin-related metabolic and cardiovascular diseases.[19] Large-scale dietary studies in malnourished children in Jordan and Nigeria confirmed the importance of adequate chromium in the diet in maintaining appropriate glucose balance and alleviation of diabetes. A study in China showed that supplementary dietary Cr(III) improved blood glucose levels, insulin sensitivity, blood cholesterol levels and haemoglobin A_1C in a

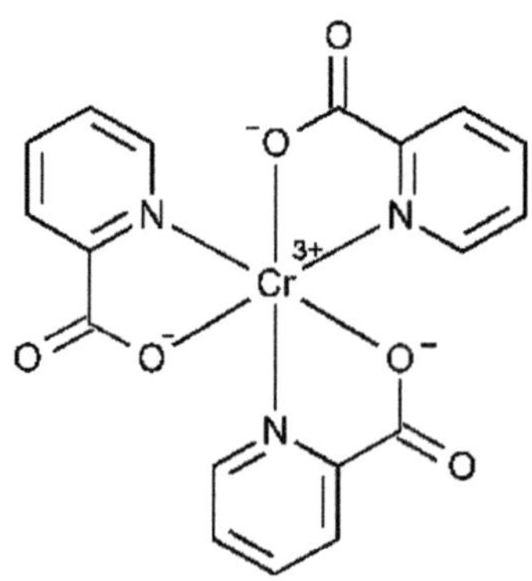

Figure 4.1 Chromium picolinate.

dose-dependent manner.[20] Experimental studies suggest that chromium acts in the form of an "auto-amplification system" for insulin signalling and promotes enhancement of insulin sensitivity. The mechanism is not understood. Hummel *et al.* indicated that the insulin receptor protein tyrosine kinase present in all cells is stimulated or activated by a low molecular weight oligopeptide (chromodulin) which selectively binds chromium.[19] This chromodulin possibly acts in a similar fashion to the calcium carrier protein calmodulin. Chromodulin is capable of binding four Cr(III) ions and performs a central function in maintaining chromium flux and possibly in preventing the onset of juvenile diabetes.

Human diets and environmental exposures contain Cr(III) and Cr(VI) compounds. There is no published evidence to suggest that Cr(III) compounds present in diets containing cereals, nuts, yeast, green vegetables, egg yolk and some wines and beers are oxidised to Cr(VI). Contrary to some published texts, recent laboratory studies in China have demonstrated that Cr(III) citrate can be transformed to Cr(VI) by a photo-oxidation process using solar light mimicked by medium pressure mercury lamps and xenon lamp.[21] Rates of photo-oxidation of two isomers of Cr(III) citrate were not sensitive to pH changes in the range 7–9, but increased significantly at higher acidity and in the presence of oxygen and the hydroxyl radial. Other studies have demonstrated that Cr(III) is oxidised to Cr(VI) in the presence of manganese oxides in sludges and environmental deposits.[22] Manganese oxides such as pyrolusite found in the soil have a high absorbency for metal ions and catalyse the oxidation process. The extent to which Cr(III) compounds found in the diet are oxidised to Cr(VI) in the presence of solar radiation of chemical contaminants such as manganese is not known, but human health implications are huge.

4.3 Absorption of Chromium

4.3.1 Dietary Intake

The chromium content of human diets varies greatly according to geographical area, cultural and social habits, and the geological profile of the country. Mean concentrations of chromium in the human body range from 5 to 6 mg.[23] Gastrointestinal absorption of Cr(III) in humans is low and ranges from less than 0.13% to 2.5% of the amount consumed.[16] Intestinal absorption is enhanced in the presence of vitamin C, amino acids and oxalates, but lowered by the dietary content of phytates, organic fibre and substances which chelate the ion. Studies in rats have suggested that absorption is a non-saturable process and not carrier mediated. Absorbed chromium is bound by the iron-binding protein transferrin which exhibits two binding sites with differing affinities for iron and chromium.[24] As with absorption of many metal nutrients, chromium receptor binding is a competitive process. Experiments in the perfused rat intestine have demonstrated that a mean level of 5.9% chromium was absorbed and that most was transported to the circulation.[25] Absorption in live animals exposed to ^{51}Cr-labelled chromium picolinate was two-fold higher than in animals given chromium chloride ($CrCl_3$), but the half-life was shorter and most was excreted within 24 hours.[26]

Metabolic studies confirm that the bioavailability and relative distribution of Cr(III) and Cr(VI) in different tissues is similar. In an industrial situation, it is expected that both ions will be absorbed [Cr(VI) > Cr(III)] gastrointestinally through consumption of contaminated food and drink, failure to wash hands, eating in contaminated areas and failure to change from contaminated clothing, *etc*. It is expected, however, that gastrointestinal absorption of chromium ions will be low in comparison with that absorbed following dermal contact or through inhalation of aerobic emissions.

4.3.2 Dermal Absorption

A large number of professionals are exposed dermally to chromium compounds in their workplace (Table 4.3). The greatest dermatological problems are experienced by builders, who are regularly exposed to chromates in wet cement and concrete, but tanners and leatherworkers, painters and colourers, pigment manufacturers and dyers, steel workers and welders are commonly in contact with chromates throughout their working lives.[6] Serious dermatological problems experienced include chrome ulcers, dermatitis and perforation of the nasal septum. Depending upon periods of exposure, and chromium concentration, these lesions may extend deeply into the skin and be associated with severe pain. The Cr(VI) compounds used in industry are appreciably stronger oxidising agents than Cr(III) compounds; many are highly soluble and capable of inducing skin burns.[6] The Cr(III) ion tends to precipitate as relatively innocuous oxides or hydroxides at near neutral pH. Cr(VI) is ulcerogenic but Cr(III) is not.

Table 4.3 Occupational exposure to chromates.[6]

Acetylene and aniline production	Automobile workers	Battery makers
Bleachers	Builders	Candle makers
Carbon printers (photography)	Cement and concrete makers	Chrome workers
Chrome platers	Colour and pigment makers	Rubber compounders
Crayon producers	Cutting oil makers	Dye makers and dyers
Decorators (wallpaper hangers)	Electroengravers/ electroplaters	Etchers
Enamellers/enamel platers	Explosives/Firework makers	Fire fighters
Foundry workers	Glass makers and colourers	Ink makers and printers
Linoleum manufacture	Lithographers	Pottery workers
Marble and stone colourers	Match factory workers	Paint workers
Paper industry workers	Paper money makers	
Pencil (coloured) workers	Photographic workers	Railway employees
Diesel workers	Rust proofers	Stainless steel producers
Chrome alloy workers	Leather and tannery workers	Television manufacturers
Textile printers	Rubber workers	Tyre manufacturers
Vulcanisers	Waterproofers	
Wax workers	Welders	Woodworkers
Wood preservers		

Cr(VI) is absorbed percutaneously through damaged skin or skin exposed to chromates used in bleaches and strong detergents. Early evidence for percutaneous penetration of chromium salts was provided by using ^{51}Cr compounds in disappearance experiments using guinea pig skin.[29] Penetration of Cr(VI) (sodium chromate) was significantly higher than Cr(III) (chromic chloride) and absorption was proportional to the concentration applied; 1–4% of that absorbed was located intra- and inter-cellularly in the epidermis.[29,30] Cr(VI) is reduced to Cr(III) on the skin and in the circulation. Even at dermal applications greatly exceeding the EPA recommended maximum dose of 100 μg l^{-1}, health risks to humans through percutaneous absorption are low.[31] This is in part owing to the fact that Cr(III)or Cr(VI) ions bind strongly to epidermal keratin and accumulate in outer regions of the epidermis, to be lost naturally by desquamation of keratinocytes.[30] Dermal exposures to 22 mg l^{-1} of Cr(VI) compounds in tap water were well tolerated in an environmental health study. In the USA, concentrations of Cr(VI) compounds in tap water rarely exceed 1–2 mg l^{-1} and are regarded as completely safe. However, clinical experience shows that concentrations as low as 2 μg are sufficient to evoke a positive skin reaction in hypersensitive subjects.

4.3.3 Inhalation

Inhalation of Cr(VI) residues in environmental dust, cigarette smoke, industrial emissions, gases and vapours constitutes a major health hazard. The US National Institute of Safety and Health estimated that 2.5 million workers were exposed to chromium and its compounds in the workplace in the period 1972 to 1974; of these, 175,000 were exposed to chromates released in processing chrome ore.[32] Typical airborne concentrations of Cr(VI) in industrial emissions range from 5–25 μg m^{-3} in chrome plating, 10–140 μg m^{-3} in ferrochrome alloys, 50–400 μg m^{-3} in stainless steel welding, 100–500 μg m^{-3} in chromate production, to 60–600 μg m^{-3} in chrome pigment production.[33] Threshold limit values (TLV) were set at 0.5 μg m^{-3} for Cr(VI) compounds, although a lower TLV of 0.05 μg m^{-3} was applied for water-soluble compounds such as lead chromate, and certain insoluble compounds.

The lungs and respiratory tract are major target organs for chromium-related carcinogenicity. Chronic exposure to chromium dusts and emissions causes bronchitis, perforation of the nasal septum, pneumonia and chronic respiratory disease. Evidence of raised chromium in blood, hair, body tissues, urine and faeces provides a convenient means of monitoring pulmonary absorption of chromium. Gastrointestinal disorders and neurological diseases are further indications of toxicity resulting from inhalation of chromium trioxide or other Cr(VI) compounds in the workplace or environmentally. Pulmonary absorption of chromium compounds is a function of the size, solubility, concentration and oxidation state of the particles, but is influenced by the protective capacity afforded by pulmonary macrophages, and the respiratory patterns of individuals. A commonly disregarded fact in occupational medicine is the fact that cigarettes contain chromium, and that this is inhaled in cigarette smoke along with other metal contaminants. Cr(VI) is reduced to the less toxic Cr(III) form by ascorbate and glutathione in the lungs. Necropsies performed on eight chromate workers demonstrated that most of the chromium inhaled is deposited in peripheral pulmonary tissue (36.7 μg g^{-1}) with lower levels in large airways (0.51 μg g^{-1}).[34] Higher concentrations of chromium present in the upper lobes of the lung suggest that pulmonary absorption reflects a difference in the rate either of chromium clearance or deposition in the tissues. Chromium content in lung tissue was proportional to the duration of exposure.

Absorption of Cr(VI) compounds following inhalation is closely related to solubility and the ability of the ion to penetrate alveolar cell membranes.[35,36] Workers exposed to Cr(VI) dusts developed higher blood and urine chromium levels than those inhaling less soluble ferrous or magnesium chromite dust.[37] Experiments in rats showed the water-soluble chromates such as zinc chromate are readily absorbed by the lungs and distributed to erythrocytes before being reduced to Cr(III).[38] Cr(VI) is reduced to Cr(III) by the alveolar fluids and pulmonary alveolar macrophages in the lower respiratory tract. The biochemical efficiency of alveolar macrophages in reducing Cr(VI) was significantly higher in tobacco smokers than in non-smokers.[39,40]

4.4 Chromium and Chromates as a Cause of Cancer

Cancer is the principle risk associated with exposure to Cr(VI) compounds in mining and manufacturing industry. Epidemiological surveys conducted in the past 50 years have shown that workers engaged in chrome plating, leather industries, chrome pigment manufacture, steel work and in chromate extraction plants in Asia, the USA and Europe are subject to excesses in lung cancer mortality following inhalation of Cr(VI) compounds, but not Cr(III) compounds.[41–48] These observations are supported by experimental studies in rodents where inhalation, injection or dietary/oral administration of soluble and insoluble Cr(VI) compounds induced and promoted pulmonary tumours under defined circumstances; Cr (III) compounds administered by oral or inhalation routes provided little evidence for carcinogenicity.[38,49] The tumour incidence in humans is influenced by lifestyle factors, *e.g.* smoking, alcohol, job specification (cigarette smoke is known to contain As, Pb, Ni, Cr and Cd).[50]

4.4.1 Current Legislation

Hexavalent Cr(VI) compounds are listed in the 12th Annual Report (RoC) and the 1990 IARC Monograph as human carcinogens on the sufficiency of evidence provided by human studies.[1,2] The US Environmental Protection Agency (EPA) has classified Cr(III) compounds as not carcinogenic to humans.[4]

4.4.2 Clinical and Epidemiological Evidence of Chromium-related Carcinogenicity

Chromium has been known to constitute a carcinogenic hazard to humans for more than 100 years, although there are many deficiencies in existing knowledge.[38,51] Epidemiological and case studies provide a strong but incomplete profile of the carcinogenic risks posed by occupational and environmental exposure to chromium in the human body. Many of the published studies are scientifically or statistically unsound and based upon retrospective analyses of *post-mortem* material. The studies are often deficient in important detail regarding the nature and duration of exposure, the circumstances under which it occurred, and lifestyle factors including smoking. Chromium levels in mainstream cigarette smoke range from 0.0002 to 0.5 mg per cigarette.[50] Importantly, the relative concentrations of Cr(III) and Cr(VI) are frequently not known. Both ions are widespread in the environment, but whereas trivalent chromium is commonly regarded as non-carcinogenic, the hexavalent form is claimed to be carcinogenic through ingestion, inhalation and possibly percutaneous absorption.[52] Much Cr(VI) absorbed from the diet or other routes is reduced extracellularly and intracellularly by a complex network of events to the relatively innocuous Cr(III) and generation of reactive oxygen species.[51]

4.4.2.1 Chromium in Food and Drinking Water

Cr(III) and Cr(VI) compounds are present in drinking water and food materials throughout the world. In North America alone, at least 74 million people drink chromium-polluted tap water in 42 states.[53] Most foods contain chromium at <100 μg kg^{-1}, with cereals and milk containing <10 μg kg^{-1}, with little geographical variation.[54] Levels of chromium in food increase with processing and cooking, and through contact with stainless steel materials. The US National Academy of Sciences established a "safe daily dietary" intake of 20–200 μg per day.[55,56] No convincing evidence of carcinogenicity has been seen at these concentrations in lifetime exposure, possibly because gastrointestinal absorption of the metal is low. It is possible that intakes of 1000 μg per day are safe. A review of nine studies evaluating the health risk associated with Cr(VI) in tap water concluded that oral exposure of 10 mg l^{-1} did not exceed the ability of gastric juices to reduce the ion, and that inhalation of this concentration during showering was without cancer risk.[31] Chromate solutions were well tolerated in dermal contact at 22 mg l^{-1}. As a further safety measure, Californian officials in 2009 proposed setting a "public health goal" for hexavalent chromium in drinking water as low as 0.06 parts per billion (ppb).

The risk of gastrointestinal or other cancers arising in humans through ingestion of chromium in tap water is difficult to assess in view of conflicting reports published in China.[57–59] Whilst a higher incidence of stomach cancer was reported in areas of high chromium in 1987, this was retracted 10 years later because the work was deemed to be scientifically and statistically unsound.[60] Beaumont concluded that a risk of stomach cancer did exist in populations exposed to high levels of Cr(VI) near a ferrochrome plant and that lung cancer was slightly higher than in "unexposed" areas, but other cancers were in line with the general population incidence.[59] No clear relationship was established in the Chinese studies between chromium in tap water and cancer incidence. More reliable evidence of the cancer risk from consuming contaminated water [Cr(VI) concentration up to 156 μg l^{-1} in 2008] is provided by a recent study in an industrialised region of northern Greece.[61] Assessment of *post-mortem* material from 474 fatalities in the resident population showed a higher standard cancer mortality rate and significant increases in liver and lung cancer rates (*p*-values of 0.001 and 0.047 respectively). Rates of lip, oral cavity, pharynx, stomach, breast, prostate cancers and leukaemia were also higher than in the general population, prompting introduction of more stringent safety guidelines than the European Union Council directive of 50 μg l^{-1}.[62]

This Greek study strongly indicates that Cr(VI) should be labelled as a multi-system carcinogen in humans, but the statistical power of the observations is seriously questioned owing to lack of information on conditions and durations of Cr(VI) exposures, lifestyle factors, smoking, *etc.*[63] Interpretation of published epidemiological studies is further questioned on the lack of information relating to the latent period between Cr(VI) exposure and diagnosis of tumours, and delays in reporting.[64,65] Despite the prevalence of Cr(VI) in drinking water, identified in USA as the "Erin Brockovich chemical", there seems to be no clear evidence that a carcinogenic risk is associated with chromates in tap water at concentrations found. A drinking water standard of 0.1 mg l^{-1} for total chromium set by the US EPA is possibly realistic.[4] This standard is based on the best available science available at the time and is enforceable by law.

Nine epidemiological studies in the USA concluded that risks of gastro-intestinal cancer arising through long-term ingestion of drinking water contaminated with Cr(VI) compounds or consumption of dust is low because most of the ions are reduced to less soluble Cr(III) in the acidic medium of the stomach.[31] A meta-analysis of 32 studies conducted in 1950–2009 has also confirmed low risks of gastrointestinal cancer in workers employed in welding, chrome plating, pigment production and chromate work who ingest Cr(VI) compounds. However, four studies did show an increase in oesophageal cancer.[66] This has not been substantiated. Concentrations of <10 ppm Cr(VI) compounds did not overwhelm the reductive and protective capacity of the stomach or blood.

4.4.2.2 Industrial Exposures

More conclusive evidence of the carcinogenicity of Cr(VI) is provided by occupational health studies, where inhalation of chromates of high and low water solubility is a potential cause of lung cancer.[43,67–69] Exposure to airborne chromium varies greatly according to the type of work, protective measures in force at the time, and awareness of the potential risks of lung cancer. A 28-year evaluation in an Ohio chromium plant[45] involving 800 measurements showed that workers were exposed to average airborne Cr(VI) concentrations ranging from 2.3 to 8.9 mg m^{-3}. High concentrations are associated with bagging chromate, but with increased awareness of cancer risk, environmental concentrations in plants have decreased over time, from 0.72 mg m^{-3} in 1940 to 0.039 in 1972. Recent national and occupational health surveys in the USA have summarised that exposure levels range from 5–25 μg m^{-3} in chrome plating to >400 μg m^{-3} in chromate pigment production and in stainless steel welding.[33] Ten occupational health studies involving workers in chromate-related industries, ferrochrome production, chromium platers (and chromium/nickel platers) and smelter workers have shown statistically significant increases in lung cancer risk, and that this risk is increased in smokers.[41,42,44,45,70–75] It is relevant to note that all these studies were based on retrospective analyses of *post-mortem* observations. As such the scientist is deprived of essential information relating to work patterns, exposure levels, other toxic contaminants in the environment and lifestyle factors. How many of the workers were smokers? Horiguchi *et al.* recorded that 7 of 265 men employed in a bicycle plant and exposed to chromium who died from lung cancer and stomach cancer were exposed to copper and nickel![73] Skin ulceration and perforated nasal septum were suggestive of high chromium exposure,[6] but it is unclear whether the tumour incidence was attributable to chromium, nickel or a co-carcinogenic mechanism involving both and promoted by copper.

Gibb *et al.* of the US EPA, following up earlier studies in the Baltimore region showing that employees in chromate plants had increased levels of tracheal, bronchial and lung cancer risk, examined health records of 2357 employees in a chromate production plant in the 1950–1974 period.[76–78] They characterised exposure to Cr(VI) and Cr(III) by job specification, cumulative chromate exposures, race and smoking habits. Table 4.4 shows that lung and prostatic cancers were higher than expected from US cancer mortality rates. The evaluation allowed a statistically valid conclusion that occupational exposure to Cr(VI) is a lung cancer risk whereas Cr(III) inhalation is more likely to be cause of pulmonary irritation. The cancer risk was proportional to cumulative exposure to Cr(VI) irrespective of whether workers smoked or not. Non-white workers showed a slightly higher incidence of prostatic cancer than the US population, but numbers were too low to draw valid conclusions. Cr(VI)-related prostatic cancer has not been validated by experiments in animal models. Langård reviewed reports of chromate-related cancer over the past 100 years and emphasised that a greater cancer risk lies in occupational exposure to water-soluble chromates rather than those of low solubility.[79]

With the knowledge that hexavalent chromium is mutagenic in cultured human cell lines, Littorin *et al.* examined cytogenetic changes in peripheral lymphocytes from 24 stainless steel workers with a mean period of 19 years'

Table 4.4 Observed (O) and observed/expected (O/E) cancer incidence in US chromate production workers (the table does not account for workers of unknown race).[77,78]

	Whites		*Non-whites*		*Total*	
Cause of death	*O*	*O/E*	*O*	*O/E*	*O*	*O/E*
All cancers	120	1.14	99	1.44	235	1.25
Lung cancer	71	1.86	47	1.88	122	1.80
Prostate cancer	5	0.71	11	2.03	16	1.22

exposure to 5–321 μg m^{-3} Cr(VI) (mean 81 μg m^{-3}) in arc welding.[80] In six industries, urine chromium levels were greatly increased in the workers but no statistically significant increase was seen in chromosomal damage, micronucleus formation or sister-chromatid exchanges. Worker smoking did not significantly influence these findings.

Lung tumours associated with inhalation of Cr(VI) are characteristically of the squamous cell type, located in the central aspects of the tissue and at the bifurcation of the bronchi where chromium deposits are highest.[44,81,82] Small cell tumours of the lung and carcinoma of the nasal cavity are also recorded. Precancerous lesions including squamous metaplasia were diagnosed, but in one study these changes regressed without therapy. Squamous metaplasia occurs as a pre-carcinogenic response, a "transformed state", in response to Cr(VI) exposure. A *post-mortem* study of eight Japanese chromate workers showed that peripheral lung tissue concentrated 99% of the total chromium inhaled and much was deposited in the upper lobes of the lungs where it persisted for long after work in the chromate plant was concluded.[70] Experimental studies in rats implanted with pellets of sparingly soluble genotoxic Cr(VI) compounds showed that the incidence of lung tumours was proportional to the solubility of the chromate implanted and duration of exposure.[81,82] Convincing epidemiological and experimental evidence has demonstrated that zinc chromate is a potent carcinogen and that calcium chromate and chromium trioxide are carcinogenic in humans.[38] Langård pointed out that, in 1988 at least, human studies had not confirmed the carcinogenic potency of other chromium compounds, including lead and strontium chromates which had been shown to be carcinogenic in rodent studies.[38] Although lead is classified as a human carcinogen,[1] lead chromate is categorised differently from other hexavalent chromium compounds because of its low solubility (*i.e.* 0.06 g l^{-1} water) and the limited data available on carcinogenicity. As discussed later, lead chromate is mutagenic in a battery of *in vitro* tests but both bacterial mutagenicity tests and studies of chromosomal genotoxicity have led to the conclusion that the chromate ion is responsible for the genotoxic properties of lead chromate. Strontium chromate is also listed as a human carcinogen and it has shown positive evidence of mutagenicity in the *Salmonella typhimurium* reverse assay. Experimental studies in rats have failed to demonstrate the carcinogenicity of high oral doses of chromium trioxide administered at high doses for 840 days, on account of its low solubility and minimal intestinal absorption.[83]

Risks of skin or other cancers arising through percutaneous penetration or ingestion of chromium compounds are low.[84,85] Although dermal exposure to chromates is expected to be high in paint and pigment manufacture, leather tanning and electroplating plating, occupationally related skin tumours are rare. A review of workers in six electroplating plants in Finland showed that dermal exposure levels of up to 28 mg Cr(VI) h^{-1} was a profound cause of sensitization, skin ulceration and dermatitis, but not skin or lung cancer.[84,85] Cr(VI) compounds are reduced in and on the skin to innocuous Cr(III).[86] This reduction is catalysed by ferrous ion, vanadium, and body fluids.

4.4.2.3 *Perspectives on the Carcinogenicity of Chromium in Humans*

Epidemiological observations relating to chromate exposure in industry are frequently difficult to interpret on account of wide variations in workplace environments, demographic and geological factors, nutritional status and exposure patterns.[74,75] Lung cancer is a major risk associated with chronic occupational exposures to soluble Cr(VI) compounds, but the true incidence is confounded by long lag phases in clinical diagnosis of tumours which may not be recognised until 20 years or more after retirement from lifetime exposures in chromate-related industries.[42] Chromium deposits persist in lung tissue for some time and progressive absorption of Cr(VI) ion and a slow metabolism to more mutagenic and genotoxic chromium species may underlie production of DNA adducts and lead to squamous metaplasia and carcinoma.[73,87] Added to that is the knowledge that that the strong oxidising capacity of chromates and dichromates, the release of reactive oxygen species, and inflammatory/immunological responses associated with release of cytokines and growth factors are manifestations of epigenetic change and alterations of signalling pathways. The Cr(VI) absorbed intracellularly through non-specific anion channels is metabolically reduced by ascorbate, glutathione and cysteine to Cr(V), Cr(IV) and ultimately Cr(III).[87]

More also requires to be learned regarding genetic factors that signal increased sensitivity to chromium and why black workers may be 40% more susceptible to lung cancer than white workers after >20 years of exposure in chromate industries.[42] Epidemiological data point to risks of chromium-related lung cancer being influenced by age, work practice and duration of exposure, and levels of environmental contamination. Crump conducted a statistical analysis of lung cancer in a chromate production plant and predicted that lung cancer risk was not significantly increased below a lifetime cumulative occupational exposures of 1 mg–year m^{-3}, and there was no excess risk[72] for workers whose average monthly exposure did not exceed permissible exposure limits of 52 μg m^{-3}. The estimated lifetime risk of cancer after 45 years' exposure to 1 μg m^{-3} was estimated to be very low. Another estimate gave lifetime cancer risk at exposures to 0.2 μg m^{-3} environmental Cr(VI) compounds as about 1 in 8, or 120 000 times higher than usual.[88]

Virtually all clinical, epidemiological and case studies seen in this review fail to recognise that, with rare exceptions, occupational and environmental to exposures to chromium are complex, and occupational environments in steel production, chrome plating, tanning, *etc.* contain a cocktail of metal and non-metal contaminants including nickel, iron, aluminium, cobalt, lead, silica, constituents of cutting and lubricating oils, cement dusts, *etc.*, any of which can influence carcinogenic action.[89] Siberian red lead ore – mined as crocoite – is composed of lead chromate but is commonly found in lead veins in rock formations with quartz, zinc, silver and copper. Lead, aluminium and magnesium are used in chromium extraction processes. Alumina-thermic production of high purity chromium metal accounts for over 95% of chromium produced.

The first step in this process requires that the chromite ore be roasted with soda and lime in air at 2000 °F (1000 °C), which creates a sodium chromate containing calcine. This can be leached away from the waste material and then reduced and precipitated as chromic oxide (Cr_2O_3). The chromic oxide is then mixed with powdered aluminium and put into large clay crucibles. Barium peroxide and magnesium powder are then spread on to the mixture and the crucible is surrounded by sand which acts as insulation.

While the studies discussed provide strong evidence for an association between chromium exposure and lung and maybe stomach cancer, the cytopathic mechanism for carcinogenic change is unlikely to be simple and is expected to involve complex biochemical interaction, mutagenic and genotoxic change, metal-carrier proteins, control by upregulation of genetic factors, release of reactive oxygen radicals and other epigenetic changes.[51,90]

4.4.3 Experimental Studies

4.4.3.1 Animal Models

Experimental studies show that certain Cr(VI) compounds are carcinogenic to rats or mice by injection, inhalation/intratracheal instillation or administration in drinking water in long-term studies.[38,78] They confirm the carcinogenicity of calcium, zinc, lead and strontium chromates and sodium dichromate. The range of species used is mainly limited to rat (F344) and mice (B6C3F1) strains preferred by the US EPA, thereby precluding an opportunity to evaluate genetic sensitivity *in vivo*. Tumour yields are frequently low even where animals have received high levels of sodium dichromate in drinking water or their diet. Despite *in vitro* evidence that Cr(III) is genotoxic, mutagenic and influences cell behaviour and gene-signalling, there is no conclusive evidence for the carcinogenicity of compounds such as chromium picolinate in laboratory animals.[39,51]

De Flora provided an authoritative view of published animal studies purporting to show the carcinogenicity of Cr(VI) compounds (chromates and dichromates) in lung or other tissues, and emphasised that most failed to mimic human exposure patterns.[51] Further, in most studies claiming a dose-related tumour incidence, animals were given massive concentrations of chromates by means that "by-passed" normal physiological defence mechanisms. This is well illustrated by early long-term studies in which rats were injected intramuscularly once weekly with doses ranging from 500 to 5000 mg calcium chromate suspension in 0.2 ml arachis oil for 20 weeks.[91] A 7% incidence of non-metastatic spindle cell or pleomorphic sarcomas was observed at injection sites in rats surviving to 400 days. The tumours were associated with massive inflammatory change, fibrosis and abnormal mitoses. Malignant tumours were demonstrated also at sites of implantation of chromite ore, calcium chromate or chromium trioxide in rats.[92,93] On the other hand, no local tumours were observed in rats given single intra-articular injections of 20 mg per rat chromium alloys with cobalt and molybdenum to mimic their use in orthopaedic implants.[94] Control animals injected with nickel subsulfide. developed pleomorphic sarcomas.

Rats fed chromic oxide baked in bread at equivalent to 1,2 3 or 5% for a total of 840 days failed to show toxic changes, suggesting that insufficient biologically active Cr(VI) ion was absorbed to elicit carcinogenic change. Even where animals were administered Cr(VI) at doses sufficient to cause 12–37% reduction in liver and spleen weight, tumour induction was not observed.[83] As in limited human experience, protective mechanisms for detoxification and reduction of the Cr(VI) in the stomach and intestine were considered to be sufficient to safeguard against carcinogenic change, *i.e.* reduction of Cr(VI) to Cr(III) or chelation of the ion to organic substrates. Inconsistencies evident in other studies where rats or mice were given high levels of Cr(VI) in drinking water are attributed to variations in the solubility of the salt used, dietary composition and intestinal absorption, and metabolism of the biologically active ion in sensitive tissues.[95,96]

The National Toxicology Program in the USA has extensively examined the metabolism, tissue deposition and mechanisms of carcinogenicity of sodium dichromate (Cr(VI)) and picolinic acid (Cr(III)) in F344 rats and B6C3F1 mice in diet or in drinking water over up to 2 years.[95–98] Rats were more sensitive than mice in 90-day studies. Tumour induction was not seen after this period but, after dosing for up to 2 years at concentrations comparable to those experienced by humans drinking contaminated water, sodium dichromate led to the production of dose-related carcinogenicity. The gastrointestinal tract was a target tissue but tumour location, incidence and tissue sensitivities differed between the two species and male mice were more sensitive to high levels of sodium dichromate than female animals. In drinking water studies, male and female rats showed a dose-related increase in malignancies in the tongue and oral mucosa. The highest dose of 516 mg l^{-1} did not affect survival but animals lost weight and showed a transient microcytic anaemia. Only 6 of 49 surviving male rats (12.2%) developed oral carcinoma and one exhibited papilloma of the tongue. Oral carcinoma was seen in 11 of 50 female rats, but only 2 of the 300 animals receiving lower doses of sodium dichromate showed carcinoma of the tongue. Mice did not develop oral tumours but exhibited a dose-related incidence of tumours involving the duodenum, jejunum and ilium. Male animals showed a combined rate of adenoma and carcinoma of 43% at 257.4 mg l^{-1} whereas females dosed at 516 mg l^{-1} showed at combined tumour incidence of 46%. At these massive doses of sodium dichromate, rats and mice exhibited profound signs of oxidative stress including depressed iron levels in bone marrow, microcytic anaemia, cytokine changes, and gastrointestinal pathology attributable to an accumulation of chromium sufficient to overwhelm inherent tissue protective mechanisms.[95] High levels of Cr(VI) accumulated in liver, kidney and glandular stomach, without tumour formation. Gastrointestinal carcinogenicity in rodent studies is attributed to the insufficiency of normal mechanisms for reduction of the high levels of Cr(VI) ion and its active absorption by transporter proteins in the intestinal mucosa. Low levels of Cr(III) ion are absorbed to satisfy nutritional requirements.

High concentrations of chromium picolinate in drinking water were not carcinogenic in rats or mice.

Intra-bronchial implantation of pellets of sparingly soluble strontium chromate led to a significant increase in squamous cell carcinomas in the lungs and respiratory tract in rats in 2-year studies.[82] Insoluble zinc or other chromates were not carcinogenic in this model. This contrasts with a second study in which rats were continually exposed to sub-micron aerosols of sodium dichromate or pyrolysed chromium oxides at concentrations of 25–100 μg m^{-3} for 2 years.[99] In the following 18 months, these animals showed an increased mortality associated with massive pulmonary deposits of chromium but the incidence of primary lung tumour (adenoma and adenocarcinoma) was low and of marginal significance.

The experimental evidence reviewed leads to the conclusion that Cr(VI) compounds such as sodium dichromate are capable of inducing tumour formation in rats or mice at high concentrations.

Other evidence points to the action of the ion as a tumour promoter. Thus, chromates have been shown to promote or otherwise enhance the production of skin tumours in nude mice induced by UV-irradiation.[69,100] At doses of 2.5 or 5 ppm chromate in drinking water, the yield of tumours induced in irradiated skin was increased three-fold in 28 weeks of study.

Hexavalent chromium should be identified as a complete carcinogen. It would be nice to see some well controlled experimental studies conducted in strains of rodent other than the F344 rat and B6C3F1 mouse, as routinely used by the NTP in regulatory style experiments. Experiments in animals have not confirmed genotypic sensitivity to chromium-induced carcinogenesis.

4.4.3.2 In Vitro Cytotoxicity, Mutagenicity and Genotoxicity

A vast number of sub-cellular and *in vitro* experiments have been conducted to evaluate molecular and genetic mechanisms of chromium-related carcinogencity. A battery of recognised studies have been published confirming the ability of Cr(VI) and Cr(III) to evoke cytotoxic, genotoxic and mutagenic changes.[101–106] Many chromates and dichromates have induced positive responses in sister chromatid exchange, chromosomal damage, micronucleus formation, inhibition of nucleic acid and protein synthesis, impairment of membrane-associated enzyme activity tests, induction of DNA-adducts, and bacterial reversion screens. Cr(VI) penetrates cell membranes more readily than Cr(III) and is reduced to Cr(III) in the presence of cell membranes, mitochondria, endoplasmic reticulum and cytosolic factors including glutathione and phosphate groups of nucleotides.[107–109] Reduction also occurs extracellularly in the presence of organic matter. Whilst *in vivo* evidence indicates that Cr(III) compounds should not be listed as carcinogenic in humans or animals, they are mutagenic, genotoxic and show cytotoxic effect in short-term *in vitro* tests.[81,110] Cr(III) compounds have not been associated with increased tumour incidence in humans or other animals on account of their lower solubility and the inability of the Cr(III) ion to cross cell membranes in sufficient quantity. Cr(VI) is

100–1000 times more cytotoxic than Cr(III) and four times more genotoxic in cultured fibroblasts.[111]

Cr(VI) compounds are cytotoxic and mutagenic in cultured human diploid fibroblasts (HDF) and evoke dose-dependent inductions of anchorage-independent colonies.[112,113] Induction of anchorage-independence can be used to study metal-related mechanisms of genotoxicity in HDF, cell-to-cell adhesions and behavioural characteristics. Cytotoxicity, mutagenicity and anchorage dependency was related to the valence state and solubility of chromium compounds. Cr(VI) compounds ($PbCrO_4$, $CaCrO_4$, Na_2CrO_4, and CrO_3) were 1000-fold more cytotoxic to HDF (average 50% lethal dose 0.5 μM) than Cr(III) compounds ($CrCl_3$, Cr_2O_3, Cr_2S_3; average 50% lethal dose 500 μM). The clastogenicity, mutagenicity and chromosomal integrity of insoluble lead chromate ($PbCrO_4$) is evident in HDF and Chinese hamster ovary cell cultures.[114] Lead and calcium chromates were cytotoxic, and induced cellular transformation in cultured C3H/10T$\frac{1}{2}$ mouse embryo cells with anchorage independence. The mutagenicity of lead chromate was related to extracellular dissolution, phagocytosis and internalization of the Cr(VI) and Pb(IV) ions, and intracellular stress responses.[115] Lead chromate is mutagenic in a battery of microbial tests including *Salmonella typhimurium* reversion assay, *Escherichia coli* survival tests and in *Saccharomyces*.[51,116]

4.5 Mechanisms of Cr(VI)-induced Carcinogenicity

Chromium compounds are subject to a vast amount of clinical and experimental research. Trivalent Cr(III) salts are essential at micro-concentrations as nutrients in the human body, with a role in insulin regulation and glucose metabolism. Hexavalent Cr(VI) compounds including chromates and dichromates are toxic and potentially carcinogenic following chronic occupational exposure. Allergenicity, profound dermal changes, hepatic and renal failure, and gastrointestinal complaints are complications of high chromate exposures in industry. Evidence is presented to show that high chromate exposure can overwhelm defensive mechanisms in the human body to induce lung and possibly gastric cancer.

Mechanisms for the carcinogenicity are complex and expected to involve mutagenic, genotoxic and epigenetic mechanisms. Cr(VI) is a complete carcinogen and the $-CrO_4^{2-}$ ion is bioactive and carcinogenic in animal models following parenteral injection, inhalation and long-term administration in the diet. Levels of toxicity and carcinogenicity are proportional to the environmental or in-body reduction of the Cr(VI) to Cr(III) ion in the presence of ascorbate or ferrous iron. Salnikow and Zhitkovich reason that, at neutral pH, Cr(VI) exists as a mixture of chromate or hypochromate ions in the ratio 3 : 1 and enters cells through sulfate channels.[104] As such, human and mammalian cell lines accumulate intracellular levels 10–20-fold higher than extracellular concentrations. Whilst these ions are unreactive with DNA at physiological pH, subsequent reduction and intracellular detoxification in the presence of ascorbate, glutathione and cysteine with production of Cr(III) is the activating event that is responsible for generation of DNA-adducts, chromosomal damage and other forms of cytotoxicity.

Cr(VI) is absorbed and penetrates cell membranes more readily than Cr(III) and studies in mammalian and bacterial cell lines show that it invokes irreversible nuclear and extra-nuclear changes, and alterations in cell membrane function predisposing to neoplastic transformation.[96,102,112,113,117] Lack of carcinogenic changes in the liver, kidney and glandular stomach are an indication of the reductive capacity of Cr(VI) in these tissues. Cr(VI) absorbed by erythrocytes is efficiently reduced to form stable complexes with haemoglobin.[11,119]

Intra-nuclear changes induced in a cultured intestinal cell line include direct DNA damage, dose-dependent translocation of transcription factors, micronucleus formation and changes consistent with oxidative damage.[117,118] In human bronchial cell cultures, Cr(VI) induced changes in membrane function and behavioural profile, thereby downregulating factors determining apoptotic change, cell-to-cell interaction, anchorage dependence and adhesion, predisposing to malignant change.[112,113,120] Cr(VI) was shown to be mutagenic in four strains of *Salmonella typhimurium* and it interacts directly with bacterial cell DNA forming frame-shift mutations and base-pair substitution at subtoxic concentrations.[121]

References

1. US Department of Health and Human Services, National Toxicology Program, *Report on Carcinogens*, Research Triangle Park, NC, 2011.
2. International Agency for Research on Cancer (IARC*)*, *IARC Monographs on the Evaluation of Carcinogenic Risk of Chemicals to Humans*, IARC, Lyon, France, 1980, vol. 23, p. 205; *Ibid.*, 1990, vol. 49.
3. Agency for Toxic Substances and Disease Registry (ATSDR), *Toxicological Profile fora Chromium*, US Department of Health and Human Services, Public Health Service; bATSDR, 2000.
4. US Environmental Protection Agency (EPA), *Chromium-6 in drinking water*. EPA, Washington, DC.
5. News in Brief, *Chemistry World*, 2011, **8**, 4.
6. A. Fisher, *Contact Dermatitis*, Lee and Febiger, Philadelphia, 1987, p. 762.
7. US Food and Drug Administration (FDA), *Metal-on-Metal Implant Systems*, FDA Executive Summary Memorandum, Gaiththersburg, MD, 2012.
8. C. G. Lewis, R. M. Belniak, M. C. Plowman, S. M. Hopfer, J. A. Knight and F. W. Sunderman, *J. Arthroplasty*, 1995, **10**, 75.
9. J. C. Smith and W. Maret, *J. Nutr.*, 2008, **138**, 247.
10. W. Mertz, *J. Am. Coll. Nutr.*, 1998, **17**, 544.
11. J. B. Vincent, *J. Am. Coll. Nutr.*, 1999, **18**, 6.
12. H. A. Schroeder, *Circulation*, 1967, **35**, 570.
13. H. A. Schroeder, *Am. J. Clin. Nutr.*, 1968, **21**, 230.

14. H. A. Schroeder, A. P. Nason and I. H. Tipton, *J. Chron. Dis.*, 1970, **23**, 123.
15. R. A. Anderson, *J. Am. Coll. Nutr.*, 1998, **17**, 548.
16. US Office of Dietary Supplements, *Chromium*, National Institutes of Health, Bethesda, MD, 2005.
17. National Research Council (USA), *Recommended Dietary Allowance*, NRC, National Academy Press, Washington, DC, 10th edn, 1989.
18. R. A. Anderson and A. S. Kozlovsky, *Am. J. Clin. Nutr.*, 1985, **41**, 1177.
19. M. Hummel, E. Standl and O. Schnell, *Hormone Metab. Res.*, 2007, **39**, 743.
20. Z. Krejpcio, *Polish J. Environ. Studies*, 2001, **10**, 399.
21. R. Dai, C. Yu, J. Liu, Y. Lan and B. Deng, *Env. Sci. Technol.*, 2010, **44**, 6959.
22. L. E. Eary and D. Rai, *Environ. Sci. Technol.*, 1987, **21**, 1187.
23. D. M. Czajka-Narins, in *Food Nutrition and Diet Therapy*, L. K. Mahon and M. V. Kraus, Sanders, Philadelphia, 1979, vol. 2, p. 114.
24. Y. Sayato, K. Nakamura, S. Matsui and M. Endo, *J. Pharmacodynam.*, 1980, **3**, 17.
25. H. J. Dowling, E. G. Offenbacher and F. X. Pi-Sunter, *J. Nutr.*, 1989, **119**, 1138.
26. K. Kottowitz, N. Laschinskky, R. Fisher and P. Nielsen, *Biometals*, 2009, **22**, 289.
27. B. Bahadori, S. Wallner, H. Schneider, T. C. Wascher and H. Toplak, *Acta Med. Austriac.*, 1997, **24**, 185.
28. C. L. Broadhurst and P. Domenico, *Diabetes Technol. Ther.*, 2006, **8**, 677.
29. J. E. Wahlberg and E. Skog, *Arch. Dermatol.*, 1965, **92**, 315.
30. R. E. Bagdon and R. E. Hazen, *Environ. Health Persp.*, 1991, **92**, 111.
31. D. J. Paustenbach, B. L. Finley, F. S. Mowat and B. D. Kerger, *J. Toxicol. Environ. Health*, 2003, **66**, 1295.
32. National Institute for Occupational Safety and Health (NIOSH), *National Occupational Hazard Survey* 1972-1974, NIOSH, Cincinnati, OH, 19.
33. *US* Agency for Toxic Substances and Disease Registry (ATSDA), *Toxic Substances Portal – Chromium*, US DHHS, ATSDR, Atlanta, GA, 2008.
34. Y. Tsunete, Y. Ohsaki, K. Kimura, H. Mikami, S. Abe and M. Murao, *Thorax*, **35**, 294.
35. W. Mertz, *Physiol. Rev.*, 1969, **49**, 163.
36. H. J. Wiegand, H. Ottenwalder and H. M. Bolt, *Toxicology*, 1984, **33**, 341.
37. H. E Stokinger, in *Patty's Industrial Hygiene and Toxicology*, G. D. Clayton and F. E. Clayton, John Wiley and Sons, New York, 3rd edn, 1981, p. 1493.
38. S. Langard, *Sci. Total Environ.*, 1988, **71**, 341.
39. A. D. Dayan and A. J. Paine, *Human Exp. Toxicol.*, 2001, **20**, 439.
40. F. L. Petrelli, G. A. Rossi, A. Camoirano, M. Romano, D. Serra, C. Bennicelli, A. De Flora and S. De Fora, *J. Clin. Chem.*, 1986, **77**, 1917.

41. T. Sorahan, D. C. Burges, L. Hamilton and J. M. Harrington, *Occup. Environ. Med.*, 1998, **55**, 236.
42. K. D. Rosenmann and M. Stanbury, *Am. J. Ind. Med.*, 1996, **29**, 491.
43. G. Kazantzis, *Ann. Occup. Med.*, 1972, **15**, 25.
44. Y. Ishikawa, K. Nakagawa, Y. Satoh, T. Kitagawa, H. Sugano, T. Hirano and E. Tsuchiya, *Br. J. Cancer*, 1994, **70**, 160.
45. D. M. Proctor, J. P. Panko, E. W. Liebig, P. K. Scott, K. A. Mundt, M. A. Buczynski, R. J. Barnhart, M. A. Harris, R. J. Morgan and D. J. Paustenbach, *Appl. Occup. Environ. Hyg.*, 2003, **18**, 430.
46. T. Birk, K. A. Mundt, L. D. Dell, R. S. Luoppold, L. Miksche, W. Steinmann-Steiner-Haldenstaett and D. J. Mundt, *J. Occup. Environ. Med.*, 2006, **48**, 426.
47. E. Halasová, T. Baska, F. Kukura, D. Mazúova, E. Bukovská, D. Dobrota, I. Poliacek and M. Halasa, *Neoplasma*, 2005, **52**, 287.
48. E. Halasová, E. Matakova, L. Kavcova, L. Musak, M. Letkova, M. Adamkov, E. E. Ondrusova, Bukovska and A. Singliar, *Neurol. Endocrinol. Lett.*, 2009, **30**(Suppl. 1), 182.
49. H. Gibb and C. Chen, *Sci. Total Environ.*, 1989, **86**, 181.
50. C. J. Smith, S. D. Livingston and D. J. Doolittle, *Food. Chem. Toxicol.*, 1997, **35**, 110.
51. S. De Flora, *Carcinogenesis*, 2000, **21**, 533.
52. A. Zhitkovich, *Chem. Res. Toxicol.*, 2011, **24**, 1617.
53. Environmental Work Group, *Cancer Causing Chromium-6 Pollution in U.S. Tap Water*, EWG, Washington, DC, 2012.
54. J. T. Kumpulainen, *Biol. Trace Elem. Res.*, 1992, **32**, 9.
55. National Research Council, Food and Nutrition Board. *Recommended Dietary Allowances*, National Academy Press, Washington, DC, 10th edn, 1989.
56. J. N. Hathcock, *Vitamins and Mineral Safety*, Council for Responsible Nutrition (CRN), 2nd edn, 2004.
57. J. D. Zhang and X. L. Li, *Zhonghua Yu Fang Yi Xue Za Zhi*, 1987, **21**, 262.
58. J. D. Zhang and S. Li, *J. Occup. Environ Med.*, 1997, **39**, 315.
59. J. J. Beaumont, R. M. Sedman, S. D. Reynolds, C. D. Sherman, L. H. Li, R. A Howd, M. S. Sandy, L. Zeise and G. V. Alexeeff, *Epidemiology*, 2008, **19**, 12.
60. A. H. Smith, *Epidemiology*, 2008, **19**, 24.
61. A. Linos, A. Petralias, C. A. Christolphi, E. Christoforidou, P. Kouroutou, M. Stoltidis, A. Veloudaki, E. Tzala, K. C. Makris and M. R. Karagas, *Environ. Health*, 2011, **10**, 50.
62. European Union, *Handbook on the Implementation of EC Environmental Legislation*, Council Directive 98/83/EC, 1998, pt. B.
63. L. McNeill, J. McLean, M. Edwards and J. Parks, *State of the Science of Hexavalent Chromium in Drinking Water*, Water Research Foundation, Denver, CO, 2012.

64. R. M. Sedman, J. Beaumont, T. A. McDonald, S. Reynolds, G. Krowech and R. Howd, *J. Environ. Sci. Health C, Environ. Carcin. Ecotoxicol. Rev.*, 2006, **24**, 155.
65. B. D. Kerger, D. J. Paustenbach, G. E. Coprbett and B. L. Finley, *Toxicol. Appl. Pharmacol.*, 1996, **141**, 145.
66. N. M. Gatto, M. A. Kelsh, D. H. Mai, M. Suh and D. M. Proctor, *Cancer Epidemiol.*, 2010, **34**, 388.
67. R. M. Park, J. F. Bena, L. T. Stayner, R. J. Smith, H. J. Gibb and P. S. J. Lees, *Risk Analysis*, 2004, **24**, 1099.
68. M. Costa, *Crit. Rev. Toxicol.*, 1997, **27**, 431.
69. M. Costa and C. B. Klein, *Crit. Rev. Toxicol.*, 2006, **36**, 155.
70. Y. Tsuneta, Y. Ohsaki, K. Kimura, H. Mikami, S. Abe and M Murao, *Thorax*, 1980, **35**, 294.
71. T. Birk, K. A. Mundt, L. D. Dell, R. S. Luippold, L. Miksche, W. Steinmann-Steiner-Haidenstaett and D. J. Mundt, *J. Occup. Environ. Med.*, 2006, **48**, 426.
72. C. Crump, K. Crump, E. Hack, R. Luippold, K. Mundt, E. Liebig, J. Panko, D. Paustenbach and D. Proctor, *Risk Analysis*, 2003, **23**, 1147.
73. S. Horiguchi, K. Morinaga and G. Endo, *Asia Pacif. J. Publ. Health*, 1990, **4**, 169.
74. E. Halasová, T. Baska, F. Kukura, D. Mazúrova, D. Bukovska, D. Dobrota, I. Poliacek and M. Halasa, *Neoplasma*, 2005, **52**, 287.
75. E. Halasova, T. Matakova, E. Kavcova, L. Musak, L. Letkova, M. Adamkov, M. Ondrusova, E. Bukovska and A. Singliar, *Neuro. Endocrinol. Lett.*, 2009, **30**(Suppl. 1), 182.
76. R. B. Hayes, A. M. Lilienfeld and L. M. Snell, *Int. J. Epidemiol.*, 1979, **8**, 365.
77. H. Gibb and C. Chen, *Sci. Total Environ.*, 1989, **86**, 181.
78. H. J. Gibb, P. S. Lees, P. F. Pinsky and B. C. Rooney, *Am. J. Ind. Med.*, 2000, **38**, 115.
79. S. Langård, *Am. J. Ind. Med.*, **17**, 189.
80. M. Littorin, B. Högstadt, B. Strömbäck, A. Karlsson, H. Welinder, F. Mitelman and S. Skerfving, *Scand. J. Work Environ. Health*, 1983, **9**, 259.
81. L. S. Levy and S. Venitt, *Carcinogenesis*, 1986, **7**, 831.
82. L. S. Levy, P. A. Martin and P. L. Bidstrup, *Br J. Ind. Med.*, 1986, **43**, 243.
83. S. Ivankovic and R. Preussman, *Fd. Cosmet. Toxicol*, 1975, **13**, 347.
84. M. Mäkinen and M. Linnainmaa, *Ann. Occup. Med.*, 2004, **43**, 277.
85. G. E. Corbett, B. L. Finley, D. J. Paustenbach and B. D. Kerger, *J Exp Anal Environ. Epidemiol.*, 1997, **7**, 179.
86. P. Apostoli, R. Cornelia, J. Duffus, P. Hoet, D. Lison and D. Templeton, *Environmental Health Criteria 234*, World Health Organization, Geneva, 2006.
87. K. P. Nickens, S. R. Patierno and S. Ceryak, *Chem. Biol. Interact.*, 2010, **188**, 276.

88. Wisconsin Dept. Natural Resources, *Consolidated Assessment and Listing Methodology (WisCALM)*, Clean Water Act, Section 305b, 314, and 303, Integrated Reporting, Wisconsin, 2010.
89. L. R. Reyes-Gutiérrez, E. T. Romero-Guzman, A. Cabral-Prieto and R. Rodriguez-Castillo, *J. Mineral. Mater. Character. Engin.*, 2007, **7**, 59.
90. T. R. Durham and E. T. Snow, in *Cancer, Cell Structures Carcinogens and Genome Instability*, L. P. Bignold, Birkhäuser Verlag, Basel, 2006, p. 97.
91. F. J. C. Roe and R. L. Carter, *Br. J. Cancer*, 1968, **23**, 172.
92. W. C. Hueper and W. W. Payne, *Am. Ind. Hyg. Ass. J.*, 1959, **20**, 274.
93. W. W. Payne, *Am. Med. Ass. Arch. Ind. Health*, 1960, **21**, 530.
94. C. G. Lewis, R. M. Belnaik, M. C. Plowman, S. M. Hopfer, J. A. Knoght and F. W. Sunderman, *J. Arthroplasty*, 1995, **10**, 75.
95. C. M. Thompson, D. M. Proctor, M. Suh, L. C. Haws, C. D. Hébert, J. F. Mann, H. G. Shertzer, J. G. Hixon and M. A. Harris, *Toxicol. Sci.*, 2012, **125**, 79.
96. B. L. Collins, M. D. Stout, K. E. Levine, G. E. Kissling, R. L. Melnik, T. R. Fennell, R. Walden, K. Abdo, J. B. Pritchard, R. A. Fernando, L. T. Buka and M. J. Hooth, *Toxicol. Sci.*, 2010, **118**, 368.
97. National Toxicology Program (NTP), *Technical Report Series*, NTP,-TR 546, NIH Publ. Research Triangle Park, NC, 08-5887.
98. M. D. Stout, R. A. Herbert, G. E. Kissing, B. J. Collins, G. S. Travlos, K. L. Witt, R. L. Melnick, K. M. Abdo, D. E. Malarkey and M. J. Hooth, *Environ. Health Persp.*, 2009, **117**, 716.
99. U. Glaser, D. Hochrainer, H. Klöppel and H. Oldiges, *Toxicology*, 1986, **42**, 219.
100. A. N. Uddin, F. J. Burns, T. G. Rossman, H. Chen, T. Klug and M. Costa, *Toxicol. App. Pharmacol.*, 2007, **221**, 329.
101. S. Venitt and L. S. Levy, *Nature*, 1974, **250**, 493.
102. V. Bianchi, R. Sal Toso, P. Debetto, A. G. Levis, S. Luciani, F. Majone and G. Tamino, *Toxicology*, 1980, **17**, 219.
103. C. B. Klein, L. Su, D. Bowser and J. Leszczynska, *Environ. Health Perspect.*, 2002, **110**, 739.
104. K. Salnikow and A. Zhitkovich, *Chem. Res. Toxicol.*, 2008, **21**, 28.
105. P. Venier, A. Montaldi, F. Majone, V. Bianchi and A. G. Levis, *Carcinogenesis*, 1982, **3**, 1331.
106. S. S. Wise, A. L. Holmes, H. Xie, W. D. Thompson and J. P. Wise, *Chem. Res. Toxicol.*, 2006, **19**, 1492.
107. F. L. Petrelli and S. De Flora, *Br. J. Ind. Med.*, 1987, **44**, 356.
108. T. Wolf, R. Kasemann and H. Ottenwalder, *Arch. Toxicol.*, 1989, **13**(Suppl), 48.
109. T. Wolf, R. Kasemann and H. Ottenwalder, *Carcinogenesis*, 1989, **10**, 655.
110. T. Norseth, *Environ. Health. Persp.*, 1981, **40**, 121.
111. S. A. Katz and H. Slem, *J. Appl. Toxicol.*, 1993, **13**, 217.
112. K. A. Biedermann and J. R. Landolph, *Cancer Res.*, 1987, **47**, 3815.
113. K. A. Biedermann and J. R. Landolph, *Cancer Res.*, 1990, **50**, 7835.
114. J. P. Wise, J. C. Leonard and S. R. Patierno, *Mutat. Res.*, 1992, **278**, 69.

115. S. R. Patierno, D. Banh and J. R. Landolph, *Cancer Res.*, 1988, **48**, 5280.
116. E. R. Nestmann, T. I. Mula, G. R. Douglas, K. C. Bora and D. J. Kowbel, *Mutat. Res.*, 1979, **66**, 357.
117. C. M. Thompson, Y. Fedorov, D. D. Brown, M. Suh, D. M. Proctor, L. Kuriakose, L. C. Haws and M. A. Harris, *PLoS ONE*, 2012, **7**, e42720q.
118. X. He, G. X. Lin, M. G. Chen, J. X. Zhang and Q. Ma, *Toxicol. Sci.*, 2007, **98**, 298.
119. S. I. Gray and K. Sterling, *J. Clin. Invest.*, 1950, **29**, 1604.
120. A. N. Costa, V. Moreno, M. J. Prieto, A. M. Urbana and M. C. Alpoim, *Mol. Carcinogenesis*, 2010, **49**, 582.
121. L. F. Petrilli and S. De Flora, *Appl. Environ. Microbiol.*, 1977, **33**, 805.

CHAPTER 5

Cobalt and Nickel

5.1 Introduction

Cobalt and nickel are multivalent transition metals and, like iron, are classified in Group VIII of the Periodic Table. Both are known to fulfil essential functions as micro-trace nutrients in the human body. They occur widely in the human environment and are found together in natural deposits with iron, copper and arsenic. Cobalt naturally occurs in nickel-bearing laterites and nickel–copper sulfide deposits and is often extracted as a by-product of nickel and copper. According to the Cobalt Development Institute, about 48% of cobalt production originates from nickel ores, 37% from copper ores and 15% from primary cobalt production. Arsenide ores are also a source of the two metals.

Cobalt and nickel resemble iron in being magnetic metals. They serve major uses in heavy metal industries and in production of tough resilient metal alloys with high tensile strength and resistance to corrosion. The alloys and super-alloys tolerate extremely high temperatures and corrosive environments and find use in jet engines, gas turbines, power generation applications, chemical plants and petroleum refining. Typically, nickel alloys or cobalt alloys and super-alloys are used in environments where temperatures are in access of 500 °C. Their resistance to corrosion is recognised in their applications in marine engineering, oil rigs, electronics and other applications where common stainless steels fail to provide adequate performance. Cobalt alloyed with iron, nickel and other metals is used form Alnico, an alloy with exceptional magnetic strength. Cobalt, chromium and tungsten are alloyed to form Stellite, which is used for high-temperature, high-speed cutting tools and dies used in metal engraving and coinage.

The 12th Annual Report (RoC) lists cobalt sulfate (cobaltous sulfate) as reasonably anticipated to be a human carcinogen on the basis of studies in experimental animals, even though no epidemiological evidence of human risk had been seen.[1] The compound is soluble in water and the resulting solution is

Issues in Toxicology No. 18
The Carcinogenicity of Metals: Human Risk through Occupational and Environmental Exposure
By Alan B. G. Lansdown

Published by the Royal Society of Chemistry, www.rsc.org

acidic. Metallic cobalt is used extensively in industry, and health risks are expected through inhalation of dusts containing cobalt and its inorganic compounds.[2,3] Cobalt dust was nominated for toxicology and carcinogenicity studies based upon widespread occupational exposure and the occurrence of such occupational diseases as hard metal disease, associated with exposure to cobalt and compounds such as cobalt–tungsten–carbide, used in metallurgy.[4] The RoC listed "certain" nickel compounds (notably nickel carbonyl, nickel sulfide) as reasonably anticipated to be human carcinogens in their initial report in 1980, but reclassified "all" nickel compounds as known carcinogens on the basis of additional information seen in their 2000 review. Metallic nickel is still regarded as reasonably carcinogenic to humans, but not on the basis of human epidemiological or case studies. Epidemiological and case studies are reported for both cobalt and nickel and their compounds (mostly inorganic), but as in many studies conducted in industrial environments, workers are exposed to several toxic metals and non-metallic substances, which can influence the interpretation and statistical significance of the observations. Cobalt and nickel residues are included amongst potentially harmful constituents of cigarette smoke and/or smokeless tobacco products.[5,6]

5.2 Cobalt

Cobalt is a brittle, hard, silver-grey transition metal with magnetic properties similar to those of iron and nickel (ferromagnetic). Twenty-seven isotopes have been identified, the stable isotope ^{59}Co and other isotopes with half-lives ranging from a few hours to >5 years (^{60}Co). ^{60}Co is used in tracer studies. Cobalt fumes and dust are black and odourless but several inorganic compounds appear coloured, *i.e.* cobalt blue [cobalt(II)oxide–aluminium oxide].

Cobalt is not found naturally. Its average abundance in the soil is low (8 ppm) although this level will be a lot higher in mining areas and in the proximity of heavy metal industries, where concentrations may reach 70 ppm. Anthropogenic sources include the burning of fossil fuels, sewage sludge, phosphate fertilizers, mining and smelting of cobalt ores, processing of cobalt alloys, and industries that use or process cobalt compounds. Natural sources of cobalt ores are influenced by erosion (wind-blown continental dusts), weathering of rocks and soil, seawater spray, volcanoes, forest fires, extraction by plants, and continental and marine biogenic emissions. Cobalt is found in meteoritic materials. At least 34 cobalt-containing minerals have been identified, including sulfides, arsenides, selenides, carbonates and sulfates, often in deposits rich in nickel, copper, arsenic, silver, iron and manganese. Ores of commercial value include the linnnaetite rocks:[7] Carollite $(Co_2Cu)_3S_4$, Linnaeite Co_3S_4, Siegenite $(CoNi)_3S_4$; and cobalt-bearing volarites: $(CoNiFe)_3S_4$. Chief deposits are in Zaire and central Africa, Canada, Australia and Morocco.

Cobalt is a stable element at room temperature and does not react directly with either hydrogen or nitrogen. It is insoluble in water. However at higher temperatures, metallic cobalt is oxidised to cobaltous oxide (CoO), and it reacts with carbon, phosphorus or sulfur. Cobalt dissolves slowly in dilute mineral acids to produce numerous bioactive salts. As a transition metal, its configuration of electrons allows for variable oxidation state, but Co(II) and Co(III) are the usual states encountered environmentally and industrially. Some environmental microorganisms may directly oxidise Co(II) to Co(III) and such biological activities may exert some control on the behaviour of cobalt in nature.[8] The National Fire Protection Association has not assigned a flammability rating to cobalt metal, dust, and fume; however, powdered cobalt (pyrophoric cobalt) is combustible and burns brilliantly in air at room temperature. Dusts of cobalt metal or cobalt compounds may form explosive mixtures in air and hence present an occupational hazard.[9]

5.2.1 Nutritional Requirements and Metabolism

Cobalt is an essential micronutrient in the human body and its uptake and metabolic role in vitamin B_{12} (cobalamine complex) are documented (Figure 5.1).[10–13] Desoxyadenosyl-cobalamine (co-enzyme B_{12}) is the predominant vitamin B_{12} analogue found in the human diet, although other analogues including methylcobalamine and hydroxycobalamine occur. They are absorbed unchanged in the lower intestine and bind to cell proteins.

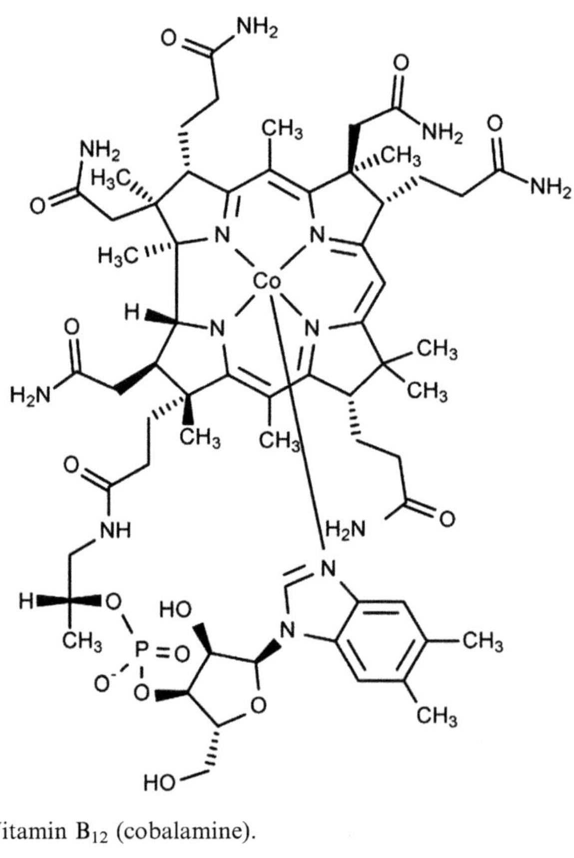

Figure 5.1 Vitamin B_{12} (cobalamine).

Vitamin B_{12}, cyanocobalamine, helps to protect nerves and is involved in the formation of red blood cells. Vitamin B_{12} is also concerned in the biosynthesis of methyl groups of choline and methionin.

Cobalt resembles iron chemically and biochemically such that intestinal absorption by carrier proteins such as transferrin is expected to be similar.[14] The uptake and metabolism of vitamin B_{12} in patients with blood disorders is possibly modulated by a physiological mechanism involving an intrinsic factor.[10,11] This intrinsic factor is a glycoprotein [molecular weight (MW) 55 000] synthesised by gastric parietal cells. Specific receptors were identified for uptake of vitamin B_{12} – intrinsic factor complexes on microvillous membranes in the ileum – and it was found that cobalt transfer involved calcium or magnesium ions. Intracellular cobalt binds to mitochondria in intestinal cells. Some is transferred to the circulation as a protein complex for mobilisation throughout the body. Cobalt may be absorbed through pinocytic mechanisms or through a specific cobalt-carrier protein, "transcobalamine".[15,16] Absorption of cobalt as vitamin B_{12} is low and related to dietary concentration (Figure 5.2). Chanarin considered that, with high pharmacological doses of vitamin B_{12}, about 1% may diffuse across the intestinal mucosa by a mechanism that does not require intrinsic factor, the limitation being related to the saturation of specific ileal cell receptor sites by the intrinsic factor–vitamin B_{12} complex.[11] Although the modulation of cobalt absorption is not fully understood, it is clear that, under normal circumstances, cells in the body are exposed to concentrations not exceeding $0.15 \pm 0.07\ \mu g\ l^{-1}$ although far higher concentrations of $9.7 \pm 7.1\ \mu g\ l^{-1}$ are seen in cobalt workers.[17]

Vitamin B_{12} is located throughout the body and functions as a co-enzyme for a critical methyl transfer that converts homocysteine to methionine, and for the

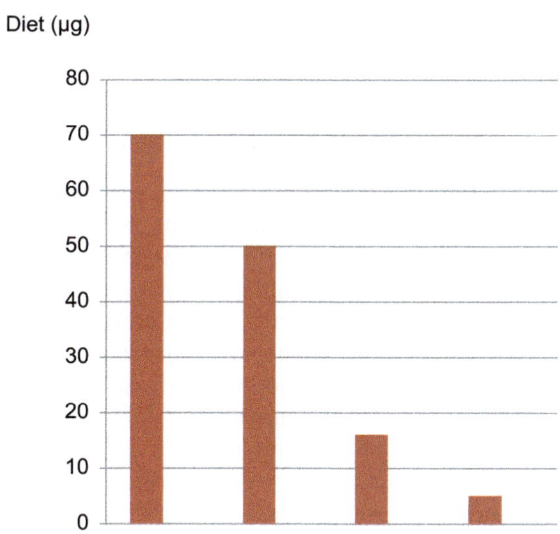

Figure 5.2 Absorption (%) of vitamin B_{12} according to dietary concentration.[1–11]

conversion of L-methylmalonyl-coenzyme A (CoA) to succinyl-CoA. Vitamin B_{12} is required as a co-enzyme in the transfer of single carbon units in nucleic acid metabolism.[18] It is essential for erythropoiesis, DNA synthesis, maintenance of the neurological tissues and body growth. Intestinal bacteria are involved in the synthesis and inter-conversion of vitamin B_{12} analogues.[11] The total amount of cobalt in the human body is low and expected to be 1.5 μg. Analyses of cobalt[17] in the plasma of control subjects found 0.15 ± 0.07 μg l^{-1} although in cobalt workers it was a lot higher at 9.7 ± 7.1 μg l^{-1}. Highest sources of cobalt in the diet include red meat, vegetables, cereals, fruit and eggs. Cobalt shares many biochemical similarities with iron in its uptake and metabolism. In cases of deficiency, coeliac diseases and genetic defect, pernicious and macrocytic forms of anaemia are diagnosed.

Uptake of cobalt and iron in the small intestine occurs in a competitive fashion with high levels of cobalt inhibiting iron uptake, and high iron impairing cobalt absorption.[19] This suggests that the metals share a common transport mechanism where cobalt exerts a regulatory influence on iron uptake, as well as interacting with and substituting for calcium and zinc metabolic systems.[19,20] Increased cobalt absorption is seen in patients with iron uptake disorders.[11,21]

Signs of cobalt deficiency include low thyroid hormone and cytological changes in neurological and other tissues. Dementia, polyneuropathy, myelopathy and blindness are reported.[22,23] Vitamin B_{12} deficiency has been shown to result in hepatocellular disorders involving cellular enzymes in glycogen metabolism and lipid synthesis. The key enzymes ATP citrate lyase and Krebs cycle enzyme citrate synthase are increased in cobalt deficiency and mitochondrial cristae membranes are disturbed.[24] Regulation of cobalt absorption is complex and probably involves intrinsic factors, intestinal bacterial balance and the composition of the diet. Excesses in vitamin B_{12} are associated with hyperthyroidism, increased haemopoiesis and raised blood volume leading to heart failure. Cardiomyopathies are reported in people who consume high levels of beer.[25–27]

5.2.2 Industrial Uses and Human Exposure

Cobalt is a relatively minor element in the human environment and is found at low concentrations in soil, waterways and oceans, particulate matter in the air and foodstuffs of animal and plant origin. It is absorbed into the human body through ingestion, inhalation and implantation of prostheses; percutaneous absorption is low.[28] Exposure to cobalt, cobalt compounds and alloys in industry and everyday life is high, particularly in metal extraction, metallurgy, heavy metal industries, battery production (with cadmium) and electrical industries, polishing and grinding, building and allied trades and metal refineries. Cobalt-based pigments are widely employed in paints, glass and ceramics production. Considerable concern is evident currently concerning worker exposures to cobalt and mixed metal dusts and fumes in hard metal production involving cobalt–tungsten carbides, and increased mortality from lung cancers.[9,29,30] Cobalt is commonly associated with contaminated detergents, cement, clays and pottery making. Industry-related respiratory problems arising from inhalation of cobalt dusts, leading to metal-related fume fevers, have led to increased environmental monitoring of airborne metal concentrations, and collation of employee health records on personal hygiene, blood and urine estimations and respiratory physiology.[4]

The largest single source of cobalt exposure for the general population is the food supply. Most of this is in inorganic form. Additionally, cobalt is a common contaminant of tobacco (<0.3–2.3 $\mu g\ g^{-1}$ dry weight), and approximately 0.5% of this cobalt is present in mainstream smoke. Occupational exposure to dust and fumes containing cobalt and other metals is a recurrent problem in hard metal industries, and fume fever and pneumoconiosis are recognised.[31] Cobalt concentrations in coal, crude oil, fuel oil and gasoline in the USA were found to be 5 mg kg^{-1}, 0.001–10 mg kg^{-1}, 0.03–0.3 mg kg^{-1} and <0.1 mg kg^{-1}, respectively.[32]

Special concerns have been raised concerning the safety of cobalt sulfate heptahydrate ($CoSO_4.7H_2O$).[1,33,34] This water-soluble compound composed of non-inflammable red crystals is acidic in solution and used in electroplating and electrochemical industries, and as a colouring agent in ceramics and glass, paints, enamels and porcelain. The Co^{2+} is biologically active and binds protein. It exhibits low toxicity when ingested and is used as a dietary supplement for cattle grazing low-cobalt pastures and in other forms of cobalt-deficiency diseases.

5.2.3 Toxicity and Carcinogenicity

5.2.3.1 General Aspects

Metallic cobalt is one of the least toxic elements in the Periodic Table. It is inert at room temperature although respiratory problems are associated with inhalation of cobalt dust. Contact with cobalt in everyday life is commonly associated with nickel-plated objects. Complications arise in identifying specific problems because cobalt and nickel are chemically similar and strongly allergenic.[35] A second non-malignant problem encountered with cobalt is that of cardiovascular diseases and cardiomyopathies.[25–27] This complication was originally noted in heavy beer drinkers in Canada in 1966, and later was associated with mild congestive heart failure, cardiomegaly, cardiogenic shock, low cardiac output and metabolic acidosis, but not pulmonary symptoms. The cardiotropic and electrophysiological effects of cobalt were confirmed in rats and dogs given repeated oral doses or parenteral injections of cobalt sulphate, or cobalt infusion.[25,36] No link has been identified between high cobalt consumption in beer or other beverage and human cancer.

Three major concerns are documented relating to the potential carcinogenicity of cobalt:

a. Long-term inhalation of cobalt in dusts and fumes
b. Chronic exposure to cobalt sulfate heptahydrate
c. Cobalt alloyed to nickel, chromium, tungsten and other metals, and use of alloys in orthopaedic prostheses.

Difficulties have been experienced in evaluating toxic risks attributable to cobalt in epidemiological and case studies, because commonly workers are exposed to metals such as chromium and nickel which, as discussed elsewhere in this volume, are appreciably more toxic, allergenic and mutagenic. More tangible information on the toxicity and carcinogenicity is provided by regulatory-style long-term experimental trials and *in vitro* studies, even though extrapolation may be difficult.

5.2.3.2 *In Vitro Cytological and Mutagenic Studies*

Several authors review the cytogenicity and mutagenicity of metallic cobalt, inorganic cobalt compounds, and cobalt-containing hard metal dusts and indicate their genotoxic properties *in vitro* and *in vivo*.[37,38] Early studies demonstrated that rat fibroblasts tolerating 5 or 7.5 μg ml^{-1} cobalt chloride developed a high incidence of aneuploidy, mitotic abnormality and cytological changes consistent with reduced oxygen uptake and aerobic and anaerobic glycolysis.[39] The Co(II) ion was assumed to affect cellular respiration, but with continued culture in non-cobalt medium, the cells developed "cobalt-tolerance". Later studies have yielded conflicting views of the mutagenicity of Co(II). Inconsistent or negative responses have been observed in tests with *Salmonella typhimurium* and *Escherichia coli*.[40,41] Léonard and Lawerys claimed that the Co(II) ion interacts with cell proteins and impairs nucleic acid synthesis, transforms Syrian hamster embryo cells and evokes chromosomal damage in plant cells, yet it is not clastogenic.[38] This is not borne out by later work.

A recent study in which human lung cells were cultured in the presence of cobalt chloride demonstrated that Co(II) ion influences cell replication and apoptosis by gene activation or suppression.[42] It established that cobalt induces intracellular hypoxic stress, and that the ion influences the upregulation and transcription of the genes for specific cobalt carrier proteins (*FBXL2, ZNT1, SLC12A5*) and tumour suppressors or transcription factors (*MAZ, DLG1, MYC, AXL*). In this model, cobalt was shown to downregulate expression of TIMP2 (tissue inhibitor of metalloproteinase 2), a peptidase involved in degradation of the extracellular matrix and possible other actions.[43]

The genotoxity of metallic cobalt and Co(II) compounds has been studied extensively by Lison and his colleagues in Belgium. They provided evidence that cobalt dust can cause some clastogenic action through DNA strand breakage, sister-chromatid exchanges and inhibition of DNA repair in human peripheral lymphocyte cultures in a concentration-related pattern, but the action is weak.[37,44–46] They attributed this genotoxicity to the ability of Cr(II) to induce reactive oxygen species and intracellular oxygen stress. Human lymphocytes incubated with metallic cobalt (0.06–6 μg l^{-1}) showed a dose- and time-related profile of DNA single strand breaks in *Comet* assays, and micronucleus formation. Peripheral lymphocytes taken from 26 workers exposed to cobalt dust in production plants showed an increase in sister-chromatid exchanges when participants were matched for age and smoking habits, and confirm the view that cobalt acts as a weak mutagen.[47] Tumour markers carcino-embryonic antigen and tissue polypeptide antigen were not significantly different from control values, except when smoking habits were taken into consideration. However, a second study of 35 workers employed in Belgian cobalt refineries showed no significant changes in genotoxic markers or mutagenic change, except where workers were smokers or were exposed to other hard metals.[48]

Additional studies designed to examine the comparative effects of cobalt and cobalt–tungsten carbide (WC) dusts have consistently shown that WC accentuates the genotoxicity of cobalt. A recent study analysed the profile of gene expression in peripheral lymphocytes and monocytes and reported an upregulation of genes involved in apoptosis and stress/defence responses with a downregulation of others determining immune response.[49,50] Later research using human lymphocytes focused upon the rapid burst of oxygen stress placed upon cells by release of reactive oxygen species. Induction of a cascade of genotoxic and translational and post-translational changes regulated gene expression, activation and stabilization. Gene expression sister-chromatid exchanges *etc.* influenced by WC-Co exposure was wider than seen with cobalt alone and may be attributable to some as yet unidentified interaction between cobalt and tungsten ions, physico-chemical properties of the ions and their ability to induce oxidative stress, uncoil chromatin loops or increased DNA sensitivity to fragmentation.[37,44]

5.2.3.3 Experimental Studies in Animals

5.2.3.3.1 Inhalation Studies. The risk of pulmonary cancer in cobalt workers following chronic inhalation of cobalt dust, fumes and inorganic Co(II) compounds[2,4,38,51,52] has been examined experimentally in numerous studies in rats, mice, rabbits, guinea pigs, Syrian hamsters and miniature pigs. Whilst they have proved useful in studying absorption, metabolism, excretion, pharmacokinetics and acute toxicity of inhaled of cobalt dust, few have provided evidence of carcinogenicity.[53] A regulatory inhalation toxicity study, conducted for the National Toxicology Program (NTP) in 1999, showed that F344N rats and B6C3F1 mice inhaling aerosols containing 0.3, 1.0 or 3.0 mg m^{-3} cobalt sulfate heptahydrate for 2 years developed dose-related profound pulmonary inflammatory change, proteinosis and non-neoplastic lesions of the respiratory tract, and increased numbers of alveolar/bronchiolar tumours and adrenal phaeochromocytomas.[33] Increased haemangiosarcomas in mice were related to concurrent *Helicobacter pylori* infection and not attributed to cobalt. Whilst pulmonary pathology was related to the concentration of cobalt in the aerosols, tumour incidence in rats was significantly increased only in female rats, and in male > female mice. Previous studies had shown that rats and mice exposed to aerosols containing cobalt sulfate heptahydrate at up to 30 mg m^{-3} developed non-malignant pulmonary disease after 13 weeks but no tumours.[34] Intratracheal administration of 2 or 10 mg Co(II) oxide in lifetime studies in rats led to a 6%

incidence of pulmonary tumours.[54] When cobalt oxide was alternated with intratracheal instillation of the carcinogen benzo[α]pyrene it greatly enhanced/promoted production of malignant lung tumours.

Rats, mice, hamsters and rabbits have been exposed to cobalt or related compounds by inhalation, intratracheal implantation or parenteral injections for subacute or lifetime studies, using concentrations greatly in excess of maximum permitted levels in factories or workplaces.[54–59] Whereas tumours were observed in rats and mice following all routes of cobalt exposure, hamsters and rabbits did not develop cobalt-related neoplasms following inhalation of cobaltous oxide (CoO), cobaltous chloride ($CoCl_2$) or cobalt dust.[33,34,58,60] Rats and mice inhaling high levels (0.3–3 mg m^{-3}) cobalt sulfate hexahydrate for 6 h per day over 2 years developed profound respiratory distress symptoms with a dose-related incidence of pulmonary adenoma and carcinoma, but survival rates in all groups were poor (40–70%). The OHSA gave permissible exposure limits for humans as 0.1 mg m^{-3}, 8 h per day Time Weighted Average (TWA).[9] Mice showed a high incidence of adrenal phaeochromocytoma. In earlier studies, these authors exposed rats to <50 mg m^{-3} for 13 weeks and reported pre-neoplastic squamous metaplasia in the respiratory epithelium, polyps in the larynx and alveolar hyperplasia at 0.3 mg m^{-3}, the larynx being the most sensitive tissue.[33,34] When cobaltous oxide (CoO) was instilled into the trachea at single doses of 2 or 10 mg kg^{-1} for total doses of 78 and 390 mg kg^{-1} respectively (100 animals per group), only two benign pulmonary tumours were seen in the lower dose group and two benign and four malignant tumours in the higher dose group, suggesting that cobaltous oxide is weakly carcinogenic.[54]

5.2.3.3.2 Intramuscular or Intra-articular Implantation. Repeated subcutaneous injection of Co(II) chloride at 4 mg per 100 g body weight for two courses of 5 days with a 9-day interval between doses led to massive fibrosarcoma in 8 of 11 rats surviving to 12 months and 6 of 16 surviving for 8 months.[57] Previous studies demonstrated subcutaneous sarcoma at injection sites following subcutaneous administration of metallic cobalt powder of varying particle size.[55,56,61,62] Implantation of cobalt particles in "bulk form" in rats led to production of pleomorphic sarcomas in five of six animals within 12 months.[59] Tumour formation probably reflected the physicochemical properties of the implant (particulate *vs.* bulk form) rather than the surface to volume ratio. In these studies, nickel particles led to visible nodules of rhabdomyosarcoma at implant sites within 4–6 months and did not differ significantly from tumours induced by Ni–Co implants. In primary rhabdomyosarcoma, induced in the rat by intramuscular implantation of powdered metallic cobalt, at least 50% of the inducing metal accumulated in the nuclei of malignant cells and in nucleoli. The remainder was distributed approximately equally between the nuclear sap and chromatin.

Steinhoff and Mohr injected massive doses of cobalt oxide (total dose 1000 mg kg^{-1}) in divided amounts in rats in a lifetime study and 19 of 20 animals developed malignant tumours at injection sites.[54] In the same laboratory, 14 of 20 animals given three intraperitoneal injections developed malignant tumours whereas those receiving similar treatment with a Co–Al–Cr complex showed tumour incidence of 2 in 20 animals, possibly on account of the lower cobalt content. Other studies have demonstrated the carcinogenic potential of injections of cobalt with chromium and/or molybdenum alloys.[63,64]

Concern has been raised over the past 30 years regarding the safety in use of metal-on-metal orthopaedic prostheses and issues relating to release and possible toxicity from cobalt fragments or interactions between metal ions leading to carcinogenicity.[63,65,66] Experimental studies have shown that cobalt alloys used in orthopaedic prostheses do produce subcutaneous sarcoma when injected parenterally or implanted in rats, and that these have been associated with intense local reaction. This was demonstrated when rats were implanted or injected with a suspension of microspheres of cobalt–chromium–molybdenum, and implant-associated tumours were found with loose implants and foreign body reactions.[67] In an attempt to reproduce in-life release of prosthetic fragments, Lewis *et al.* injected powders of cobalt–chromium–molybdenum alloy directly into the articular cavity of the knee or hip joints in rats (20 mg per animal).[68] They did not observe tumours in test or negative control groups, yet in a group of 12 animals treated with nickel sulfide, 10 developed pleomorphic sarcomas.

5.2.3.3.3 Oral or In-diet Administration. No data are available in the open literature on the potential carcinogenicity of cobalt following exposure *via* the oral route either in humans or in experimental animals. The exposure of consumers to total dietary cobalt has been considered.[69]

5.2.3.4 Epidemiological and Human Case Studies

Sufficient evidence exists from epidemiological evidence that inhalation of cobalt and hard metal dusts is a cause of profound respiratory diseases and fume fever,[2,34,51,70–72] but the limited database on exposures to cobalt and its inorganic compounds makes it difficult to implicate them as human carcinogens in occupational or environmental exposures.[3,73,74] Pneumoconiosis is a common feature of many cases recorded and, from limited information available, lung diseases including cancer are likely to depend upon the amount and the physico-chemical properties of the particles inhaled, and on patterns of exposure. Other difficulties relate to the low numbers of cases of lung cancer in exposed populations. Nevertheless cobalt dust and cobalt sulfate heptahydrate salts are classified as being "possible" or "reasonably anticipated to be human carcinogens", mainly on the basis of experiments in animal models.[1,75,76] Lung cancer is identified as the principal risk associated with cobalt dust but numbers of cases are small and clinical evidence rests upon studies conducted in Scandinavia and France. Studies based on lung cancer mortality in hard metal industries are difficult to interpret because workers are exposed to several metals, notably tungsten carbide and chromium. Cigarette smoking was a confounding factor in evaluating cobalt as a cause of pulmonary diseases in several studies.[2,31]

A Danish study of 874 women exposed to cobalt in porcelain factories showed a slightly higher incidence of lung cancer than expected (67 observed *vs.* 56 expected), but although the number of cases was slightly higher than in all national surveys, the incidence was too low to be of statistical significance.[52,71] An industry wide case-control study involving 10 factories in the French hard metals industry, over the period 1968 to 1991, showed only 63 cases of lung cancer mortality among 5777 males and 1682 females (*i.e.* 0.84%) exposed to cobalt dust (with or without tungsten carbide).[29] The cancer mortality was significantly higher than in national averages. When they ignored exposures in the previous 10 years, the clinicians claimed a two-fold increase in lung cancer in workers exposed to cobalt and tungsten carbide, and the risk increased with cumulative exposure. In a second French study, four cases of lung cancer mortality were reported in a cohort of 1143 workers exposed to cobalt dust in an electrochemical plant producing cobalt and sodium.[70] However, the occupational risk was not authenticated because tobacco consumption could not be excluded from the analysis.

Hogstedt and Alexandersson[77] evaluated reports on 3163 male workers with at least 1 year of occupational exposure to hard metal dust at one of three hard metal manufacturing plants in Sweden in 1840–1882, and who were followed during the period 1951–1982. Allowing for different levels of cobalt exposure, they observed 73 cancer deaths of which 17 were from lung cancer. Comparing the high *vs.* low exposure categories, standard mortality rates were similar. However, among workers with at least 10 years of exposure and more than 20 years since first exposure, there was a significant increase in lung cancer risk even though numbers were low, and this was unrelated to smoking habit.

The possible co-carcinogenicity between cobalt, tungsten carbide (WC) and chromium in patients implanted with metal-on-metal prostheses is complex and unresolved. A nationwide survey conducted in Sweden[78] examined cancer statistics on 116 727 patients who underwent hip surgery in the period 1965 to 1974. The study reported no elevated cancer risk relative to the general population and that hip implant patients had similar rates of most types of cancer to those in the general population. Berkenstock, in 1992, reviewed the subject and related incidences of primary neoplasia to the physico-chemical properties of implants, ionic valences of the metals, electrochemical properties of implants and presumed metabolic pathways of ions released into the circulation.[66] Twenty cases of primary tumours were recorded in a study in 1975, but Berkenstock considered that, in his 50 years of practice, cobalt–chromium implants had a safe and efficacious record. This view is substantiated by a recent survey showing that available data do not support a causal link between total hip or knee arthroplasty and the development of cancer. Although it is biologically plausible for the materials used in total joint replacement to induce malignant degeneration, this relationship has not been demonstrated. There is ongoing concern about an association between haematopoietic cancer and total joint replacement.[79]

No regulatory authority or epidemiological study has implicated cobalt with or without tungsten carbide as a cause of childhood leukaemia. However, it is noteworthy that in the vicinity of the towns of Sweet Home and Fallon in Nevada, a cluster of 17 children were diagnosed with leukaemia in areas of high airborne cobalt and tungsten carbide dust.[80] Whilst statistics are claimed to show a temporal relationship between the onset of childhood leukaemia and environmental cobalt and tungsten carbide concentrations, further clinical and experimental study is required to confirm the risk. Scientists from the University of Arizona found high levels of tungsten and cobalt around a tungsten smelter. While tungsten is not a proven cause of leukaemia, researchers have found that, combined with cobalt, it causes mutations and tumours in mice.

5.2.4 Conclusions on the Carcinogenicity of Cobalt

Epidemiological studies in cobalt-related industries provide limited evidence for the carcinogenicity of the metal and the ability of chronic cobalt dust inhalation to cause lung cancer. The few studies available are confounded by worker smoking, exposure to other potentially carcinogenic elements such as chromium and nickel, or are deficient in essential details concerning patterns and durations of worker exposure.[3,37,38] Evidence provided by one study indicated that a long latent period may exist between working in the hard metal industries and the appearance of fatal cancers, but this remains to be qualified.[77] Anxiety that release of metal fragments or metal ions following implantation of cobalt-containing orthopaedic prostheses may be a cause of local sarcoma formation is presently not justified. Tharani *et al.* reviewed 25 cases of malignant disease occurring in association with a total hip or knee prosthesis and concluded that, on the basis of this small number of case reports, there is no apparent relationship between any specific implanted material and cancer.[79]

At least 10 experimental studies, mainly in rats and mice, have substantiated that cobalt is a potential cause of broncho-alveolar tumours and/or tracheal metaplasia following inhalation or intratracheal administration. Adrenal tumours were seen in mice following subcutaneous cobalt sulfate injection, but the mechanism is unclear.[34] Carcinogenic risk has not been associated with dermal exposure to cobalt or its inorganic compounds, although subcutaneous sarcoma has been reported in rodents injected or implanted with metallic cobalt or related compounds in long-term studies,[34,54,57,59] as with many other metal and non-metal compounds.[81]

Numerous *in vitro* studies in which human peripheral lymphocytes or other cell lines have been cultured in the presence of cobalt, Co(II) compounds or cobalt–tungsten carbide complexes indicate that cobalt has a weak mutagenic effect and can exert a clastogenic effect though induction of sister-chromatid exchanges, micronucleus formation and chromosomal aberration, but useful information has been provided to show that cobalt can upregulate or downregulate key genes involved in cell replication, apoptosis and transformation. Many of these studies identify production of intracellular oxygen stress through elaboration of reactive oxygen radicals. The effect seems to be amplified by tungsten carbide. Cobalt is not a cumulative nutrient, but it does bind MT and is known to interact with other elements.[82]

Available studies indicate that Co(II) ion should be listed as a low-grade carcinogen. It has been shown to induce tumours in experimental animals and has a direct genotoxic effect. Very limited experimental information is available to show that it promotes tumours induced by experimental carcinogens.

5.3 Nickel

Nickel was originally identified in niccolite (nickel arsenide) ores in Cronstedt in Germany in 1751 and was named from the German word, *Nickel*, for Satan or "Old Nick". The nickel-containing ore was called *kupfernickel*, meaning Old Nick's copper or false copper. Nickel is a transitional ferromagnetic metal with extensive uses in hard metal alloys. It is a silver–white lustrous metal isolated from niccolite, pentlandite ($Fe_9Ni_9S_{16}$), magnetic nickeliferous pyrites and chalcopyrite pyrrhotite with principle commercial sources in Canada and the USA, Australia, Cuba, Russia and Indonesia. New Caledonia in the South Pacific is the world's fourth-largest producer of mined nickel and the largest producer of ferro-nickel, and nickel products dominate the economy.

Nickel is the fifth most abundant element on our planet and occurs in concentrations of up to 50% (*ca.* 13 000 ppm) in meteoritic rocks. Nickel concentrations in the Earth's crust are estimated[83] to be 60–90 mg kg^{-1}. Metallic nickel is insoluble in water but reacts with oxygen; the pure metal is seldom found on the Earth's surface. Laterite ores containing nickel and cobalt are found in tropical climates and by the end of 2012 are expected to yield 50% of the world's nickel production. A wide variety of other nickel ores are found in different parts of the world, with workers being exposed during extraction and purification to iron, chromates, antimonides, arsenides, selenides, tellurides, and complexes containing cobalt, cadmium, copper, precious metals and silicates. Niccolite ore contains 43.9% nickel and 56.1% arsenic. Nickel is separated during "roasting" and reduction, and smelting. Through erosion of rocks, geochemical activity and in metal industries, nickel dust is widely distributed in the atmosphere. Nickel is a major component of meteorites and settles as minute particles (<1 μm diameter) from meteoritic dust from the stratosphere.

Nickel is a hard, ductile and malleable metal with modest ability to conduct heat and electricity. It exhibits three main oxidation states, Ni(II), Ni(III) and Ni(IV), and 31 isotopes of which five are stable: ^{58}Ni (68.1%), ^{60}Ni (26.2%), ^{61}Ni (1.1%), ^{62}Ni (3.6%) and ^{64}Ni (0.9%). Common nickel compounds include the oxides (NiO, Ni_2O_3), sulfides (NiS, NiS_2), subsulfide (Ni_3S_2) and nickel chloride ($NiCl_2$). Nickel has many uses in industry but its chief value lies in the production of hard, corrosion-resistant alloys and in stainless steel production. Common nickel alloys include Invar®, Monel®, Iconel® and Hastelloy®, Nichrome®, Permalloy® and constantin. Nickel-steel is used for armour plate and burglar-proof vaults. Resistant copper–nickel alloys have proved efficient in desalination plants for conversion of sea water to fresh drinking water. Nickel is extensively used in the world's coinage as a consequence of its durability.

Nickel and nickel compounds are toxic in humans and other animals, and many if not all are now recognised as carcinogens.[1,84] Fisher described nickel as the ubiquitous contact allergen, possibly causing more cases of dermatitis than any other metallic element.[35] Nickel allergy is encountered through sensitisation to cheap jewellery, coinage, nickel buttons on clothing, spectacles, hair products, dyes, electrical and metallurgical work, paints and pigments, glass manufacture and plastics. In industry, nickel solutions readily penetrate rubber gloves, and heavy vinyl gloves are recommended. Nickel occurs in tap water, and surveys in the USA in 1969–1970 revealed levels of 4.8 μg l^{-1}, leading to an expected human daily intake of 10 μg per day. Soil concentrations of up to 1000 ppm of nickel have been recorded in the region of mining sites.

5.3.1 Nickel as a Nutrient

Nickel is present in the diet and drinking water and is recognised as an essential micro-trace nutrient in the human body, but its functions are still largely unclear. The total nickel content in the human body is estimated to be less than 10 mg,[85] with mean blood levels being 3–8 μg l^{-1} and urinary excretion at 10–100 μg l^{-1}. Absorption of nickel from food and drinking water in four subjects given ^{62}Ni tracer was shown to be 29–40%, with excretion over 5 days in the range 51 to 82%.[86] Peak plasma concentrations were achieved within 2.5 hours after ingestion. Anke and his colleagues[87,88] monitored daily nickel intake in 56 persons aged 20–60 years who were fed diets containing 0.36 to 0.68 mg kg^{-1}. They recorded that 111–256 μg of nickel was consumed daily but this varied greatly according to lifestyle and geographical area. They presumed that dietary requirements for nickel in humans and other animals as low as 500 μg kg^{-1} body weight precluded the possibility of primary nickel deficiency! The Food and Nutrition Board are still to set a recommended daily intake of Ni(II) for optimal human health (?$<$1 mg per day), although signs of nickel deficiency are recognised.[89,90] Nickel is not cumulative in animals or humans.[91]

Ni(II) is absorbed across the intestinal brush border by passive diffusion as well as by a competitive carrier protein mechanism involving transferrin.[92] In the blood, nickel binds serum proteins including albumins and macroglobulins for distribution to all organs, notably blood, liver, brain, liver, bone, pancreas, adrenal glands and skin. Higher levels are reported in cancer patients and in cases of thyrotoxicosis and myocardial infarction.[21] Nickel binds RNA in preference to DNA and in tracer studies radio-labelled Ni(II) was located in the nuclear fraction in homogenates of liver, kidney, spleen and lungs, bound to RNA but not to DNA or proteins.[93] Various functions have been ascribed to nickel in the human body including hormonal action, regulation of cell membranes, enzyme cofactors, RNA synthesis and lipid metabolism. Much of this information derives from experimental observations of nickel deficiency in rats, sheep, pigs and chickens fed highly refined diets. (Nickel residues leach readily into food and drinking water from nickel alloys used in containers, processers, piping and utensils and the atmosphere, such that provision of a true nickel-free diet is very difficult to achieve.) These studies point to the view that nickel may act as a carcinogen by "controlling gene expression post-transcriptionally". Nielsen reviewed much of the early experimental work on nickel deficiency in animals and noted marked interspecies differences in response, ranging from a failure to grow, impaired reproductive performance, skin and liver pathology, to haematological change.[94] He later concluded that nickel is needed for the fixation of carbon dioxide to propionyl-CoA to form D-methylmalonyl-CoA in mitochondria.[95] Ultrasructural changes in the liver in Ni-deficient rats and chickens were consistent with reduced oxidation of α-glyceryl phosphate, increases in total lipids, phospholipids and cholesterol, and changes in the ribosomal profile. Nickel-deprived lambs and piglets exhibited reduced growth, serum lipids and cholesterol, protein and erythrocyte counts, serum transaminase, dietary nitrogen utilisation and liver copper concentrations.[96,97] Nielsen also suggested a role for nickel in modulating prolactin secretion.[98] High nickel was shown to enhance production of prolactin inhibiting factor (PIF) in isolated pituitary glands. Experimental evidence points also to the liver as a target organ in nickel deficiency, with ultrastructural degeneration, reduced oxidative action and alterations in polysome formation being reported in nickel-deficient rats.[94] Watt and Ludden identified a class of seven nickel-dependent metalloenzymes: urease, a hydrogenase, carbon monoxide dehydrogenase, methyl-coenzyme M reductase, Ni-superoxide dismutase, glyoxalase I and *cis–trans* isomerase.[99] Accessory or Ni-carrier proteins are required for binding and transport of the Ni(II) ion across cell membranes and its incorporation into enzymes.

5.3.2 Industrial Use and Human Exposure

Nickel is found in every sphere of human activity, in the atmosphere, food and drinking water and things we touch. Nickel is absorbed mainly as Ni(II), following ingestion of food and drinking water, but some is taken up through inhalation of particles in the atmosphere and following dermal contact. It is now known that nickel fragments and ions are released from metal-on-metal orthopaedic prostheses in sufficient quantities to induce metal allergy.[100] Various figures have been suggested for dietary nickel intake but, in test studies, ingestion of 22.4 g of nickel led to transitory increases in plasma nickel concentrations, rising from 48.8 μg l^{-1} after 1 hours, peaking at 80 μg l^{-1} after 3 hours and declining to 53.3 μg l^{-1} after 4 hours.[101] These responses were suppressed by addition of ascorbic acid. At a mean daily intake suggested for normal health of 168 μg per day (food) and 2 μg per day (drinking water), nickel has no observed toxicological effect.[102] Nickel residues enter human food chains from soil, and ground and sea water sources. Food taken from nickel mining areas, reclamation sites, combustion and waste sites, and in the vicinity of coal mining, smelters and heavy metal industries can be expected to contain greatly increased levels of nickel in comparison with unpolluted extra-urban sites. Seafood will concentrate nickel residues from waters draining from mining sites.[103]

Levels of nickel in the air vary greatly according to the time of sampling and geographical area. Assays in the USA in 1977–1982 ranged from 7 to 12 ng m^{-3}, although lower levels of 2.2 ng m^{-3} are recorded nowadays. Average concentrations[104] of nickel in indoor air in 1993 were 10 ng m^{-3}. Greatly increased aerobic concentrations of nickel particles are expected in the region of occupational sources listed in Table 5.1, possibly as high as 150 ng m^{-3} or more near point sources. About 20–35% of the nickel inhaled in non-industrial situations is absorbed into the circulation and eliminated in the urine without ill effect. Nickel is not a cumulative toxin but adverse effects can be expected where exposures exceed tolerated levels and where concentrations exceed normal protective mechanisms. In an industrial situation, toxicity resulting from excessive inhalation of nickel residues relates to the duration of exposure, the type of work involved, and safety measures in force at the time of the study.

The International Commission on Nickel Carcinogenesis in Man (INCO) conducted a survey of occupational nickel exposures in mining, smelting and refining in Canada and Scandinavia in 1990 and showed an unequivocal relationship between the incidence of lung and nasal cancer, and environmental nickel concentrations and professional responsibilities.[105] Greatest risk lies with exposures to oxidic and sulfidic ore in smelter plant operation, ore roasting, mixed electrolysis, sinter plants, metallurgy and with soluble nickel compounds. Cancer mortalities were higher following long-term working in these industries and cumulative exposures.[106] Risks of nickel-related cancers are also recorded in other professions, listed in Table 5.1. Combustion of fuels, diesel oil, coal and nickel waste are additional risk factors. Nickel residues are found in tobacco and in mainstream cigarette smoke.[5] The chemical forms of nickel differ greatly in physicochemical properties, biological effects and bioavailability of Ni(II) in body fluids and target cells. Whilst there is contradictory epidemiological evidence, Grimsrud and Anderson argue strongly that material evidence demonstrates a strong association between water-soluble nickel

Table 5.1 Industries associated with increased environmental nickel concentrations.

Chemical and catalyst production
Cement manufacture
Ceramics and glass production
Disinfectant and detergent manufacture
Electroplating industries
Battery production
Ink and dye manufacture
Jewellery and textiles
Metal mining, smelters, refining
Production of nickel alloys and hard metal industries, stainless steel
Paint and varnish making
Petroleum and gas industries

Table 5.2 Nickel compounds present in industrial environments.

Compound	*Chemical formula*	*Solubility in water*	*Exposure in industry*
Metallic nickel	Ni	Insoluble	Metallurgy, coins, alloys, steels, *etc.*
Oxide	NiO	Insoluble	Nickel salts, ceramics, glass, pottery, sinters Ni-refining processes, catalysts
Hydroxide	$Ni(OH)_2$	Insoluble	Catalyst intermediate, Ni–Cd batteries
Carbonyl	$Ni(CO)_4$,	Liquid at room temperature	Intermediate in carbonyl nickel-refining, process catalyst, Ni-plating, production of Ni-alloys
Subsulfide	Ni_3S_2	Insoluble	Pyrometallurgy
Sulfide	NiS	Insoluble	Catalyst in petroleum hydrogenation, intermediate in nickel industries, processing of Ni ores
Disulfide	NiS_2	Insoluble	Intermediate in Ni production
Carbonate	$NiCO_3$		Manufacture of Ni catalysts, pigments, glass, *etc.*, electroplating
Nitrate	$Ni(NO_3)_2$	Soluble	Manufacture of catalysts, Ni–Cd batteries
Chloride	$NiCl_2$		Intermediate in catalyst production
Sulfate	$NiSO_4$	Soluble	Electroplating, Ni-electrorefining, catalyst, electroless Ni-plating
Acetate	$Ni(OCOCH_3)_2$	Soluble	Electroplating, catalyst
Ammonium sulfate	$Ni(NH_4)_2.(SO_4)_2$	Soluble	Dye mordant, electroplating metal finishing compositions

compounds and risks of lung cancer and nasal cancer.[107] Nickel compounds encountered in industrial settings are listed in Table 5.2.

5.3.4 Toxicity and Carcinogenicity

5.3.4.1 General

The Ni(II) ion is bioactive and binds or otherwise interacts with RNA, chromosomes/chromatin, mitochondria and proteins in all cells, but not with DNA. It interacts with iron, calcium and zinc, and manganese acts as a cofactor in some nickel-modulated enzyme systems.[87,108,109] Dehydrogenases, transaminases and α-amylase levels are reduced in Ni-deficiency states and levels of intracellular carbohydrate metabolism, lipids and glycogen as energy sources are lower than usual. In view of its strong allergenic properties many studies have been undertaken to investigate cellular and subcellular changes in lymphoid cells that are induced by nickel as an integral part of its allergenic action.[35,110–114]

5.3.4.2 *In Vitro Cytological and Mutagenic Studies*

Nickel is a micro-trace metal in human cells, but at supra-optimal levels it induces a well documented sequence of cytochemical and morphological changes predisposing to carcinogenicity. Cytotoxic and genotoxic effects vary greatly according to the solubility of the nickel compound, ionisation and the absorption of Ni(II). Much of the nickel absorbed locates in the nucleus, but as seen in cultured A549 cells, the distribution balance (nucleus *vs.* cytoplasm) varies greatly from >0.5, in cells exposed to insoluble black nickel oxide (NiO), to <0.18 for nickel chloride. Costa was of the opinion that water-insoluble compounds such as nickel sulfide or subsulfide enter cells by a highly efficient phagocytic mechanism, with bioactive Ni^{2+} being released through dissolution of the compound in the cytosol.[84] Nickel ions released from water-soluble compounds are probably absorbed by an active process through iron or calcium channels with transferrin or calmodulin respectively as carrier proteins. Manganese, copper and zinc ions inhibit calcium-dependent Ni–calmodulin binding.

Nickel influences events in the cell cycle, including expression of p53 protein and induction of cyclin D1 and cyclin E.[115,116] In human bronchial epithelial cells and hamster ovary cells, nickel advanced cells in the S-1 phase but blocked cells in G2 and M phases leading to mitotic arrest and apoptosis. Overexpression of cyclin D1 led to arrest of cells in M phase, rendering them sensitive to errors in DNA transcription and protein expression. In cultured human dermal keratinocytes, nickel was shown to regulate 134 specific genes, of which 97 were induced and 37 suppressed.[117] Functional categories of regulated genes suggest that nickel inhibits apoptosis, promotes specific steps in the cell cycle and induces synthesis of extracellular matrix proteins and proteases. Nickel is implicated also in upregulation of growth factors such as vascular endothelial growth factor (VEGF) and placental growth factor (PGF), whilst suppressing the cytokines interleukin (IL)-18, galectin-3 and lipopolysaccharide-induced tumour necrosis factor (TNF). In addition to their possible role in Ni-induced allergic reactions, these changes may be of significance in promotion of carcinogenic change. Insoluble nickel subsulfide absorbed by HeLa cells exerted oxidative changes in DNA, whereas insoluble nickel oxide and water-soluble nickel sulfate did not.[118] In HeLa cells at least, nickel subsulfide was shown to evoke a sharp increase in 8-hydroxydeoxyguanosine (8-OH,dG), a commonly used marker of oxidative stress-derived DNA damage. In contrast, when the three nickel salts were instilled intratracheally in rats, all induced 8-OH,dG in lung cells. This may be interpreted as a direct form of oxidative damage through interaction of Ni^{2+} and endogenous or induced hydrogen peroxide, or as a result of undefined epigenetic changes (inflammation-related cytokines *etc.*).

Biederman and Landolf studied the ability of carcinogenic metals to alter behaviour and transformation of cultured cells, and the molecular and genetic basis for induction of anchorage independence.[119,120] They demonstrated that human foreskin fibroblasts cultured in the presence of nickel subsulfide, acetate and sulfate were not resistant to cytotoxic change, and that nickel subsulfide did not induce mutations at concentrations that induced anchorage independence. Later work showed that nickel sulfide induced anchorage independence in immortalised osteoblastic cell lines whereas the soluble nickel chloride did not.[121] Isolated anchorage-independent cells exhibited changes in cell morphology, shape and growth in multi-layers, and continued proliferation consistent with carcinogenic potential.

Nickel does not bind DNA but promotes DNA hypermethylation.[84,122] Nickel compounds with high carcinogenicity failed to exhibit mutagenic change in cultured cell lines. Mutagenicity tests in nickel bacterial reversion tests have mostly proved negative, but nickel is known to impair bacterial and viral DNA polymerase and its transcription.[123,124] Results with insoluble nickel subsulfide were negative, but the compounds induced clastogenic effects, chromosomal aberrations and a marked increase in micronuclei in mouse erythrocytes.[125] Nickel cytotoxicity is manifest by chromosomal aberrations, sister-chromatid exchanges, DNA-strand breaks, alterations in gene expression and silencing and breakage of DNA–protein cross-links, as a possible consequence of lipid peroxidation.[84,126] Nickel carbonyl inhibited DNA-dependent RNA polymerase in cultured liver cells, leading to transcriptional errors in RNA synthesis and inhibition of ^{14}C-orotic acid incorporation.[109] Other studies using radio-tracers and X-ray diffraction have demonstrated that nickel can impair uptake of ^{3}H-thymidine as a means of causing chromosomal aberrations, somatic mutations and morphological transformation in cultured cells. Messenger RNA-dependent induction of enzyme synthesis in cultured liver and lung cells was depressed in the presence of excess nickel.

Costa and Klein emphasised that nickel-induced cytotoxic changes were attributable to high levels of the ion, implying a saturation of endogenous protective mechanisms such as metallothioneins (MT) and other metal-binding proteins.[84] Nickel induces and binds MT as a protective mechanism and at the same time increases intracellular zinc levels.[127] Zinc is a relatively non-toxic metal and is known to have a mitigating influence on Ni-related cell damage, possibly through induction of a metal transcription factor. Studies in cultured human epithelial cells and in the murine lung have shown that MT was critical in limiting nickel-related injury. Soluble nickel chloride ($NiCl_2$) was shown to increase MT in cultured hepatocytes by eight times within 17 hours in a concentration-dependent manner.[128] Nickel induced expression of the *MT2A* gene in cultured lung cells.

Metallic nickel nanoparticles (average size 80 nm) have been shown to elicit higher cytotoxicity and apoptosis induction than fine particles (3 μm) in a JB6 cell line,[129] suggesting a surface phenomenon. Metallic nickel induced JB6 cell death through apoptosis, but not necrosis after 24 h treatment in a dose range of 0.1–20 μg cm^{-2}. Apoptotic changes were mediated by a sequence of molecular changes, activation of pro-apoptotic factors and expression of key proteins including a *Fas*-associated protein with a "death domain", caspase-8 and a death receptor-3. Caspases are a family of cysteine proteases which play essential roles in apoptosis, necrosis and inflammation. Cytochrome *c* is released into the cytoplasm from the mitochondria in response to pro-apoptotic stimulation and is an important factor in the intrinsic apoptotic pathway. Apoptotic induction factor (AIF), but not cytochrome *c*, was released from mitochondria into the cytoplasm in the apoptotic process in JB6 cells induced by metallic nickel particles without changes of mitochondrial membrane permeability. These researchers claimed that nickel induced cell death through a caspase 8/AIF-mediated cytochrome *c*-independent pathway.

5.3.4.3 Experimental Studies in Animals

Experimental studies on the toxicity and carcinogenicity of metallic nickel and compounds commonly encountered in industrial environments have been conducted in a limited range of mouse and rat strains. Other species, including hamsters, rabbits, guinea pigs and dogs, have been used in some specialised research. The European Commission Scientific Committee on Health and Environmental Risks was sceptical about the use of experimental animals in studying nickel carcinogenicity and emphasised that many published studies were scientifically inadequate or inappropriate in assessing carcinogenicity of the metal following inhalation, the principle route of human exposure.[130] Animals show different patterns of sensitivity to metallic nickel and nickel compounds, and dose-related tumours have been recorded in animals at sites not previously recorded in humans (adrenal and pituitary tumours, rhabdomyosarcoma, keratoacanthoma, *etc.*).[109,131]

Experimental studies have provided useful mechanistic information and demonstrated the risks associated with exposure to insoluble (metallic nickel, nickel sulfides and oxides) compared to water-soluble compounds such as nickel sulfate, chlorides and nitrate. Three features emerge from animal studies:

1. Nickel is not a complete carcinogen. It acts more as a promoter than inducer.[109,132]
2. Ni(II) exhibits low mutagenic activity and evidence suggests that it is a slow mutagen whatever its molecular mechanism of action.
3. Low carcinogenicity for soluble nickel compounds is attributed to insufficient biologically active Ni(II) ion in the proximity of the nucleus of target cells to interact with RNA and DNA.[118,136]

Experimental evidence has established that insoluble nickel salts are carcinogenic in rats (mostly of the F344 strain) following subcutaneous or intramuscular injection, inhalation and intratracheal instiallation.[109,132–135]

5.3.4.3.1 Inhalation. The carcinogenicity of soluble (sulfate) and insoluble (oxide and subsulfide) nickel compounds was examined in a 2-year inhalation study in F344F rats and B6C3F1 mice in a regulatory style study for the US National Institute of Environmental Health.[131] The study has a number of shortcomings and extrapolation of the results deserves caution. Mortality in nickel-exposed and control groups of both strains was unacceptably high, at

23–67%. Animals receiving nickel compounds showed a dose- and time-related increase in lung weight consistent with severe pulmonary distress, particularly in animals receiving the insoluble compounds. Nickel sulfate inhalation (0.25 or 0.5 mg m^{-3}) did not lead to significant increases in lung tumours in either strain, but insoluble nickel oxide (0.15 or 1.00 mg m^{-3}) or nickel oxide (1.25 or 2.5 mg m^{-3}) was associated with an increased incidence of alveolar/bronchogenic carcinomas in rats but not mice (control mice exhibited <20% lung tumours). Male and female rats also showed a high incidence of adrenal phaeochromocytoma. The study showed a relationship between the nickel content of the compounds and their carcinogenicity:

- Nickel sulfate.6H_2O – 22.3% Ni – not a pulmonary carcinogen
- Nickel subsulfide – 73.3% Ni – pulmonary carcinogen
- Nickel oxide – 78.6% Ni – pulmonary carcinogen.

In a second, chronic dosing, study, male mice were not susceptible to pulmonary carcinogenesis following chronic nickel subsulfide inhalation whereas F344 rats exposed to 0.97 mg Ni m^{-3} nickel sulfide for 78 weeks showed an increased incidence of bronchial or broncho-alveolar tumours.[137] Pulmonary inflammation and mortality were high in nickel-exposed and control groups.

In an attempt to understand the susceptibility of the respiratory tract to nickel-related carcinogenicity, tracheal isografts were implanted with pellets of 1 or 3 mg nickel subsulfide in gelatine.[138] Nickel content in the isografts declined over 9 months during which time the tracheal epithelium developed transitional hypertrophic changes, inflammation, squamous metaplasia and some atrophy. Sarcomas were evident after 10.5 months, all of which developed in connective tissue surrounding the grafts and were locally invasive and not metastatic; all were sarcomas rather than carcinomas of tracheal epithelialtissue (fibrosarcoma, leiomyosarcoma, fibromyosarcoma, rhabdomyosarcoma and fibromyyxosarcoma). In 20 months, 45 of 64 tracheal isografts treated with 3 mg nickel subsulfide had developed malignancies. This study confirms a clear relationship between nickel exposure and tracheal carcinogenicity but emphasises also that nickel is a slow mutagen, and that a long latent period exists between initial exposure and manifestation of neoplasia.[139]

5.3.4.3.2 Parenteral Administration. With the exception of nickel alloys used in orthopaedic prostheses, humans are unlikely to be exposed to metallic nickel and nickel compounds by parenteral injection or intratracheal implant. However at least seven experimental studies are published where soluble (sulfate) or insoluble (oxides, sulfide or subsulfide) nickel have been injected intramuscularly, intravenously or subcutaneously with high single or repeated doses, and observed for periods of up to 2 years.[109,133–135,138–140] Experiments are reported where animals have been given single injections of nickel compounds into the salivary gland, kidney, tracheal mucosa, gastrointestinal tract, testis or cheek pouch (Syrian hamsters only).[141–143] Following intramuscular or subcutaneous administration of nickel subsulfide in mice, sarcomas developed from 6 months with metastases to lung, liver and regional lymph nodes, but using ^{63}Ni tracer, phagocytosed nickel dust was seen to be translocated to liver, kidney, spleen, intestine and other tissues, none of which developed primary tumours.[133] In Fischer F344 rats, local sarcoma was observed at or close to injection sites following administration of nickel subsulfide (α-Ni_3S_2) within 2 years, with intense local reaction, but injections of a partially converted nickel–iron sulfide matte, nickel copper matte and nickel-containing powder were less effective; amorphous nickel sulfide (NiS) did not evoke tumours.[134] Nickel was located within the body of tumours and was transported to liver, spleen and lymph nodes by the action of reticuloendothelial cells. Tumours were transplantable. In contrast, no tumours were induced when 5 mg nickel subsulfide pellets were implanted intratracheally in Wistar rats, but epithelial hyperplasia was observed when 2 mg of benzpyrene was administered concurrently.[143] Oskarsson reported that his analytical studies with ^{63}Ni showed that α-Ni_3S_2 was converted within tumours to α-Ni_7S_6.

Sunderman (1990, 1984) provided additional evidence to show that nickel-related tumours were promoted by iron, cobalt and sulfur through enhancement of reactive oxygen species (ROS) and through production of Ni(III) ion, but inhibited in the presence of manganese dust.[135,140] In an attempt to expand on the range of experimental models available, Sunderman administered single injections of nickel subsulfide to Fischer F344 rats and Syrian hamsters by intramuscular injection (5 or 10 mg).[142,144] Malignant tumours were not induced in hamster cheek pouches, oral cavity or gastrointestinal tract even after repeated injections as high as 1.1 g, yet when they injected the compound intratesticularly in rats, sarcomas developed. Malignancies were observed following injection of nickel subsulfide into the submaxillary gland or *via* the portal system into the liver. In his rat experiments Sunderman claimed a dose–response relationship for tumour induction following intra-renal and intramuscular injection.

5.3.4.3.3 Oral or In Diet Administration. Experimental evidence that high doses of nickel compounds are carcinogenic when administrered in the diet is limited and inconclusive. A 2-year study in which F344 rats were given 50 mg kg^{-1} nickel sulfate showed reduced body weights and higher mortality in female animals.[145] Several animals dosed with 10 mg kg^{-1} developed the unusual dermal keratoacanthoma but observations led to the inevitable conclusion that nickel sulfate was non-carcinogenic in lifetime studies in rats fed maximum tolerated doses. Even nickel subsulfide, which is carcinogenic by inhalation or parenteral injection, failed to exhibit carcinogenic effects in hamsters by oral intake.[146] Hamsters are not suitable models for studying the carcinogenicity of nickel compounds.

5.3.4.3.4 Transplacental Exposures. Limited experimental evidence is published to show that nickel can induce tumours in the offspring of rats through the transplacental route.[147–149] Renal tumours developed in male offspring of rats given soluble nickel acetate intraperitoneally in late pregnancy and post-natal sodium barbital during weeks 4–85 of age. Sodium barbiturate given in drinking water was not carcinogenic in neonates. Female offspring did not develop renal tumours, but both male and female animals developed malignant pituitary tumours with or without barbital. Sunderman *et al.* failed to observe transplacental carcinogenesis in rats given 25 mg nickel subsulfide in early pregnancy. Administration of nickel carbonyl by inhalation to Syrian hamsters in early pregnancy was a cause of severe fetal distress and malformation but not neoplasia.[148–150]

5.3.4.4 Epidemiological and Human Case Studies

Numerous occupational health studies have been conducted in cohorts of workers exposed to nickel in mining, metal extraction and smelting, and to nickel processing in North America and Europe. Early evidence indicated that the lung cancer incidence was 2.2 to 16 times higher in nickel factory workers than in unexposed populations, and that the risk of nose and paranasal sinus cancers was up to 196-fold higher,[109] and subsequent epidemiological studies in large cohorts have confirmed that workers in nickel-related industries are at risk of lung or nasal cancer. Whilst there are many variables in terms of working conditions, the nickel compounds/ores to which workers were exposed, and environmental contamination with other toxic and carcinogenic materials, such as arsenic, radon and uranium breakdown products, asbestos sulfuric acid mists and cobalt, a cancer risk is indicated for workers in mining, milling refining, smelting and electrolysis. Whilst Grandjean was of the view, in 1988, that only one study demonstrated an exposure-related tumour response, even that was not statistically significant.[151] Current views emphasise that a direct relationship exists between the period of work in the industry and risk of cancer,[107,152] and that cumulative exposures over 10–15 years or more in mining present a major risk of lung cancer.[153] Shen and Zhang reviewed cancer risks in 140,888 workers exposed to nickel oxides and sulfides for 6 months to 6 years, and exposed to nickel compounds at concentrations ranging from <2 mg m^{-3} to >10 mg m^{-3}, and listed 274 cases of lung cancer, many of which showed high concentrations of nickel residues in lung tissue.[154] The authors were unclear as to whether the nickel content of lung tissue is a useful guide to cancer risk in nickel exposed workers.

An increased risk of lung cancer and nasal cancers in electrolysis workers exposed chronically to water-soluble nickel compounds in Norway up to the 1980s was presumed to be due largely to chronic exposure to soluble nickel compounds.[107] However, recent examination of environmental exposure patterns and lifestyle factors now implicates lifetime smoking, arsenic, contamination and other occupational carcinogens to which workers are exposed "outside the refinery". Cigarette smoking and alcohol consumption are major confounding lifestyle factors in evaluating epidemiological studies conducted in nickel and other metal industries.[5,6,155] Alcoholic beverages are not known to contain significant amounts of nickel but cigarettes and tobacco are known to contain 2.2 to 4.9 mg kg^{-1} nickel regardless of their origin.[156] Blood nickel concentrations are significantly higher in smokers (0.01–0.42 μg l^{-1}) than non-smokers (0.01–0.26 μg l^{-1}) and nickel levels in soft tissues and urine are significantly increased. Nasal and lung cancers are influenced by smoking, alcohol consumption and exposure to other workplace carcinogens.[151,153,155,157–159]

Other variables illustrated by occupational health evaluations in Ontario and Denmark include differences in the professional role of workers in mining, refining and manufacture, health and safety regulations in force, and duration of employment.[151,159] Mining, smelting refining, electrolysis work, calcining and sintering operations present the greatest cancer risk to workers, but levels of significance vary greatly among reports and national trends. Many rely on the accuracy and details given in past records, several of which failed to disclose salient details on the magnitude of nickel exposure, nickel speciation, and lifestyle factors. A large Canadian study of 54,509 nickel miners followed for 35 years showed a three-fold increase in lung cancer mortality which rose steeply with duration of service, peaking at 943 with 10 to 15 years in one pit and 806 after 20 to 25 years service in a second mine.[152] Nasal cancer was also increased but other cancers including those of larynx and kidney were comparable to national averages. A large study of 28,261 workers employed in 12 plants in West Virginia recorded 25–50% excess lung cancer mortality in maintenance categories when the health data were analysed according to occupational groupings.[160] No conclusion was reached as to the causation of excess cancers of the liver and the large intestine. A more comprehensive study in the USA involving 31,165 workers in 13 nickel alloy plants exposed to metallic nickel or oxidised forms of nickel in the period 1948 to 1988 indicated an elevated risk of 13% for lung cancer in men employed in grinding, allocated services and production of high nickel alloys, exposed to a mean level of 0.298 mg m^{-3} nickel, but the risk was not significantly higher than in local populations.[160,161] The study did not confirm risks of kidney or colonic cancer following nickel exposures.

Several epidemiological studies fail to confirm that nickel exposure presents a significant cancer risk. Thus, an extended follow-up survey of 11,567 Canadian nickel workers, engaged in nickel mining, milling and smelting in 1950–1976 and 1977–1984, showed a low incidence of nasal cancer.[153] Although lung cancer mortality rates beyond 15 years from first exposure and cumulative exposure increased overall, trends were not significant or consistent with occupational background. Similar inconclusive observations were reported in New Caledonia in the South Pacific in 1978–1984, where lung or upper respiratory tract cancers did not correlate well with presumed nickel exposures.[155,158] A retrospective analysis of 1959 workers in a Chinese nickel mine and in nickel smelting and mining also failed to reveal significant results even though the overall incidence of lung cancer mortality in 17 years was increased.[162]

A broader occupational survey in England and Wales in the period 1963–1967 showed a higher incidence of nasal cancers amongst cabinet makers, wood machinists, boot and shoe operatives, nickel smelters in South Wales, coal miners and furnace men in coke, gas and chemical industries, without any clear association between a particular type of tumour and occupation.[163] Asbestosis, disintegration products of radon, combustion fumes, cobalt and sulfuric acid occurring as contaminants in many industries and mining operations are associated with high risks of lung and nasal cancers. Additionally, workers engaged in electroplating, steel industries and other nickel-related industries are exposed to varying environmental levels of chromium, arsenic, iron, manganese, copper and elements that act as tumour inducers or promoters. Kusiak examined the causation of high risks of lung cancer in a cohort of 54,128 men working in Ontario mines in the period 1955 to 1986.[164] They emphasised that gold miners showed an increased risk of lung cancer, and their statistics showed no increase in lung carcinoma mortality in men who commenced mining nickel after 1936, presumably reflecting changes in working practice.

Whilst there is equivocal evidence that inhalation of metallic nickel dust *per se* is carcinogenic in humans, risks associated with inhalation of compounds such as nickel carbonyl [$Ni(CO)_4$] gas released in the extraction process, insoluble nickel subsulfide (Ni_3S_2), or nickel oxide found in refinery dusts should not be discounted.[165] A second difficulty in diagnosis of cancer in nickel-related industries relates to the long lag phase between initial exposure to nickel residues and diagnosis of tumours. In a Chinese nickel mine, men with an average age of 54.3 years and exposed to nickel sulfide showed a latent period of 22.3 years from presumed first employment until diagnosis of cancer.[162]

The IARC re-assessed the epidemiological evidence for the carcinogenicity of nickel and nickel compounds in 2009 and in their opinion the weight of evidence suggested that some exposures to soluble nickel compounds could promote lung cancer, but not as complete carcinogens.[106,132,166] If these exposures are carcinogenic then the mode of action, probably *via* epigenetic pathways, is unclear. Goodman considered that soluble nickel compounds "would not deliver sufficient levels of bioactive Ni(II) ion to nuclei of target cells in the lung to evoke genotoxic change".[132]

Cytogenetic analyses of peripheral lymphocytes from 21 workers exposed occupationally to nickel showed chromosomal abnormalities following occupational exposure to nickel oxide ($9.5 \pm 3.2\%$) or nickel sulfate ($5.2 \pm 1.9\%$).[167] No chromosomal abnormalities were seen in a group of 24 manual metal arc stainless steel welders.[168]

5.3.5 Conclusions on the Carcinogenicity of Nickel

Exposure to insoluble nickel compounds in mining, milling, smelting and electrolysis has to be listed as a potential carcinogenic risk. A latent period of 10 or more years can be expected between leaving the industry and diagnosis of cancers of the lung, bronchus, larynx or nasal passages. This review does not allow a distinction to be made, however, between cancers that are related to nickel exposure *per se* and those that are more correctly identified as an "industry-related" risk. Clearly, occupational environments in mining, electrolysis, alloy production, *etc.*, contain many metals other than nickel, together with combustion products and non-metallic toxins, many of which are listed carcinogens.[1] If the latter is true, it is unclear whether the nickel ion should be regarded as tumour initiator, or as *in vitro* studies in isolated cells predict, as a promoter or co-carcinogen. Clinical studies do not allow an explanation and there is only limited information in animal models to show that nickel promotes neoplastic change induced by carcinogens such as 7,12-dimethylbenz[α]anthracene.[132,167] Experimental studies suggest that Ni(II) acts as a "slow mutagen" whatever its molecular mechanism of action. Sunderman provided additional evidence to show that nickel-related tumours were promoted by iron, cobalt and sulfur through enhancement of reactive oxygen species (ROS), and through production of Ni(III) ion.[135]

The predictive value of animal studies in the study of nickel-related carcinogenicity is viewed with caution. The animal species used show different patterns of sensitivity to metallic nickel and nickel compounds, and dose-related tumours have been recorded in animals at sites not previously recorded in humans (adrenal and pituitary tumours, rhabdomyosarcoma, keratoacanthoma, *etc.*).[84,109,131] The number of species and strains used is very limited, but even so, they do provide useful information on mechanisms of nickel-related carcinogenesis, not possible with human studies.

Lack of sufficient Ni(II) ion release in the region of the nucleus and production of changes in RNA binding and transcription may be an explanation as to why soluble nickel compounds such as the sulfate and chloride, and ammonium sulfate are non-carcinogenic. Nickel subsulfide is phagocytosed into cells and free ion is released to interact with RNA. Carcinogenicity seems to be related to the solubility of the nickel compound and the ability of free intracellular nickel ion to interact with RNA and DNA.[168] Preliminary work suggesting that soluble nickel acetate might be a transplacental carcinogen has not been ratified or confirmed in human studies.[147–149]

References

1. US Department of Health and Human Services (USDHHS), *Report on Carcinogens*, National Toxicology Program, Research Triangle Park, NC, 12th edn, 2011.
2. V. Verougstraete, A. Mallants, J.-P. Buchet, B. Swennen and D. Lison, *Am. J. Resp. Clin. Care Med.*, 2004, **170**, 162.
3. G. Nordberg, *Sci. Total Environ.*, 1994, **150**, 201.
4. S. Masten and K. E. Haneke, *Cobalt Dust: Review of the Toxicological Literature*, NTP, Research Triangle Park, NC, 2002.
5. S. S. Hecht, *J. Natl. Cancer Inst.*, 1999, **91**, 1194.
6. H. Oliveira, E. A. N. Fernandez, M. A. Bacchi and G. A. Sarriés, *J. Radioanalytic. Nuclear Chem.*, 2000, **244**, 299.
7. J. F. Riley, *Mineralogical Magazine*, 1980, **43**, 733.
8. Y. Lee and B. M. Tebo, *Appl. Environ. Biol.*, 1994, **60**, 2949.

9. Occupational Health and Safety Administration (OHSA) *Occupational Safety and Health Guidelines for Cobalt Metal, Dust and Fume*, US OHSA, Washington DC, *www.OHSA.gov.*
10. I. Chanarin, in *Megaloblastic Anaemias*, Blackwell, Oxford, 1969.
11. I. Chanarin, *J. Clin. Pathol.*, 1971, **24**(Suppl 5), 60.
12. F. H. Nielsen, *J. Nutr.*, 1996, **126**, 2377S.
13. J. Neve, *J. Pharm. Belg.*, 1991, **46**, 271.
14. T. A. D. Smith, *Bioorg. Med. Chem.*, 2006, **12**, 4578.
15. H. A. Huebers and C. A. Finch, *Physiol. Rev.*, 1987, **67**, 520.
16. C. A. Hall and A. E. Finkler, *J. Lab. Clin. Med.*, 1965, **65**, 459.
17. I. Andersen and A. C. Høgetveit, *Fresenius Zeitschr. Anal. Chem.*, 1984, **318**, 41.
18. N. S. Scrimshaw and V. R. Young, *Sci. Am.*, 1976, **235**, 51.
19. A. B. Thomson, L. S. Valberg and D. G. Sinclair, *J. Clin. Invest.*, 1971, **50**, 2384.
20. G. K. Alamada, J. L. Evelhoch, J. L. Sudmeier and R. R. Birge, *Biochem.*, 1985, **24**, 1757.
21. D. M. Czajka-Narins, in *Food and Nutrition Diet Therapy*, ed., L. K. Mahan and M. V. Kraus, Saunders, Philadelphia, 1999.
22. E. J. Fine and E. D. Soria, *South. Med. J.*, 1991, **84**, 1475.
23. M. I. Behrens, V. Diaz, C. Vásquez and A. Donoso, *Rev. Med. Chile*, 2003, **131**, 915.
24. E. P. Frenkel, A. Mukherjee, C. R. Hackenbrock and P. A. Srere, *J. Biol. Chem.*, 1976, **251**, 2147.
25. H. C. Grice, T. Goodman, I. C. Monro, G. S. Wiberg and A. B. Morrison, *Ann. N.Y. Acad. Sci.*, 1969, **156**, 189.
26. C. S. Alexander, *Am. J. Med.*, 1972, **53**, 395.
27. J. M. Duckham and H. A. Lee, *Lancet*, 1976, **i**, 1350.
28. J. J. Hostýnek, R. S. Hinz, C. R. Lorence, M. Price and M. Guy, *Crit. Rev. Toxicol.*, 1993, **23**, 171.
29. J. J. Moulin, P. Wild, S. Romazini, G. Lesfargues, A. Peltier, C. Bozec, P. Deguerry, F. Pellet and A. Perdrix, *Am. J. Epidemiol.*, 1998, **148**, 241.
30. National Toxicology Program, *Report on Carcinogens, Cobalt-Tungsten Carbide; powders and hard metals*, NTP, Research Triangle Park, NC, 2009.
31. P. Kelleher, K. Pacheco and L. S. Newman, *Environ. Health. Perspect.*, 2000, **108**(Suppl. 4), 685.
32. I. C. Smith and B. L. Carson. *Trace Metals in the Environment. Cobalt.* Ann Arbor Science Publ. Inc., Ann Arbor, MI, vol. 6, 1981.
33. J. R. Bucher, J. R. Hailey, J. R. Rycroft, J. K. Haseman, R. C. Sills, S. L. Grumbein, P. W. Mellick and B. J. Chou, *Toxicol. Sci.*, 1999, **49**, 56.
34. J. R. Bucher, M. R. Elwell, M. B. Thompson, B. J. Chou, R. Renne and H. A. Ragan, *Fundam. Appl. Toxicol.*, 1990, **15**, 357.
35. A. A. Fisher, *Contact Dermatitis*, Les and Febiger, Philadelphia, 1987.
36. G. E. Sandusky, M. P. Crawford and E. D. Roberts, *Toxicol. Appl. Pharmacol.*, 1981, **60**, 263.

37. M. De Boeck, M. Kirsch-Volders and D. Lison, *Mutat. Res.*, 2003, **533**, 135.
38. A. Léonard and R. Lauwerys, *Mutat. Res.*, 1990, **239**, 17.
39. M. Daniel, J. T. Dingle, M. Webb and J. C. Heath, *Br. J. Exp. Pathol.*, 1963, **44**, 163.
40. D. A. Pagano and E. Zeiger, *Environ. Mol. Mutagen.*, 2006, **19**, 139.
41. H. I. Ogawa, Y. Ohyama, Y. Ohsumi, K. Kakimoto, Y. Kato, Y. Shirai, T. Nunoshiba and K. Yamamoto, *Mutagen.*, 1999, **14**, 249.
42. V. Malard, F. Berenguer, O. Prat, S. Ruat, G. Steinmetz and E. Quemeneur, *BBC Genomics*, 2007, **8**, 147.
43. A. H. Baker, D. R. Edwards and G. Murphy, *J. Cell Sci.*, 2002, **115**, 3719.
44. D. Anard, M. Kirsch-Volders, A. Elhajouji, K. Belpaeme and D. Lison, *Carcinogenesis*, 1997, **18**, 177.
45. D. Lison, M. De Boeck, V. Verougstraete and M. Kirsch-Volders, *Occup. Environ. Med.*, 2001, **56**, 619.
46. F. Van Goethem, D. Lison and M. Kirsch-Volders, *Mutat. Res.*, 1997, **392**, 31.
47. J. P. Gennart, C. Baleux, C. Verellen-Dumoulin, J. P. Buchet, R. De Meyer and R. Lauwerys, *Mutat. Res.*, 1993, **299**, 55.
48. M. De Boeck, S. Lardau, J. P. Buchet, M. Kirsch-Volders and D. Lison, *Environ. Mol. Mutagen.*, 2000, **36**, 151.
49. N. Lombaert, D. Lison, P. Van Hummelen and M. Kirsch-Volders, *Toxicol. Appl. Pharmacol.*, 2008, **227**, 299.
50. N. Lombaert, E. Castrucci, I. Decordier, P. Van Hummelen, M. Kirsch-Volders, E. Cundari and D. Lison, *Arch. Toxicol.*, 2013, **87**(2), 259.
51. M. Demedts and J. L. Ceuppens, *Chest*, 1989, **95**, 2.
52. A. A. Jensen and F. Tüchsen, *Crit. Rev. Toxicol.*, 1990, **20**, 427.
53. IARC Working Group, *Cobalt and Cobalt Compounds*, IARC Working Group, Lyon, France, 1991.
54. D. Steinhoff and U. Mohr, *Exp. Toxicol.*, 1991, **41**, 169.
55. J. C. Heath, *Br. J. Cancer*, 1960, **14**, 478.
56. J. C Heath and M. Daniel, *Br. J. Cancer*, 1962, **16**, 473–478.
57. A. A. Shabaan, V. Marks, M. C. Lancaster and G. N. Duffeu, *Lab. Anim.*, 1977, **11**, 14.
58. A. Johansson, T. Curstedt and P. Camner, *Environ. Res.*, 1991, **64**, 24.
59. T. Hansen, G. Clermont, A. Alves, R. Elroy, C. Brochhausen, J. P. Boutrand, A. M. Gatti and C. J. Kirkpatrick, *J. Roy. Soc. Interface*, 2006, **3**, 767.
60. A. P. Wehner, R. H. Busch, R. J. Olson and D. K. Craig, *Am. Ind. Hyg. Assoc. J.*, 1977, **38**, 338.
61. J. C. Heath, *Nature*, 1954, **173**, 822.
62. J. C. Heath, *Br. J. Cancer*, 1956, **10**, 668.
63. J. C. Heath, M. A. R. Freeman and S. A. V. Swanson, *Lancet*, 1971, **297**, 564.
64. S. A. V. Swanson, M. A. R. Freeman and J. C. Heath, *J. Bone Joint Surg.*, 1973, **55**, 759.

65. F. W. Sunderman, S. M. Hopfer, T. Swift, W. N. Rezuke, L. Ziebka, P. Highman, B. Edwards, M. Folcik and H. R. Gossling, *J. Orthop. Res.*, 1989, **7**, 307.
66. O. L. Berkenstock, *Contemp. Orthop.*, 1992, **24**, 265.
67. P. R. Bouchard, J. Black, B. A. Albrecht, R. E. Kaderly, J. O. Galante and B. U. Pauli, *J. Biomed. Mater. Res.*, 1996, **32**, 37.
68. C. G. Lewis, R. M. Belniak, M. C. Plowman, S. M. Hopfer, J. A. Knight and F. W. Sunderman, *J. Arthroplasty*, 1995, **10**, 75.
69. European Food Safety Authority (EFSA), Scientific opinion on the use of cobalt compounds as additives in nutrition, *EFSA J.*, 2009, **7**, 1383.
70. J. M. Mur, J. J. Moulin, M. P. Charruyer-Seinerra and J. Lafitte, *Am. J. Ind. Med.*, 1987, **11**, 75.
71. F. Tüchsen, M. V. Jensen, E. Villadsen and E. Lynge, *Scand. J. Work Environ. Health*, 1996, **22**, 444.
72. B. Nemery and J. L. Abraham, *Am. J. Resp. Crit. Care Dis.*, 2007, **176**, 2.
73. J. Siemiatycki, L. Richardson, K. Straif, B. Latreille, R. Lekhani, S. Campbell, M.-C. Rousseau and P. Boffetta, *Environ. Health Perspect.*, 2004, **112**, 1447.
74. M.-C Rousseau, K. Straif and J. Siemiatycki, *Environ. Health Perspect.*, 2005, **113**, A580.
75. NTP, *Cobalt sulphate heptahydrate* $CoSO_4.7H_2O$ NTP, Research Triangle Park, NC, 1998.
76. International Agency for Research on Cancer (IARC), *Monographs on the Evaluation of Carcinogenic Risks to Humans*, IARC, Lyon, France, 2006, p. 86.
77. C. Hogstedt and R. A. Andersson, *Arbete. Och. Halsa*, 1990, **21**, 1.
78. L. B. Signorello, W. Ye, J. P. Fryzek, L. Lipworth, J. F. Fraumeni, W. J. Blot, J. K. McLaughlin and O. Nyrén, *J. Nat. Cancer Inst.*, 2001, **93**, 1405.
79. R. Tharani, *J. Bone Joint Surg.*, 2001, **83**, 774.
80. P. R. Sheppard, R. J. Speakman, G. Ridenour and M. L. Witten, *Environ. Health. Persp.*, 2007, **115**, 715.
81. P. Grasso, S. D. Gangolli, L. Golberg and J. Hooson, *Food Cosmet. Toxicol.*, 1971, **9**, 463.
82. M. Vasák and J. H. Kägi, *Proc. Natl. Acad. Sci.*, 1981, **78**, 6709.
83. *CRC Handbook of Chemistry and Physics*, ed. W. M. Haynes, Taylor and Francis, Boca Raton FL, 9th edn, 2013.
84. M. Costa and C. B. Klein, *Environ. Health Perspect.*, 1999, **109**, A438.
85. W. Mertz, *Proc. Nutr.*, 1974, **33**, 307.
86. M. Patriarca, T. D. Lyon and G. S. Fell, *Am. J. Clin. Nutr.*, 1997, **66**, 616.
87. M. Anke, B. Groppel, H. Kronemann and H. Grün, *IARC Sci Publ.*, 1984, **53**, 339.
88. M. Anke, B. Groppel, U. Krause, W. Arnhold and M. Langer, *J. Trace. Elem. Electrolyte Health Dis.*, 1991, **5**, 69.
89. World Health Organization (WHO), *Nickel in Drinking Water*, WHO, Geneva, 2007.

90. Institute of Medicine, Food and Nutrition Board, *Dietary Reference Intakes for Vitamin A, Vitamin K, Arsenic, Boron, Chromium, Copper, Iodine, Iron, Manganese, Molybdenum, Nickel, Silicon, Vanadium, and Zinc*, National Academy Press, Washington, DC, 2001, 1.
91. D. G. Barceloux, *Clin. Toxicol.*, 1999, **37**, 239.
92. S. A. S. Gropper and J. L. Smith, *Advanced Nutrition and Human Metabolism*, Wadsworth Publ. Co., Belmont, CA, 2012.
93. P. Kaur and H. M. Dani, *J. Environ. Pathol. Toxicol. Oncol.*, 2003, **22**, 29.
94. F. H. Nielsen and H. H. Sandstedt, *Am. J. Clin. Nutr.*, 1974, **27**, 515.
95. F. H. Nielsen, *Fed. Am. Soc. Exp.Biol. J.*, 1991, **5**, 2662.
96. J. W. Spears, E. E. Hatfield, R. M. Forbes and S. E. Keonig, *J. Nutr.*, 1978, **108**, 313.
97. J. W. Spears, E. E. Jones, L. J. Samsell and W. D. Armstrong, *J. Nutr.*, 1978, **114**, 845.
98. F. H. Nielsen and D. A. Ollerich, *Fed. Proc.*, 1974, **33**, 1767.
99. R. K. Watt and P. W. Ludden, *Cell Mol. Life Sci.*, 1999, **56**, 604.
100. D. J. Gawkrodger, C. W. McLeod and K. Dobson, *Br. J. Dermatol.*, 2012, **166**, 82.
101. N. W. Solomans, F. Viteri, T. Shuler and F. H. Nielsen, *J. Nutr.*, 1982, **112**, 39.
102. R. B. Hayes, *Cancer Causes Control*, 1997, **8**, 371.
103. B. Jezierska and M. Witeska, in *Soil and Water Pollution Monitoring Protection and Remediation*, ed. I. Twardowska, H. E. Allen, M. M. Haggblom and S. Stephaniak, Springer, New York/Heidelberg, 2006, **33**, pp. 3–23.
104. Agency for Toxic Substances and Disease Registry (ATSDR), *Toxicological Profile for Nickel*, ATSDR, Atlanta, GA, 2005.
105. Report of the International Committee on Nickel Carcinogenesis in Man, *Scand. J. Work Environ. Health*, 1990, **16**(Suppl. 1), 1.
106. A. R. Oller, *Environ. Health Perspect.*, 2002, **110**(Suppl. 5), 841.
107. T. K. Grimsrud and A. Andersen, *J. Occup. Med. Toxicol.*, 2010, **5**, 7.
108. F. H. Nielsen, R..A. Poellot and E..O. Uthua, *J. Trace Elem. Exp. Med.*, 1995, **7**, 167.
109. F. W. Sunderman, *Dis. Chest.*, 1968, **54**, 527.
110. L. A. Van Loon, P. W. Van Elsas, J. D. Bos, H. C. ten Harkel-Hagenaar and C. L. Davidson, *J. Oral Pathol.*, 1988, **17**, 129.
111. F. Sinigaglia, *J. Invest. Dermatol.*, 1994, **102**, 398.
112. S. Silvennionen-Kassinen, I. Ikäheimo, J. Karvonen, M. Kauppinen and M. Kalloinen, *J. Allergy Clin. Immunol.*, 1992, **89**, 794.
113. M. K. Robinson and D. L. Sneller, *Toxicol. Appl. Pharmacol.*, 1990, **104**, 106.
114. D. J. Gawkrodger, E. McVittie and J. A. Hunter, *Dermatologica*, 1987, **175**, 171.
115. Y. H Shiao, S. H. Lee and K. S. Kasprzak, *Carcinogenesis*, 1998, **19**, 1203.

116. J. Ding, G. He, W. Gong, W. Wen, W. Sun, B. Ning, S. Huang, K. Wu, C. Huang, M. Wu, W. Xie and H. Wang, *Cancer Epidemiol. Biomarkers Prev.*, 2009, **18**, 1720.
117. A. Gazel, M. Rosdy, C. Tornier, A. de B. De Fraissinette and M. Blumenberg, *J. Cell. Pathol*, 2008, **217**, 686.
118. S. Kawanishi, S. Oikawa, S. Inoue and K. Nishino, *Environ. Health Perspect.*, 2002, **110**(Suppl. 5), 789.
119. K. A. Biedermann and J. R. Landolph, *Cancer Res.*, 1987, **47**, 3815.
120. J. R. Landolph, A. Verma, J. Ramnath and F. Clemens, *Environ. Health Perspect.*, 2002, **110**(Suppl. 5), 845.
121. X. Lin and M. Costa, *Environ. Health Perspect.*, 1994, **102**(Suppl.3), 289.
122. M. Costa, in *Toxicology of Metals*, CRC Press, Boca Raton, FL, 1996, p. 245.
123. F. W. Sunderman, *Environ. Health Perspect.*, 1981, **40**, 131.
124. A. Reith and A. Bragger, *IARC Sci. Publ.*, 1984, **53**, 175.
125. F. Z. Arrouijal, H. F. Hildebrand, H. Vophi and D. Marzin, *Mutagenesis*, 1990, **6**, 583.
126. H. Savolainen, *Rev. Environ. Health*, 1996, **11**, 167.
127. A. A. Nemac, G. D. Leikauf, B. R. Pitt, K. J. Wasserloos and A. Barchowsky, *Am. J. Respir. Cell Mol. Biol.*, 2009, **41**, 69.
128. F. W. Sunderman and C. B. Fraser, *Ann. Clin. Lab. Sci.*, 1983, **13**, 489.
129. J. Zhao, L. Bowman, X. Zhang, X. Shi, B. Jiang, V. Castranova and M. Dong, *J. Nanotechnol.*, 2009, **7**, 177.
130. European Commission Scientific Committee on Health and Environmental Risks (SCHER), *Risk Assessment Report on Nickel-Human Health*, SCHER, Brussels, 2006.
131. R. Dunnick, M. R. Elwell, A. E. Radovsky, J. M. Benson, F. F. Hahn, K. J. Nikula, E. B. Barr and C. H. Hobbs, *Cancer Res.*, 1995, **55**, 5251.
132. J. E. Goodman, R. L. Prueitt, D. G. Dodge and S. Thakali, *Crit. Rev. Toxicol.*, 2009, **39**, 365.
133. A. Oskarsson, Y. Andersson and H. Tjälve, *Cancer Res.*, 1979, **39**, 4175.
134. F. W. Sunderman and R. M. Maenza, *Res. Comm. Chem. Pathol. Pharmacol.*, 1976, **14**, 319.
135. F. W. Sunderman, S. M. Hopfer, M. C. Plowman and J. A. Knight, *Res. Comm. Chem. Pathol. Pharmacol.*, 1990, **70**, 103.
136. J. P. Berry, M. F. Poupon, J. C. Judde and P. Galle, *Ann. Clin. Lab Sci.*, 1985, **15**, 109.
137. A. D. Ottolenghi, J. K. Haseman, W. W. Payne, H. L. Falk and H. N. MacFarland, *J. Natl. Cancer Inst.*, 1975, **54**, 1165.
138. T. Yarita and P. Nettesheim, *Cancer Res.*, 1978, **38**, 3140.
139. G. D. Lumb and F. W. Sunderman, *Sci. Total Environ.*, 1994, **148**, 185.
140. F. W. Sunderman, K. S. Kasprzak, T. J. Lau, P. P. Minghetti, R. M. Maenza, C. Onkelinx and P. J. Goldblatt, *Cancer Res.*, 1976, **36**, 1790.
141. G. Jasmin and B. Solymoss, *Adv. Exp. Med. Biol.*, 1978, **91**, 69.
142. F. W. Sunderman, R. M. Maenza, P. R. Allpass, J. M. Mitchell, I. Damjanov and P. J. Goldblatt, *Adv. Exp. Med. Biol.*, 1977, **91**, 57.

143. K. S. Kasprzak, L. Marchow and J. Breborowicz, *Res. Comm. Chem. Pathol., Pharmacol.*, 1973, **6**, 237.
144. F. W. Sunderman, *Ann. Clin. Lab. Sci.*, 1984, **14**, 93.
145. K. E. Heim, H. K. Bates, R. E. Rush and A. R. Oller, *Toxicol. Appl. Pharmacol.*, 2007, **224**, 126.
146. F. W. Sunderman, *Basic Life Sci.*, 1983, **24**, 107.
147. B. A. Diwan, K. S. Kasprzak and J. M. Rice, *Carcinogenesis*, 1992, **13**, 1351.
148. H. Guan, F.-Y. Piao, X.-W. Li, Q.-J. Li, L. Xue and K. Yokohama, *Biomed. Environ. Sci.*, 2010, **23**, 458.
149. F. W. Sunderman, K. S. McCully and L. A. Rinehimer, *Res. Comm. Chem. Pathol. Pharmacol.*, 1981, **31**, 545.
150. F. W. Sunderman, S. K. Shen, M. Reid and C. Allpass, *Teratol. Carcinogen. Mutagen.*, 1980, **1**, 223.
151. P. Grandjean, O. Anderson and G. D. Nielsen, *Am. J. Ind. Med.*, 1988, **3**, 193.
152. R. S. Roberts, J. A. Julian, D. C. Muir and H. S. Shannon, *Toxciol. Ind. Health*, 1989, **5**, 975.
153. H. S. Shannon, *Toxicol. Ind. Health.*, 1991, **7**, 277.
154. H. M. Shen and Q. F. Zhang, *Environ. Health Perspect.*, 1994, **102**(Suppl. 1), 275.
155. M. Goldberg, P. Goldberg, A. Leclerc, J. F. Chastang, R. Fuhrer, J. M. Brodeur, N. Segnan, J. J. Floch and G. Michel, *Int. J. Cancer*, 1987, **40**, 300.
156. D. Stojanović, D. Nikić and K. Lazarević, *Cent. Eur. J. Public Health*, 2004, **12**, 187.
157. A. Andersen, S. R. Berge, A. Engeland and T. Norseth, *Occup. Environ. Med.*, 1996, **53**, 708.
158. M. Goldberg, P. Goldberg, A. Leclerc, J. F. Chastang, M. J. Mame and D. Dubourdieu, *Cancer Causes Control*, 1994, **5**, 15.
159. N. Lightfoot, C. Berriault and R. Semenciw, *Occup. Med.*, 2010, **60**, 211.
160. C. K. Redmond, *IARC Sci. Publ.*, 1984, **53**, 73.
161. V. C. Arena, N. B. Sussman, C. K. Redmond, J. P. Costiano and J. M. Trauth, *J. Occup. Environ. Med.*, 1998, **40**, 907.
162. C. Qisheng, *Ind. Health Occup. Dis.*, 1992, **3**, 8.
163. E. D. Acheson, R. H. Cowdell and E. H. Rang, *Br. J. Ind. Med.*, 1981, **38**, 218.
164. R. A. Kusiak, J. Springer, A. C. Richie and J. Muller, *Br. J. Ind. Med.*, 1991, **48**, 808.
165. M. Cempel and G. Nikel, *Polish J. Environ. Stud.*, 2006, **15**, 375.
166. International Agency for Research on Cancer (IARC) Monograph, Nickel and Nickel Compounds, IARC, Lyon, France, 2009.
167. V. Senfta, F. Losan and M. Tucek, *Mutat. Res.*, 1992, **279**, 171.
168. M. Littorin, B. Höstedt, B. Strömbäk, A. Karlsson, H. Welinder, F. Mitelman and S. Skerfving, *Scand. J. Work. Environ. Health*, 1983, **9**, 259.

CHAPTER 6

Calcium, Strontium, Magnesium and Copper

6.1 General Introduction

Calcium, copper and magnesium are recognised nutrients in the human body and serve well defined roles in cell physiology, the structure and function of intercellular matrices and as co-factors in numerous enzymes. Strontium is listed as a micro-trace element although its putative functions in bone development are less well defined. Strontium exhibits similar chemical properties to calcium and mimics it in bone formation.[1,2] The four elements interact in human physiology and morphogenesis, and with the possible exception of strontium, serve key roles in the cell cycle and in cell replication in the skin and intestinal epithelia, as well as in tissue repair following injury.[3] The stable isotopes of all four elements are not carcinogenic in the mammalian body but there are numerous reports on the health risks faced by populations exposed to ^{90}Sr from radioactive fallout in nuclear weapons test areas.[4,5]

6.2 Calcium

Calcium is a divalent metal ion and the most abundant element in the body. It is found throughout the world, and limestone is one the commonest of terrestrial ores known to man. Calcium is a reactive element and not found naturally in its metallic form. Twenty-four isotopes have been identified, three of which are stable. The commonest isotope of calcium (96.94%), ^{40}Ca, has a half-life of $>3.0 \times 10^{21}$ years. Common natural forms of calcium include limestone (oxide), carbonate (chalk, marble, calcite), stalactites and stalagmites in subterranean caves, silicates, together with fossilised skeletons and shells dating back many millennia. Lime kilns are a feature of many rural landscapes with early records

Issues in Toxicology No. 18
The Carcinogenicity of Metals: Human Risk through Occupational and Environmental Exposure
By Alan B. G. Lansdown

Published by the Royal Society of Chemistry, www.rsc.org

showing a kiln in Mesopotamia dating from about 2450 BC. Archaeological evidence shows that lime burning in Britain dates from Roman times. To the biologist, calcium is an essential component of bones, teeth, shells, horn and claws but minute quantities are found in every living cell in the body.

Calcium is recognised as a central regulator for the cell cycle, post-mitotic differentiation and functional maturation with key roles in apoptosis and tissue homeostasis. Research in the past century has focused upon the role of calcium in gastrointestinal absorption, protein binding, tissue stability and metabolism, regulated by vitamin D (1,25-dihydroxyvitamin D), parathyroid hormone and dietary phosphate. Current research shows that calcium absorption, cellular regulation and transport are regulated by more than 100 calcium-binding proteins (CaBP) of which the EF-Hand series, including 16 S100 proteins, has been catalogued.[6,7] Their structural configuration resembles an outstretched hand (Figure 6.1). Several high-resolution structures containing different metal ions bound to EF-hand sites have given new insight into the modulation of their binding affinities. The EF-hand calcium-binding proteins with a helix–loop–helix Ca2þ binding motif constitute one of the largest protein families, and are involved in numerous biological processes including cell–cell adhesion, transmission of calcium-dependent cell regulatory signals, cell migration and motility, membrane function, cell growth and extracellular cytokine-like action.[7,8]

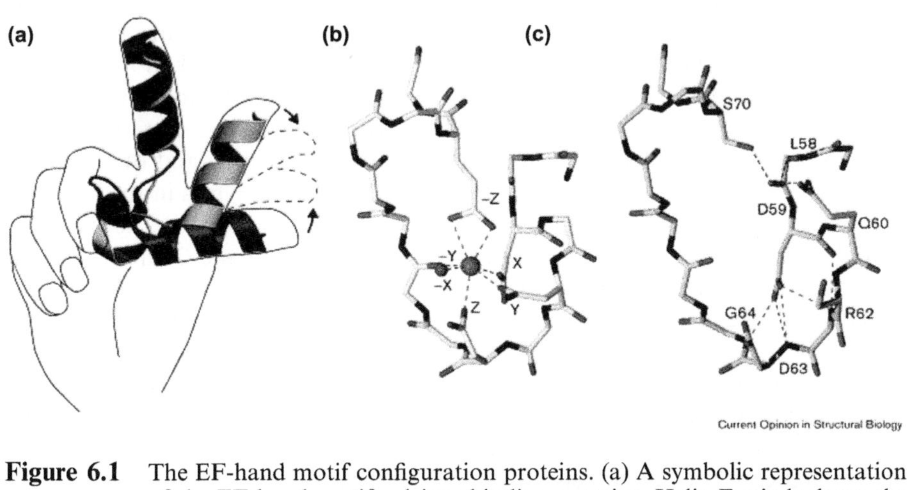

Figure 6.1 The EF-hand motif configuration proteins. (a) A symbolic representation of the EF-hand motif calcium-binding proteins. Helix E winds down the index finger, whereas helix F winds up the thumb of a right hand. When the calcium ion binds, helix F moves from the closed (apoprotein, light grey) to the open (holoprotein, dark grey) conformation. (b) The geometry of the calcium ligands. At positions X and Y, we usually find the side chains of aspartic acid or asparagine; the side chains of aspartic acid, asparagine or serine are found at Z and a peptide carbonyl oxygen lies at –Y. –X is usually a water molecule and –Z is a conserved bidentate ligand, glutamic acid or aspartic acid. (c) The modified EF-hand loop (S100A10), showing the network of stabilising hydrogen bonds.[8]

Calcium has not been shown to evoke neoplastic change *in vitro* or *in vivo* but cahederins (a family of calcium-dependent glycoproteins) and other CaBP are now known to influence cell adhesion, apoptosis and intercellular anchorage through over-expression or inhibition of expression in a variety of tumours. In certain cases profiles of CaBP may be important as diagnostic markers in tumour progression and invasiveness.[9,10] Thus, disabled or aberrant E-cahederin was shown to advance progression of well differentiated adenoma to invasive carcinoma whereas well differentiated squamous carcinomas of the head and neck expressed high levels of E-cahederin.

6.2.1 Human and Animal Nutrition

Calcium is a macronutrient. The human body requires calcium with phosphate throughout life for a wide range of physiological and morphological functions including haemostasis, tissue repair, musculo-skeletal function, growth, pregnancy and lactation.[3] Of the average body content of 1500 g or 1.5–2% of human body weight, 99% is contained within skeletal tissues and teeth, the remainder being distributed amongst soft tissues and body fluids. Normal concentrations of calcium in blood plasma range from 8.5–10.5 mg dl^{-1}, half as the free ion and the remainder bound to albumin. The recommended daily intake for a healthy adult is approximately 800 mg. Calcium balances are regulated by age and physiological requirements for growth and maturation, parathyroid hormone (PTH), androgen–oestrogen balance, dietary phosphate, and vitamin D_3. Dietary vitamin D is fat soluble and is absorbed in the small intestine. Vitamin D from the diet or dermal synthesis is biologically inactive and requires enzymatic conversion to active metabolites, notably to 25-hydroxyvitamin D, the major circulating form of vitamin D, and then to 1,25-dihydroxyvitamin D, the active form of vitamin D, by enzymes in the liver and kidney.[11] The half-life of 1,25-dihydroxyvitamin D is four to six hours, compared with two to three weeks for 25-hydroxyvitamin D and 24 hours for the parent compound. The First Food World Congress on Calcium and Vitamin D in Human Life in 1996 discussed the requirements for the metal in health and disease with particular reference to growth and development in childhood and adolescence, and diseases such as osteoporosis, hyperparathyroidism, dietary and vitamin C deficiencies, cardiovascular complaints and age-related conditions.[12] Cushing's disease, Paget's disease and hormonal imbalances are additional causes of bone deformity or calcium loss.

Calcium in bone is mostly bound in a hydroxyapatite complex in the form of a calcium carbonate and calcium phosphate crystal structure embedded in a polysaccharide intercellular matrix. This polysaccharide matrix binds other metals including strontium, zinc, magnesium, iron, lead and chromium which may be released into the circulation. Lead is known to impair calcium uptake and metabolism.[13] Some of the calcium content of bone is in a labile form and is available to maintain homeostatic balances in body fluids and tissues, whilst the remainder is in a stable state and mobilised only in a state of severe hypocalcaemia. Bone and to a lesser extent tooth calcium exist in a state of flux to respond to changing body demands.

Calcium is absorbed through the mucosa of the small bowel in a relatively inefficient manner with only about 30% taken up into the circulation. Absorption is proportional to "available" calcium ion in the intestinal contents and free binding sites on the enterocyte membranes. The efficiency of absorption declines with advancing age and in relation to reduced body requirements. Intestinal absorption of calcium is an active process involving an electrochemical gradient controlling calcium channels, but a passive trans- or intercellular diffusion may occur. In the former case, calcium uptake depends upon CaBP such as calbindin and sequestration *via* endocytic or lysosomal vacuolation. Once absorbed, calcium ion is transferred to mitochondria, secondary lysosomes and Golgi-vesicles before entering the peripheral circulation *via* the basolateral membrane.[14–16]

Calcium absorbed intracellularly binds calbindin-D as a "ferry protein" and is taken up into mitochondria where it stimulates Ca-sensitive dehydrogenases of the Krebs cycle.[17] Calbindin also modulates levels of intracellular calcium as an endogenous protective system. In the circulation, calcium is mobilised to intra- and intercellular sites elsewhere in the body where the ions activate calcium channels, pumps and ion exchange mechanisms involving magnesium, sodium, potassium, other CaBP and Vitamin D_3.[18–20] PTH may function in a similar manner to vitamin D_3 although its principal action is in regulation of extracellular calcium levels and mobilisation of labile calcium reserves. Mutations in calcium-sensing receptors have been implicated in familial and diet-related hypercalcaemia.[21,22]

6.2.2 Occupational and Industrial Exposure

Calcium is encountered in food and drinking water in all parts of the world and in many industries including agriculture, chemical syntheses and water treatments. Calcium oxide as lime is perhaps the most important commercial compound, with uses in treatment for drinking water, agriculture and building. It exists in several forms for use in building:

- Hydrated lime
- Natural hydraulic lime
- Lime putty: an aqueous suspension of calcium hydroxide
- Hydraulic lime mortar: a mixture of aggregate and natural hydraulic lime
- Fine lime plaster: an aqueous suspension of calcium hydroxide and aggregate.

Global consumption of lime probably exceeds 100 million tons annually. Calcium silicates are components of Portland cement. The environmental effects of calcium oxide and calcium hydroxide are minimal when properly used. Calcium oxide (quicklime) reacts promptly with water to form calcium hydroxide (hydrated lime). Recarbonation by air or bicarbonate ion in surface waters converts calcium hydroxide to calcium carbonate, the same mineral as in limestone. Calcium oxide is an effective agent for soil stabilisation and aggregate modification as well as drying up muddy construction sites and in a variety of alkaline chemical neutralisation applications. Calcium hydroxide [$Ca(OH)_2$] as hydrated lime is effective as an additive to hot mix asphalt to improve adhesion between the cement and the aggregate, to reduce stripping, rutting, moisture damage and age hardening and to increase the retained tensile strength.

Hydrated lime has an irritant effect on moist skin (due to transpiration or humidity) after prolonged contact and calcium oxide as quicklime, caustic lime or burnt lime is corrosive and known to "burn" the skin when moistened, but is not a cause of allergenic sensitisation.[23] Many international suppliers and trade organisations publish health and safety guidelines on handling of these products but none specify lung cancer as a potential concern. Inhalation of industrial dusts and fumes containing silica, products of combustion and certain other elements is a well documented cause of chronic respiratory disease.[24–26] Frequent inhalation of large quantities of hydrated lime dust over a long period of time can also increase the risk of developing lung diseases. Pneumoconiosis is a rare occupational risk in limestone quarry workers but in a recorded case of a 43-year-old man, granulomatous foreign body reactions were associated with birefringent nodules containing calcium, aluminium and silica.[27] Calcinosis is a second uncommon hazard experienced by workers inhaling calcium carbonate dust with magnesium oxide, silica dioxide and aluminium oxide.[28] Correct diagnosis of calcinosis requires an accurate histological study and an appropriate occupational history. The nodules show a foreign-body granulomatous response with several foreign-body cells. Light microscopy reveals the presence of numerous birefringent crystals with the chemical composition of limestone.

Eye contact with hydrated lime (dry or wet) may cause serious and potentially irreversible injuries. Many dermatological problems are reported in the literature following chronic exposure to wet cement, concrete and mortar, but it is not possible to apportion risks of contact allergy or dermatitis to calcium absorption. Cement contains chromium and nickel residues which are far more toxic and allergenic.

6.2.3 Cell Biology and Homeostatic Regulation

Recent research using new antibody technology, fluorescent probes and analytical chemistry has provided great insight into the role of calcium in cellular physiology and behaviour under the influence of dietary deficiencies, hormonal disturbances and disease states.[29,30] Using marker CaBP or Ca(II)-dependent enzymes (*i.e.* ATPases), calcium can be located in the cell membranes, nucleus, endoplasmic reticulum, mitochondria and in cytosol.[31–33] Intracellular calcium acts as a signal transmitter for mitosis, differentiation, terminal differentiation and apoptosis, and local distribution of CaBP can serve as a diagnostic and prognostic marker in tissue behaviour and in tumour biology.[3,34,35] Calcium is identified as a key intracellular signaller or messenger modulating cellular function and gene expression.[36] Calcium concentrations in epithelial tissues occur in waves determined by physical and/or chemical injury, disease, or toxicological change. Under normal circumstances, calcium concentrations in tissue are tightly regulated through the action of PTH, peptide hormones such as calcitonin, and vitamin D or its metabolised form, vitamin D_3 (1,25-dehydroxy-cholecalciferol).[32,36,37] Experimental studies indicate that zinc, iron, magnesium, copper and manganese interact with calcium in sequentially regulated events of the tissue repair cascade.[34,38]

The CaBP are critical in understanding the functional role of Ca(II) ions in health and disease.[39] Melanie Wilson of the Scripps Research Institute compiled a comprehensive list of CaBP of the EF-Hand configuration and summarised their role in cellular transformation, gene transcription and location of CaBP-related genes.[40] Of particular relevance, in discussing the promotional action or apoptotic mechanisms involving calcium in tumour growth or progression, are proteins of the S100 series, notably calmodulin, calbindin, cahederin and calpain (Table 6.1). They have been defined as a family of 10–14 kDa EF-Hand CaBP that function to transmit calcium-dependent regulatory signals to target cells.[41] They have no enzymatic action but bind in a calcium-dependent manner to target proteins

Table 6.1 Calcium binding proteins.

CaBP	*Chemical*	*Location*	*Presumed action*	*Comment*
Calmodulin	Protein	Cytosol	Intracellular calcium homeostatsis; Activates Ca-mediated events, controls calcium channels; activates signal transduction channels; promotes cell proliferation	4 Ca^{2+} binding sites; binds zinc
Cahederin	Glycoprotein	Cell membrane	Cell–cell adhesion, motility, migration, Ca-sensors, gene transcription, cytoskeletal function	4 Ca^{2+} binding sites
Calbindin	Protein		Calcium sensor protein; buffer for intracellular calcium; regulates physiological functions	Mainly research in neurological tissue
Calpain	Protease	Intracellular	Calcium pump, calcium extrusion protein; interacts with calmodulin	
Calcyclin	Peptide	Cytosol	Upregulated in proliferating and differentiating cells	
Calretinin	Protein	Cytosol	Calcium buffer or sensor	Binds zinc

Evidence from keratinocyte cultures suggests that S-100 CaBP target and modify enzymes, such as transaminases, involved in protein synthesis and regulate cell motility through their action on the actin microfilaments of the cytoskeleton.[42] Low levels of S-100 proteins have been demonstrated in basal cells of the epidermis, whereas higher levels were demonstrated in post-mitotic differentiating cells of the stratum spinosum and granular layers, in line with calcium concentrations. Cell motility is inversely proportional to S-100 and calcium concentrations. Calcyclin is upregulated in proliferating and differentiating cells.[43] Nagy showed that expression of S-100A2 is downregulated in epithelial tumour cells and might be identified as a tumour suppressor gene.[42] It exhibited an inhibitory effect on cell motility whilst modifying certain kinetic factors. The S-100A2 protein was hardly expressed at all in non-epithelial tissues.

Observation that calcium-dependent events in the cell cycle are sensitive to and modulated by cytoplasmic zinc has prompted the view that the two metals exhibit reciprocity.[44] Whereas calcium activates some intracellular processes, zinc inhibits calcium–calmodulin complexing. It also inhibits expression of calmodulin in mice following parenteral injection. In these studies, calmodulin and cyclic adenosine monophosphate (CAMP) levels exhibited an inverse relationship with tissue zinc levels.

Cahederins are a large family of transmembrane glycoprotein CaBP having a critical role in morphogenesis, maintenance of tissue architecture, and cell-to-cell and desmosomal adhesion in virtually all multicellular systems through a homotypic action.[45–47] They are expressed on the cell surface of all basal epithelial cells, within the plasma membranes. They have been labelled as "prototypic calcium sensors" present on the surface of all eukaryotic cells. They have been shown in *in vitro* systems to modulate intracellular process including general signal transduction and transcription, ion conductivity, vesicular fusion and cytoskeletal function.[48,49] Cahederins exhibit four calcium-binding sites and are identified either as epithelial (E-type) or placental (P-type) cahederins. The former are potentially important in promotion of neoplastic changes in that they regulate intercellular adhesions and motility in post-mitotic cells.[50,51] E-cahederins act at cell surfaces to achieve membrane stability, with individual molecules co-operating in the form of a linear cell adhesion zipper effect. They possibly interact with intracellular catenins (α,β and γ) on epithelial cells to form a link with actin filaments of the cytoskeleton.[52,53] Catenins are polypeptides that bind to the conserved cytoplasmic tail of cadherins and are required for cadherin function.[54] Cell surface expression of E-cahederins is reduced in neoplastic tissue and this may prove a useful prognostic factor in disease progression and metastasis.[10,55] Keratinocytes exposed to drugs that inhibit calmodulin activity have been shown to inhibit proliferative activity in rapidly dividing cells.[56,57] In contrast, epidermal keratinocyte cultures exposed to high concentrations of cahederin expressed a high level of adhesiveness and a strong tendency to form cell sheets and differentiate. Any disruption of catenin activity had been shown to abolish the adhesive tendency of E-cahederins and can be expected to retard cell differentiation and maturation.[53]

6.2.4 Neoplastic Modulation and Promotion

No experimental or human clinical studies are reported to show that elemental calcium is carcinogenic, although compounds such as calcium chromate and calcium arsenate are listed as human carcinogens owing to the action of the chromate and arsenate anions; much of the relevant information is reviewed in the IARC Monographs of 1980 and 1987.[57,58] Thus, intratracheal administration of a pesticide mixture containing calcium arsenate formerly used in vineyards in rats led to production of a high incidence of pulmonary adenocarcinoma.[59,60] In two lifespan studies in hamsters, weekly intratracheal administration of calcium arsenate induced significant increase in lung tumours (adenomas and an adenocarcinoma).[56,61] Pulmonary carcinomas were observed in rats in which pellets of calcium chromate were implanted intrabronchially.[62]

Although metallic calcium cannot be regarded as a human carcinogen *per se*, profiles of certain CaBP as shown by highly pure and specific antibodies are potentially useful guides in the clinical diagnosis, prognosis and behaviour of malignancies.[7] At the moment few useful comparative studies have been published and much of the information available is widely scattered and incomplete. Expression of genes of the S-100 complex has been best researched, and information on their properties in cell growth, motility, cell cycle regulation and transcription in epithelial cells is available. Most of the genes regulating expression of the S-100 family are located in a cluster on human chromosome 1q21. Each member of the S-100 family shows a specific and characteristic pattern of distribution in normal cells, and this may be deregulated in cancerous tissue. The cahederins for example are a family of transmembrane glycoproteins with a critical role in morphogenesis, cell binding and maintenance of cellular architecture and have contributions in certain forms of cancer.[45,63,64] The S-100A2 protein is expressed marginally if at all in non-epithelial proliferating tissues, but is highly expressed in epithelial tissues such as the skin.[42] S-100A2 is dramatically reduced in neoplastic tissue, where it locates in the nucleus and cytoplasm. Addition of the protein to culture medium inhibits cell proliferation, possibly through its action on the actin of the cell cytoskeleton. Blood levels of S-100A have proved useful in the prognosis and clinical management of tumours of the brain, gastrointestinal cancers and breast carcinomas, whereas S-100B has been used as a monitor for malignant melanoma and metastases.[65–68] Filipek and Kuznicki demonstrated that calcyclin expression is upregulated in proliferating and differentiating cells.[69]

Drugs acting as antagonists to CaBP, such as calmodulin, have been investigated as antiproliferative agents in experimental studies.[57] Calmodulin levels are naturally high in fetal epithelia and in psoriasis where they activate hyperproliferation and/or differentiation.[30,70] They have been researched also as markers for the central role of calcium in tissue repair and remodelling following injury.[3,34] While studies have identified miconazole, dequalinium chloride, bromocryptine and tamoxiphen as potentially useful calmodulin antagonists for use in treatment of psoriatic conditions, they have potential relevance in cancer therapy. Tamoxiphen, of course, has long-standing value as an efficacious anti-oestrogenic in breast cancer treatment and gynaecomastia therapy.[71] Elsewhere, human calmodulin-like protein has been shown to be downregulated in primary tumours or established tumour cell lines, suggesting that this may serve as a marker for terminal differentiation.[72]

The role of cahederins as cell surface CaBP markers of differentiation has been investigated in histological tissue taken from 136 patients with squamous cell carcinomas of the head and neck, oesophagus or skin.[10] Thus, E-cahederin was expressed at very high levels in epithelia adjacent to or distant from tumour sites as well as in most well differentiated tumours, as an indication of the level of squamous cell differentiation. In contrast, in poorly differentiated tumours, E-cahederin levels were 60% lower. All-*trans*-retinoic acid, which induces differentiation in epidermal tissues, upregulated E-cahederin expression, thereby confirming the value of the protein as a marker for tumour prognosis. Other evidence suggests that loss of this cahederin has a role in transition of epithelial cells from a benign to an invasive state.[33] N-cahederin expressed in highly invasive cell lined deficient in E-cahederins served as a marker for increased migration and invasion of gel culture media, and showed adhesion in monolayers.[73] N-cahederin upregulates matrix metalloproteinase-9 (MMP-9) as a means of advancing invasiveness. This was further demonstrated by injection of the tumour cell line into immunodeficient nude mice where tumours actively metastasised to liver, pancreas, lung, lymph nodes and muscle.

6.3 Strontium

Strontium is a bivalent metal with close chemical and biological properties to calcium. It is classified with calcium and magnesium in Group IIa of the Periodic Table and is the 16th most abundant element in the Earth's crust. Strontium reacts vigorously with water and quickly oxidises readily in air, and is not found naturally in its elemental state. Finely powdered strontium metal will ignite spontaneously in air to produce both strontium oxide and strontium nitride. Strontium occurs commonly in nature, and comprises an average of 0.034% of all igneous rock. The two principal minerals of strontium available in sufficient quantities to make recovery practical are strontianite ($SrCO_3$) (named after the Scottish town of Strontian, near where it was first identified) and celestite ($SrSO_4$). The principal mining sites for strontium in China, Spain and Mexico produce approximately 98% of celestite. The celestite deposits in the USA have not been mined since 1959. Strontium deposits occur naturally with other metals including barium, iron and vanadium. Rocks, soil, dust, coal, oil, surface and underground water, air, plants, and animals all contain varying amounts of strontium. Strontium, as strontium sulfate, is an essential element in the mineralisation of the shells of certain deep-sea creatures including stony corals. Strontium residues are eluted into water courses from the soil by rainwater and through weathering of rocks. Strontium enters food chains at low levels and is presently classified as a minor non-essential nutrient in the human body. Its physiological role is debatable at present.

Strontium is a soft yellowish shiny metal, best known through the health risks attributable to emissions from its radioactive isotope ^{90}Sr in the proximity of nuclear weapons testing sites. More than 25 isotopes of strontium have now been identified, only four of which are stable. Others exhibit half-lives ranging from a few milliseconds to 10^{25} years. As in the case of calcium, stable isotopes of strontium are of low toxicity and are classified as non-carcinogenic. Greatest biological interest lies in the action of strontium as a bone-seeking element.[1,74–76] Highest concentrations in the human body are found in the skeleton and in teeth. Clinical and experimental evidence shows that strontium interacts and impairs calcium metabolism through suppression of CaBP and possible responses to vitamin D_3-induced gastrointestinal absorption.[2] Strontium enhances the content of magnesium and potassium in bone ash. Clinical studies in young children growing up in areas where food was grown on soil rich in strontium and low in calcium have shown an increased incidence of rickets ("strontium rickets").[77] This has been re-examined experimentally in rats and chickens. Dental caries is an additional complication of excess strontium consumption in drinking water.[78–80] Research in avian embryology shows that strontium concentrates in egg shells and is transferred to developing embryos, and is a potential cause of structural abnormality and mineralisation defects.[81–84] Radioactive ^{90}Sr in nuclear fallout is teratogenic, mutagenic and carcinogenic. Bone tumours result from emissions of ^{90}Sr.[5,85–87]

6.3.1 Industrial Use and Human Exposure

Strontium has a limited range of industrial uses and is not an element of great commercial importance. Its uses are similar to those of calcium and barium, but it is rarely employed because of its higher cost. Principal uses of strontium compounds are in pyrotechnics, for the brilliant reds in fireworks and warning flares and in greases. Some is used as a "getter" in vacuum tubes to remove the last traces of air. Most strontium is used as the carbonate in special glass for television screens and visual display units. Although ^{90}Sr is a dangerous radioactive isotope, it is a useful by-product of nuclear reactors from whose spent fuel it is extracted. Its high-energy radiation can be used to generate an electric current, and for this reason it can be used in space vehicles, remote weather stations and navigation buoys.

Human exposure to stable strontium isotopes occurs through ingestion in food and drinking water. Strontium, as ^{90}Sr and other radioactive isotopes, is absorbed by inhalation and transported to bones and teeth. Rare cases of hazard relating to strontium exposure are reported in the manufacture and combustion of pyrotechnics and fireworks. Strontium is absorbed into the body by ingestion, inhalation and through percutaneous absorption.

Other commercial uses of strontium include cosmetics and depilatory agents (strontium chloride), prescription therapy for osteoporosis (strontium ranelate), and in production of coloured glass for use in television tubes. Strontium aluminate oxide doped with "rare earth" impurities such as europium is used to provide a photo-luminescent pigment for use in watch and clock dials. Phosphorescent pigments usually appear a pale green or beige in normal light, but emit green, red and blue pigments in the dark.

6.3.2 Strontium in Human Nutrition

Strontium is listed as a micro-trace element in human nutrition, although its specific functions are ill-defined.[88] The typical adult body burden of strontium is about 0.3–0.4 g, with 99% found in the skeleton. Normal dietary intake of strontium in children of 4–14 years of age is about 1 mg but patterns of uptake and excretion at higher levels are controlled largely by calcaemia.[89] Renal clearance of strontium and calcium is far lower in children than for adults, indicating the importance of both elements in the growth and mineralisation of bone, and in tooth growth. Whilst some early workers claimed that addition of strontium to the drinking water of rats had no obvious effect on growing bone,[90] later workers provided irrefutable evidence that added strontium in the diet enhances the structure and strength of bone, especially in childhood.[91,92] Clinically, strontium ranelate (Protelos®) is an efficacious therapy for osteoporosis but is associated with serious side effects of thromboembolism and hypersensitivity reactions.[92–95] MacCaslin and Janes noted that osteoporotic patients given strontium lactate or strontium lactate with oestrogen and testosterone supplement showed marked improvement in their osteoporotic symptoms.[91] Similar observations were made at McGill University in 1985 where patients given strontium carbonate therapy showed marked improvement in bone strength and density in hip biopsies.[92] The results showed a 172% increase in bone formation. Other confirmatory evidence of the relevance of strontium in bone formation and induction of osteogenesis has been provided by experimental studies in rodents and chickens.[74,93]

Clinical and experimental studies show highest levels of strontium and calcium accumulation in epiphysial growth regions, possibly indicating cooperation between the elements in initiation of osteogenesis. Uptake and absorption of strontium in bone is possibly regulated by levels of calcaemia.[89] In pregnancy, strontium levels in fetuses are possibly controlled through inherent protective mechanisms in the mother or through a "saturation" state in fetuses that blocks further transplacental transfer.[96]

High levels of strontium, leading to saturation of protective mechanisms, are known to inhibit calcium uptake and mineralisation in the epiphyses of growing bone and to induce a condition of strontium rickets.[97,98] It is proposed that strontium exhibits a regulatory influence on calcium metabolism, possibly through an action on proteoglycan or glyosaminoglycan synthesis in the bone matrix. On the other hand, early experimental work suggests that the body discriminates against strontium in transfer of calcium from blood to bone and that the balance changes as the skeleton matures in adolescence and adulthood.[89,99,100]

6.3.3 Cytology and Intracellular Management

Stable isotopes of strontium generally appear to be non-toxic or mutagenic in standard laboratory *in vitro* assays whereas the genotoxicity and mutagenicity of radionuclides such as ^{90}Sr are well documented.[101] A single study has reported a dose-related incidence of chromosomal aberrations in bone marrow cell metaphase preparations from mice dosed with strontium chloride for 6–24 hours.[102] Female animals were more susceptible than males. Strontium ranelate was not cytotoxic in cultured human periodontal ligament fibroblasts exposed to 2.5, 5, 10 or 20 mg ml^{-1} and cell survival was not impaired after 8 days of exposure.[103] Proliferation in osteoblasts and pre-osteoblastic cells derived from rat calvariae was increased by up to 60% after 96 hours and DNA synthesis and bone formation were induced.[104] Other studies have shown that bone marrow cell stimulation enhanced by strontium ranelate is sensitive to cyclo-oxygenase-2 (COX-2) and is markedly reduced in the presence of COX-2 inhibitors or disruption of COX-2 gene expression.[105]

Strontium exhibits a specific predilection for osteoblasts and is known to stimulate proliferation and increase DNA and mRNA levels.[106] Experiments with bone marrow stromal cells taken from rats treated orally with strontium for 12 weeks following ovariectomy, or human bone marrow extracts, showed enhance osteoblastic activity but a depressed level of adipogenic gene expression and lower adipogenic mRNA expression. In cultured keratinocytes, added strontium substituted for calcium ion and was shown to enhance proliferation, cell viability and differentiation significantly.[107] This was reflected in sequential changes in the cell cycle, that is an increase of cells in G_2–M phase and a lower level of cells in G_0–G_1. No significant changes were seen in cell cycle profile in melanocyte or fibroblast cultures, which must be presumed to be non–responsive to Sr. Other changes reported in response to strontium include promotion of fibronectin expression in keratinocyte cytoplasm and on cell membranes.

Soluble strontium compounds have shown a lack of mutagenic change in several strains of *Salmonella typhimurium* reverse mutation assays or other bacterial tests.[108–110] Strontium sulfate did not evoke chromosomal aberrations or polyploidy in cultures of Chinese hamster lung cells. Radioactive isotopes of strontium have been shown to induce evidence of clastogenicity and mutagenicity in standard *in vitro* assays. Human lymphocytes exposed to ^{90}Sr-β radiation at doses of 0.2 to 5.0 Gy from a source emitting 0.2 Gy min^{-1} showed chromosomal aberrations including acentric fragmentation, double minutes and dicentrics.[111] However, irradiation in the range used exhibited no effect on the model chromosome number of irradiated cells or on cell cycle kinetics. In Comet assays, there was an increase in DNA migration as a function of radiation dose as evaluated by image analysis or visual classification. The ^{90}Sr irradiation induced DNA damage and dose-related cell death.

6.3.4 Carcinogenicity

Stable strontium isotopes are not known to cause cancers in humans following ingestion, inhalation or percutaneous absorption. Studies in rats have mainly shown that chronic or sub-chronic dosing with high levels of soluble strontium salts induces changes in the ossification of long bones and changes consistent with hypocalcaemia. Stable strontium can cause cancer in animals when administered as strontium chromate or strontium arsenate, but the carcinogenicity is attributed to chromate and arsenate anions, as discussed elsewhere in this volume, and not to Sr(II). No co-carcinogenicity or other interaction has been demonstrated between strontium and chromate or strontium or arsenate in human or chronic animal studies.

Radioactive strontium isotopes are carcinogenic through emissions of β-radiation.[4,112] Strontium-90 (^{90}Sr) does not occur naturally and has a half-life of 28.7 years. Environmental ^{90}Sr comes from three main sources:

1. Fallout from above-ground explosions of nuclear weapons testing worldwide from 1963 to 1980.
2. Radioactive emissions from the 1986 Chernobyl nuclear power plant accident in the Ukraine.
3. Radioactive releases from nuclear power plants into the environment.[113]

The principal carcinogenic risks include bone cancers and childhood leukaemias. These cancers were significantly higher in the USA during the years of atomic bomb research (1951–1964), but ^{90}Sr contributed only part of the radiation to which people were exposed at the time. In retrospective surveys of cancer deaths, ^{90}Sr exposure was validated by analysing emissions from deciduous teeth. However, in view of wide variations in the duration and levels of irradiation attributable to ^{90}Sr, useful statistics are hard to find.[4] Calculations suggest that latent periods of less than 10 years occurred between the exposure period and identification of tumours.[4,114–116] Other analyses of decedents showed that exposure to ^{90}Sr during the nuclear test years led to fatal malignancies in line with exposure, and that ^{90}Sr emissions were responsible for many tumours other than those of bone and bone marrow, including brain, bladder, connective tissue, gastrointestinal tumours, melanomas and Hodgkin's type lymphomas.[117] Current views also point to *in utero* exposures being a cause of childhood leukaemias.

Elevated rates of bone and other cancers have been observed in rodents, rabbits, pigs, dogs and humans exposed to ^{90}Sr in experimental studies. Radiation-induced bone tumours commonly occur in the region of active endochondrial ossification in long bones at the interface between the epiphyses and diaphysis.[118] In early pre-carcinogenic phases, normal osteogenesis is disturbed and retarded but with progressive neoplastic transformation, bizarre bone formation overshadows resorption and dysplastic changes progress to frank neoplasia with abnormal osteoid and trabeculae. Changes are prevalent in spongy bone where metabolically active tissues around Haversian canals are sensitive to the toxic and mutagenic effects of the β-emissions. In a study in which young Wistar rats were exposed to 3.7 MBq ^{90}Sr, pathological changes consistent with medullary osteogenic sarcoma developed within 8–10 months of intramuscular injection.[86] Whereas bone tumours induced by α-irradiations tend to predominate in vertebral regions, those attributable to β-emitters such as ^{90}Sr are mainly found in growth regions (diaphyses) of long bones.[119,120] On rare occasions, murine sarcomas associated with ^{90}Sr have been associated with a chondroblastic tissue, which is rare in the case of tumours induced by β-emitters. Other studies have demonstrated that intravenous injection of ^{90}Sr in the ear vein led to local sarcomas as well as bone tumours.[121,122]

6.4 Magnesium

Magnesium is an essential trace element found in every living cell in the body. It is an alkaline earth metal with many similarities to calcium and strontium and has three stable isotopes and five radioisotopes with short half-lives. It one of the commonest metals found in the human environment and its compounds are found throughout the world in the soil, sea water, plants and animals. Magnesium is present as a chelate in the porphyrin molecule of plant chlorophylls A and B. In both cases the magnesium atom is bound centrally in the molecule. Magnesium is a highly reactive element and oxidises strongly in water to release hydrogen gas. It is extremely inflammable in air. As such, magnesium is not found naturally, its main mineral salts being carbonate (magnesite), hydrated potassium magnesium chloride (carnallite, $KMgCl_3 \cdot 6H_20$), calcium magnesium carbonate [dolomite, $Ca{:}Mg(CO_3)_2$]. Carnallite is an important commercial source of potash for fertilisers. Talc or basic magnesium silicate [$Mg_3Si_4O_{10}(OH)_2$] is a soft natural source of magnesium. It is a soft (soapstone), either colourless, white, pale green, bright emerald-green, dark green, brown or grey mineral found as a metamorphic mineral in veins often associated with dolomite, magnesite and calcium carbonate (calcite), and the magnesium silicates serpentine, tremolite and forsterite. Common impurities include nickel, calcium, iron and aluminium. Talc is mined in all parts of the world, although highest production is from the USA and China. Epsom salts ($MgSO_4$) occur naturally in springs near the Surrey town and elsewhere and the spa waters have been used for many years as a remedy for aches and pains.

6.4.1 Magnesium as a Nutrient

Magnesium is an essential nutrient in plants and animals. It is an essential component, co-factor or activator of at least 300 enzyme systems, notably all those enzymes involved in the transfer of phosphate to and from ATP, and the glycolytic pathway.[123,124] Additionally, magnesium-related enzymes are essential in protein, lipid and carbohydrate metabolism, stabilising long ribonucleic and deoxyribonucleic acid side chains in nucleic acid synthesis, muscle contraction and neuronal transmission.[125] Early studies indicated that magnesium exerts a curare-like action at neuromuscular junctions.[126] Other early reports claim that injection of magnesium sulfate depresses central nervous system activity and may serve as an anaesthetic in human patients, although the side effects of high doses used are prohibitive.[127]

The human body is expected to contain 21–28 mg of magnesium, much of which is contained in bone and tooth matrix (60%), and in muscle (20%); 20–30% of the magnesium bound in bone is exchangeable and available to maintain serum levels in the range 19–25 mg l^{-1}. Factors regulating magnesium metabolism and mobilisation from labile sources are listed in Table 6.2.[128]

Table 6.2 Factors regulating serum magnesium levels and renal tubular resorption.

Factors
Plasma magnesium concentration/magnesium status
Glomerular filtration rate
Volume status
Parathyroid hormone
Calcitonin
Calcium levels in serum and tissue
Antidiuretic hormone
Glucagon
Insulin
Hypophosphataemia
Acid–base status
Diuretics

Efficient endogenous homeostatic mechanisms keep serum magnesium levels and magnesium–protein complexes within narrow limits.[123] Magnesium is absorbed mainly in the small intestine, at levels ranging from 40 to 60% of the dietary content, but this is inhibited by high blood concentrations of calcium and lead and the dietary content of phytates, plant fibres and materials that chelate the Mg(II) ion. Mucosal absorption may involve passive diffusion but is more likely to occur through an active process involving metal binding proteins and transporter proteins. Dietary requirements for magnesium are slightly higher for males than females and increase with age, *i.e.* 39–75 mg per day for infants, 240 mg per day for 9–13 year olds, to 420 mg per day for adults.

6.4.2 Industrial Uses and Human Exposure

Principal uses of magnesium include production of light alloys with aluminium for use in the automobile industry, de-sulfuring of pig iron and steel, and the production of cast iron. Magnesium oxide, chloride, citrate and sulfate have medical applications, magnesium sulfate as Epsom salts being used widely as an antacid for gastrointestinal complaints. Since magnesium is a requirement for plant growth, salts such as magnesium sulfate, oxide, citrate, potassium–magnesium concentrates and phosphate are used as fertilisers, particularly for magnesium-poor soils. Of the commercially used magnesium salts, talc is perhaps the best known and possibly the most toxic. It is widely used in dusting powders and personal care products, but has many industrial applications including animal feed, asphalt, bleaching agents, ceramics, cereal polishing, confectioneries, detergents, dry fire extinguishing powder, floor wax, leather treatment, paper, paints, pharmaceuticals, plastics, printing ink, putties, shoe polish, water filtration and welding rod coating. A contentious use of talc is in dusting powder for surgical operating gloves.

Whereas magnesium and most of its inorganic compounds are not classified by the RoC or the IARC as being carcinogenic to humans, inhalation of talc powder or use of dusting powder for hygiene purposes may be associated with cancer. In different industries, mining, milling, polishing, powder preparation and manufacturing, workers are exposed to fine talc dust, and talcosis or massive pulmonary fibrosis is reported.[129–131] Smoking is a complicating factor in these studies, but observation of black dust and birefringent crystals in lung biopsies confirmed the main cause of the condition. Electron microscopy has demonstrated talc particles within pulmonary macrophages. Ellis surveyed problems associated with exposure to the talc used to provide easy donning of surgical gloves in operating theatres.[132] By 1943, 50 cases of granuloma had been reported in patients contaminated with the talc used in surgical gloves. Many of these talc-granulomas were located internally in the peritoneum or omentum, some following operations for appendicitis or gynaecological conditions. Other problems arising through the use of talc in surgical gloves include granulomas in open wounds and incisional wounds made during surgery, and talc residues left in the abdominal cavity. A number of case–control studies in the USA have suggested that the frequent use of body powders containing talc by women in the genital area may be a risk factor in ovarian cancer, but none were proved.

Considerable confusion is noted in the lay press relating to the carcinogenicity of talc in industrial exposures and in consumer products. Survey of the literature shows that some of this confusion relates to the misidentification of granulomas (essentially foreign body reactions) as neoplastic. A second source of error concerns the mistaken notion that talc is closely related to the potent carcinogen asbestos. The granule size is similar. There is further concern that talcum powder produced from natural magnesium trisilicate if contaminated with asbestos can cause cancer. The IARC re-evaluated the clinical and experimental information and the nature of the exposures that people experienced in industry or through exposure to consumer product.[133] Their conclusion was that a potential risk of ovarian cancer exists through application of talc-based body powder, but that inhaled talc "not containing asbestos fibres" is not classifiable as a human carcinogen. The rationale for this decision is discussed later under silicates, but the working party found considerable inconsistency in the reports available.

6.4.3 Cell Biology

Magnesium is found in all living cells in the body and Paunier was of the opinion that intracellular levels are a more accurate guide to hypomagnesaemia than serum concentrations.[123] Intracellular concentrations and bimolecular functions of magnesium are closely interlinked with calcium such that dietary deficiencies or exogenous agents such as lead that impair calcium uptake and metabolism also influence magnesium-dependent actions.[134,135] Most of the magnesium in the body exists within the cellular compartment, 1–3% as free Mg(II) ion and the remainder bound to macromolecules (*i.e.* ATPases, enzyme proteins), membranes, endoplasmic reticulum, mitochondria and nuclear proteins.[136] Intracellular concentrations range from 5 to 20 mmol but wide variations are reported according to the analyses used, and the cell type studied, *i.e.* bone > skeletal muscle > cardiac muscle.

Magnesium is now known to be a co-factor, activator or catalyst for more than 300 enzymes involved in overall cellular metabolism, membrane function, growth and macromolecular biosynthesis. These perform key functions in the glycolytic pathway and glucose utilisation in energy generation, insulin signalling, transfer of phosphate to and from ATP, and a multitude of reactions affecting lipid and protein metabolism, nucleic acid synthesis and oxidative phosphorylation.[123,124,128,137] Wacker emphasised that magnesium is necessary for stabilising long RNA and DNA chains and for the structural integrity of ribosomes.[125,138] Other intracellular functions ascribed to Mg–enzyme complexes include mitochondrial oxidative phosphorylation and regulation of membrane calcium channels, in each case competing with calcium for intracellular metal binding proteins. Mitochondrial enzymes utilize the ability of the Mg(II) ion to chelate important intracellular anionic ligands of ATP and adenosine diphosphate (ADP) as substrates for phosphate transfer reactions.[125] Intracellular magnesium concentrations are low in patients with Type II diabetes mellitus or hypertension.[139] An inverse association exists between the plasma magnesium level and insulin resistance, due to intracellular changes.

Experiments with cultured human mammary cells (HC-11) show clearly that addition of magnesium to the culture medium regulates epithelial cell proliferation (Figure 6.2).[140] The figure shows that cells exposed to magnesium sulfate at a higher level (H, 45 mM $MgSO_4$) did not significantly affect accumulation of cells in any phase of the cell cycle compared with controls (0.05 mM $MgSO_4$), whereas the low concentration (L, 0.8 mM $MgSO_4$) increased the accumulation of cells in the G_0/G_1 phase. This was matched by a corresponding fall in cells in the S and G_2/M phases of the cell cycle. This effect was abolished when cells were given a subsidiary dose of magnesium sulfate after 4 hours (L + 4h Mg).

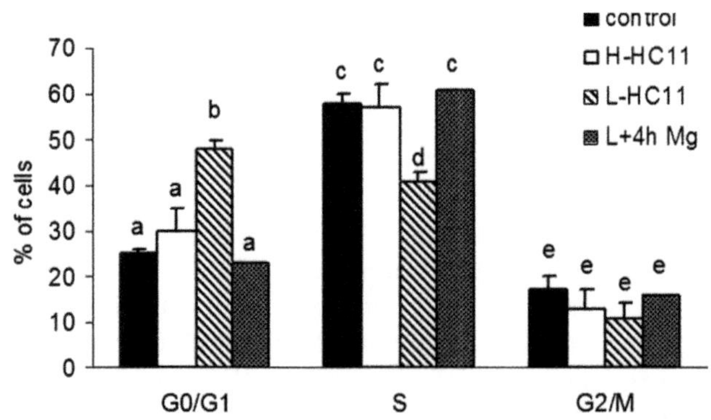

Figure 6.2 Cell cycle distribution in human mammary cell cultures (HC-ll) as influenced by low (L) or high (H) concentrations of magnesium sulphate (0.8, 45 mM $MgSO_4$), or Controls (0.05 mM $MgSO_4$), and the effect of addition of 0.8 mM Mg after 4 hours.[140]

6.4.4 Toxicology and Carcinogenicity

Magnesium is absorbed into the body through the intestinal mucosa and regulated by serum concentrations, phosphate and calcium levels, hormones (parathyroid hormone, calcitonin, *etc.*) and idiopathic factors. Hypermagnesaemia is rare except in cases of renal insufficiency, where affected patients exhibit depression of the central nervous system, impaired neuromuscular action, respiratory distress and cardiac disorders.[141,142] Exposures to magnesium and most of its inorganic compounds are not listed by the IARC as being likely to cause cancer in humans. However, it is necessary to examine two important aspects of magnesium in human cancer: first, the possibility that, since magnesium is a known prerequisite for normal cell proliferation and differentiation,[140,143,144] it should be considered as a possible tumour promoter in much the same way as zinc. Secondly, concerns are publicised in the literature that talc as magnesium silicate [$Mg_3Si_4O_{10}(OH)_2$] is an occupational carcinogen or is harmful following its use in talcum powder in cosmetics and hygiene preparations. The carcinogenic risk associated with occupational exposure to silicates is discussed in Chapter 8, but it is clear from the literature that contaminants such as asbestos fibres and other carcinogenic materials encountered in the workplace are major factors in epidemiological studies.[133]

Evidence that magnesium may act as a promoter in tumour growth is not substantiated by clinical or epidemiological evidence. A meta-analysis of eight population studies totalling 338,979 patients and 8000 cases with colorectal cancer showed that the relative risk for highest *vs.* lowest magnesium was 0.89, with little evidence of heterogeneity.[145] (Similar observations were made when comparing high and low calcium intakes.) Whereas hypomagnesemia is a complication in many critically ill cancer patients,[146,147] epidemiological studies have substantiated that people living in areas with higher magnesium actually show a lower risk of cancers than those living elsewhere.[148] This is seen in Egypt where the soil and water are rich in magnesium and the national cancer rate is about 10% of that reported in Europe and America. Researchers from Japan's National Cancer Center in Tokyo have found that high daily intake of magnesium (327 mg per day) reduced the risk of colonic cancer in men by about 52%.[149] No such relationship was seen in women. A follow-up study of 300 US patients with prostatic cancer showed that, with adjustments for lifestyle factors such as smoking, a statistically sound inverse relationship occurred between magnesium intake and pancreatic cancer.[150] Similar observations have been reported in relation to breast and prostate cancers, and tumours of the gastrointestinal tract.

Talc as a cause of cancer is a highly equivocal issue. In its pure form, plate-like particles are white, silvery-white or grey in colour, but natural forms of talc are pale to dark green or greenish grey or a brownish colour, according to the source and contamination with other minerals. Talc is insoluble in water or body fluids because the molecule does not cleave in biological fluids to release Mg(II) and silicate. The insoluble granules are associated with granulomas or foreign body reactions both when talc is used as a lubricant in surgeon's gloves,

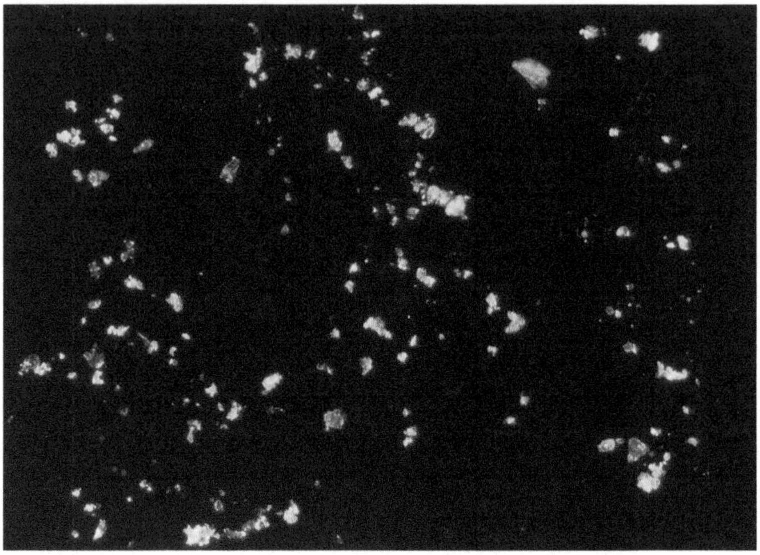

Figure 6.3 Birefringent particles of talc in lung tissues following inhalation.

as well as in occupationally related lung diseases where workers inhale talc particles in the course of their work.[132] The talc particles are phagocytosed, induce fibrogenic change and are toxic to macrophages *in vitro*, but have not been associated with carcinogenic change under these conditions.[151] Talc crystals embedded in fibrous tissue are revealed as birefringent particles in histological preparations from lung biopsies in cases of pulmonary distress in miners and millers exposed to talc dust occupationally (Figure 6.3). Pneumoconiosis and profound fibrosing lung conditions are occupational risks in talc miners and millers, but lung cancer is not presently regarded as a hazard except where individuals are exposed to asbestos fibres, nitrosamines or other toxic/carcinogenic contaminants in the course of their work.[152–154]

Experimental studies have yielded inconsistent evidence that inhalation of non-asbestiform talc is a potential carcinogenic risk.[155] Talc is phagocytosed by pulmonary macrophages in laboratory animals and clearance mechanisms are presumed to be much the same as in humans.[156] In short-term (20 days) exposures, no clinical signs were reported in mice or rats inhaling <20.4 mg.talc m^{-3} and <17 mg.talc m^{-3} respectively, but lung clearance patterns were "delayed"; by this stage normal protective mechanisms are saturated. A major evaluation of cosmetic grade talc conducted in F344 rats and B6C3F1 mice by the National Toxicology Program showed that animals inhaling concentrations of talc sufficient to overwhelm pulmonary clearance mechanisms and impair lung function resulting in 80% mortality showed massive pulmonary fibrosis and inflammation.[155] Carcinogenic responses were inconsistent. Male and female mice did not show dose-related pulmonary tumours when exposed to talc at 6 or 18 mg m^{-3} for up to 18 months. Male and female rats exhibited increased incidence of adrenal gland tumours, but only female animals exhibited a significantly higher incidence of benign and malignant alveolar/bronchiolar carcinomas.

A second feature of talc toxicity is emphasised by use of talcum powder for hygiene purposes. Numerous epidemiological studies, mostly conduced in the USA, have concluded that "a lifetime pattern" of use of talc in perineal areas may increase the risk of epithelial ovarian and endometrial cancer. Ovarian cancer in the USA accounts for more than 20% of all female cancers and a recent review by the National Institutes of health indicates that talc may be one of numerous contributory causes, but that the available information is conflicting.[157] An overview of published studies shows that the association between use of talcum powder in undergarments, sanitary napkins and diaphragms, or direct application of body talcum, and ovarian cancer is largely based upon subjective questionnaires or past records and provides little accurate information as to the amount of talc applied, duration or mode of application. The studies differ greatly in the number of women assessed, cancer incidence and duration of talc usage.

One prospective study, conducted in 1976–2000, evaluated 78,630 female nurses, 307 of whom developed ovarian cancers.[158] No association was substantiated between use of talc and incidence of ovarian cancer; 40.4% of the nurses denied using talc in the observation period and only 14.5% recorded using it on a daily basis. This study did show however that perineal use of talcum powder was associated with an increased risk of invasive serous ovarian cancer. The mechanism is unclear. Other studies, also based upon the validity of questionnaires and medical records, claimed that a 24% increase in endometrial cancer was related to regular perineal use of talcum-based powders, particularly among post-menopausal women.[159] This too awaits ratification using more definitive data. The IARC concluded that "there is limited evidence in humans for the carcinogenicity of perineal use of talc-based body powder" and that, as an overall evaluation, perineal use of talc-based body powder should be viewed as being possibly carcinogenic to humans.[160]

In summary, if talc in talcum powders is to be viewed as a potential carcinogen, then further data are required on the route of transfer of the particles to the target tissues and the mechanism of carcinogenesis. There is no evidence to implicate Mg(II) as a carcinogen and no work has been reported to my knowledge to show that talc particles are phagocytosed by ovarian tubal or stromal tissue and lead to preneoplastic changes predisposing to malignancy in experimental animals or humans. The results are at variance with observations made by Ellis and others, who allude to the ability of talc particles derived from surgeons' gloves to evoke foreign body granulomas in the peritoneal cavity through a "direct action",[132] yet talcum powder particles applied to the genital area are seemingly able to migrate *via* the genital tract to the ovaries or other sites and set up carcinogenic change. We have no histological or pathological evidence of talc-related birefringent bodies in the ovary or endometrium of any of the numerous epithelial cancer cases reported. No data have been seen on the genotoxic effects to humans of exposure to talc. The results of the few *in vitro* studies available on the genetic toxicology of talc are negative.

6.5 Copper

Copper is best known through its distinctive yellow–red colouration and its widespread use in electrical wire and conductors, copper pipes and water containers, and numerous alloys, bronzes and brasses. Copper has been known to the human race for more than 10,000 years and has been found in ornaments, jewellery and artefacts unearthed in Mesopotamia, China, Egypt and North America. Copper coinage in Britain dates back at least to Roman times (Figure 6.4).

Worldwide copper resources are estimated to be 2.6 trillion kilograms, of which only a small fraction has been mined in 5000 years. Massive copper deposits in the form of chalcocite (Cu_2S), chalcopyrite ($CuFeS_2$), malachite [$2CuCO_3 \cdot Cu(OH)_2$], azurite (Cu_2O), and native copper are found in North and South America, central Africa and numerous sites in Asia and the Pacific regions. About 200 years ago, copper mined in Cornwall and Wales placed the UK as the world's largest source of the metal. Copper ores are often found in association with other elements including arsenic, antimony, zinc, iron and silicon. Tennantite [$(Cu,Fe)_{12}As_4S_{13}$] comprises 47.5% copper and 20.4% arsenic. World copper output in 1995 was 9.8 million tons and demand is still high.

Human exposure to copper occurs through food and drinking water as an essential nutrient, and *via* utensils used in food preparation. Dermal contact with metallic copper and with more than 45 alloys with nickel, zinc silver, aluminium, tin and beryllium is expected to be high. Probably the most notable contact is with copper alloys in coinage. Copper forms alloys more freely than most metals and is useful through its corrosion resistance, thermal conductivity and durability. Pure copper has the best electrical and thermal conductivity of any commercially available metal.

Copper is classified as a transitional Group 1b metal of the Periodic Table with two oxidation states, Cu(I) and Cu(II), and two naturally occurring stable

Figure 6.4 Copper coins from the Roman period of the Emperor Hadrian AD 119–120 (British Museum).

isotopes, ^{63}Cu (69.2%) and ^{65}Cu (30.8%). Nine radioactive isotopes of copper have been identified. Copper is an interesting element, partly through its bioactivity and essential role in human and animal physiology and in plant nutrition, but it exhibits strong interactions with many other elements in the biosphere. In human medicine, inherited defects in copper uptake and homeostasis are now well researched and clinical problems associated with Menke's disease, Wilson's disease and neurological conditions such as swayback are well documented.[161–166] Study of these conditions has shed much light on the importance of the metal and cuproenzymes in human physiology and nutrition. Bremner and Beattie have reviewed the biochemical role of the numerous cupro-enzymes in cell biology, connective tissue biosyntheses, neurological conduction and growth.[163] At the moment, minimal evidence exists to show that metallic copper or its compounds are carcinogenic in animals or humans. Mention is made however of the anxiety that was raised a few years ago relating to the use of copper in intrauterine contraceptive devices.[167]

6.5.1 Copper in Human Nutrition

Copper has been recognised as an essential nutrient in humans and animals since 1928, when Hart demonstrated its co-operative role with iron in haemoglobin synthesis in weanling rats fed a milk-based diet.[168] Subsequent studies have emphasised that, in human nutrition, copper balance and the function of many enzymes based on copper in normal physiology are regulated by zinc and iron.[163,164,169] Principle cupro-enzymes are listed in Table 6.3. Copper enzymes, notably oxidases are involved in cross-linking and strengthening of connective tissue fibres including collagen and elastin.[163,169] Other cuproproteins include erythrocuprein which constitutes about 60% of erythrocyte copper content, cerebrocuprein which has been isolated from the brain and contains 0.3% copper by weight, hepatocuprein and hepatomitochondrocuprein.[170] The functions of these proteins are imperfectly understood.

Table 6.3 Principle cuproenzymes in the mammalian body.

Enzyme	*Enzyme function*
Cytochrome-*c* oxidase	Electron transport, terminal oxidase
Superoxide dismutase	Superoxide dismutation
Catechol oxidase	Synthesis of melanin
Lysyl 6-oxidase	Collagen and elastin cross-linking
Ceruloplasmin	Ferroxidase
Amine oxidases	Deamination of primary amines
Dopamine-b-monooxygenase	Dopamine > norepinephrine
Peptidylglycine monooxygenase	α-Amidation of neuropeptides
Tyrosinase	A monooxygenase catalysing the formation of melanin from tyrosine

Much information on the role of copper in human nutrition has been gained from the study of children in nutritionally deprived areas of South America, and through the two inherited copper-related conditions, Menke's disease and Wilson's disease.[164,171] Thirty young children in Peru showing low serum copper and reduced levels of the copper carrier protein caeruloplasmin responded well to copper supplementation given orally. Normally, human demands for copper are low, with total body levels expected to be 90–120 mg, with highest concentrations in brain, liver, heart, kidney and bone.[88] Demands for copper are highest in infants and young children, in keeping with body growth and development. Copper uptake and metabolism are closely linked with zinc and iron, and biochemical evidence suggests that the elements are complementary in tissue homeostasis and repair following injury.

The two rare inherited conditions – Wilson's disease and Menke's syndrome – provide useful information relating to systemic overload of copper and impaired copper transport within the body. Wilson's disease is an autosomal recessive trait characterised by impairment in biliary excretion and a massive build-up of copper, leading to hypercuprinaemia and tissue saturation.[166,172] Affected infants exhibiting copper overload in liver, brain and other tissues are subject to oxidative damage in mitochondria and degenerative change in the tissues.[165,166,173] Hepatolenticular progressive degeneration is a feature of Wilson's disease, which is also associated with degenerative changes in the brain and nervous system, and increased fatality. Zinc, acting as a copper antagonist ion, and other anti-copper therapies are employed to treat infants with Wilson's disease, with limited success.[162] Menke's disease is a rare, X-linked recessive disorder characterised by neuronal degeneration, abnormal hair, malformed connective tissue, mental retardation, bone and tooth disorders and a shortened lifespan.[174] Menke's disease is best known through abnormal or "kinky" hair patterns, but it is dominated by profound neurological pathology and connective tissue disorders resulting from deficiency in cupro-enzymes.[161] The mechanisms of impaired copper transport and intracellular management of copper in Menke's syndrome are not known but may be expected to reflect defects in copper-transport P-type ATPases, carrier proteins (caeruloplasmins) and copper chaperones which are believed to be critical in copper homeostasis.[175] Copper deficiency may be manifest by underdevelopment of myelin, as shown by experiments in copper-deficient guinea pigs.[176] An experimental study in the transgenic "crinkled" mouse has shown that copper supplementation early in life mitigates the effects of the mutant gene.[177]

Copper absorption in the stomach or small intestine is an inefficient process with approximately 30% of that provided in the diet being absorbed into the circulation. Absorption is influenced by the solubility of copper compounds in the diet, and blood levels. Studies with radio-labelled copper have shown that most copper absorbed is bound in caeruloplasmin complex, with some bound to albumins and amino acid fractions. Principal sites for copper accumulation include bone marrow for haemopoiesis, liver and brain. Copper uptake, and inter- and intracellular metabolism, is modulated by metal-binding proteins, notably caeruloplasmin, transcuprein and albumins.[163,178] Copper induces and binds metallothioneins and competes with and can displace zinc from MT-binding sites.[179]

6.5.2 Industrial Uses and Human Exposure

Since Roman times or before, copper has been an important metal in human civilisations, and artefacts from 5000 years ago show copper in coinage, weapons, decorations, jewellery and containers. These days, greatest use is made of copper in alloys and metal plating with aluminium, nickel, chromium, tin and zinc; of these, aluminium bronze, tin bronze, phosphor bronze, brass, gunmetal, cupro-nickel, beryllium copper and coinage copper are well known. Delayed contact allergy and dermatoses are documented through exposure to these alloys, which have a large range of commercial and domestic applications.

Greatest occupational risks are expected in copper mining, smelting and refining, copper plating industries and in production of alloys. In most cases, workers will be exposed to other metals/metalloids some of which, such as arsenic, lead and beryllium, are acknowledged carcinogens. Arsenic is held to be an occupational hazard in copper mining where inhalation of dust is common.[180,181] Problems associated with these metals present major environmental and human health problems in countries such as Chile which depend extensively on copper mining and exportation.

Principal routes of exposure to copper, copper alloys and copper compounds are through dermal contact, inhalation of dust, and ingestion in food and drinking water. Exposure through ingestion may lead to about 30% of copper being absorbed, but this has not been associated with lasting health risk or disfigurement. As much as 80% of copper absorbed into the body is excreted *via* the kidneys and through the hair or toenails, and excretion patterns can give a useful indication of body copper burden.[182,183] Green hair is a common manifestation of the excretion of excess copper.[184] Hostynek and Maibach reviewed dermal uptake and responses to copper compounds and noted that metallic copper is readily oxidised in the presence of sebum and sweat and that resulting soluble salts penetrate the stratum corneum.[185] Percutaneous penetration of solutions of copper sulfate or copper acetate is low, however (10^{-6} and 10^{-5} cm h^{-1} respectively), since most of the ion is strongly bound in keratin complex.

6.5.3 Cell Biology

Copper is found at low concentrations in all living cells in the body where homeostasis is achieved through interaction and binding of membrane transport proteins, intracellular receptors and soluble peptides.[179] Caeruloplasmin, a multicopper ferroxidase, has a central role in copper metabolism, and mutations in its expression impair copper metabolism but also distort iron and zinc metabolism with implications in the causation of Wilson's disease.[186,187] The copper ATPases are a family of transporter proteins with integral roles as membrane proteins. They show specific receptors for attaching to cell membranes and co-inducting ATP-dependent movement of copper ions. The action of these ATPases extends into the cytosol. Intracellular copper metabolism is a function of vesicles derived from the Golgi membrane and copper chaperones or copper-binding peptides. The incorporation of copper into key enzymes, such as lysyl oxidase with an integral role in the cross-linking and stabilisation of collagen fibres, is effected through the production of a prolysyl oxidase fraction which binds free Cu(II) before transformation to functional lysyl oxidase.[188] Lysyl oxidase is highly specific in oxidising lysyl residues in elastin and collagen, thereby initiating formation of covalent cross-linkages to stabilise these fibrous proteins. Kagan and Li have emphasised that lysyl oxidase functions within the cell to modulate cellular activity.[189]

6.5.4 Toxicity and Carcinogenicity

Copper is not listed as a carcinogenic metal by the IARC or other international bodies. Long-term intake of high copper diets has been associated with disturbances in copper homeostasis with possible adverse effects on immune status, but not increased risk of cancer.[190] However, in his review, Cordano did refer to increased levels of serum copper in cases of leukaemia and lymphosarcoma.[170] Secondly, concern was shown several years ago in relation to use of copper in intrauterine contraceptive devices or coils (IUD).[167,191] Although reports of increased lung cancer are reported in copper miners and smelters in several sites in USA and Canada, the epidemiological evidence strongly points to inhalation of arsenic or nickel contaminants as the principal cause.[192–194]

In an early pilot study, Hrgovcic reported increased serum copper concentrations in 70 patients with lymphoma and leukaemia, including 28 patients with Hodgkin's disease and 23 patients with acute lymphocytic leukaemia.[195] They concluded that levels of serum copper were not a cause of the malignancies but reflected the increased proliferative activity in the bone marrow and might be a useful indicator of "the general activity of Hodgkin's disease" and of value in the differential diagnosis of the active or inactive status of the disease. Demir *et al.* also reported increased levels of copper in the serum of 42 patients with leukaemia but these normalised following remission of the disease.[196] These and similar studies should not be viewed as an indication of the causative role of copper in the induction of cancer, but they are a reflection of the metabolic state of the tissue and the cellular and physiological demands for cupro-enzymes. An evaluation of 5100 women on the Island of Guernsey failed to establish a significant link between serum copper levels and breast cancer.[197] Serum copper in cancer cases was 1.31 mg l^{-1} compared to 1.26 mg l^{-1} in control patients. No major changes were observed in risk estimates when statistical evaluation was restricted to cases diagnosed more than 10 years after collection of blood samples

Copper in IUDs is not now considered to present a risk of endometrial cancer in women. An analysis of 815 women with a 48-month follow-up with regular smears of ectocervix, endocervix and endometrial aspirates failed to find conclusive evidence of a link between copper IUD exposure and cancer.[191] A similar lack of evidence was provided in a retrospective study of 405 women with endometrial cancer in the USA.[167] Notable observations indicate that risk did not vary with the number of years for which IUDs had been used or with years since first use, even though the devices did provoke endometritis.

Experimental studies in animals have failed to show an association between high copper levels and cancer in any tissue. However in one study, Fischer F344 rats given *N*-nitrosodiethylamine to induce liver tumours followed by high concentrations of copper gluconate in the diet (10, 300 or 6000 ppm) and partial hepatectomy showed that copper did promote cell proliferation and apoptosis.[198,199] Hepatic mRNA expression of genes relating to copper metabolism, inflammation and apoptois was significantly elevated suggesting that, in this medium-term assay, copper acts as tumour promoter, possibly acting through oxidative stress or by influencing inflammatory or pro-apoptotic signalling events.

References

1. J. Samachson, *Nature*, 1967, **216**, 193.
2. A. R. Johnson, *Calc. Tiss. Res.*, 1973, **11**, 215.
3. A. B. G. Lansdown, *Wound Rep. Regen.*, 2002, **10**, 271.
4. J. J. Mangano and J. D. Sherman, *Int. J. Health Serv.*, 2012, **42**, 47.
5. J. M. Gould, E. J. Sternglass, J. D. Sherman, J. Brown, W. McDonnell and J. J. Mangano, *Int. J. Health Serv.*, 2000, **30**, 515.
6. M. Nelson, *The EF-Hand Calcium Binding Proteins Library*, Structural Biology Program, Vanderbilt University, Nashville, TN, 1997.
7. C. W. Heizmann, G. Fritz and B. W. Schäfer, *Front. Biosci.*, 2002, **7**, 1356.
8. A. Lewit-Bentley and S. Réty, *Curr. Opin. Struct. Biol.*, 2000, **10**, 637.
9. K. Vieminickx and R. Kemler, *BioEssays*, 1999, **21**, 211.
10. H. Wu, R. Lotan, D. Menter, S. M. Lippman and X. C. Xu, *Anticancer Res.*, 2000, **20**, 1385.
11. Z. S. Agus, in *Metabolism of Vitamin D*, ed. M. K. Dresner and J. E. Mulder, Wolters Kluwer Health, Amsterdam, 2012.
12. FAO/WHO, *First World Congress on Calcium and Vitamin D in Human Life*, FAO and the World Health Organization (WHO) and the Italian National Institute of Nutrition, Rome, 1996.
13. C. S. Fullmer, *Neurotoxicol.*, 1992, **13**, 7926.
14. R. H. Wasserman, J. S. Chandler, S. Meyer, C. A. Smith, M. E. Brindik, C. S. Fullmer and R. Kumar, *J. Nutr.*, 1992, **122**, 662.
15. D. Pansu, C. Bellaton and C. Roshe, *Diabete. Metab.*, 1984, **10**, 106.
16. F. Bronner, *Miner. Electrol. Metabol.*, 1990, **16**, 94.
17. A. W. Norman, *Am. J. Clin. Nutr.*, 1990, **51**, 290.
18. W. D. Stein, *J. Nutr.*, 1992, **122**(Suppl.), 651.
19. T. Mauro, D. B. Dixon, I. Komuves, K. Hanley and P. A. Pappone, *J. Invest. Dermatol.*, 1997, **108**, 864.

20. X.-M. Xia, B. Fakler, A. Rivard, G. Wayman, T. Johnson-Pais, J. E. Keen, T. Ishil, B. Hirschberg, C. T. Bond, S. Lutsenko, J. Maylie and J. P. Adelman, *Nature*, 1998, **395**, 503.
21. D. E. C. Cole, V. D. Peltekova, L. A. Rubin, G. A. Hawker, C. C. Reinhold Vieth, D. M. Hwang, J. Evrovski and G. N. Hendy, *Lancet*, 1999, **353**, 112.
22. J. Singh, N. Moghal, S. H. S. Pearce and T. Cheetham, *Arch. Dis. Child.*, 2003, **88**, 403.
23. A. A. Fisher, *Contact Dermatitis*, Lea and Febiger, Philadelphia, PA, 1987.
24. A. D. Oxman, D. C. F. Muir, H. S. Shannon, S. R. Stock, E. Hnizdo and H. J. Lange, *Am. Rev Respir Dis.*, 1993, **148**, 38.
25. P. S. Burge, *Eur. Resp .J.*, 1994, **7**, 1032.
26. E. Hnizdo and V. Vallyathan, *Occup. Env. Med.*, 2003, **60**, 237.
27. F. Crummy, I. Carl, C. H. S. Cameron and L. G. Heaney, *Occup. Med.*, 2004, **54**, 497.
28. L. Flors, M. L. Domingo, C. Leiva-Salinas, M. Mazón, E. Roselló-Sastre and J. Vilar, *Am. J. Roentgen.*, 2010, **194**, W20.
29. J. A. Fairley, *Semin. Dermatol.*, 1991, **10**, 225.
30. J. A. Fairley, C. L. Marcelo, V. A. Hogan and J. J. Voorhees, *J. Invest. Dermatol.*, 1985, **84**, 195.
31. R. Rizzuto, P. Pinton, W. Carrington, F. S. Fay, K. E. Fogarty, L. M. Lifshitz, R. A. Tuft and T. Pozzan, *Science*, 1998, **289**, 1763.
32. R. Talmage, *Clin. Orthop.*, 1969, **67**, 210.
33. J. F. Whitfield, R. P. Bird, B. R. Chakravarthy, R. J. Isaacs and P. Morkey, *J. Cell. Biochem.*, 1995, **59**, 74.
34. A. B. G. Lansdown, *J. Anat.*, 1999, **195**, 375.
35. B. Clapa, D. Pesando, M. Wilding and M. Whitaker, *Nature*, 1994, **368**, 875.
36. O. P. Hornstein and Z. Hautkr, 1984, **59**, 1125.
37. D. Fraser, *Calcium Metabolism*, in *The Contribution of Nutrition to Human and Animal Health*, ed. E. M. Widdowson and J. C. Mathers, Cambridge Univ. Press, Cambridge, 1992, p. 141.
38. A. B. G. Lansdown, *CRC Crit. Rev. Toxicol.*, 1995, **25**, 397.
39. R. H. Kretsinger, *Ann. Rev. Biochem.*, 1976, **45**, 239.
40. M. Wilson, *The EF-Hand Calcium-Binding Proteins*, Data Library, Scripps Research Institute, La Jolla, CA, 1997.
41. M. Ruse, A. Lambert, N. Robinson, D. Ryan, K. J. Shon and R. L. Eckert, *Bochem.*, 2001, **40**, 3167.
42. N. Nagy, C. Brenner, N. Markadieu, C. Chaboteaux, I. Camby, B. W. Schäfer, P. Pochet, C. W. Heizmann, I. Salmon, R. Kiss and C. Decaestecker, *Lab. Invest.*, 2001, **81**, 599.
43. M. Nowotny, S. Bhattacharya, A. Filipek, A. Krezel, W. Chazin and J. Kuznicki, *J. Biol. Chem.*, 2000, **275**, 31178.
44. M. K. Heng, M. K. Moon and M. C. Y. Heng, *Br. J. Dermatol.*, 1993, **129**, 280.
45. A. W. Koch, D. Bozic, O. Pertz and J. Engel, *Struct. Biol.*, 1999, **9**, 275.

46. K. A. Knudson and A. P. Soler, *Methods Mol. Biol.*, 2000, **137**, 409.
47. F. Furukawa, K. Fijii, Y. Horiguchi, N. M. Matsuyoshi, N. Fujita, K. Toda, S. Imamura, H. Wakita, S. Shirahama and M. Takigawa, *Micros. Res. Technol.*, 1997, **38**, 343.
48. L. J. Van Eldik and D. M. Watterson, *Calmodulin and Signal Transduction*, Academic Press. New York, 1998.
49. C. D. DeMaria, T. W. Song, B. A. Alseikhan, R. S. Alvania and D. T. Yue, *Nature*, 2001, **411**, 484.
50. R. Kemler, *Semin. Cell Biol.*, 1992, **3**, 149.
51. H. Aberle, H. Schwartz and R. Kemler, *J. Cell. Biochem.*, 1996, **61**, 514.
52. M. Amagai, *J. Invest. Dermatol.*, 1995, **104**, 146.
53. S. H. Kee and P. M. Steinert, *Mol. Biol. Cell.*, 2001, **12**, 1983.
54. B. M. Gambiner and P. D. McCrea, *J. Cell Sci. Suppl.*, 1993, **17**, 155.
55. H. Shiozaki, H. Oka, M. Inoue, S. Tamura and M. Monden, *Cancer*, 1996, **77**, 1605.
56. G. Pershagen and N. E. Bjorklund, *Cancer Lett.*, 1985, **27**, 99.
57. IARC, *Monographs on the Evaluation of Carcinogenic Risk of Chemicals to Humans*, IARC, Lyons, vol. 23, 1980.
58. IARC, Arsenic, in *Overall Evaluations of Carcinogenicity. An Updating of IARC Monographs 1 to 42*, IARC, Lyons, 1987, Suppl. 7, p. 100.
59. S. MacNeil, R. A. Dawson, G. Crocker, W. F. G. Tucker, B. Bittiner, J. G. Singleton, T. Hunter and D. F. Tierneys, *Br. J. Dermatol.*, 1993, **128**, 143.
60. S. Ivankovic, G. Eisenbrand and R. Preussmann, *Int. J. Cancer*, 1979, **24**, 786.
61. A. Yamamoto, A. Hisanaga and N. Ishinishi, *Int. J. Cancer*, 1987, **40**, 220.
62. L. S. Levy, P. A. Martin and P. L Bidstrup, *Br. J. Ind. Med.*, 1986, **43**, 243.
63. A. S. Yap, W. M. Brieher and B. M. Gumbiner, *Ann. Rev. Cell Dev. Biol.*, 1997, **3**, 119.
64. A. P. Kowalczyk and B. A. Nanes, *Mol. Mech. Tiss. Dev. Dis, Subcell. Biochem.*, 2012, **60**, 197.
65. Y. Yonemura, Y. Endou, K. Kimura, S. Fushida, E. Bandou, K. Taniguchi, K. Kinoshita, I. Ninomiya, K. Sugiyama, C. W. Heizmann, B.W. Schafer and T. Sasaki, *Clin Cancer Res.*, 2000, **6**, 4234.
66. L. Lauriola, F. Michetti, N. Maggiano, J. Galli, G. Cadoni, B. W. Schäfer, C. W. Heizmann and F. O. Ranelletti, *Int. J. Cancer*, 2000, **89**, 345.
67. I. Ninomiya, T. Ohat, S. Fushida, Y. Endo, T. Hashimoto, M. Yagi, T. Fujimura, G. Nishimura, T. Tani, K. Shimizu, Y. Yonemura and C. W. Heizmann, *Int. J. Oncol.*, 2001, **18**, 715.
68. D. Liu, P. S. Rudland, D. R. Sibson, A. Platt-Higgins and R. Barraclough, *Br. J. Cancer*, 2000, **83**, 1473.
69. M. Nowotny, S. Bhattacharya, A. Filipek, A. M. Krezel, W. Chazin and J. Kuznicki, *J. Biol. Chem.*, 2000, **275**, 31178.
70. A. M. Al-Ani, A. G. Messenger, J. Lawry, S. S. Bleehen and S. MacNeil, *Br. J. Dermatol.*, 1988, **119**, 295.

71. Cancer Research UK, *Tamoxiphen, data sheet*, London.
72. M. S. Rogers and E. E. Strehler, *J. Biol. Chem.*, 2001, **276**, 12182.
73. R. B. Hazan, G. R. Phillips, R. F. Qiao, L. Norton and S. A. Aaronson, *J. Cell. Biol.*, 2000, **145**, 779.
74. N. S. MacDonald, L. Braden, E. James, R. A. Wood, *Proc. Soc. Exp. Biol. Med.*, 1967, **124**, 69.
75. J. T. Triffitt and A. Sutton, *Calc. Tiss. Res.*, 1969, **4**, 174.
76. M. E. J. Curzon and P. C. Spector, *J. Dental Res.*, 1980, **59**, 1988.
77. S. Ozgür, H. Sume and G. Kocoglu, *Arch. Dis. Child.*, 1996, **75**, 524.
78. C. Meyerowitz, P. C. Spector and M. E. Curzon, *Caries Res.*, 1979, **13**, 203.
79. T. M. Athanasssouli, D. S. Papostathopoulos and A. X. Astopoulos, *J. Dent. Res.*, 1983, **62**, 989.
80. M. E. J. Curzon, *J. Dental Res.*, 1985, **64**, 1386.
81. F. R. Mraz, *Poultry Sci.*, 1970, **69**, 310.
82. A. R. Doberenz, C. W. Weber and B. L. Reid, *Calc. Tiss. Res.*, 1969, **4**, 180.
83. C. R. Creger, L. B. Colvin and J. R. Crouch, *J. Exp. Biol. Med.*, 1967, **124**, 445.
84. C. R. Creger and L. R. Colvin, *Radiat. Res.*, 1967, **32**, 131.
85. W. J. Clarke, R. F. Palmer, E. B. Howard and P. L. Hackett, *Science*, 1970, **169**, 598.
86. B. Graf, J. Lafuma, C. Parmentier and N. Parmentier, *Bull. Cancer*, 1970, **57**, 381.
87. J. J. Mangano and J. D. Sherman, *Int. J. Health Serv.*, 2011, **41**, 137.
88. D. M. Czujka-Narins, in *Food, Nutrition and Die Therapy*, ed. L. K. Mahon and M. V. Kraus, Saunders, Philadelphia, 1979.
89. A. Sutton, H. Shepherd, G. E. Harrison and D. Barltrop, *Nature*, 1971, **230**, 396.
90. International Life Sciences Institute, *Nutr. Rev.*, 1959, **17**, 312.
91. F. E. McCaslin and J. M. Janes, *Proc. Staff Meetings Mayo Clin.*, 1959, **34**, 32.
92. P. J. Marie, S. C. Skoryna, R. J. Pivon, G. Chabot, F. H. Glorieux and J. F. Stara, *Trace Subst. Env. Health.*, 1985, **19**, 193.
93. J. D. Ringe, A. Dorst and P. Farahmand, *Arznein. Mittelforschung*, 2010, **60**, 267.
94. The Medicines Healthcare and Products Regulatory Agency (MHRA), *Drug Safety Update*, 2012, **5**, Issue 10, A3.
95. R. Rizzoli, J.-Y. Reginster, S. Boonen, G. Bréart, A. Diez-Perez, D. Felsenberg, J.-M. Kaufman, J. A. Kanis and C. Cooper, *Calc. Tiss. Int.*, 2011, **89**, 91.
96. A. B. G. Lansdown, *Experientia*, 1972, **28**, 558.
97. E. Storey, *Austral. Ann. Med.*, 1965, **47B**, 145.
98. F. P. Reinholt, B. Engfeldt, D. Heinegård and A. Hjerpe, *Coll. Relat. Res.*, 1985, **5**, 41.
99. S. G. Kshirsagar, E. Lloyd and A. Vaughn, *Br. J. Radiol.*, 1966, **39**, 131.

100. P. J. Marie, *Mol. Intervent.*, 2010, **10**, 305.
101. Agency for Toxic Substances and Disease Registry (ATSDR), *Toxicological Profile for Strontium*, Atlanta, GA, 2010.
102. S. Ghosh, G. Talukder and A. Sharma, *Biol. Trace Elem Res.*, 1990, **25**, 51.
103. K. Er, Z. A. Polat, T. Taşdemir, U. Sezer and S. H. Soso, *J. Formos. Med. Ass.*, 2008, **107**, 609.
104. E. Canalis, M. Hott, P. Deloffre, Y. Tsouderos and P. J. Marie, *Bone*, 1996, **18**, 517.
105. S. Choudhary, P. Halbout, C. Alander, L. Raisz and C. Pilbeam, *J. Bone Mineral. Res.*, 2007, **22**, 1002.
106. S. Peng, X. S. Liu, T. Wang, G. Zhou, K. D. Luk, X. E. Guo and W. W. Lu, *J. Orthop. Res.*, 2010, **28**, 1208.
107. F. Furukawa, J. C. Huff, M. B. Lyons, M. Weston and D. A. Norris, *J. Invest. Dermatol.*, 1988, **90**, 690.
108. H. Tateno and Y. Kamiguchi, *Mol. Repro. Dev.*, 1997, **47**, 72.
109. Servier Laboratories, *Protos® Strontium Ranelate Data Sheet*, Servier Laboratories, Victoria, Australia, 2012.
110. L. A. Loeb, M. A. Sirover, L. A. Weymouth, D. K. Dube, G. Seal, S. S. Agarwal and E. Katz, *J. Toxicol. Environ. Health*, 1977, **2**, 1297.
111. E. M. de Oliviera, M. F. Suzuki, P. A. do Nascimenta, M. A. da Silva and K. Okazaki, *Mutat. Res.*, 2001, **476**, 109.
112. A. B. G. Lansdown, in *General and Applied Toxicology*, ed. B. Ballantyne, T. C. Marrs and T. Sylversen, John Wiley, Chichester, 2009, vol. 3, p. 1491.
113. United Nations Scientific Committee on the Effects of Atomic Radiation (UNSCEAR), *Sources and Effects of Ionizing Radiation, Vol. 1: Sources.* United Nations, New York, 2000.
114. P. J. Baker and D. G. Hoel, *Eur. J. Cancer Care*, 2007, **16**, 355.
115. C. Spix, S. Schmiedel, P. Kaatsch, R. Schulze-Rath and M. Blettner, *Eur. J. Cancer*, 2007, **44**, 275.
116. J. J. Mangano and J. D. Sherman, *Int. J. Health Serv.*, 2011, **41**, 137.
117. US Congress, *Radiation Exposure Compensation Act Amendments*, Public Law, 2000, 105–245.
118. N. N. Litvinov and J. N. Soloviev, in *Pathology of Tumours in Laboratory Animals*, ed. V. Turasov and U. Mohr, IARC Publ., Lyon, 1990, vol. 1, p. 659.
119. A. Nilsson, *Strontium-90 Induced Bone and Bone Marrow Changes*, Almquist and Wikell, Uppsala, 1962.
120. A. Nilsson, *Acta Radiol., Therap., Phys. Biol.*, 1970, **13**, 107.
121. H. A. Sissons and J. Vaughn, *Nature*, 1960, **185**, 399.
122. S. Kshirsagar, J. Vaughan and M. Williamson, *Br. J. Cancer*, 1965, **19**, 777.
123. L. Paunier, *Monatsschr. Kinderheilkd.*, 1992, **140**(Suppl. 1), S17.
124. P. D. Mier and D. W. K. Cotton, *Br. J. Dermatol.*, 1970, **82**, 142.
125. W. E. C. Wacker and A. F. Paresi, *N. Engl. J. Med.*, 1968, **278**, 884.
126. A. H. Ryan, F. V. Guthrie and C. C. Guthrie, *Proc. Soc. Exp. Biol. Med.*, 1909, **7**, 239.

127. C. H. Peck and S. J. Meltzer, *J. Am. Med. Assoc.*, 1916, **67**, 1131.
128. R. Swaminathan, *Clin. Biochem. Rev.*, 2003, **24**, 47.
129. A. C. Hunt, *Thorax*, 1956, **11**, 287.
130. G. W. H. Schepers and T. M. Durkan, *Arch. Ind. Health*, 1955, **12**, 317.
131. K. Reijula, P. Pääkko, P. Kerttula, O. Taikina-Aho, T. Tuuponen and J. Hassi, *Br. J. Ind. Med.*, 1991, **48**, 140.
132. H. Ellis, *Ann. Roy. Coll. Surg., Engl.*, 1994, **76**, 5.
133. International Agency for Research on Cancer (IARC), *Monographs on the Evaluation of Carcinogenic Risks to Humans*, IARC, Lyon, 2010, vol. 93.
134. E. B. Dawson, D. R. Evans, R. Kelly and J. W. Van Hook, *Biol. Trace Elem. Res.*, 2000, **74**, 107.
135. T. Todorovic, D. Vujanovic, I. Dozic and A. Petkovic-Curcin, *Magnesium Res.*, 2008, **21**, 43.
136. J. T. McCarthy and R. Kumar, in *Atlas of Diseases of the Kidney*, ed. R. Schrier, Wiley-Blackwell, Philadelphia, PA, vol. 4, 1999.
137. L. Garfinkel and D. Garfinkel, *Magnesium*, 1985, **4**, 60.
138. W. E. C. Wacker, *Ann. N.Y. Acad. Sci.*, 1969, **162**, 717.
139. J. Takaya, H. Higashino and Y. Kobayashi, *Magnesium Res.*, 2004, **17**, 126.
140. F. I. Wolf, V. Trapani, M. Simonacci, A. Boninsegna, A. Mazur and J. A. M. Maier, *Nutr. Cancer*, 2009, **61**, 131.
141. J. R. Schelling, *Clin. Nephrol.*, 2000, **53**, 61.
142. J. Ferdinandus, J. A. Pederson and R. Whang, *Arch. Int. Med.*, 1981, **141**, 669.
143. J. J. Grzesiak and M. D. Pierschbacher, *J. Clin. Toxicol.*, 1995, **95**, 22.
144. S. Banai, L. Haggroth, S. E. Epstein and W. Casscells, *Circulation Res.*, 1990, **67**, 645.
145. G.-C. Chen, Z. Pang and Q.-F. Liu, *Eur. J. Clin. Nutr.*, 2012, **66**, 1182.
146. E. D'Erasmo, F. S. Celi, M. Acca, S. Minisola, G. Aliberti and G. F. Mazuoli, *Biomed. Pharmacother.*, 1991, **45**, 315.
147. H. Lindsey, *Oncol. Times*, 2007, **29**, 21.
148. P. A. Wark, R. Lau, T. Norat and E. Kampman, *Am. J. Clin. Nutr.*, 2012, **96**, 622.
149. E. Ma, S. Sasazuki, M. Inoue, M. Iwasaki, N. Sawada, R. Takachi and S. Tsugane, *J. Nutr.*, 2010, **140**, 779.
150. Y. Kesavan, *Am. J. Epidemiol.*, 2010, **171**, 233.
151. R. Davies, J. W. Skidmore, D. M. Griffiths and C. B. Moncrieff, *Fd. Chem. Toxicol.*, 1983, **21**, 201.
152. P. Wild, *Occup. Envir. Med.*, 2006, **63**, 4.
153. Y. Honda, C. Beall, E. Delzell, K. Oestenstadt, I. Brill and R. Matthews, *Ann. Occup. Med.*, 2002, **46**, 575.
154. M. Coggiola, D. Bosio, E. Pira, P. G. Piolatto, C. La Vecchia, E. Ngri, M. Michelazzi and A. Bacaloni, *Am. J. Ind. Med.*, 2003, **44**, 63.
155. National Toxicology Program (NTP), *Natl. Toxicol. Program Tech. Rep. Ser.* 1993, **421**, 1.

156. J. A. Pickrell, M. B. Snipes, J. M. Benson, R. L. Hanson, R. K. Jones, R. L. Carpenter, J. J. Thompson, C. H. Hobbs and S.C. Brown, *Environ. Res.*, 1989, **49**, 233.
157. National Cancer Institute (NCI), *Ovarian Cancer Screening (PDQ®)*, National Institutes of Health, Bethesda, MD, 2012.
158. D. M. Gertig, D. J. Hunter, D. W. Cramer, G. A. Colditz, F. E. Speizer, W. C. Willett and S. E. Hankinson, *J. Natl. Cancer Inst.*, 2000, **92**, 249.
159. S. Karageorgi, M. A. Gates, S. E. Hankinson and D. V. Immaculata, *Canc. Epidemiol. Biomarkers Prev.*, 2010, **19**, 1269.
160. IARC, *Monographs on the Evaluation of Carcinogenic Risks to Humans*, IARC, Lyon, 2006, vol. 93.
161. N. Horn, T. Tønessen and Z. Turner, *Brain Pathol.*, 1992, **2**, 351.
162. G. J. Brewer, V. Yuzbasiyan-Gurkan, V. Johnson, R. D. Dick and Y. Wang, *J. Am. Coll. Nutr.*, 1993, **12**, 26.
163. I. Bremner and J. H. Beattie, *Proc. Nutr. Soc.*, 1995, **54**, 489.
164. A. Cordano, *Am. J. Clin. Nutr.*, 1998, **67**(Suppl. 5), 1012S.
165. A. Ala, A. P. Walker, K. Ashkan, J. S. Dooley and M. L. Shilsky, *Lancet*, 2007, **369**, 397.
166. R. F. Pfeiffer, *Seminar Neurol.*, 2007, **27**, 123.
167. S. R. Sturgeon, L. A. Brinton, M. L. Beerman, R. Mortel, L. B. Twiggs, R. J. Barrett, G. D. Wilbanks and J. R. Lurain, *Int. J. Epidemiol.*, 1997, **26**, 496.
168. E. B. Hart, H. Steenbock, J. Waddell and C. A. Elvehjem, *J. Biol. Chem.*, 1928, **77**, 797.
169. B. L. O'Dell, *Philos. Trans. Roy. Soc. Lond.*, 1981, **294**, 91.
170. A. Cordano, J. M. Baertil and C. G. Graham, *Blood*, 1964, **28**, 280.
171. M. Schaefer and J. D. Gitlin, *Am. J. Physiol*, 1999, **276**, G311.
172. I. Sternlieb, *Gastroenterol.*, 1980, **78**, 1615.
173. P. Günther, W. Hermann, H. J. Kühn and A. Wagner, *Trer. Umsch.*, 2007, **64**, 57.
174. F. S. Walker, C. Goho and R. B. Schanzer, *ASDC J Dent Child.*, 2002, **69**, 70/12.
175. J. Camakaris, I. Voskoboinik and J. F. Mercer, *Biochem. Biophys. Res. Comm.*, 1999, **261**, 255.
176. C. H. Gallagher, J. D. Judah and K. R. Rees, *Proc. Roy. Soc. B*, 1956, **144**, 134.
177. L. S. Hurley and L. T. Bell, *Proc. Soc. Exp. Biol. Med.*, 1975, **149**, 830.
178. J. Ma and N. M. Betts, *J. Nutr.*, 2000, **130**, 2838.
179. E. D. Harris, *Nutr. Rev.*, 2001, **59**, 281.
180. T. A. Hysong, J. L. Burgess, M. E. Cebrián Garcia and M. K. O'Rourke, *J. Exp. Anal. Environ. Epidemiol.*, 2003, **13**, 211.
181. P. Richtera, R. Seguel, I. Ahumada, R. Verdugo, J. Narváez and Y. Shibatac, *Chil. Chem. Soc.*, 2004, **49**, 333.
182. M. Wilhelm, F. K. Ohnesorge, L. Lombeck and D. Hafner, *J. Anal. Toxicol.*, 1989, **13**, 17.

183. M. Wilhelm, D. Hafner, I. Lombeck and F. K. Ohnesorge, *Sci. Total Environ.*, 1991, **103**.
184. A. B. G. Lansdown, Metal ions affecting the skin and eyes, in *Metal Ions Life Sciences*, ed. A. Sigel, H. Sigel and R. K. O. Sigel, 2011, vol. 8, p. 187.
185. J. Hostynek and H. I. Maibach, *Toxicol. Mech. Method.*, 2006, **16**, 245.
186. T. I. Mzhel'skaya, *Bull. Exp. Biol. Med.*, 2000, **130**, 719.
187. N. E. Hellman and J. D. Gitlin, *Ann. Rev. Nutr.*, 2002, **22**, 439.
188. T. Kononen, J. Y. Urin-Hare, M. S. Clegg, C. L. Keen and R. B. Rucker, *Biochem. J.*, 1997, **327**, 283.
189. H. M. Kagan and W. Li, *J. Cell. Biochem.*, 2003, **88**, 660.
190. J. R. Turnlund, R. A. Jacob, C. L. Kewen, J. J. Strain, D. S. Kelley, J. M. Domek, W. R. Keyes, J. L. Ensunsa, J. Lykkesfeldt and J. Coulter, *Am. J. Clin. Nutr.*, 2004, **79**, 1037.
191. U. K. Luthra, A. B. Mitra, A. K. Prabhakar, P. Bhatnagar and S. S. Agarwal, *Ind. J. Med. Res.*, 1980, **72**, 659.
192. A. Lee-Feldstein, *J. Natl. Cancer Inst.*, 1983, **70**, 601.
193. J. H. Lubin, L. E. Moore, J. F. Fraumeni and K. P. Cantor, *Environ Health Perspect.*, 2008, **116**, 1661.
194. N. E. Lightfoot, M. A. Pacey and S. Darling, *Chronic Dis. Canada*, 2010, **29**(Suppl. 2), 101.
195. M. Hrgovcic, C. F. Tessmer, T. M. Mingkler, B. Mosier and G. H. Taylor, *Cancer*, 1968, **21**, 743.
196. G. Demir, H. Demir, R. Esen, A. Sehitogullari, M. Atmaca and M. Alay, *Asian Pacific J. Cancer Prev.*, 2011, **12**, 347.
197. K. Overvad, D. Y. Wang, J. Olsen, D. S. Allen, E. B. Thornling, R. D. Bulbrook and J. L. Hayward, *Am. J. Epidemiol.*, 1993, **137**, 409.
198. M. Abe, N. Suzuki, M. Yoshida, K. Usida, S. Furukawa, L. R. Juneja, T. Okubo and D. Nakai, *Fd. Chem. Toxicol.*, 2008, **46**, 1760.
199. M. Abe, K. Usuda, S. Hayashi, I. Ogawa, S. Furukawa, M. Igarashi and D. Nakae, *Arch. Toxicol.*, 2008, **82**, 563.

CHAPTER 7

Minor Trace Elements: Manganese, Vanadium, Molybdenum, Tin

7.1 General Introduction

These four elements are recognised as micro-trace metals in the human body and fulfil criteria set by Underwood and others as being essential for normal growth, maturation and adult health, as co-factors in enzymes or components of intercellular matrices.[1] The minerals are normally found in the body in ionic form at concentrations of less than 0.02 g. All are absorbed from food but mechanisms of absorption and daily requirements are unknown. Presumed dietary intake is shown in Table 7.1. The four elements are dissimilar chemically and metabolic pathways are different. They are transition metals and biologically active. Although none of them is identified as a suspected or proven human carcinogen in its ionic state, certain compounds or related products have been a cause for concern.

By definition, macro- and micronutrients should exhibit defined roles in normal human physiology, biochemical or biosynthetic processes, or structural components of cells or intercellular matrices, and in states of deficiency the body exhibits tell-tale signs of failure to thrive.[1–4] Only limited evidence is presently available for the importance of tin in human nutrition whereas manganese, vanadium and molybdenum are recognised as components or co-factors of essential enzyme systems even though patterns of uptake, homeostatic regulation and metabolism are imperfectly understood. Manganese and possibly vanadium are linked with regulation of glucose metabolism, insulin binding and prevention of diabetes but their sites and modes of action are unclear.[5,6] Both elements are presumed to act as enzyme co-factors.

Issues in Toxicology No. 18
The Carcinogenicity of Metals: Human Risk through Occupational and Environmental Exposure
By Alan B. G. Lansdown

Published by the Royal Society of Chemistry, www.rsc.org

Table 7.1 Dietary requirements for trace elements.

Metal	*Concentration in diet (mg per day)*	*Dietary source*	*Concentration in human body*
Manganese	1–8	Tea, cereals, vegetables	10–20 mg
Molybdenum	0.5–2	Green vegetables	0.009
Tin	3.5–17	Vegetables and meat (especially canned)	0.17
Vanadium	0.000025	Grain, seeds, meat, seafood	0.18

Manganese-related enzymes include xanthenes and aldehyde oxidases, phosphohydrolases and phospho-transferases in proteoglycan and mucopolysaccharide synthesis and bone formation. Vanadium-related enzymes may be important in energy metabolism. As with better known metal nutrients such as zinc, calcium, iron and magnesium, it is expected that dietary requirements for molybdenum, manganese, vanadium and tin will vary according to age, sex and physiological state. Afridi *et al.* have shown that concentrations of manganese, chromium and zinc are lower in diabetic women and their infants, suggesting a role for these nutrients in the pathogenesis of aberrant carbohydrate metabolism.[7] Animal models including rodents and pigs have provided some information relating to dietary deficiency and the toxicity of excesses, but the reliability of extrapolation to humans is questioned.[3,8]

7.2 Manganese

Manganese is a hard, brittle light grey coloured metal of Group VIIb of the Periodic Table. It is the 12th most abundant element in the Earth's crust and has 23 known isotopes, of which only ^{55}Mn is stable. Manganese is an essential element in photosynthesis and, without it, there would be no free oxygen on Earth. Like other transition metals, manganese exhibits several oxidation states but Mn(II) is most stable. Other oxidation states include Mn(IV) as in manganese dioxide, Mn(VII) as found in the purple permanganate anion MnO_4^-, and Mn(VI) which exists in the form of a green oxidation state. Manganese is a biologically active metal and readily forms strong ferromagnetic alloys with many industrial uses. Principal sources of manganese include the minerals pyrolusite (MnO_2) romanechite (psilomelane), hausmannite and rhodochrosite ($MnCO_3$). Manganite is an opaque mineral of medium hardness, ranging in colour from steel grey to iron black and having a dark, reddish-brown streak. Manganese is found also in manganese nodules on the ocean floor. Commercially viable manganese deposits occur mainly in Australia, Brazil, China, Gabon, Ghana, India, South Africa, Scandinavia and Ukraine; some deposits contain chromium, silver and other metals.

Manganese is recognised as an essential nutrient with physiological and biochemical roles in glucose metabolism, insulin regulation and in synthesis of mucopolysaccharides, bone development and otolith formation.[9–11]

7.2.1 Industrial Uses and Human Exposure

Manganese is a hard metal and estimates suggest that at least 95% of annual production is used in hard metal alloys and steel production. Advantage is taken in steel production of the ability of manganese to fix sulfur, its strong oxidising capacity and alloying properties. Manganese imparts resilience, corrosion resistance and hardness to alloys for use in heavy industry, tool making and construction. Alloys with manganese include, iron, nickel, copper, aluminium and silicon, *e.g.* ferromanganese, silicomanganese. Manganese dioxide is used in dry-cell batteries. Other uses relate to the strong oxidising properties of manganese dioxide which finds extensive use in analytical chemistry and as an antiseptic in medical practice, purification of drinking water and in treating waste water. Several new manganese compounds have been introduced recently as components of fungicides (*e.g.* Maneb), antiknock agents in petroleum and as contrast agents in nuclear magnetic resonance tomography.[12] The distinctive colouration of manganese salts, pink-reds, greens, *etc.*, has for many centuries been used in glass production and ceramics. Industrial exposure to manganese is a known cause of manganic pneumonia and manganism. These diseases mainly concern workers occupied in manganese ore mills, smelting works, battery factories and manganese mines. Manganic pneumonia is a form of lobar pneumonia that is unresponsive to antibiotic treatment. Industrial exposures to manganese residues in mines and surrounding areas of Mexico, Canada and Germany have pointed to neurological symptoms as a principal cause for concern.[13–15] Whilst the route of uptake and amount of free ion absorbed into the circulation is not known, it is evident that in the occupational environment workers are exposed to 2–3 times higher levels of manganese than in urban areas and that blood levels reach 88 $\mu g\ l^{-1}$.

The commonest exposure to inorganic manganese salts is through food and drinking water. Manganese absorption is low and greatly influenced by the dietary content of iron and zinc and competitive binding of receptors on the intestinal mucosa.[16] High manganese inhibits iron absorption, whereas a five-fold excess of Zn : Fe reduced iron uptake by 56% in human volunteers. Manganese uptake is higher in young children, in keeping with the role of the Mn(II) ion in glucose metabolism, cartilage and bone development and lipid metabolism.

Other routes of exposure to manganese in industry and in use of manganese-related products in hygiene include dermal exposure and inhalation. The latter route is of importance in mining, steel production and in processing facilities. Limited available evidence suggests that even under sustained heavy industrial exposures and close contact with manganese-containing compounds, percutaneous absorption is very low with much being bound by epidermal keratins and hair.[17] It is notable that skin, sweat and hair are principal routes of excretion of manganese. Exposure to strongly oxidising solutions of permanganate is liable to cause corrosive damage to skin and eyes. Rodier surveyed risks associated with exposure to manganese in Moroccan mines in 1955 and reported neurological symptoms, muscular weakness, paraplegia and pneumonopathies in 150 miners as occupational risks.[18,19] Whilst patterns of manganese absorption following inhalation are not well documented, there is increasing evidence that exposure to the metal in mining at least can be a cause of neurological diseases including Parkinson-like symptoms.[20–22] Experimental studies in rats have shown that intravenous injections of manganese did lead to increased brain concentration and that this was accentuated in iron deficiency.[23]

7.2.2 Mutagenicity and Carcinogenicity

The mutagenicity and carcinogenicity of manganese are incompletely known. No clinical or experimental evidence is seen to implicate manganese as an animal or human carcinogen although manganese chloride ($MnCl_2$) is cytotoxic, inhibits cells in mitosis and induces chromosomal damage without polyploidy in human lymphocyte and A549 cell cultures at high concentrations.[12,24,25] It arrests cells in a time–dose-related manner and leads to apoptosis. G_0/G_1 cells obtained by serum starvation and exposed to $MnCl_2$ showed a disruption of cell cycle progression with mitotic arrest in G_0/G_1 followed by block in the S phase, reflecting a decrease in cyclin-dependent kinases. Absence of polyploidy in mitosis suggests that manganese does not influence formation of the mitotic spindle. Mn(II) may be weakly mutagenic and possibly clastogenic *in vivo* but its genotoxicity in humans is not known.[26]

Case studies conducted in welders exposed to high manganese emissions indicate that workers may be subject to neurological and respiratory problems dependent upon the size and concentration of the particles in the atmosphere.[15,26,27] At manganese concentrations well above recommended threshold limit values recommended by the American Conference for Governmental Occupational Hygienists (0.24 mg m^{-3}),[28] none of the 10 workers assessed exhibited evidence of neoplastic change in any tissue. Other studies in Germany have collated manganese exposure in welders and respirable concentrations but failed to report lung or other cancers or increased cancer mortality.[27,29] Environmental toxicity due to manganese exposure in 150 Moroccan miners showed that risks of neurological damage and pneumonopathies far outweighed any potential carcinogenic risk associated with exposure to manganese.[18] Hobbesland examined cancer incidence among 6363 male workers in Norwegian ferromanganese and silicomanganese production plants and concluded that furnace and non-furnace workers may have exposures that increase the incidence of several cancers, but a causal relationship between manganese exposure and cancer was not established in humans.[30]

Experimental studies paint a confused picture of manganese as a mutagen or carcinogen. Relatively high concentrations influence DNA replication and repair, but Ames tests in bacteria were largely negative.[12] Four chronic toxicity studies in experimental animals have failed to show the carcinogenic potential of inorganic manganese compounds. Suspensions of manganese dioxide or manganese powder in trioctanoin were not carcinogenic when injected intramuscularly in Fischer F344 rats or Swiss albino mice, although rats injected with the organomanganese compound manganese acetylacetonate did show a statistically significant increase in injection site fibrosarcomas.[31] Witschi *et al.* injected female A/J mice intraperitoneally with 80 mg kg^{-1} methylcyclopentadienyl manganese tricarbonyl and reported pulmonary cell proliferation but no increase in tumour incidence.[32] Two other chronic studies in mice injected subcutaneously or intraperitoneally with manganese chloride for up to 6 months showed an increased number of lymphosarcomas and pulmonary adenomas but the increase in pulmonary tumours was only significant at the high dose of 30 mg kg^{-1} thrice weekly for 7 weeks.[33,34] Observation of high numbers of tumours in control animals minimises the evidence that manganese compounds are carcinogenic.

It should be concluded that available evidence from clinical and case studies in humans and experimental studies in animals fails to show that inorganic manganese compounds present a carcinogenic risk. Experimental evidence showing that several organic derivatives of manganese are associated with carcinogenicity in some organs is not supported by evidence in humans. Further clinical study is required to show the contribution if any played by manganese in the production of bone sarcomas at the site of prosthesis implantation. An old Russian study refers to the unusual ability of human and rabbit bone tumours to concentrate manganese and copper.[35] There is no evidence to show that manganese induces neoplastic change in bone, but it is known that manganese-related enzymes are important in chondrogenesis and early ossification of long bones and may be involved in bone physiology and in alkaline phosphatase activity.[36,37]

7.3 Vanadium

Vanadium is a transitional metal in Group V of the Periodic Table. It is a reactive metal within a variety of brightly coloured compounds. Vanadium is found in many parts of the world, its principal minerals being carnotite ($K_2O \cdot {}_2UO_3 \cdot V_2O_5 \cdot 3H_2O$), roscoelite [$K(V_3,Al,Mg)_2AlSi_3O_{10}(OH)_2$], vanadinite [$Pb_5(VO_4)Cl_3$], mottramite [$(Cu \cdot Zn)PbVO_4 \cdot OH$] and patronite ($V_5S_4$). Vanadium has two stable isotopes, ^{50}V and ^{51}V, and eight radioactive isotopes. Vanadium exhibits four main oxidation states, of which the V(II) and V(III) salts have similar properties to iron, and are ionic. Vanadium oxides of V(IV) and V(V) oxidation states are amphoteric; vanadium pentoxide (VPO) is more acidic than basic. Vanadium salts are toxic and levels of risk relate to the chemical formulation of the compound, its solubility, oxidation state, organic *vs.* inorganic form, and the route and duration of exposure.[38] VPO is probably the most toxic of the inorganic vanadium compounds investigated.

Principal mining areas for vanadium include South Africa, the USA, Mexico, Canada, Russia and China. In 2007, the last three countries produced more than 95% of the 58,600 tonnes of the world output of vanadium. Vanadium mining is a risky profession, on account of the uranium, lead and other toxic metals found in mineral deposits. Uranium has to be separated from the vanadium residues in processing. A bulletin from South Africa recorded that "the impacts of vanadium mining are devastating the miners and their families who suffer from kidney and organ failures, a wide array of pulmonary diseases and ailments and ultimately death".[39]

7.3.1 Industrial Use and Human Exposure

A large proportion of the vanadium (80–90%) produced annually is used in production of hard metal alloys with iron, copper, aluminium, nickel, cobalt and molybdenum. The most important use of vanadium is as an alloying element in the steel industry where it is added to produce grain refinement and hardening in steels. Vanadium is used in a range of ferrous and non-ferrous alloys with specific roles in automobile manufacture, die castings, ball bearings, foil, jet engines and the aircraft industry. Vanadium–titanium alloy (6% Al : 4% V : 90% Ti) is important in supersonic aircraft where strength-to-weight ratio is a primary consideration. Ferrovanadium is a well known alloy. Other uses of vanadium alloys include electrical appliances, semiconductors, batteries, pigment production, mercury safety lamps and resistors. The strong oxidising properties of vanadium pentoxide make it useful in chemical analyses and as catalysts.

Human exposure to vanadium compounds occurs through diet and drinking water, dermal contact and inhalation of dust in mining, extraction and manufacturing, and exposure to atmospheric particles. Fuel and heating oil, lubricants and asphalt also contain significant quantities of vanadium. Natural sources of atmospheric vanadium include continental dust, marine aerosol, industrial effluents and volcanic emissions. The concentrations of vanadium in workplace air (0.01–60 mg m^{-3}) are much higher than those in the general environment. Vanadium in drinking water is generally low, at 0.001 mg l^{-1}, but this can be expected to rise greatly in areas close to mining and steel production plants. A study of natural background levels of airborne vanadium in Canada in 1974 recorded concentrations in rural areas ranging from a few nanograms to tenths of a nanogram per m^3, and in urban areas from 50 ng m^{-3} to 200 ng m^{-3}, but appreciably higher levels of up to 2000 ng m^{-3} were recorded in cities during the winter, when fuel oil with a high vanadium content was used for heating.[40] Air pollution by industrial plants may be less than that caused by power stations and heating equipment. The concentrations of vanadium in workplace air (0.01–60 mg m^{-3}) are much higher than those in the general environment. Kucera *et al.*[41] examined the uptake and excretion of vanadium from factory dust and environmental vanadium pentoxide gas and reported that exposures to vanadium led to urine concentrations of 3.03–762 ng ml^{-1} (mean 33.0 ng ml^{-1}) compared with median control levels of 2.53 ng ml^{-1}.

Fragmentary information suggests that absorption of vanadium compounds from the intestine is as low as possibly 6–10% of that consumed; less than 1.0% metallic vanadium is absorbed. Percutaneous absorption of vanadium and its inorganic salts is also minimal because free vanadium ions actively bind strongly to epidermal keratin. The greatest intake of vanadium occurs following inhalation. Absorption depends on particle size, the solubility of the compound and respiratory profiles, but available studies suggest that 25% of soluble inorganic vanadium compounds inhaled may be absorbed. VPO is absorbed readily through the pulmonary epithelium.[42] Vanadium bound to transferrin or other proteins is distributed throughout the body but much is eliminated rapidly in urine. Inhalation of VPO is a major industrial risk and clinical and experimental studies have demonstrated its ability to evoke pulmonary and bronchial inflammation, oedema, and systemic effects through action on mitochondrial enzymes and disruption of cell membrane function, but not carcinogenicity.[41] A study of 52 workers in a vanadium plant showed that inhalation of vanadium pentoxide led to a seven-fold higher blood vanadium concentration (31 $\mu g\ l^{-1}$) than in corresponding controls and DNA damage in peripheral leucocyte preparations, micronuclei, nucleoplasmic bridges and nuclear buds as an indication of its clastogenic effect.[43]

7.3.2 Vanadium as a Nutrient

Although vanadium is now recognised as a minor trace element much speculation exists concerning its physiological and biochemical role in the human body. Uptake from the intestine is low and possibly occurs in a similar manner to iron, and involving iron-related carrier proteins such as transferrin. In the circulation, vanadium is largely present in the form of vanadate [V(v)] bound to plasma transferrin or albumins. Vanadium is absorbed intracellularly by an anion transport system and then reduced to the V(IV) form via a non-enzymatic reaction.[44] Most intracellular vanadium becomes bound to proteins such as ADP, nicotine adenine dinucleotide (NAD), glutathione, ribosides, haemoglobin, catecholamine, transferrin and glucose-related enzymes, with less than 1% remaining in a free form. There is increasing evidence from experimental studies in rats and *in vitro* that vanadium acts to lower blood glucose levels and to enhance or mimic the action of insulin in Type I and Type II diabetes. The mechanism is not known but it is thought that vanadium suppresses glucose production through inhibition of gluconeogenic enzymes or through action on signalling pathways.[6]

Information gained from dietary deficiency studies in chicks and rats indicate that vanadium has a role in lipid metabolism, bone growth and as a catalyst in oxidation–reduction reactions.[2] Rats subject to vanadium deficiency exhibited reduced growth patterns, lowered plasma cholesterol and raised triglycerides. Other views suggest that the metal is involved in regulating the Na–K ATPase or sodium pump process and acts as a diuretic agent.

7.3.3 Cytotoxicity and Mutagenicity

Cytological evidence indicates that vanadium is biologically active in the intracellular compartment and binds strongly to a range of intracellular proteins, cell membranes and mitochondria in key tissues such as the liver. It is also cytotoxic through its ability to evoke production of reactive oxygen species.[45] Suggestions have been made that it may act as an anticancer agent but this awaits verification.[46] Studies in rat hepatocyte cultures have shown that vanadium exposure leads to a sequence of degenerative changes reflecting mitochondrial and lysosomal membrane damage, cytochrome *c* release, oxidation and depletion of glutathione and oxidative stress.

Léonard and Gerber reviewed evidence for the cytotoxicity of vanadium and its compounds and concluded that V(V) is not clastogenic and only weakly mutagenic but has the capacity to evoke mitogenic change, chromosomal change and aneuploidy in cultured cells.[47] Evidence is fragmentary and although there seems to be a lack of evidence that vanadium and its compounds, including vanadium pentoxide gas, are mutagenic in Ames *Salmonella typhimurium* reverse transcription assay,[48] several studies allude to the genotoxicity of soluble vanadium compounds, and their ability to induce "mitotic gene conversion" and point reverse mutations, aneuploidy, sister-chromatid exchanges and micronuclei has been demonstrated in yeast and cultured Chinese hamster cells.[49–51] Experiments in a mouse cell line emphasised that the cytotoxicity and clastogenicity of vanadium are related to its oxidation state.[52] Pentavalent vanadium compounds are reduced *in vivo* to the less toxic V(IV) through the action of intracellular glutathione (GSH). Fifty-two workers in a VPO factory showed evidence of chromosomal damage attributable to the strong oxidation of DNA bases, without clinical evidence of VPO-related pulmonary neoplasia.[53] Further epidemiological evidence is necessary to confirm this observation.

7.3.4 Carcinogenicity

There is no satisfactory evidence to show that exposure to vanadium or its compounds is carcinogenic in humans. I agree entirely with Professor Duffus, who criticised the classification of VPO as a possible human carcinogen on the basis of experimental studies which showed inherent weaknesses in experimental design and procedure.[54] Extension of the conclusions on the basis of two studies on VPO to cover all vanadium compounds is viewed as irresponsible. A retrospective case study of 196 men working in a South African vanadium mining operation provided no indications of a link between exposure to vanadium pentoxide and any form of cancer.[55] A larger survey of vanadium miners in Montrose County in Colorado also failed to show an increase in cancer incidence in 1877 workers compared to the local population.[56] The authors concluded that many industrially related cancers could have resulted from prior exposure to other occupational carcinogens including radon and cigarette smoking. Underground miners extracting vanadium and other metals such as gold, copper, *etc.* are exposed to arsenic, chromium, nickel and radioactive isotopes of uranium, and presumably their radioactive breakdown products.

The American Conference of Governmental Industrial Hygienists (ACGIH) assigned respirable VPO a threshold limit value (TLV) of 0.05 mg m^{-3} as an 8-hour time-weighted average.[57] In a study conducted for the National Toxicology Program, F344/N rats and B6C3Fi mice were exposed to VPO at concentrations greatly exceeding ACGIH recommended levels.[58] At concentrations of 0.5, 1 or 2 mg m^{-3} (rats) or 1, 2 or 4 mg m^{-3} (mice) in inhalation chambers for 2 years, animals exhibited profound respiratory stress and reduction in body weight gain but lung tumour incidence in male and female rats was marginally higher than in control groups. Lung tumours were significantly higher in mice of both sexes, but control male mice showed a 44% incidence of pulmonary adenoma/carcinoma. All VPO-treated animals exhibited a dose-related respiratory pathology indicating the irritancy of the compound. A more recent study conducted in four strains of mice treated with the tumour initiator 3-methylcholanthrene followed by 5 weekly aspirations of VPO (4 mg kg^{-1}), demonstrated that VPO can promote development of lung tumours induced by a standard carcinogen, but that the action is strain-specific.[59] Lung tumour incidence was significantly increased in three of the four strains after 20 weeks but not in a C57BL/6J (B6) strain. Tumour incidence correlated closely with levels of pulmonary irritancy, release of inflammatory cytokines, upregulation of transcription factor activity, and presumably hyperpermeability of pulmonary membranes.

Major flaws in the design of both experimental studies and lack of confirmation of observations make validation of the carcinogencity of VPO on the basis of two animal experiments extremely difficult.[57,60] Assignment of VPO to a non-carcinogenic risk category should be viewed with caution, however, in the light of *in vitro* evidence that it can generate reactive oxygen species, and chromosomal and mitotic changes leading to aneuploidy in cultured human cells. VPO is viewed as being weakly mutagenic and not clastogenic, but there is a potential risk of induction of pulmonary tumours as a consequence of its action in inducing "DNA instability" and related changes.[47,53,59]

7.4 Molybdenum

Molybdenum is a hard silvery white transitional metal in Group VI of the Periodic system with an atomic configuration close to those of chromium and tungsten. It is biologically very different from those metals and exists naturally in the form of seven stable isotopes and numerous radionuclides with half lives ranging from a few milliseconds to several years. Natural deposits comprising 1.1 ppm of the Earth's crust include molybdenite (MoS_2), wulfenite ($PbMoO_4$) and powellite ($CaMoO_4$). Molybdenum is also obtained as a by-product of copper mining. Principal mining areas are in the USA, Canada, Russia, Mongolia, Peru, Chile and China.

Molybdenum exhibits six oxidation states but Mo(VI) is the most stable and of greatest biological importance. Molybdenum forms a large range of compounds including several complex organic salts, few of which have been fully evaluated biologically and toxicologically. Molybdenum is a minor trace element in the human body with presumed function in a least six enzyme systems as a co-factor.[61–63] Since the body requirements are normally low, molybdenum deficiency due to dietary deficiency or inherited defect is rare. Defective molybdenum metabolism occurs as a result of a rare autosomal condition arising from abnormalities in the expression of the genes *MoCS1*,

MoCS2 and *MoSC3*.[64,65] Affected children die of neonatal seizures and neurological symptoms due to an inability to synthesise molybdenum co-factor (MoCO) and deficiencies in the key enzymes sulfite oxidase, xanthene oxidase and aldehyde oxidase.

Molybdenum and its inorganic compounds exhibit a low order of toxicity in the human body and are not presently classified by National Institutes of Health or IARC as suspected or proven human carcinogens. Studies in cattle feeding on molybdenum-rich pastures have shown that ruminants are especially prone to developing diarrhoea or scours, anaemia, joint diseases and failure to thrive, attributable to impaired copper metabolism.[66] Copper and molybdenum interact in the body and, in cattle, high intakes of molybdenum increase the protein-bound copper in liver and induce copper deficiency syndrome. Anaemia, reduced red blood cell counts and joint deformities are characteristic of molybdenum toxicity in animals, whereas several older Russian studies show that factory workers exposed to high levels of molybdenum dust developed high serum caeruloplasmin (copper-binding protein), joint pains, skin reactions, reduced fertility, dermatological and immunological changes characterised by defects in T-cell function.[67,68]

7.4.1 Molybdenum as a Nutrient

Molybdenum requirements for normal human health are low at 200–500 μg per day. Population and genetic studies have established that molybdenum is essential as a co-factor in several oxidases, notably xanthene oxidase (and dehydrogenase), aldehyde oxidase and sulphite oxidase, with functions in the transformation of xanthene to uric acid, detoxification of aldehydes, pyrimidines, pteridines and related compounds, and catalysis of the transformation of sulfite to sulfate respectively.[61,62] Sulfite oxidase is a mitochondrial enzyme. Diets low in molybdenum are associated with poor growth and survival in infants. Molybdenum deficiency in childhood has been associated with cardiac abnormalities, neurological symptoms, night-blindness and coma.[61] Johnson studied inborn errors of molybdate metabolism in infants, and associated deficiencies in sulfite oxidase and xanthene dehydrogenase and oxidoreductases with severe neurological abnormality, dislocated ocular lenses and mental retardation and associated the conditions with the level of the molybdenum cofactor in the enzymes.[63]

Homeostatic mechanisms for molybdenum in the human body are not known but like that of many other metals, gastrointestinal absorption is probably determined by plasma concentrations, metal-binding proteins and tissue requirements. Iron and copper interact with molybdenum metabolism and Cu : Mo ratios are important in haemopoiesis and prevention of iron-deficiency anaemia.[69] Tungsten acts as an antagonist of molybdenum uptake and metabolism. Absorption of molybdates in liquid form is rapid, with the ion being metabolised in the form of plasma protein complexes in liver, kidneys and adrenal glands for storage.[70] Molybdenum is excreted in the bile and urine.

7.4.2 Industrial Uses and Human Exposure

The greatest human exposure to molybdenum is through mining, processing of the metal, and the steel and hard metal industries. In mining, workers are exposed to other metals including lead, copper, arsenic, silver and cadmium, several of which are associated with toxic signs. The vast proportion of molybdenum produced internationally is used in the production of steels and hard metal alloys. Molybdenum is a heat-tolerant metal and can withstand extremely high temperatures without changing shape, expanding or softening significantly. This property lends molybdenum alloys to use in manufacture of furnaces, missiles, armour plating, aircraft parts, industrial motors, filaments and electrical contacts. Increasing use has been made in recent years of molybdenum alloys in orthopaedic prostheses and modular hip systems; other metals included are chromium, nickel, iron, titanium, cobalt and tantalum. Considerable scrutiny has been given to the safety and physical efficiency of these implants in recent years and much clinical and experimental study conducted to evaluate health risks associated with liberation of metal ions into the circulation and local tissue reactions due to metal fragments.[71–74] Other therapeutic uses proposed for molybdenum compounds include nutritional additives and therapy for iron-deficiency anaemia, prevention of dental caries and anticancer treatments. The latter use has been subject to considerable experimental investigation but has not achieved clinical acceptance.[75–77] Whilst there is experimental and clinical evidence to support claims for efficacy in these uses, much requires to be understood concerning mechanisms of action and long-term safety issues.

Workers exposed to molybdenum and its compounds in dusts and residues in mining, refining and in hard metal and steel industries absorb the ion through inhalation. A Russian study of 352 workers showed that 19.6% developed occupationally related skin conditions, eczema and reduced immunological responsiveness.[67] Other routes of exposure include dermal contact, contact with the eye, inhalation and ingestion of contaminated air, but there is little information relating to absorption and metabolism of bioactive ion from any site other than ingestion, which is presumed to vary from 30 to 80% depending upon the solubility of the material and the age of the person involved. Children absorb molybdenum more readily than adults. Reviews of the literature conducted by the International Molybdenum Association show that, whilst there were no reported deaths attributable to molybdenum poisoning in industry, at levels of occupational exposure molybdenum is rarely a serious source of environmental pollution. Chronic exposure, as seen in Russian workers, may be a source of physiological change, hepatic, renal and bone pathology, and neurological damage,[78,79] but these reports are incomplete and fail to reveal useful details regarding exposure patterns, lifestyle factors, job specifications and the occupational dust concentrations to which workers were exposed. Importantly, workers engaged in molybdenum ore crushing and extraction processes were subject to neurological effects but carcinogenic changes were not reported. Molybdenum dusts are a known cause of dermal irritancy in occupational exposure but it is unclear how much molybdenum ion is actually absorbed percutaneously.[17]

7.4.3 Mutagenicity and Carcinogenicity

7.4.3.1 *In Vitro Studies*

Very inconsistent evidence is seen in standard laboratory *in vitro* tests to show that molybdenum or its salts are mutagenic. Many studies employ high concentrations of the molybdenum compounds and inconsistency, in part, seems to relate to the laboratory involved and the solubility of the compound. This makes determination of the true mutagenic risk of molybdenum difficult to appreciate (Table 7.2).[80] Virtually insoluble molybdenum trioxide was not mutagenic in five strains of *Salmonella typhimurium* with or without metabolic activation, and negative observations were made in *Bacillus subtilis,* whereas of the soluble salts ammonium molybdate was mutagenic in two of three strains of *Escherichia coli,*[81] but gave negative results in *a Saccharomyces cerevisiae* reverse mutation test and gene conversion assay.[82] The soluble sodium molybdate was neither mutagenic nor recombinogenic in this test but did show a modest effect in reducing viability and increasing micronucleus formation in

Table 7.2 Mutagenicity of molybdenum compounds in bacterial assay.[80]

Test schedule	*Compound tested*	*Dose range*	*Observation*
Salmonella typhimurium	MoO_3		–ve
	$(NH_4)_6Mo_7O_{24}$		–ve
			–ve
E. coli	$(NH_4)_6Mo_7O_{24}$	2–10 mM	+ve
			–ve
	Na_2MoO_4		–ve
	$MoCl_6$		–ve
Bacillus subtilis	MoO_3		–ve
	$(NH_4)_6Mo_7O_{24}$		+ve
	$MoCl_6$		–ve
	MoS_2		–ve
Chinese/Syrian hamster ovary cell transformation test	MoO_3		+ve
Cell transformation Balb/c-3T3 cells	MoO_3	2.3–11 mM	–ve
Micronucleus test, human lymphocytes	MoO_3	250–750 μg ml^{-1}	+ve
	Na_2MoO_4	0.1–5 mM	+ve
	$(NH_4)_6Mo_7O_{24}$	0.1–2 mM	+ve
Saccharomyces cerevisiae (meiosis)	Na_2MoO_4	40–150 mM	+ve diploidy
Gene conversion in yeast	$(NH_4)_6Mo_7O_{24}$		–ve
	Na_2MoO_4		–ve
Sister-chromatid exchanges	MoO_3		–ve
Bacteriophage test	Na_2MoO_4		±ve
Cell transformation assay (SA7/SHE)	MoS_2	62–1000 μg ml^{-1}	–ve

binucleated lymphocytes in culture and in murine bone marrow cells.[83] In this test, male mice injected intraperitoneally with sodium molybdate at high doses of 200–400 mg kg^{-1} for two days and then mated with untreated female animals showed modest increases in post-implantation loss. Sodium molybdate induced dominant lethality at the post-meiotic stage of spermatogenesis. Additionally, sodium molybdate was shown to invoke micronuclei in bone marrow with chromosomal breakage in male mice 24 hours following intraperitoneal injection. In a "wing-spot" test using *Drosophila melanogaster* larvae, molybdenum trichloride evoked positive changes indicative of mutagenicity.[81,84]

7.4.3.2 Experimental Animal Studies

Experimental evidence is available to show that certain molybdenum compounds are capable of inducing tumours in rodents and act as promoters of tumours induced by known carcinogens. Other indications of the possible role of molybdenum in carcinogenesis derive from recent observations showing that mammary carcinomas in dogs contained higher zinc and molybdenum levels than in non-malignant tissues.[85] Whereas higher levels of zinc as zinc-metalloenzymes should be expected in rapidly proliferating tissues, the role of molybdenum is unclear. Molybdenum is known to interact with copper in biological systems and tetrathiomolybdate is shown to release copper from metallothionein and caeruloplasmin complexes in experimental and farm animals and is a possible therapy for copper-metabolic disturbances such as Wilson's disease.[76,86–88]

Stoner *et al.* provided the first evidence that a molybdenum compound (MoO_3) can induce cancer in laboratory animals, by giving mice a total of 19 intraperitoneal injections at maximum tolerated doses.[89] The Strain A mice injected with a massive total dose of 4.75 g kg^{-1} MoO_3 developed a significant increase in lung tumour incidence within 30 weeks; other tumours were not seen. A cumulative dose of half this amount did not cause a significant increase the number of tumours. The survival of the molybdenum-treated mice at the end of the experiment was reported to be 75%. The study was scientifically unsound, unethical and its validity is questioned. In a larger-scale study in which F344/N rats and B6C3F1 mice were exposed to MoO_3 (10, 30 or 100 mg m^{-3}) by inhalation for two years in a standard NTP-style evaluation, male but not female animals developed dose-related degenerative changes in the nasal epithelium, precancerous squamous metaplasia in pharyngeal epithelium and a marginally significant increase in pulmonary adenoma/carcinoma.[90] In mice exposed to MoO_3 at similar levels, all animals exhibited hyperplastic and metastatic changes in the respiratory tract and the pulmonary cancer incidence was significantly increased. Whereas the National Toxicology Program considered that these results provide some evidence that molybdenum trioxide is carcinogenic under extreme conditions, its relevance to human risk is questionable. In each of these studies, the concentrations of molybdenum to which animals were exposed were many times higher than the recommended maximum exposure for humans set by the OHSA of 15 mg m^{-3} for molybdenum and insoluble molybdenum compounds averaged over an 8-hour shift.[91] The American Conference of Governmental Occupational Hygienists[92] recommended a threshold limit value of 10 mg m^{-3}.

Four experimental studies have been seen illustrating the possible anti-carcinogenic role of molybdenum compounds on experimental tumour formation in rodents treated with well-known carcinogens. In the model in which mammary tumours were induced in rats by single injections of *N*-nitroso-*N*-methyl urea (NMU), molybdenum administered in the diet significantly reduced the onset, development and progression of tumour growth.[93] Similar observations were reported by Luo *et al.*, who exposed weanling rats to low levels of molybdenum in drinking water and then administered *N*-nitrososarcosine ethyl ester by intragastric intubation twice weekly for 2–8 consecutive weeks.[94] Molybdenum at 2 or 20 ppm significantly inhibited oesophageal and fore-stomach tumours. In the third model, rats receiving 200 ppm molybdenum as ammonium molybdate in drinking water were injected repeatedly *via* the intraperitoneal route with 1,2,dimethyl hydrazine (DMH) carcinogen.[95] The incidence, location and number of intestinal tumours recorded were significantly lower than in control groups given DMH only. The animals exhibited an accumulation of molybdenum in their livers. In all three experiments, administration of tungsten, which acts as a molybdenum antagonist, promoted tumour growth. A further study conducted in rats given dietary molybdenum demonstrated the protective role provided by the molybdenum-related enzyme xanthene oxidase which increased in gastrointestinal tissues in a dose-related fashion.[96] The animals in this case were significantly protected from the carcinogenic effect of *N*-methyl-*N*-benzylnitrosamine, which induces malignancies in the fore-stomach and oesophagus. In this model, dietary molybdenum as low as 2 ppm resulted in an accumulation of the metal in the oesophagus, stomach, liver and serum and a corresponding increase in enzyme levels, and a significantly lower incidence of carcinomas.

7.4.3.3 Epidemiological and Human Case Studies

The carcinogenicity of molybdenum and its compounds in humans is a complex subject and one that is subject to a lot of speculation based upon limited clinical information. Little good evidence exists to show that molybdenum is carcinogenic in humans through long-term exposure to industrial dusts and fumes, or through lifetime exposure to foods grown on molybdenum-rich soils. Davies in 1975 reviewed scattered evidence from studies in South Africa, China and Russia and was perhaps first to suggest an inverse correlation between the molybdenum content of the soil and food plants and cancer incidence.[97] In his view, molybdenum offered significant protection against oesophageal cancers in Bantu women in South Africa and exhibited low toxic risk. Food plants grown under such conditions were supposed to be prone to fungal infections including *Aspergillus flavus,* extracts of which are known to induce liver neoplasms in mice. Based upon earlier experimental evidence that depletion in the molybdenum-dependent enzyme xanthene oxidase is consistent with tumour growth in mice and that supplementation resulted in a regression of the tumours, Davies conjectured from earlier information that molybdenum acts as an antitumor agent.[98,99] Whilst a study in the Nigata prefecture of Japan reported an inverse correlation between molybdenum levels (3.82 ppm) and the mortality of women from oesophageal and rectal cancers, it also showed that molybdenum concentrations "correlated with mortality in ladies from pancreatic cancer".[100] The protective effect of molybdenum (and xanthene oxidase or other related enzymes) is not supported by two large-scale trials in the Linxiang region of northern China where very high levels of gastric and oesophageal cancer are associated with chronic deficiency in many key nutrients, including molybdenum and zinc.[101] Dietary repletion over 5.25 years in 30,000 people did lead to a small reduction in the cancer incidence, but molybdenum supplementation was not implicated.

Chronic occupational exposure to molybdenum dusts and effluents has not been associated with increased lung cancer mortality in metallurgy plants, although severe respiratory distress is recorded. In these situations, workers in steel melting shops, metal casting, rolling mills, *etc.* are exposed to a wide range of pollutants, probably with molybdenum concentrations of <0.004 mg m^{-3}, such that the effects observed are likely to be of multifactorial causation.[102] A similar explanation is offered for reports of neurological changes reported by workers in a Russian electrochemical plant[79,103] following chronic exposure to molybdenum disulfide at excessive levels of about 5 mg m^{-3}. Fuller details of these studies are unavailable for further comment. Only one study is seen of an investigation into possible pulmonary carcinogenesis in welders, shipyard workers, dockers, *etc.* exposed to molybdenum and other metals occupationally.[104] Questionnaires in a case–control study of 478 employees from various industries in the Antwerp region of Belgium purported to show a correlation between lung cancer incidence and molybdenum exposure, but workers were also exposed to chromium, smoking and lifestyle factors, any of which could contribute to the lung cancer incidence reported. Thus two epidemiological studies conducted in Colorado in 1971 failed to provide evidence for molybdenum-related cancer in the lungs or other tissues,[105] and progress reports in other mines in the USA at least have failed to show cancer risk.

Cancer incidence has not been linked with release of molybdenum particles from hard metal complexes used in orthopaedic implants, although local reactions, granulomas and foreign body reactions seem to be a recurrent problem.[72,74,106] The pathogenicity of metals and hard metal alloys is discussed in more detail under orthopaedic implants later in this publication.

7.5 Tin

Tin is a soft pliable silvery white metal which has been known to humankind for over 5000 years, at least from the Bronze Age. The world famous tin mines in Cornwall existed since Roman times. The Bantu peoples of central Africa mined tin from the 11th to the 15th century, but much of the world's demand these days is provided by Southeast Asia, Bolivia and Australia. Early demands for tin meant that ores had to be transported many miles, and trade routes across Europe are documented. Cassiterite (SnO_2), which contains up to 78.8% Sn, is the most valuable tin ore; stannite (Cu_2FeSnS_4, 27.5% Sn) also contains good industrial concentrations of Sn but is not very widely used because of the numerous difficulties involved in dressing the ore and extracting tin. Nordenskioldine ($CaSnB_2O_6$, 43.5% Sn) is a calcium–tin borate of secondary commercial importance. Tin ores often occur as complexes that contain W, Be, Li, Bi, Ta, Cu, Pb, Zn and Ag, to which miners will be exposed in mining and extraction processes. Tin is found in many parts of Asia, Europe, Australia and South America but many of these deposits have become depleted over centuries.

Tin is classified with lead in Group IV of the Periodic Table. It is a transitional metal with 22 known isotopes including 9 stable isotopes and 13 radionuclides, of which ^{117m}Sn has medical value in treating bone cancer. Tin occurs as Sn(II) and Sn(IV) oxidation states. Several organic complexes, including trimethyl and trimethyl tin, and aryl tin compounds are highly toxic. Tin and its inorganic compounds are not presently classified as suspected human carcinogens but the US Environmental Protection Agency (EPA) has assigned tributyltin oxide to group D, for which there is "inadequate information to assess carcinogenic potential," according to updated guidelines.[107–109]

Tin is well known in the canning and preservation of foods and as an anticorrosion coating for other metals but its value as an essential nutrient in the mammalian body is questioned. It is present in many tissues but is rarely a cause of human toxicity. Tin is found in quantities of less than 100 ppm in canned foods and some tin may elute into the food product for human consumption.[109–111]

7.5.1 Tin as a Nutrient

The human body contains approximately 0.017 g of tin but little is known concerning its essentiality in the human diet or the effects of tin deficiency.[109] Tin exhibits an atomic configuration, a strong tendency for co-ordination reactions, and two oxidation states favouring its interaction with almost any biological material, and participation in biological charge-transfer reactions.[2,3] Estimated daily intake in humans is 3.5 to 17 mg, but intestinal uptake is possibly less than 5% of this. Plasma levels are 140 $\mu g\ l^{-1}$ but the nature of protein binding of tin in human blood is not known. Tin is distributed throughout the body with higher levels in bone, lung and kidney, and excreted in the urine.[110] Amounts in excess of 130 mg per day have been shown to lead to increased accumulation in liver and kidney.

Experimental studies in tin-deficient rats have shown that general body growth is retarded and animals show defective skin and hair growth, neurological symptoms and decreased food efficiency.[112] Addition of tin as stannic sulfate to the diet in part restored growth rate, but the mechanism of action is not known. Tin as a protoporphyrin complex is known to suppress haem-oxygenase activity and bilirubin formation and possibly modulates this enzyme in the brain and other tissues.[113–115] Dietary supplementation with tin in rats leads to an increase in renal haem-oxygenase. Dietary deficiency of tin in rats has also been associated with an imbalance of key metal ion nutrients including calcium, zinc, copper and manganese.[116] On diets containing very low tin (17 ng g^{-1}), Wistar rats exhibited an increase in calcium in the lung with a concurrent depletion in magnesium, increased iron in kidney and spleen but with reductions in femoral muscle, and a reduction in zinc, manganese and copper in heart and tibial muscle. The biochemical implications are not clear, but it is suggested that tin activates and binds key metalloproteins involved in trace metal metabolism. This is in part illustrated by the observation that renal haem-oxygenase in rats was greatly induced by tin and nickel, but substantially reduced when manganese was administered simultaneously in equimolar amounts.[114] Also, haem-oxygenase induced by tin or nickel was markedly reduced when the animals were pre-treated with manganese, which was ineffective alone in preventing haem-oxygenase production in isolated microsomes. Zinc also has been shown to inhibit renal haem-oxygenase production when administered in equimolar levels with tin or nickel.

7.5.2 Industrial Uses and Human Exposure

From the Bronze Age, the principal use of tin has been in the production of alloys with silver, copper, nickel, bismuth, iron, aluminium, manganese, tellurium, cadmium and antimony. More recently tin alloys with indium and/or gallium have been used in the semiconductor industry and have given cause for concern regarding their potential carcinogenic risk. Tin salts are occasionally sprayed on to glass to produce electrically conductive coatings. Electroplating, solder, fusible metals, pewter, Babbit metal (contains copper and antimony), bell metal and die casting are additional uses of tin.

Tin is still used extensively as an anticorrosive coating for other metals, notably steel plate in the food canning industry, although this use has declined in recent years as plastics and other synthetic materials have been developed. Tin does migrate from tin cans into food and in some instances produces gastrointestinal disturbances.[110] The glass industry uses molten tin in the production of window glass, and lead–tin yellow has been employed as a colorant in glass and ceramics since mediaeval times. Lead–tin yellow II is found in Florentine, Venetian and Bohemian paintings and was used in the workshops of Giotto, di Cione, Veronese, and Tintoretto. Stannous chloride ($SnCl_2$) is a light-green pigment used in dying textiles. Stannous fluoride and stannous pyrophosphate have been used in some toothpastes on account of their antimicrobial action and ability to prevent caries and plaque formation, but dental staining and toxic risks preclude their value in these products.[117]

Inorganic tin compounds are poorly absorbed on account of their low solubility, but the stannous ion [Sn(II)] binds strongly to proteins (*e.g.* the β-chain of haemoglobins) and is slowly excreted. Humans chronically exposed to inorganic tin (*e.g.* stannic oxide dust or fumes) develop a benign form of pneumoconiosis known as stannosis, which involves mainly the lower respiratory system. In stannosis the lung surfaces become coated with radiographically visible tin oxide deposits or as profuse small nodules.[118] It may be complicated by inhalation of other metals in the work environment (*e.g.* lead and copper), but has not been associated with pulmonary cancers. Stannosis may also occur in hearth tinning where molten tin is poured into heated ironware or where materials are dipped by hand into molten tin as a coating. Other forms of stannosis develop from grinding, briquette-making and smelting the metal. Inorganic tin compounds such as stannous chloride can be toxic through their action as mitochondrial and membrane poisons. Immunosuppression, neurodegeneration, hepatocellular and pancreatic atrophy, and gastrointestinal disturbances have been reported in rats following subacute and chronic feeding.[119,120]

Recent increases in the plastics industry have seen an expansion in the use of organotin compounds. In many cases, these compounds are appreciably more toxic than the inorganic salts. Uses are extensive and diverse, with commercial uses of alkyltins (methyltins, butyltins, octyltins), phenyltins and cyclohexyltins in polyvinyl chloride (PVC), heat stabilisers, biocides, catalysts, agrichemicals and glass coatings. Tricylohexyltin hydroxide is registered by the US EPA as a miticide. These outlets account for an annual use 20,000 tons of organotin and major toxicological, environmental and ecotoxicological concerns. Inhalation of organotin (alkyl- and aryl-compounds) dusts and fumes is a cause of benign pneumoconiosis in exposed workers.[121] Organotin compounds such as trimethyl- and triethyltin are absorbed more readily from the intestine and are toxic to the liver and central nervous system.[110] They are the most toxic of the group with respect to effects on the central nervous system. Most other aryl- and alkyltin compounds are poorly absorbed and of lower toxic risk.

7.5.3 Cytotoxicity and Carcinogenicity

Limited clinical and experimental evidence exists to show that tin or its inorganic compounds are carcinogenic.[110] Mutagenicity studies show that stannous chloride was not mutagenic in the Ames test, while SnF_2 was slightly mutagenic in one strain of *Salmonella typhimurium*.[122,123] Stannous chloride induced DNA damage in *E. coli*, Chinese hamster ovary cells and human white blood cells.[124–127] Stannic chloride ($SnCl_4$) induced chromosomal damage in human lymphocytes.[128,129] Trimethyl tin administered at concentrations of 0.25, 0.4 or 5 mg kg^{-1} intraperitoneally into mice led to damage in somatic chromosomal preparations including breaks, dicentrics, mitotic change and fragmentation.[130]

Studies in laboratory animals have provided inconclusive evidence of the carcinogenicity of sodium chlorostannate or stannous-2-ethyl hexoate (rats) or stannous oleate (mice) with equivalents of 5000 ppm of tin.[130–133] Studies by the National Toxicology Program failed to find statistically significant evidence for carcinogenicity in rats (F344) or mice dosed for 78 weeks with dibutyltin acetate or stannous chloride for 105 weeks.[134,135] Wester *et al.* reported an increase in pituitary adenoma, phaeochromocytoma of the adrenals and parathyroid glands, and a low non-significant incidence of pancreatic carcinoma in Wistar rats receiving 5 mg kg^{-1} bis(tri-*n*-butyl)tin oxide for 106 weeks, but failed to identify a carcinogenic mechanism.[136]

Stannosis is classified as a benign pneumoconiosis in tinners exposed for long periods to tin dust; this has not been associated epidemiologically with increased risk of pulmonary tumours or symptoms of impaired pulmonary function or foreign body reactions, but dust-laden pulmonary macrophages were numerous.[118,137] A survey of 7855 tin workers in four Chinese mines showed that inhalation of cigarette smoke, arsenic, radon and silica were major factors implicated in lung cancer and respiratory distress.[138] It is recognised, however, that in occupational health studies, tinners, tin miners and others working in tin-related industries are rarely exposed to tin residues alone but will inhale, ingest or be subjected to dermal exposures to a wide range of other metals and materials which may modify, accentuate or otherwise alter reactions to tin. This is a large subject which has not been adequately addressed in epidemiological studies.

References

1. E. J. Underwood, *Trace Elements in Human and Animal Nutrition*, Academic Press, New York, 4th edn, 1977.
2. W. Mertz, *Proc. Nutr. Soc.*, 1974, **33**, 307.
3. P. J. Aggett, *Proc. Nutr. Soc.*, 1980, **39**, 241.
4. N. S. Scrimshaw and V. R. Young, *Sci. Am.*, 1976, **235**, 51.
5. D. L. Baly, J. S. Schneiderman and A. L. Garcia-Welsh, *J. Nutr.*, 1990, **120**, 1075.
6. K. H. Thompson, *Biofactors*, 1999, **10**, 43.
7. H. I. Afridi, T. G. Kazi, J. A. Baig, M. K. Jermali, M. B. Arain, R. A. Safraz, H. U. Shiekh, G. A. Kandhro and A. Q. Shah, *Arch. Gynec. Obstet.*, 2009, **280**, 415.
8. C. T. Whittemore and J. M. Manson, *Vet.Rec.*, 1995, **317**, 218.
9. R. M. Leach and A.-M. Muenster, *J. Nutr.*, 1962, **78**, 51.
10. R. M. Leach, A.-M. Muenster and E. Wien, *Arch. Biochem. Biophys.*, 1969, **133**, 22.
11. N. Purichia and L. C. Erway, *Dev. Biol.*, 1972, **27**, 395.
12. G. B. Gerber, A. Léonard and P. Hantson, *Crit. Rev. Oncol/Haematol.*, 2002, **42**, 24.
13. B. Pesch, T. Weiss, B. Kendzia, J. Henry, M. Lehnert, A. Spickenheuer, R. Van Gelder, M. Berges, J.-U. Hahn, H. Kafferlein and T. Bruening, *Occup Environ. Med.*, 2011, **68**, (Suppl.1), A95.
14. C. Santos-Burgoa, C. Rios, L. A. Mercado, R. Arechiga-Serrano, F. Cano-Valle, A. Eden-Wynter, J. L. Texcalae-Sagrador, J. P. Villa-Barragan, Y. Rodriguez-Agudelo and S. Montes, *Exp. Res.*, 2001, **85**, Sect. A, 90.

15. A. Smargiasse, M. Baldwin, S. Savard, G. Kennedy, D. Mergler and J. Zayed, *Appl. Occup. Environ. Hyg.*, 2000, **15**, 746.
16. L. Rossander-Hultén, M. Brune, B. Sandström, B. Lönnerdahl and L. Hallberg, *Am. J. Clin. Nutr.*, 1991, **54**, 152.
17. J. J. Hostýnek and H. I. Maibach, *Crit. Rev. Toxicol.*, 1993, **25**, 171.
18. J. Rodier, *Br. J. Ind. Med.*, 1955, **12**, 21.
19. J. Rodier and M. Rodier, *Bull. Inst. Hyg. Maroc.*, 1949, **9**, 3.
20. J. D. Wang, C. C. Huang, Y. H. Hwang, J. R. Chiang, J. M. Lin and J. S. Chen, *Br. J.Ind. Med.*, 1989, **46**, 856.
21. Chin-Chang Huang, *Chand. Gung. Med. J.*, 2007, **30**, 385.
22. E. Hell, R. Molina, T. Donaghey, J. D. Brain and M. Wessling-Resnik, *Am. J. Physiol.*, 2005, **288**, L877.
23. V. A. Fitsanakis, N. Zhang, J. G. Anderson, K. M. Erikson, M. J. Avison, J. C. Gore and M. Aschner, *Toxicol. Sci.*, 2008, **103**, 116.
24. P. D. L. Lima, M. C. Vasconcellos, R. C. Montenegro, C. O. Pessoa, L. V. Costa-Lotufo, M. O. Moraes and R. R. Burbano, *Toxicol. in Vitro*, 2008, **22**, 1032.
25. P. Zhao, W. Wong, X. Ying, Z. Yuan, J. Fu and Z. Zhou, *Toxicology*, 2008, **250**, 39.
26. F. L. Assem, P. Holmes and L. S. Levy, *J. Toxicol. Environ. Health B Crit. Rev.*, 2011, **14**, 537.
27. A. Hobson, N. Seixas, D. Sterling and B. A. Racett, *Ann. Occup. Hyg.*, 2011, **55**, 113.
28. American Conference of Governmental Industrial Hygienists (ACGIH), *Toxicological Pofile for Manganese*, ACGIH, 2007.
29. B. Pesch, T. Weiss, B. Kendzia, J. Henry, M. Lehnert, A. Lotz, E. Heinze, H. I. J. Käfferlein, R. Van Gelder, M. Berges, J. U. Hahn, M. Mattenldott, E. Punkenberg, A. Hartwig and T. Brüning, *J .Exp. Sci. Environ. Epidemiol.*, 2012, **22**, 291.
30. A. Hobbesland, H. Kjuus and D. S. Thelle, *Occup. Envir. Health*, 1999, **56**, 618.
31. A. Furst, *J. Natl. Cancer Inst.*, 1978, **60**, 1171.
32. H. P. Witschi, P. J. Hakkinen and J. P. Kehrer, *Toxicology*, 1981, **21**, 37.
33. J. A. DiPaulo, *Fed. Proc.*, 1964, **23**, 393.
34. G. D. Stoner, M. B. Shimkin, M. C. Troxell, T. L. Thompson and L. S. Terry, *Cancer Res.*, 1976, **36**, 1744.
35. A. Shevchenko and E. Y. Pankov, *Bull. Éksper. Biol. Med.*, 1970, **69**, 75.
36. H. E. Parker, F. N. Andrews, C. W. Carrick, R. D. Creek and S. M. Hauge, *Poult. Sci.*, 1955, **34**, 1154.
37. J. M. Leibolz, V. C. Speer and V. W. Hayes, *J. Anim. Sci.*, 1962, **21**, 772.
38. K. Rydzynski and D. Pakilska, Vanadium, niobium and tantalum, *in Patty's Toxicology*, ed. B. E. Bingham and B. Cohrsson, John Wiley and Son, Hoboken, NJ, 2012, vol. 1, pp. 511–564.

39. P. Tuomi, *Humanities and Social Sciences, South Africa's Hazardous Vanadium Mines*, Symposium and Multimedia Symposium, Barnard College, NY, 2004.
40. Committee on Biological Effects of Atmospheric Pollutants, *Vanadium*, National Academy of Sciences, Washington, DC, 1974.
41. J. Kucera, J. Lener and J. Mnuková, *Biol. Trace Elem. Res.*, 1994, **43–45**, 327.
42. R. Cooper, *Ind. J. Occup. Environ. Med.*, 2007, **11**, 97.
43. V. A. Ehrlich, A. K. Nerseyan, K. Atefie, C. Hoelzi, F. Ferk, J. Bichler, E. Valic, A. Schaffer, R. Schultz-Hermann, M. Fenech, K.-H. Wagner and S. Knasmüller, *Environ. Health Perspect.*, 2009, **11**, A15.
44. L. Marzban and J. H. McNiel, *J. Trace Elem. Exp. Med.*, 2003, **16**, 253.
45. M.-J. Hosseini, N. Seyedrazi, J. Shahraki and J. Pourahmad, *Adv. Biosci. Biotech.*, 2012, **3**, 1096.
46. I. Kostova, *Anticanc. Agent. Med. Chem.*, 2009, **9**, 827.
47. A. Léonard and G. B. Gerber, *Mutat. Res/Rev. Genet. Toxicol.*, 1994, **317**, 81.
48. D. F. Lewis, C. Ioannides and D. V. Parke, *Environ. Health Perspect.*, 1996, **104**(Suppl. 5), 1011.
49. A. Galli, R. Vellosi, R. Fiorio, C. Della Croca, R. Del Carratori, E. Morichetti, L. Giromini, D. Rosellini and G. Bronzetti, *Terat., Carcin. Mutag.*, 1991, **11**, 175.
50. B. Z. Zhong, Z. W. Gu, W. E. Wallace, W. Z. Wong and T. Ong, *Mutat. Res.*, 1994, **321**, 35.
51. J. Owusu-Yaw, M. D. Cohen, S. Y. Fernando and C. I. Wei, *Toxicol. Lett.*, 1990, **50**, 327.
52. E. Sabbione, L. Clerici and A. Brazzelli, *J. Toxicol. Environ. Health*, 1983, **12**, 737.
53. V. A. Ehrlich, A. K. Nersesyan, K. Atefie, C. Hoelzi, F. Ferk, J. Bichler, E. Valic, A. Scharrer, R. Schulte-Hermann, M. Fenech, K. H. Wagner and S. Knasmüller, *Environ. Health Persp.*, 2008, **116**, 1689.
54. J. H. Duffus, *Regulat. Toxicol. Pharmacol.*, 2007, **47**, 110.
55. M. H. Fourie, MSc. Thesis, University of Pretoria, 2010, ETD 08032010-181254.
56. J. D. Boice, M. T. Mumma and W. J. Blot, *Radiat. Res.*, 2007, **167**, 711.
57. American Conference of Governmental Industrial Hygienists (ACGIH), Vanadium pentoxide, in *Threshold Limit Values for Chemical Substances and Physical Agents and Biological Exposure Indices*, ACGIH, Cincinnati, OH, 2008.
58. N. B. Ress, B. J. Chou, R. A. Renne, J. A. Dill, R. A. Miller, J. H. Rycroft, J. R. Hailey, J. K. Haseman and J. R. Boucher, *Toxicol. Sci.*, 2003, **74**, 287.
59. E. A. Rhondini, D. M. Walters and A. K. Bauer, *Particle Fibre Toxicol.*, 2010, **7**, 9.
60. F. L. Assem and L. S. Levy, *J. Toxicol. Environ. Health B Crit. Rev.*, 2009, **12**, 289.

61. N. N. Amrud, *Bull, N.Y. Acad. Sci.*, 1984, **60**, 163.
62. J. L. Johnson and K. V. Rajogopalan, *J. Clin. Invest.*, 1976, **58**, 551.
63. J. L. Johnson, W. R. Waud, K. V. Rajogopalan, M. Duran, F. A. Beemer and S. K. Wadman, *Proc. Soc. Acad. Sci.*, 1980, **77**, 3715.
64. J. L. Johnson, M. M. Wuebbens, R. Mandell and V. E. Smith, *J. Clin. Invest.*, 1989, **83**, 897.
65. J. Reiss, *Human Genet.*, 2000, **106**, 157.
66. N. F. Suttle, *Ann. Rev. Nutr.*, 1991, **11**, 121.
67. L. A. Dueva and S. S. Stepanian, *Vestn. Dermatol. Venereol.*, 1989, **10**, 47.
68. A. Vyskočil and A. Viau, *J. Appl. Physiol.*, 1999, **19**, 185.
69. M. S. Seelig, *Am. J. Clin. Nutr.*, 1973, **26**, 657.
70. A. Giussani, A. M. Arogunjo, M. C. Cantone, F. Tavola and I. Veronese, *Appl. Radiat. Isot.*, 2006, **64**, 639.
71. D. A. Pulco and W. W. Huh, *J. Appl. Biomater.*, 1995, **6**, 109.
72. C. P. Case, L. Ellis, J. C. Turner and B. Fairman, *Clin. Chem.*, 2001, **47**, 275.
73. B. Daley, A. T. Doherty, B. Fairman and C. P. Case, *J. Bone Joint Surg.*, 2004, **86B**, 59.
74. T. Watanabe, N. Yamada, Y. Yoshida and O. Yamamoto, *Br. J. Dermatol.*, 2008, **158**, 402.
75. B. Fisher, *Inorgan. Chim. Acta*, 1975, **279**, 136.
76. G. J. Brewer, *Curr. Cancer Drug Targets*, 2005, **5**, 195.
77. S. Rafique, M. Idrees, H. Nasim, Akbar and A. Athar, *Biotechnol. Mol. Biol. Rev.*, 2010, **5**, 38.
78. H. A. Schroeder, J. J. Balasa and I. H. J. Tipton, *J. Chron. Dis.*, 1970, **23**, 481.
79. S. L. Eolyan, *Zh. Eksp. Klin. Med.*, 1965, **5**, 70.
80. J. Montelius, *Scientific Basis for Swedish Occupational Standards*, University of Gothenberg, 2010, **44**, 1.
81. H. Nishioka, *Mutat. Res.*, 1975, **31**, 185.
82. I. Singh, *Mutat. Res.*, 1983, **117**, 149.
83. N. Titenko-Holland, J. Shoo, L. Zhang, L. Xi, H. Ngo, N. Shang and M. T. Smith, *Environ. Mol. Mutagen.*, 1998, **32**, 251.
84. H. I. Ogawa, T. Shibahara, H. Iwata, T. Okada, S. Tsuruta, K. Sakata, Y. Kato, H. Ryo and T. Itoh, *Mutat. Res.*, 1994, **30**, 133.
85. E. M. Skibniewska, T. Kośla and M. Skibniewska, *Trace Elem. Electrol.*, 2012, **29**, 42.
86. S. Haywood, Z. Dincer, J. Holding and N. M. Perry, *Br. J. Nutr.*, 1998, **79**, 329.
87. Y. Ogra and K. T. Suzuki, *J. Org. Biochem.*, 1998, **70**, 329.
88. K. T. Suzuki and Y. Ogra, *Yakaugaku Zasshi J. Pharm. Soc. Japan*, 2000, **120**, 899.
89. M. B. Shimkin, G. D. Stoner and J. C. Theiss, *Adv. Exp. Med. Biol.*, 1977, **91**, 85.

90. P. C. Chan, R. A. Herbert, J. H. Rycroft, J. K. Haseman, S. L. Grumbein, R. A. Miller and B. J. Chou, *Toxicol. Sci.*, 1998, **45**, 58.
91. Centers for Disease Control and Prevention, *Molybdenum*, CDC, Atlanta, GA, 2011.
92. American Conference of Governmental Industrial Hygienists (ACHIH), *Molybdenum as Mo: Documentation of Threshold Limit Value for Substances in Workroom Air*, ACGIH, Cincinnati, OH, 3rd edn, 1974.
93. H. J. Wei, X. M. Luo and X. P. Yang, *Zhonghua Zhong. Liu Za Zhi.*, 1987, **9**, 204.
94. X.-M. Luo, H.-J. Wei and P. Yang, *J. Natl. Canc. Inst.*, 1983, **71**, 75.
95. M. A. Montenegro, M. S. Negrette, E. J. Gimeno and J. T. Borda, *Biocell.*, 2002, **26**, 339.
96. H. Komada, Y. Kise, M. Nakagawa, M. Yamamura and K. Hioki, *Cancer Res.*, 1990, **50**, 2418.
97. I. J. T. Davies, *Intake (British Medical Journal, Advertisers Suppl.)*, 1975, **39**, 4.
98. R. C. Bray, in *The Enzymes*, ed. P. D. Hardy, L. Hardy, K. Myrback, Academic Press, New York, 1963, vol. 7, p. 533.
99. C. A. Pritsos and D. L. Gustafson, *Cancer. Chemotherap. Oncol. Res.*, 1994, **6**, 477.
100. H. Nakadaira, K. Endoh, M. Yamamoto and K. Katoh, *Arch. Environ. Health*, 1995, **50**, 374.
101. W. J. Blot, J. Y. Li, P. R. Taylor, W. D. Guo, S. M. Dawsey and B. Li, *Am. J. Clin. Nutr.*, 1995, **62**, S1424.
102. M. Huvinen, J. Uitti, P. Oksa, P. Palmroos and P. Laippala, *Occup. Med. Oxford*, 2002, **52**, 203.
103. IOG, *Referativny Zhur. Metallurg.*, 1965, **153**, 154.
104. J. H. Droste, J. J. Weyler, J. P. Van Meerbeeck, P. A. Vermeire and M. P. Van Sprundel, *Occup. Environ. Med.*, 1999, **56**, 322.
105. F. W. Breise, in *Molybdenum and the Environment*, ed. W. R. Chappell and K. K. Petersen, Marcel Dekker, New York, 1976, vol. 1, Ch. 19.
106. IARC, *Monographs on the Evaluation of Carcinogenic Risks to Humans*, IARC, Lyon, 1999, **74**.
107. J. B. Greig and J. A. Pennington, in *WHO Food Additives Series* (addendum), FAO/WHO Expert Committee on Food, 2003, p. 46:TIN.
108. Integrated Risk Information System (IRIS), *Tributyl Tin Oxide (CAS-56-35-9)*, EPA, Washington, 2012.
109. V. W. Sardesai, *Introduction to Clinical Nutrition*, CRC Press, London, 2003.
110. K. A. Winship, *Adverse Drug React. Acute Poisoning Rev.*, 1988, **7**, 19.
111. G. W. Monier-Williams, in *Trace Elements in Food*, John Wiley & Sons, New York, 1949, pp. 138–161.
112. K. Schwartz, D. B. Milne and E. Vineyard, *Biochem Biophys. Res Commun.*, 1970, **40**, 22.
113. J. A. Mark and M. D. Maines, *Pediat. Res.*, 1992, **32**, 324.

114. G. S. Drummond and A. Kappas, *Proc. Natl. Acad. Sci.*, 1979, **76**, 5331.
115. G. S. Drummond and A. Kappas, *J. Clin. Invest.*, 1986, **77**, 971.
116. K. Yokoi, M. Kimura and Y. Itokawa, *Biol. Trace Elem. Res.*, 1990, **24**, 223.
117. W. Wade, M. Addy, J. Hughes, S. Milsom and F. Doherty, *J. Clin. Periodontol.*, 1997, **24**, 81.
118. G. K. Sluis-Cremer, R. G. Thomas, B. Goldstein and A. Soloman, *S. Afr. Med. J.*, 1989, **75**, 24.
119. H. C. Dreef-van der Meulen, V. J. Feron and H. P. Til, *Path. Europ.*, 1974, **9**, 185.
120. P. J. M. Jansssen, M. C. Bosland, J. P. Van Hees, B. J. Spit, M. I. Willems and C. F. Kuper, *Toxicol. Appl. Pharmacol.*, 1985, **78**, 19.
121. K. D. Kimbrough, *Environ. Health Perspect.*, 1976, **14**, 51.
122. E. Gocke, M. T. King, K. Eckhardt and D. Wild, *Mutat. Res.*, 1981, **90**, 91.
123. M. J. Prival, V. F. Simmon and K. E. Mortelmans, *Mutat. Res.*, 1991, **260**, 321.
124. M. Bernardo-Filho, M. Cunha, I. Valsa, A. Araujo, F. Silva and A. Fonseca, *Fd. Chem. Toxicol.*, 1994, **32**, 477.
125. F. J. Dantas, M. O. Moraes, E. F. Carvalho, J. O. Valsa, M. Bernardo-Filho and A. Caldeira-de-Araujo, *Fd. Chem. Toxicol.*, 1996, **34**, 959.
126. F. C. Silva, A. S. Fonseca, A. S. Correa, C. C. Lee, A. C. De Araujo, J. O. Valsa, M. Bernardo-Filho and A. Favre, *Microbios*, 1994, **79**, 241.
127. J. R. McLean, D. H. Blakey, G. R. Douglas and J. G. Kaplan, *Mutat. Res.*, 1983, **119**, 195.
128. G. Talukder, B. B. Ghosh and A. Sharma, *Environ. Mol. Mutagen.*, 1989, **4**, 197.
129. B. B. Ganguly, *Biol. Trace Elem. Res.*, 1993, **38**, 55.
130. B. B. Ganguly, *Mutat. Res. Envir. Mutagen. Rel. Subj.*, 1994, **312**, 9.
131. F. J. C. Roe, E. Boyland and K. Millican, *Fd. Cosmet Toxicol.*, 1965, **3**, 277.
132. E. Boyland and F. J. C. Roe, *Fd. Cosmet.Toxicol.*, 1963, **1**, 161.
133. D. K. Gulati, K. Witt, B. Anderson, E. Zeiger and M. D. Shelby, *Environ.Mol. Mutagen.*, 1989, **13**, 133–193.
134. National Toxicology Program, *Natl. Toxicol.Progr. Tech. Rep. Ser.* 1979, **183**, 1.
135. National Toxicology Program, *Natl. Toxicol.Progr. Tech. Rep. Ser.* 1982, **231**, 1.
136. P. W. Wester, E. I. Krajnc, F. X. van Leeuwen, J. G. Loeber, C. A. van der Hheijden, H. A. Vaessen and P. W. Hellman, *Fd. Chem. Toxicol.*, 1990, **28**, 179.
137. C. W. D. Cole, J. V. S. A. Davies, M. D. Kipling and G. L. Richie, *Br. J. Industr. Med.*, 1964, **21**, 235.
138. W. Chen and J. Chen, *Occup. Environ. Med.*, 2002, **59**, 113.

CHAPTER 8

The Metalloid Elements, Selenium and Silicon

8.1 General Introduction

Metalloid elements by definition combine the chemical properties of metals and non-metal elements. After oxygen, silicon is the most abundant of the elements on planet Earth and seventh in the universe. In contrast, selenium ranks 34th amongst the elements in the Earth's crust. Selenium exists in numerous allotropes (natural forms) and at different times has been classified as a metal or a non-metal with borderline properties. Both elements are now classified as essential human nutrients, but selenium is better documented as an antioxidant (as in the enzyme glutathione peroxidase) in human metabolism. Selenium displays biochemical similarities to oxygen and sulphur, with which it is placed in Group VIa of the Periodic system.

Elemental selenium is not viewed as a carcinogenic risk to humans but the 12th Report on Carcinogens classified selenium sulphide, as used in anti-dandruff shampoos, as reasonably anticipated to be a human carcinogen on the basis of limited experimental evidence.[1] Silicon does not occur in its elemental form, in view of its strong capacity to oxidise to silica, but particles of silica (SiO_2) of so-called "respirable" size are unquestionably carcinogenic following chronic inhalation. Until recently, safety concerns have not been expressed widely in relation to the extensive use of silicon-related compounds including silicones and silicone oils in medical prostheses and implants. Although silicones have been in use for breast enlargement and cosmetic surgery since the early 1960s, clinicians are now aware that the materials are not as inert as at one time thought and under some circumstances may induce foreign body reactions with the possible risk of mammary cancer.[2] A third contentious aspect relates to the possible carcinogenicity of certain silicates.[3,4] The widespread use of

Issues in Toxicology No. 18
The Carcinogenicity of Metals: Human Risk through Occupational and Environmental Exposure
By Alan B. G. Lansdown

Published by the Royal Society of Chemistry, www.rsc.org

magnesium silicate [$Mg_3Si_4O_{10}(OH)_2$], or talc, as in talcum powders, dusting powders for surgical gloves, cosmetic products for hygiene purposes, and numerous industrial applications has been associated with a 1.4% relative risk of ovarian cancer.[5–7] The expression "talc" is used broadly to denote both mineral talc and industrial products that are marketed under the name talc and contain proportions of mineral talc from about 35% to almost 100%.

8.2 Selenium

Selenium is an amorphic or monoclinic crystalline metalloid element which exists in a number of allotropic forms ranging from deep red to grey or black in colouration. Selenium-containing deposits occur in North America, China, Mongolia, Russia, Australia and South Africa, with the principal minerals being:

- Berzelianite (*copper selenide*)
- Clausthalite (*lead selenide*)
- Eucairite (*silver copper selenide*)
- Hakite (*copper mercury silver antimony selenium sulfide*)
- Klockmannite (*copper selenide*)
- Palladseite (*palladium selenide*)
- Penroseite (*nickel selenide*)
- Selen-tellurium (*selenium tellurium*)
- Tiemannite (*mercury selenide*)
- Umangite (*copper selenide*)
- Crooksite (*copper thallium selenide*).

Selenium is frequently found in copper deposits and elemental selenium is obtained as a by-product of copper mining. The average selenium content in soils is about 4000 ppb but levels may be as high as 6000 to 28,000 ppb in the alkaline Cretaceous shales of North America where the climate is arid, or even higher in the Avonian shales of Ireland and Wales. Environmental distribution of selenium is closely related to the nature of the parent rock, geochemical activity, the alkalinity of the soils and annual rainfall.[8] Se(IV) and Se(VI) are mobile and Se is bio-available in both forms. Atmospheric deposition of selenium is high in the vicinity of coal-fired or industrial or power-generating plants. Annual emissions of selenium into the soil increase from coal to ash.[9] Selenium readily accumulates in ground and water plant-life to be consumed by livestock grazing in these areas. Selenium is a component of food chains and is plentiful in cereals, nuts and certain vegetables. Fresh and marine watercourses around the world contain between 0.1 and 0.1 ppb selenium although much higher levels are seen in water draining from seleniferous soils.

Selenium was first identified as an element by Jöns Jacob Berzelius in 1817, although selenium poisoning had been reported in horses and cattle grazing selenium-rich pastures in China and Tibet in the 13th century by the great explorer Marco Polo. Selenium has six naturally occurring isotopes and three main oxidation states, including –2, +4 and +6. It is an unusual element in that it exhibits a photovoltaic action and is able to convert light energy directly into electricity, and has photoconductive action where electrical resistance decreases with increased illumination. These unique properties render selenium of particular value in photovoltaic and photo-cells, xerography, electrical rectifiers and in semiconductors.

8.2.1 Industrial Uses and Human Exposure

Humans are exposed to selenium and its inorganic and organic compounds through mining, extraction processes, and exposure to a large range of industrial, chemical and manufacturing processes, food and drinking water, environmental contamination and numerous industrial sources.[10] Combustion of coal and use of coal-burning stoves is a cause of increased atmospheric selenium and selenium in foods. Selenium is readily absorbed in humans in the expected dietary range of 0.071–0.152 mg per day, with rare evidence of systemic toxicity.[11,12] Endemic selenium intoxication was reported in the Hubei province of China in 1961, where high morbidity was associated with intoxication from food crops grown in the area, as well as through inhalation of smoke in the burning of stony coal with high selenium content.[13,14]

Major industrial uses of selenium include production of electrical components, rectifiers, photoconductors, semiconductors, colorants in paints, inks, enamels, glass and ceramics, rubber and plastics and in steel production, all of which will involve exposure to other elements. In semiconductor production and in rubber industries, selenium is often used in combination with tellurium, which is toxic, although not presently known to be carcinogenic in humans.[15] Selenium semiconductors are used in solid state electronics as well as rectifiers to convert alternating electric currents into direct current electricity. In the glass and ceramics industries, selenium imparts a ruby-red colour. Selenium is used in xerography to copy documents and as a photographic toner. Other major uses occur in the chemical industry as catalysts, solvents for metals, and in organic syntheses and stainless steel production. Selenium diethyldithiocarbonate is used as a catalyst in chemical syntheses and as a trace element in animal feeds. Fisher summarised that many selenium compounds are strong irritants and that selenium oxychloride, used as a metals solvent and chlorinating agent, is a powerful vesiculant.[16] Other selenium compounds, particularly the more soluble selenates, are well absorbed through the gastrointestinal tract, inhalation of dusts and fumes, and percutaneous absorption through damaged skin. Selenomethionine, selenium dioxide and selenious acids are known to penetrate intact human and animal skins and are skin irritants.[17–19] Experimental studies in mice have shown that oral selenomethionine accumulates in the skin following oral dosage and protects against skin damage through UV light exposure.[20] In general, uptake of selenium compounds in the industrial environment correlates well with solubility and chemical properties, the type and duration of exposure. Absorption is influenced also by the presence of other metals such as lead, copper, antimony, lead, thallium, *etc.*, to which workers are exposed in selenium mining and extraction.[12,21]

Attention is drawn to the use of "selenium sulphide" in proprietary anti-dandruff shampoos, for treatment of seborrhoeic dermatitis and associated infections with *Staphylococcus*, *Propionibacterium* and *Malassezia* spp.[22] The original patent cited "stabilized selenium disulphide" (CAS 7488-56-4) as the active component with the clause stating that this compound has great therapeutic and fungicidal value.[23] However, major difficulties are experienced in interpreting the validity of published toxicological reports which refer to the active ingredient of the shampoos as selenium sulfide (SeS) (CAS 7446-34-6). This view is reiterated in a statement made by the Hazardous Substances Data Bank,[24] which reads:

> "There is confusion in the literature between this compound and selenium disulfide (CAS7488-56-4). Users with an interest in this chemical substance are strongly encouraged to consult the record named SELENIUMCOMPOUNDS for general information on the toxicity and environmental fate of selenium ions and selenium compounds. Reference materials are often unclear as to whether the authors were discussing selenium sulphide or selenium disulfide".

Both compounds are insoluble in water and exist as yellow–orange or red–orange powders at room temperature, but they are synthesised in different manners and display different biochemical properties.[25,26] Selenium sulfides are absorbed through damaged skin, but human exposure is anticipated also through inhalation during use, in manufacturing processes and following accidental ingestion.[1] Some selenium sulfide is retained within superficial regions of the skin following topical use of shampoos, thereby giving lasting protection against skin infections, dandruff and seborrhoea. Selenium sulfide probably controls dandruff by exhibiting an antimitotic effect rather than though any antimicrobial action. Experiments using tritiated thymidine (^{3}H.Tdr) tracer showed a decrease in radio-labelled epidermal cells following application of 1% selenium sulfide (SeS).[27]

8.2.2 Selenium in Nutrition

The human body contains about 0.013 g of selenium, most of which is obtained through consumption of nuts, cereals and vegetables. Intestinal absorption through passive diffusion or *via* a sodium-mediated carrier transport mechanism is high, but varies according to the compound (*i.e.* selenates > selenites) and the composition of the diet.[12,17,28] The maximum safe daily intake of selenium in humans is not known but 50–200 μg per day is adequate for persons over 7 years of age.[29] More recent estimates show that dietary requirements for selenium increase with age and that 55 μg per day is adequate for adults; the maximum contaminant level in drinking water should be 0.05 mg l^{-1}.[30] Selenium absorbed into the body as selenites in the diet, or through parenteral supplementation, is readily metabolised, with hydrogen selenide (H_2S) being a key metabolite in the production of seleno-proteins.[28,31] Much of this is transported to the liver, lungs, kidney, heart and skeletal muscle, eye lens and thyroid gland.[32] The regulatory mechanisms for selenium absorption are not known but vitamins B_6, C and E and certain metals (copper, iron, zinc, magnesium and tin) in food can act as modulating factors.[12] Selenium is a metabolically active element and interacts with these nutrients as well as several non-nutrient elements including cadmium, mercury, lead and arsenic.

Selenium is an essential trace element in the mammalian body and a component of at least 30 proteins of which glutathione peroxidase, thyroid dehydrogenases (*e.g.* thioredoxin reductase), factor-3 and seleno-protein P are documented.[30,31,33,34] Seleno-enzymes regulate a number of essential metabolic processes and Se, acting as an antioxidant and with vitamin E, protects cell membranes and lipid-dependent intracellular structures against the detrimental effects of oxidative stress and reactive oxygen species created by certain toxic materials.[12] Vitamin E is a strong selenium synergist whereas vitamins B_6 and C may be involved in its homeostasis. Glutathione peroxidise catalyses the breakdown of hydrogen peroxide and certain endogenous peroxides.[35–38] Selenium inhibits cellular oxidation/reduction reactions, the cell cycle, and replaces sulfur in essential amino acids including cysteine and methionine. The chemistry of selenium suggests that it is present in the body as a selenol (*i.e.* selenomercaptan), or as Se-ether analogues to sulfur in methionine. The major metabolic pathway for selenium compounds involves methylation to dimethyl selenide, dimethyl diselenide or trimethyl selenolium ion in the liver, kidney, spleen and heart. Selenium also appears to be incorporated into proteins *via* post-transcriptional modification of polypeptides.

Selenium deficiency, as in Keshan's disease and Kashin-Beck syndrome, prevalent in certain parts of Asia, is recognised by cardiomyopathies, mental retardation, musculoskeletal abnormalities, osteoarthritis or "big-joint disease" (in children) and dermatological lesions.[12,39,40] Other selenenium deficiency conditions include haemolytic anaemia, cataracts, reduced immune responsiveness, depressed thyroid function and low levels of the selenium-dependent enzyme glutathione peroxidase.[41] Ultrastructural studies on biopsies from patients with Keshan showed myocardial necrosis and fibrosis and remarkable granular changes in mitochondria not seen in other forms of myocarditis. A relationship has been seen between selenium deficiency, with low vitamin E and zinc, and cystic fibrosis and Crohn's disease.[34,42] Twenty-two patients with Crohn's disease showing depletion of plasma and erythrocyte selenium and glutathione peroxidase were subject to bowel resection and reduced absorption. Selenium deficiency syndromes in cattle and other farm animals grazing selenium-rich pastures in Asia, parts of Ireland, North and South America and elsewhere are well documented.[8,13,43] Animals exhibit lameness, horizontal grooves or cracks and occasional sloughing of hooves, long periods of recumbency, loss of body condition, hair loss, reproductive failure and fatalities.

High levels of selenium are toxic to humans, cattle, pigs and wild animals living near selenium-rich soils or in the vicinity of mines. Characteristic signs of selenosis include severe gastrointestinal and neurological symptoms, fatigue and dermatological abnormalities with high blood, urine and nail selenium.[44,45]

Koller and Exon estimated that symptoms of selenosis develop in humans following a chronic ingestion of 1000–1500 μg per day, but the condition is rare.[35] The maximal tolerable level for selenium in humans is not known but selenium in hair and urine is used as a measure of the total body burden. In rural areas of China, people have been reported with toxic signs following inhalation of smoke from the burning of low quality coal with a 300–80 000 $\mu\ g^{-1}$ selenium content.[8,13] Affected individuals exhibited loss of hair and nails, dental deformities, bone and joint diseases, neurological problems and failure to thrive, following an estimated daily selenium intake of 3.2–6.7 mg.[13,14] Nuttall reviewed reports on selenosis and indicated that serum concentrations range from 400 to 30,000 $\mu g\ l^{-1}$ in acute cases and 500–1400 $\mu g\ l^{-1}$ in chronic toxicity.[46] Ingestion of inorganic compounds such as selenous acid found in gun-bluing agents can lead to acute fatality with serum selenium concentrations of >1400 $\mu g\ l^{-1}$. Other haematological changes include reduced fibrinogen, glutathione and prothrombin activity, raised alkaline phosphatase and succinic dehydrogenase.

Selenium toxicosis is better documented in cattle and horses grazing selenium-rich pastures where "blind staggers" and alkali disease are regarded as tell-tale signs of selenium toxicity, although sulfate toxicity may be contributory.[45,47] Alkali disease results from selenium substituting for sulfur in essential amino acids such as cystine and cysteine required for synthesis of mucopolysaccharides in keratinising and connective tissues. Affected animals display abnormal cartilage, bone and tooth development, neurological lesions, lameness and reduced reproduction. Weakened hooves tend to fracture when subjected to mechanical stress.

The nutritional and metabolic roles of selenium in the mammalian body are modulated by vitamins, trace metal nutrients and hormones.[12,36,38] Silver, arsenic, cadmium, mercury, lead, iron, fluoride, tin, copper and manganese antagonise selenium uptake and protein binding. In contrast, magnesium and cobalt promote selenium uptake and synthesis of the seleno-enzyme glutathione peroxidase.[48–50]

8.2.3 Cytological Effects and Mutagenicity

For *in vitro* and *in vivo* experiments considerable attention has been given to developing models in which to study the anticarcinogenic effects of selenium. Evidence for the mutagenicity and cytotoxicity of selenium compounds is mixed. Selenates and selenites have shown negligible signs of mutagenicity in several bacterial screens including mutant strains of *Salmonella typhimurium*, but levels of clastogenicity are low. Thus, base-pair substitution, sister-chromatid exchanges and unscheduled DNA synthesis were reported in human and hamster ovary cells cultured in the presence of glutathione.[51–53] The suggested mechanism implicated reduced glutathione in converting selenite/selenite/selenide to "a form capable of inducing the clastogenic changes". This is not further clarified. Experiments with the organo-selenium antioxidant Selol (selenitetraglycerid) synthesised at the Medical University of Warsaw showed lack of mutagenicity in five strains of *Salmonella* with or without the S9 fraction, but the stability of this material and the toxicity of metabolites in the presence of biological material is not known.[54]

In anticarcinogenicity studies, sodium selenite has been shown to inhibit S9-dependent mutagenic events and metabolism of benz[a]pyrene (BaP) in a T100 strain of *Salmonella,* and to invoke regressive changes in BaP-treated malignant cell lines.[55,56] Sodium selenite significantly reduced clastogenic effects induced by aflatoxin B_1 in human leucocyte cultures and mitigated the capacity of aflatoxin B_1 to cause DNA damage.[57] Sodium selenite or selenocystine alter the characteristics of human leukaemia cell cultures by depleting intracellular glutathione and stimulating glutathione reductase and oxygen consumption.[58] The mechanism of the protective action of selenite is shown in experiments in guinea pig macrophages exposed to the cytotoxic effects of quartz particles.[59] Addition of sodium selenite to the culture led to stabilization of cell membranes and prevention of lipid peroxidation, thereby promoting cell survival, intercellular adhesiveness and differentiation. This protective action is further illustrated by experiments in which cultured rabbit alveolar macrophages were exposed to selenium particles or selenous acid, which are commonly released during combustion of fossil fuels.[60] Neither selenium nor selenious acid caused cell lysis after 20 hours of exposure at concentrations which reduced cell viability. Selenious acid was appreciably more toxic than elemental selenium. Rat dermal fibroblasts were similarly unresponsive to selenium particles at concentrations approaching inhibitory levels.[61]

8.2.4 Carcinogenicity

8.2.4.1 Experimental Studies

The carcinogenicity of selenium in animals and humans is equivocal.[62–64] Whereas early experiments in animals suggested that dietary consumption of selenite or selenites could be a cause of hepatocellular tumours, these experiments were scientifically unsound and the tumour incidence not statistically significant.[65,66] Schroeder and Mitchener demonstrated that 2 ppm selenium as selenite or selenate in drinking water for 105 weeks rising to 3 ppm for up to 24 months was severely toxic to Long Evans rats, but the number of a malignant tumours was significantly higher following selenate (41.7%) than following selenite exposure (12.5%).[66] Control animals exhibited unacceptably high incidences of malignancies (16.9%) and tumour patterns, comprising many mammary and connective tissue lesions, were inconsistent across selenium-treated and control groups, casting serious doubt on the validity of the work. None of the observations made in farm and wild animals exposed to high levels of selenium in China and other eastern countries has provided any evidence of the carcinogenicity of selenium at very high levels, at which the animals exhibited severe neurological and musculoskeletal disease.

Concern over the possible carcinogenic risk afforded by 1.0% selenium sulfide used in antidandruff shampoos led the National Toxicology programme to conduct bioassays in 1980.[27,67–69] In an initial study in which 0.5 or 1.0 mg of selenium sulfide (SeS) was applied topically three times weekly for 86 weeks, the dose given was toxic and most animals died without evidence of carcinogenicity. In contrast, in a comparable study in which 25 or 50% dilutions of Selsun antidandruff shampoo were applied for up to 88 weeks, many tumours were recorded in male and female animals but the incidence was not dose related. The background incidence of pulmonary tumours was very high. In a 2-year experiment in which F344 rats were gavaged daily with 5 mg kg^{-1} of selenium sulfide in 0.5% carboxy-methylcellulose, a significant increase in hepatocellular tumours (adenomas and carcinomas) in male and female animals was reported at the high dose level, but the incidence of other tumours, including lymphomas, reproductive tract and endocrine neoplasms, was high in both test and control groups (>43%). In mice given massive doses of 20 or 100 mg kg^{-1} per day selenium sulfide (*recommended human dietary intake (RDI) is <70 μg*) for a similar period, female animals in the high-dose group developed a significant number of liver tumours, but male animals in both test and control groups exhibited an unacceptably large number of lung and liver tumours.

More recently, considerable attention has been given to the possible therapeutic value of selenium supplements as anticancer therapies for malignant tumours resulting from the action of idiopathic or chemical factors.[70–75] Administration of selenite to rats treated with known carcinogens suppressed progression of neoplasia and led to apoptosis and regression of established tumours, possibly through blocking of critical phases in the cell cycle or downregulation of enzymes integral to the carcinogenic process. Selenium supplementation had no influence on induction of hepatocellular carcinoma in rats treated with diethyl nitrosamine or *N*-acetyl-aminofluroene,[76] or the induction of oesophageal or bladder cancers by nitrosamines.[77] Oral selenium supplements led to an increase in ring-hydroxylation and an increase in *N*-hydroxylation reflecting a shift in the balance towards detoxification pathways.[78] Other experimental studies have demonstrated the capacity of selenite in the diet to reduce susceptibility to 7-12-DMBA-induced mammary carcinogenesis.[79,80] The organo-selenium compound 1,4-phenylenebis(methylene)-seleno-cyanate was shown to protect mice against development of lung and forestomach tumours induced by benz[a]pyrene or 4-[methyl-nitrosamino-1-(3-pyridyl)-1-butanone] found in cigarette smoke, possibly by inhibiting enzymes involved in the activation of the experimental carcinogens.[81]

8.2.4.2 Human Experience

No tangible evidence has been seen to implicate selenium or its compounds as human carcinogens following ingestion in food and drinking water, inhalation or dermal exposure. This is in part due to the paucity of clinical reports citing selenium as a cause of mortality or other serious conditions, and partly due to the infrequent analysis of blood silicon in clinical cases.[82] Greater emphasis has been placed on the action of selenium in preventing degenerative diseases and as an anticancer therapy.[37] Selenium toxicosis is now well acknowledged in certain regions of central and northern China where ingestion of high selenium in the diet and inhalation of smoke from low quality coals are problems, but reports fail to recognise selenium as a carcinogen.[8,13,14] Observations in the Kentucky area of the USA where groundwater contains 0.006–5.37 mg l^{-1} show evidence of increased selenium intake in local populations measured in hair and nails, but these did not correlate with increased cancer risk.[83] Nuttal examined in detail reports of human exposure spanning the range of 400 to 30,000 μg l^{-1} associated with acute exposure and 500 to >1400 μg l^{-1} in chronic exposure.[46] He reported that acute poisoning leading to blood selenium levels of >1400 μg l^{-1} involved ingestion of inorganic compounds such as selenious acid, used in gun-bluing agents.

The wider implications of selenium exposure in various industries including electrical and semiconductor manufacture, pigments in paints and glass has been examined.[44] Inhalation and dermatological exposure to selenium dioxide, formed whenever selenium is boiled in air, seem to present the greatest problems. The red fumes emitted contain up to 50% selenium dioxide and this is a severe pulmonary irritant, and a cause of bronchospasm and possibly systemic injury. Dermatitis and skin burns, severe eye irritation and systemic changes are additional acute problems associated with selenium compounds in industry. In longer-term exposure haematological and neurological problems are important, possibly complicated by damage to the liver and kidneys. Selenium-related cancer was not a problem.

Evidence that selenium compounds will protect against or act as antitumour agents in humans derives from environmental observations and a limited number of epidemiological trials. In the first instance, claims are made that people living in low-selenium areas and eating diets low in the element are more prone to cancers of the lung, colorectal region and prostate, but the incidence of non-melanoma skin cancer was higher.[75,84,85] Other studies suggest that selenium with vitamin E is beneficial in protecting against prostatic cancer in patients with low baseline plasma selenium, and is particularly effective in prostatic cancer in smokers with low vitamin E.[86] However, in an attempt to ratify these observations, the Nutritional Prevention Council Trial (NPC) in 1983–1996 conducted a major trial in the eastern USA involving 1312 patients.[87,88] They demonstrated that a daily dose of 200 μg of selenium as selenised yeast failed to protect against basal cell tumours of the skin, and that risks of squamous cell and non-melanoma tumours were significantly higher. Is this evidence that selenium is carcinogenic to the skin in humans?[63] In a reply to queries on their observations in the NPC trial, the authors emphasised that the carcinogenicity, toxicity and anticancer properties of selenium compounds vary greatly according to chemical formula and their metabolism. In the selenised yeast formulation, selenium exists in the form of selenomethionine, which is well absorbed and has a low toxicity, much like methionine.[88,89] It releases metabolites of selenocyteine, the oxidant hydrogen selenide, and methylselenol, which in their opinion exhibits protective action against cancers. However, the effect was only significant at the lower baseline of prostate specific antigen

(PSA) and plasma selenium levels. In a later sub-study, they administered 400 μg per day to 424 subjects and observed that, whereas the 200 μg per day dose decreased total cancer incidence by a significant 25%, 400 μg per day had no such effect on total cancer incidence.

The National Cancer Institute assembled a clinical trial to test further whether selenium has a preventive effect on prostatic cancer.[90] However, this trial was aborted on the grounds that selenium with or without vitamin E supplementation did not prevent prostatic cancer and that a small but not statistically significant increase in prostatic cancer cases was observed among 35,524 men over 50 years of age. A small increase was noted in the number of men with diabetes. No mention was made in the evaluation of the effects of the treatment on the incidence of other cancers.

More positive observations relating to the action of selenium in mitigating or preventing cancer are provided by a clinical trial of 73 patients in North Carolina.[72] Lower plasma levels were associated with a significantly higher incidence of colorectal adenomas, although the size and location of adenomas were not influenced. Later work with 276 colorectal cancer patients from a large multicentre adenoma prevention trial failed to substantiate that plasma selenium and total selenium levels exhibit a protective effect against adenoma recurrence.[85]

8.2.5 Summary

The carcinogenicity of selenium and related compounds, including the selenium sulfides, as presented through clinical and epidemiological studies in human patients and in laboratory models is complex. Whereas a single long-term study in rats and mice provided some indication that selenium sulfide (SeS) might evoke hepatocellular tumours in rats and a significantly increased number of lung tumours in female mice, the study is not satisfactory on the grounds that the concentrations of selenium used were unrealistically high and the total number of tumours in all groups was unacceptably high. Observations in the two species were inconsistent and the work has not been ratified. Other rodent studies reviewed are scientifically unsound. In clinical studies, limited information is provided by two clinical trials that high plasma selenium is associated with a low but increased incidence of prostatic and non-melanoma skin cancers. This is countered by large-scale population studies in China and elsewhere which indicate that high chronic selenium consumption is associated with general ill health, bone and joint diseases and failure to thrive, without evidence of increased cancers of any type. Selenium sulfide in Selsun anti-dandruff and antifungal shampoos is still widely available, and after 50 years since its original patent was issued, no cases of human cancer of any type have been associated with its use. The carcinogenicity of selenium and related products is not established. Second, whilst numerous experimental studies in animals treated with known carcinogens have demonstrated that the antioxidant effect of selenium can mitigate or prevent neoplastic changes, this is not proven in human patients.

8.3 Silicon

Silicon is a metalloid element classified in the carbon family of elements with germanium, tin and lead in Group IVb of the Periodic Table. It is a quadrivalent element with three naturally occurring isotopes. Silicon is the second most abundant element in the Earth's crust, exceeded only by oxygen, with which it combines at high temperature to form silicon dioxide (SiO_2). Silicon never occurs as a free element but is combined with one or more other elements. Substantial quantities of silicon dioxide are found worldwide in the form of sand, sandstone, granite, quartz, flint, cristobalite, tridymite and gemstones (amethyst, opal, mica, feldspar, garnet, tourmaline, asbestos, talc, zircon, emerald, and aquamarine). Crystalline silica forms a significant proportion of clay, shale and slate, but may occur also at low levels in chalk, limestone and other rock. Crystalline silica takes many forms with varying dimensions and diffraction patterns under a petrographic microscope (*i.e.* micro-crystalline or cryptocrystalline).[91] Quartz is the commonest form of silica found in concrete and mortar, which also contain crystalline silica. Asbestos fibres are naturally occurring hydrated silicates that have long been mined and used for their fire-retardant and insulating properties as construction materials. Silica occurs as a natural component of many materials used or encountered in construction site activities. Silica dust exists in the atmosphere in all parts of the world, distributed by air turbulence, volcanic action and geochemical activity.[9]

There a few situations when humans are not exposed to silica in the air or through the things we touch. Few problems have been encountered with dermal contact or ingestion of silica and most silicates and silicones are free from allergenic activity.[16] However inhalation of fine silica dust of "respirable" particle size is a cause of chronic lung disease and lung cancer. Additionally, in the past 100 years considerable health problems have been associated with inhalation of asbestos fibres. Asbestos is a generic term for a group of six naturally occurring fibrous silicate minerals that have been widely used in commercial products. There are two main groups: amphibole comprising the minerals amosite, crocidolite, tremolite, anthophyllite and actinolite; and serpentine minerals, notably chrysotile, a hydrous magnesium silicate mineral with the chemical composition of $Mg_3Si_2O_5(OH)_4$ that possesses relatively long and flexible crystalline fibres capable of being woven. Chrysotile accounts for about 95% of all asbestos in commercial use, but is distinct from other types including the fibrous forms of the amphibole group of minerals. The carcinogenicity of asbestos is well documented.[92–98] The first recorded fatal case in England concerned a woman who developed a massive pulmonary fibrosis through inhaling dust in asbestos manufacturing works in Rochdale in 1924.[99] Since then, many hundreds of patients have been diagnosed with pulmonary fibrosis, pneumoconiosis, pulmonary carcinoma and mesothelioma, some leading to litigation against employers. The Health and Safety Executive in the UK recorded that the number of mesothelioma deaths had increased from 153 in 1968 to 2347 in 2010, and over 80% of the fatalities involved men exposed to asbestos in the workplace.[100] Diagnosis is frequently complicated by a long lag phase of 20–40 years after asbestos exposure before mesothelioma is identified.

Silicon, silica and silicates occur naturally in plants and foodstuffs, bird egg shells and some sea foods. Until recently, silicon had not been recognised as a human nutrient and many early textbooks on nutrition fail to list it. Silicon is now established as an essential component of connective tissues although much remains to be understood regarding its uptake mechanism, metabolism and interaction with other metals in metal-binding proteins. No major health concerns are associated with excessive consumption of inorganic silicon compounds (including silica) in the diet.

In the past 50 years three major issues have become the centre of public attention in connection with the use of silicate and silicones in industrial environments, mining, manufacture and use of products:

1. Occupational and environmental exposures to silica dust of respirable particle size.
2. Use of magnesium silicate as talc for alleviating the adhesiveness in surgeons' gloves, and talcum powder for personal hygiene and cosmetic purposes.
3. Silicone oil, gels and implants used in cosmetic surgery, notably augmentation mammoplasty.

Each presents a different profile of exposure patterns, human tissue response and physio-pathological consequences. Each has been subjected to extensive national and international evaluation and regulation, and whilst chronic inhalation of silica dust and chrysotile (white asbestos) is an established cause of pulmonary carcinoma and mesothelioma, the carcinogenic risks associated with exposure to talc during mining, clinical or cosmetic exposure are questionable. Use of silicones in cosmetic surgery has been an emotive issue since the procedures for mammoplasty were developed in the 1960s, but as discussed, present views are that they present low risk (Figure 8.1).[101–103]

In recent years, silicones have been developed successfully as dressings and therapies for skin wounds, scars and keloids.[104,105] Long chain silicone polymers used to cover chronic wounds and burns are not thought to be bioactive but "tissue friendly" and with physico-chemical properties enabling painless application and removal. Other dermatological applications include treatment of exudating and infected wounds, skin graft sites, pilonodal cyst surgery and acniform lesions. Silver-containing silicone dressings have been developed as a therapy for infected wounds with a commendable safety record. None of these therapies implicates silicones or related products in skin cancer although on rare occasions foreign body reactions result.

8.3.1 Silicon as a Human Nutrient

Contrary to many texts, elemental silicon should be recognised as an essential micronutrient for humans and animals, but only recently have its

Figure 8.1 Silicone-gel filled implant for mammoplasty.[102]

bioavailability and absorption become appreciated.[106–109] In his survey of the "newer" human nutrients, Mertz commented that silicon may play an important role in the structural configuration of mucopolysaccharides of bone and cartilage as part of a very stable molecular structure with the proposed configuration shown below.[110]

$$\mathbf{—R_1—O—Si—O—R_2 \text{ or } —R_1—O—Si—O—Si—O—R_2—}$$

Chemical analyses show that glycosaminoglycans such as hyaluronic acid contain 1892 μg g^{-1} of silicon and that chondroitin sulfate in cartilage contains high concentrations. Recent investigations associated with the Framlingham Study, and experimental studies in rats, have shown that silicon is absorbed intestinally as orthosilicic acid and plays a central role in collagenesis in skin and wound healing and in bone formation, where it interacts with other trace elements including calcium, zinc, copper and iron.[105,107,111–115]

Carlisle was the first to identify the essentiality of silicon in the human diet and with the rudimentary analytical methods available for silicon at the time, showed that highest concentrations were present in the epidermis and connective tissues generally.[116] Her preliminary analyses showed that normal human blood concentrations were 0.5 mg l^{-1} and that these increased followed administration of silicon as sodium metasilicate or organic silicates. Blood silicon levels did not rise following inhalation of silica dust, suggesting minimal pulmonary absorption. Although the exact definition of the silicon deficiency state in humans is not known, early experiments in rats and chicks established that silicon is necessary for normal growth and development, collagenesis in tooth, bone and cartilage growth, and vascular structures.[117,118] Experimental studies in calves and rats showed that silicon supplementation in the diet enhanced levels of collagenesis and increased the calcium content in bone, whereas low silicon led to a dose-related reduction in overall body growth, lower bone development and strength, cranial abnormalities and aortic thinning with lower calcium, copper, potassium and zinc.[115,118–120] X-ray analysis of growing bone suggests that silicon is located intracellularly in the mitochondria and cytoplasm of osteoblasts, and is found specifically in the nucleus and mitochondrial granules of fibroblasts, but not other cells.[121–123] Silicon contents of tissue decline with advancing age, particular in vascular tissue such as the aorta.

The human body contains approximately 1 g of silicon, mostly in bone, cartilage and certain soft tissues where it appears to act as a cross-linking agent in maintaining the structure and resilience of connective tissue. Reference values for optimal dietary silicon derived from study of 1325 healthy individuals show serum silicon of 9.5 μmol l^{-1} in men (18–59 years) declining to 8.5 μmol l^{-1} later, whereas in women aged 18–29 years, median serum silicon was 10 μmol l^{-1} rising to 11.10 μmol l^{-1} at 30–44 years and then falling to 9.23 μmol l^{-1} in the 45–59-year age group.[124] In both men and women silicon levels declined further with advancing age. Tamada estimated that the ratio of silicon between serum and erythrocytes was in the range 0.15 to 0.39.[125]

Silicon is well absorbed gastrointestinally (50%) from the bioactive form orthosilicic acid (H_4SiO_4), with excess excreted in the urine.[126] Studies in human volunteers have shown a significant correlation between creatinine clearance and silicon levels in serum; urine renal clearance of 82–96 ml min^{-1} silicon suggests a high renal filterability. Silicon is known to protect against the toxic action of xenobiotic elements such as aluminium.[126] Poor skin quality, osteoporosis and cardiovascular deformities are manifestations of long-term silicon deficiency [118,127] Homeostasis of silicon in the human body is imperfectly understood but clinical studies showing that median silicon concentrations are age and sex dependent suggest that hormonal factors may act in a regulatory capacity. Silicon and silicates in diet seem to be well tolerated and without serious adverse effect.[128,129]

8.3.2 Silicon, Silica and Silicates in Industry and Human Exposure

There is no clear evidence that elemental silicon has any commercial value but most humans regularly come into contact with silica, silicates and silicones though building and construction industries, plastics, electronic components (transistors, integrated circuits, computer chips, solar cells, *etc.*), fire-resistant asbestos sheeting, silicon rubbers, oils, medical wound dressings, talc in surgery, food additives and in prosthetic implants. Each exposure pattern presents a different toxicological and analytical challenge and mode of exposure, through dermal contact, inhalation of particles in the atmosphere, or ingestion in contaminated food or drinking water.

Silica has a well defined physico-chemical structure with constituent atoms bound by strong covalent bonds, but its molecular configuration varies according to crystal type. Thirty-five distinct crystal types of silica have been identified, each with characteristic physico-chemical properties and presumably toxicological profiles.[130] Few have been fully investigated toxicologically. Three major forms are recognised, each with sub-divisions. Different forms of silica react variously with water. Whereas silica is commonly regarded as an insoluble and inert compound, it is marginally soluble in water (0.12 g l^{-1}) (and presumably body fluids) to form weak silicic acid in the equilibrium:

$$SiO_2 + 2H_2O \Leftrightarrow H_4SiO_4$$

This is absorbed by the human intestine. The influence of this water-solubility on alveolar absorption following inhalation of silica dust is unclear. I have seen no information relating to the per-alveolar uptake of silicic acid formed in alveolar surfactants. Silicic acid forms also during the hydrolysis of silicates. The US DHHS and IARC have indicated that crystalline particles of silica of respirable size are a potential cause of lung cancer and mesothelioma.[1,4] Workers are exposed to contaminated air in the natural environment, near quarries and gravel pits, factory sites and in professions involving grinding and abrasion of quartz and cristobalite, and in building sites. Risks also apply to workers in shale, clay, mining, glass and pottery manufacture. A frequently overlooked hazard relates to workers who are exposed to silica particles occupationally and then convey dust to their homes on clothing, in their hair or on their bodies. Risks associated with secondary contamination in children are recognised.[131] In each case, absorption and health risks relate closely to the size and density of the silica particles/fibres, frequency of exposure and use of protective measures, including observation of factory recommendations for occupational health and safety. Approximately three million workers in the USA are estimated to be exposed to silica, human-made mineral fibres, and asbestos.[132] The lung is the primary target organ of concern. Human exposure to silica and silicates in building involves a large element of dermal contact but inhalation of dust is a potential cause of pneumoconiosis, respiratory distress and cancer.[1]

Exposure to talc in mining, refinement, industry and through medical/surgical practice is an identified health risk. Soapstone is a grey to green, soft and typically "soapy" metamorphic rock composed mostly of the mineral talc (sometimes also called talc-schist). As mentioned earlier, talc used to prevent adhesion in surgical gloves and in dusting powders for cosmetic and hygiene purposes has been implicated in the induction of ovarian and other tumours even though the fine powder is insoluble in water, stable at body temperatures and biologically inert.

Silicones comprise a large group of straight chain polymers with the general formula shown in Figure 8.2.[133] The polymeric chains and networks, based on repeating –Si–O–Si– units, have widespread use in oils, greases, rubber-like materials, clothing, surgical dressings, electrical insulators, hydraulic fluids and

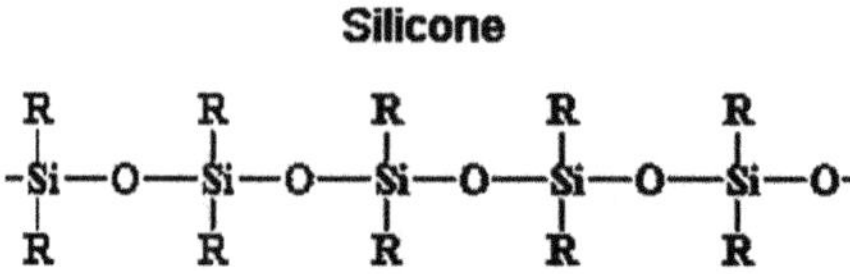

Figure 8.2 Molecular structure of silicones (R = CH_3, C_2H_5, *etc.*).

in moisture proofing. They are water repellent, heat stable and resistant to chemical attack. Silicones were introduced as breast implants and prostheses in reconstructive surgery in the 1960s, and for use in arthroplasty in rheumatoid disorders.[134] Mammoplasty implants composed of a silicone rubber shell (elastomer) and filled with a silicone gel were claimed to be a cause of mammary and other cancers, and hence became subject to intensive scrutiny and controversy.[135]

8.3.3 Cytotoxicity and Carcinogenicity

8.3.3.1 *Microcrystalline Silica*

The 12th RoC document recognises that respirable crystalline silica, and primarily quartz dusts occurring in industrial and occupational settings, constitute a human carcinogenic risk based upon sufficiency of evidence for carcinogenicity in human epidemiological studies.[1] The IARC and US DHHS emphasised that a direct link exists between silicosis and elevated lung cancer rates, and that highest risks lie with exposure to respirable quartz and cristobalite dust, but not to amorphous silica.[1,7,94,101]

The comparative toxicology of the different crystalline forms of silica is poorly understood, nanoparticles exhibit novel kinetic properties and unusual biochemical reactivity when compared with micron-sized bulk materials.[136] Crystalline silica seems to be appreciably more active than amorphous forms. Nano-particulate silica (1–100 nm range) was not mutagenic in cultured mammalian cells,[137] but in rats subjected to sub-chronic inhalation of amorphous or crystalline silica, neutrophil concentrations were significantly increased.[138] They showed evidence of cytotoxic change and mutations in the *HPRT* (hypoxanthine phospho-ribosyltransferase) gene. Other studies in cultured mammalian cells have shown that silica generates reactive oxygen species and upregulates nuclear transcription and growth factors, with oncogene expression.[139] Amorphous and crystalline silica were shown to induce pulmonary inflammation in rats, but whereas amorphous silica cleared readily from the tissue, presumably through the action of alveolar macrophages, crystalline silica persisted and was associated with greater inflammatory response and mutagenicity. Other studies have demonstrated that quartz crystals instilled intratracheally influence inflammatory cell accumulation and pulmonary surfactant profiles as a preliminary to fibrotic change and mutagenicity at low concentrations, but that changes were directly related to particle size (1.8 μm > 0.7 μm).[140,141] Pulmonary macrophages provide some protective

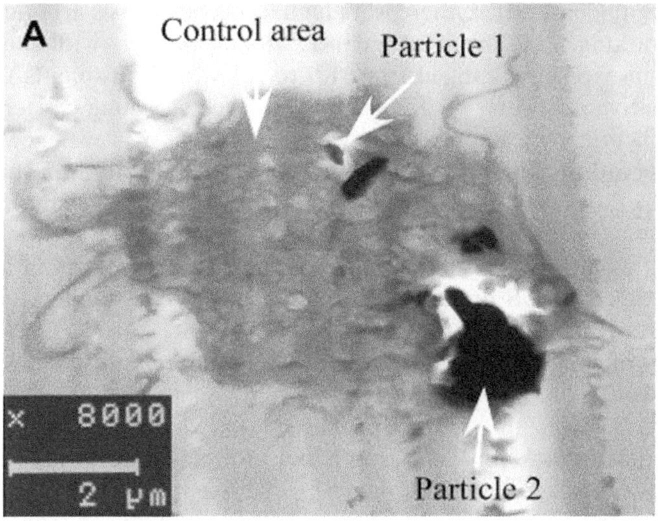

Figure 8.3 Scanning transmission electron micrograph of a macrophage exposed to silica, showing the presence of silica particles of varying sizes.[142]

function by phagocytosing silica particles. Time-lapse imaging has shown that mouse macrophages avidly bind particles that settle onto the cell surface and that cells also extend protrusions to capture distant particles.[142,143] Cultured macrophages bind crystalline silica particles, phagocytose them, and rapidly undergo apoptotic and necrotic death. Using confocal optical sectioning, silica particles were shown to be present within the cytoplasmic volume of live cells (Figure 8.3). In addition, electron microscopy and elemental analysis showed silica in internal cellular sections. Cell death was directly related to time and to silica concentration.

Recent epidemiological and occupational health studies have reviewed silica exposure as a cause of lung cancer in workers in mines, stone quarries, granite production, pottery, glass and ceramics manufacture, steel working and in semiconductor industries.[144,145] Whilst the majority of studies claim a statistically significant dose-related trend in cancer incidence and mortality in relation to exposure level, duration of exposure and work patterns, there is considerable inconsistency in the 50 or more studies published since 1997, when the IARC produced its last Monograph.[4] Silicosis and chronic obstructive lung disease are major causes of death in miners, pottery workers and quarrymen, but some surveys involving several thousand workers have failed to associate silica exposure with lung cancer.[146–148] Major causes of inconsistency seen in clinical and retrospective trials of silica workers include cigarette smoking and exposures to other contaminants including radon, arsenic and coal dust.[149–151] A recent estimate showed that, in 8000 industrial hygiene measurements in the German porcelain industry, exposure to crystalline silica for more than 4 mg m^{-3} years or more than 0.15 mg m^{-3} (mean) was strongly associated with silicosis, rather than lung cancer.[148] Chen and Chen evaluated cancer incidence in tin mines in China where there were high levels of silicosis arsenic, radon and tobacco smoking.[149] Risks of lung cancer among workers with short, medium or long term exposures ranged from 1.9 to 2.3 with adjustments for smoking and cumulative exposures to arsenic, but no excess of lung cancer was found amongst silicotic subjects in a mine where there was a prevalence of silicosis. These observations challenge the commonly held view that respirable silica particles are a cause of lung cancer; in fact, arsenic and smoking present greater risks.

Diagnostic features of silicosis (also known as miner's asthma and potter's rot), as detailed in US autopsy records of 3365 former miners (1971–1980), include massive nodular fibrosis, birefringent silicate particles, mineralisation in tracheobronchial and hilar lymph nodes and inflammatory changes.[146] Ulm *et al.* also concluded, from their survey of 247 German quarry workers, that exposure to 0.15 mg m^{-3} respirable silica dust (which was the current occupational exposure level in Germany) is not a direct cause of lung cancer, and that risks of lung cancer in the absence of silicosis are small or non-existent.[147] In their view, the odds ratio of developing lung cancer ranged from 0.85 to 1.02 with no significant trend in relation to cumulative silica exposure. Exclusion of subjects with silicosis may have reduced the statistical power to detect small risks of lung cancer. Additional studies are warranted to substantiate the risks of skin and oesophageal cancers which have been recorded in some subjects exposed occupationally to crystalline silica in some studies.[150,152,153]

Numerous experimental studies are reported where strains of rodent have been exposed to silica dust by inhalation or intratracheal instillation. Wide variations in response are recorded according to the strain and species of animal used, and details of experimental procedures.[154] Although Wistar rats are possibly a suitable experimental model in which to demonstrate development of silicotic nodules, rat strains generally seem to be susceptible to silica (quartz or cristobalite)-induced lung cancer. However, on anatomical and physiological grounds, the rat should not be regarded as a suitable surrogate model in which to investigate human risks attributable to silica dust inhalation. Marked interspecies differences in patterns of particle clearance from lungs, the protective role performed by alveolar macrophages, profiles of alveolar macrophage responsiveness and differences in sensitivity to inhaled substances are expected.[155–158] These features preclude direct extrapolation of experimental studies in terms of human risk.

8.3.3.2 Silicates

Silicates are the most widespread Si compounds, and possible comprise more than 90% of the weight of the minerals found in the Earth's crust and in meteoric debris. The main structural unit of silicates is a tetrahedral cluster containing one silicon atom and four oxygen atoms. This SiO_4^{4-} anion is relatively stable and the strength of the sulfur–oxygen bonds provides the thermal stability and chemical resistance of the majority of silicate compounds. Silicates can exist in crystalline or non-crystalline forms, each with differing physico-chemical properties. Quartz for example has a solubility of 6 ppm, in contrast to silica which is soluble in water at 25 °C at 70 ppm. As a group, silicates represent a large and chemically varied group of compounds with a wide range of crystal and molecular structures, many with one or more metal cations *e.g.* Mg+ in talc as $Mg_3Si_4O_{10}(OH)_2$; Na^+, Al^{3+} in Albite ($NaAlSi_3O_8$); K^+, Al^{3+} in Microcline ($KAlSi_3O_8$) and Ca^{2+}, Al^{3+} in Anorthite ($CaAl_2Si_2O_8$). Principal commercial applications include ceramics and hydraulic cements (crystalline synthetic silicates), microchips and conductors in computer technology, oscillators in watches, radios and pressure gauges, glass, cosmetics, dental fillers, ant caking devices, fillers and stabilizers, soaps and detergents.[3,159,160] Few have been adequately evaluated toxicologically, other than talc, sodium silicate and aluminium silicate.

8.3.3.2.1 Talc. Talc is a soft mineral powder composed of microscopic platelets (Figure 8.4). It exists in various forms but is insoluble in water, does not react with air and exhibits low reactivity with body fluids. Adhesion between the platelets is low, enabling them to slide by one another, hence the soft and greasy feel of talc or talcum powder as used in cosmetic products. Talc is used to replace graphite as an industrial lubricant. In cosmetics, fine talcum powder adsorbs sweat and perspiration, but talc is not absorbed by body fluids. The toxicity and carcinogenicity of talc is one of the more contentious of the silicates, as discussed in Chapter 7 under magnesium.

Figure 8.4 Talc: a white pseudomorphous powder [$Mg_3Si_4O_{10}(OH)_2$] of hexagonal crystals (*Mineral News*).

The IARC evaluated talc as a possible human carcinogen in 1998 and concluded that there was inadequate evidence for carcinogenicity except where the powders were contaminated with asbestiform fibres in mining or preparative phases.[161] Muscat and Huncharek reviewed studies conducted in the years 1982–2007 and concluded that the 1.4% risk suggested by some work could not be substantiated because most of the studies failed to establish whether the talc entered the female body following external application, implying that the results were scientifically unsound.[5] Nevertheless, the association between talcum powder used for hygiene purposes and ovarian cancer is still a contentious issue.[6] Experimental studies in animal models have provided inadequate evidence for carcinogenicity.[162] A study in which seven different types of talc particles of respirable size were exposed to isolated mouse peritoneal macrophages *in vitro* showed that all induced cytotoxic change was manifest by enzymatic and morphological change, not consistent with mutagenic change or clastogenicity.[163] There seems to be little evidence to suggest that talc miners are at risk of developing lung cancer or mesothelioma following chronic exposure in mining, milling or related talc industries. This is well illustrated by a survey in an Italian mine where health records of 1795 men exposed to talc dust in the period between 1946 and 1995 showed no excess in cancer mortality but noted that the workers were exposed also to diesel fumes and radon in the course of their work.[164] In this cohort, there was an excess total mortality due chiefly to non-neoplastic respiratory diseases (mainly classified as silicosis), liver cirrhosis, other digestive tract diseases, and suicides. Pneumoconiosis, bronchitis and bronchiolitis are complications of "talcosis" in mining, milling, talc loading and bagging.[165] Combined exposure to talc and asbestos was a major cause of gastric cancer in a study of 11,633 German rubber workers.[166]

8.3.3.2.2 Other Silicates. A wide range of silicates occur in the human environment. Elmore reviewed the safety of 17 silicates commonly used in cosmetic formulations and emphasised that at least attapulgite fibres, clinoptilite, phillipsite, mordenite, non-fibrous Japanese Zeolite and synthetic Zeolite were not classifiable as human carcinogens.[3] They are not mutagenic in the Ames bacterial reversion test and failed to show unscheduled DNA synthesis or changes consistent with clastogenicity in primary hepatocytes or other mammalian cell cultures. Six silicates are used as ant caking agents in food production:

- Aluminium calcium silicate
- Calcium silicate
- Magnesium silicate
- Sodium aluminosilicate
- Sodium calcium aluminosilicate
- Tricalcium silicate.

They are not known to be contact allergens.[16] Human carcinogenicity data have not been seen.

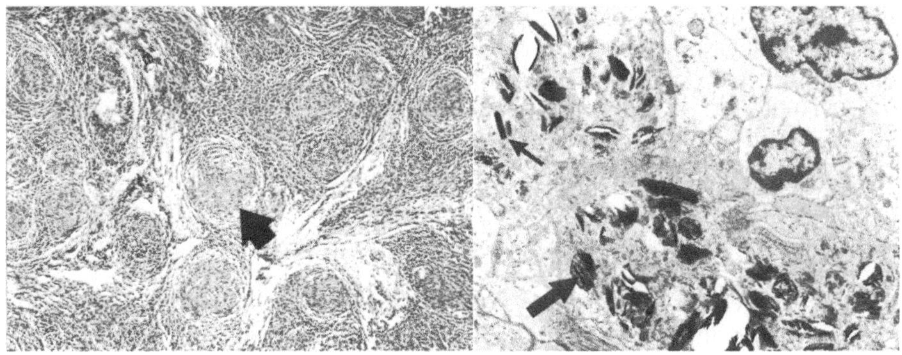

Figure 8.5 (left) Granulomata (arrow) replace the normal lymph node architecture (x50); (right) large quantities of needle-like particles are free (small arrow) or arranged in spherical bodies (large arrow) in the cytoplasm of macrophages. Holes in the tissue are artefacts of preparation (x5400). (From Pearson *et al., Veterinary Record*, 1986, **118**, 450).

Zeolites constitute an important group of silicates with applications in paper making, food manufacture and processing, detergent manufacture and agriculture. Zeolites have a rigid three-dimensional crystalline structure (similar to a honeycomb) consisting of a network of inter-connected tunnels and cages. Water moves freely in and out of these pores but the Zeolite framework remains rigid. Another special aspect of this structure is that the pore and channel sizes are nearly uniform, allowing the crystals to act as a molecular sieve. The porous Zeolite is host to water molecules and ions of potassium and calcium, as well as a variety of other positively charged ions. No evidence is seen to show that any of more than 50 Zeolites occurring naturally or synthesised is carcinogenic in humans, and a limited number of animal studies have failed to show carcinogenicity by oral or parenteral administration, or intratracheal instillation.[4,167,168]

An interesting veterinary case highlights a potential risk associated with accidental intradermal implantation of aluminosilicate particles in dogs.[169] Two Springer spaniels were referred with granulomatous lymphadenopathy associated with birefringent needle-like particles in macrophages and histiocytic cells in peripheral lymph nodes (Figure 8.5). The owner of one animal was a builder who regularly used building blocks manufactured from volcanic ash. Electron microscopy showed particles free within the cytoplasm or as lysosomally bound bodies. The lesions were originally identified as granuloma but later reclassified as lymphosarcoma on the basis of cytology and metastatic behaviour.

8.3.4 Silicones

Silicones are a family of polymers that comprise alternating silicon–oxygen atom backbones. In addition to their oxygen linkage to form polymeric chains, silicon atoms are bonded to organic groups such as methyl groups. They are formulated into materials with varying structural characteristics and physical properties including emulsions, resins, elastomers, rubbers, oils, gels, sheeting, *etc.* with a wide range of medical applications of which in-dwelling catheters, colostomy bags, sealants, waterproof adhesives and wound dressings are well known. They even have use in the aeronautics industry where their light weight, temperature stability and durability are advantages.

8.3.4.1 Mammoplasty and Cosmetic Surgery

Silicones were introduced into cosmetic surgery in 1962 when no specific statutory guidelines were in place to control the production, marketing or implantation of medical devices in the USA, UK or Europe. Four main types of implant have been developed:

- Silicone rubber envelope filled with saline
- Silicone rubber envelope filled with silicone oil (see Figure.8.1)
- Silicone rubber envelope covered with a polyester-polyurethane foam coating
- Dual silicone rubber envelopes with an inner containing silicone gel and an outer filled with saline.[170]

Whereas silicones were once thought to be inert in the presence of body tissues and fluids, there is now good evidence to show that some silicones used in mammary implants, intravenous catheters, *etc.* and other implantable medical devices are bioactive and lead to increased blood silicone levels or local tissue foreign body reactions and/or granulomas.[170–175] Silicones released into the circulation or into tissues surrounding the implantation site through seepage or rupture of implant envelopes can give rise to pathological complications including sclerodermatous skin lesions and arthritic pain.[176] Connective tissue disease is now well documented as a result of silicone gel seepage from implants, accidental rupture or permeability of implant envelopes.[177] Tissue reactions ranging from local inflammation, oedema, fibrosis to severe granulomas are reported in the capsular region surrounding an implant rupture and at more distant sites including the abdominal wall, limbs or liver, in each case associated with local increases in silicon concentration.[175,178,179] To some extent macrophages provide a defensive role in phagocytosing the gel in areas of granulomatous reaction. A case of a 45-year-old woman injected with silicone during cosmetic surgery to repair disfigurement following a motor cycle accident revealed granuloma fragments in her elbow resulting from migration of silicone.[180] Silicone fragments were identified using polarised light and X-ray microanalysis. Reactions to these silicone implants may not develop for months or even years following implantation of prosthetic breast material, and some instances persist after their removal. Phlederer and Garrido examined 15 patients with silicone gel-filled implants and reported silicone resonances in the liver 3–4 years after implantation; the levels identified were indicative of whether the implants were intact or had ruptured.[178]

Blood levels of silicone or silicon, as revealed by modern techniques of atomic absorption spectrometry, direct current plasma emission spectrometry or inductively coupled plasma emission spectroscopy, do not necessary correlate with obvious tissue reaction. In a five-year study, women with silicone gel breast implants had markedly elevated silicon levels, ranging from 9980 to 14,390 μg g^{-1} dry weight, compared with average control levels of 30 to 209 ng g^{-1}.[181] On the other hand, a study of 29 patients with silicone implants and showing breast silicon levels of 4–444 μg g^{-1} showed no association between silicon concentration and connective tissue disease.[174]

The association between silicone implantation, silicone gel release or even injection of liquid silicone oil into the breast for mammoplasty and human cancer is not proven. However a variety of tumours have been reported in the literature, associated with implantation of silicone materials. They include cases of mammary squamous carcinoma and "siliconomas" associated with invasive papillary micro-carcinomas following injection of silicone oil, on occasion many years after surgery.[182–185] Of these cases, a siliconoma in the breast was diagnosed in one woman 42 years following augmentation mammoplasty, which then progressed to an invasive micro-capillary carcinoma at multiple sites.[183] Rare cases of primary anaplastic large cell lymphomas (non-Hodgkins type) of the breast are reported following mammoplasty or elective breast augmentation and implants of a saline-filled implanter.[186–188] In each case the tumours developed in the periprosthetic capsular region 9–14 years following surgery. On occasion, multiple silicone granulomas can manifest as masses with suspicious morphology and positron emission tomography (PET) and magnetic resonance imaging (MRI) scans characteristic of neoplasia.[189]

In one of the largest clinical studies on the long-term effects of silicone breast implants, the National Cancer Institute (NCI) found no association between silicone implants and mammary cancer, with 12.9 years of follow-up.[189–192] Louise Brinton, the principal investigator in these NCI trials, found no association between the implants and immunological diseases, neurological problems or other systemic diseases in 13,500 women in the period between 1962 and 1989. In her view, epidemiological studies provide no support for an increased risk of sarcoma or multiple myeloma among breast implant patients. Increased incidence of cervical, vulvar and pulmonary cancers in some groups of breast cancer patients were attributed mainly to lifestyle factors including smoking and sexual behaviour rather than to silicone carcinogenicity.

Limited clinical evidence points to the possibility that silicone implants in mammoplasty may mitigate or even lower the risk of breast cancer. A population study of 25,230 women with or without silicone breast implants in Canada reported that 41 women with implants developed breast cancer and that this was 47.6% lower than predicted, within the 10 years of the study.[193] The average time from mammoplasty to cancer diagnosis was 7.5 years. This observation may indicate that the participants were drawn from a population with low cancer risk or that silicone did not substantially enhance the cancer risk. A larger scale study in Connecticut, linked with the tumour registry in

34 hospitals, demonstrated a modest reduction in the incidence of breast cancer (0.59 *vs.* 0.88%) in implanted *vs.* control patients, but a more significant reduction in non-breast cancers (0.59 *vs.* 2.7%).[194] Other studies using computerised records of 2033 patients over a 30-year period showed a 15% failure rate but that 85% implants were intact, and no increase in malignancies or expected immune disorder was indicated.[195] Subjective views on the acceptability and satisfaction with silicone mammary implants provided in interviews or questionnaires from patients indicated that the implants were beneficial and that mammoplasty was successful, yet 27% of 174 randomly selected women would be unlikely to choose silicone implants again in light of the potential complications.[196,197] Current views are that silicones used in mammary surgery are without carcinogenic risk.[190,192,197]

If silicones in breast implants or other long-term implanted devices are a potential cause of neoplasia, the mechanism is unclear. The pattern of cancers is not consistent and not specifically linked to any form of implant.[198] Experimental studies in animals provide little information concerning the carcinogenicity of silicone implants. In fact, rats and mice implanted with silicone two weeks before administration of carcinogens such as methylnitrosourea showed a reduction in mammary tumour incidence.[199] Exposure to silicone prostheses, silicone gel or silicone sheet alone did not increase tumour incidence in any tissue. Extracts of silicone gel were shown to be non-mutagenic in Ames *Salmonella typhimurium* tests and they failed to induce micronuclei in polychromic erythrocytes or chromosomal aberrations in bone marrow cells.[200]

References

1. US Department of Health and Human Services, Public Health Service, *12th Report on Carcinogens*, NIH, Washington, DC, 2011.
2. A. B. G. Lansdown, *Chemistry World*, 2012, **9**, 1.
3. A. R. Elmore, *Int. J. Toxicol.*, 2003, **22**(Suppl.1), 37.
4. International Agency for Research on Cancer (IARC), *IARC Monographs on the Evaluation of Carcinogenic Risks to Humans*, IARC, Lyon, 1997, vol. 68.
5. J. E. Muscat and M. S. Huncharek, *Eur. J. Cancer Prev.*, 2008, **17**, 139.
6. D. M. Gertig, D. J. Hunter, G. A. Colditz, F. E. Speizer, W. C. Willett and S. E. Hankinson, *J. Natl. Can. Inst.*, 2000, **92**, 249.
7. IARC, *IARC Monographs on the Evaluation of Carcinogenic Risks to Humans*, IARC, Lyon, 2010, vol. 93, p. 1.
8. R. G. Burau, *Calif. Agric.*, 1985, **39**, 16.
9. I. Thornton, *Metals in the Global Environment*, International Council on Metals and the Environment, ICME, Ontario, 1995.
10. Agency for Toxic Substances and Disease Registry (ATSDR), *Toxicological Profile for Selenium*, Public Health Service DHHS, Atlanta, GA, 1996.
11. World Health Organization (WHO) *Vitamin and Mineral Requirements in Human Nutrition*, WHO, Geneva, 2004.

12. D. L. Watts, *J. Orthomolec. Med.*, 1994, **9**, 111.
13. P. D. Whanger, *J. Nutr.*, 1985, **119**, 1236.
14. G. O. Yang, S. Z. Wang, R. H. Zhou and S. Z. Sun, *Am. J. Clin. Nutr.*, 1983, **37**, 872.
15. M. C. Yarema and S. C. Curry, *Pediatr.*, 2005, **116**, e319.
16. A. A. Fisher, *Clinical Dermatitis*, Lea and Febiger, Philadelphia, PA, 1987.
17. C. H. Lin, C. L. Fang, S. A. Al-Suwayeh, S. Y. Yang and J. Y. Yang, *Acta Pharmacol. Sin.*, 2011, **32**, 1181.
18. R. H. Guy, *Int J Pharm.*, 1999, **188**, 121.
19. J. Hostýnek, R. S. Hinz, C. R. Lorence, M. Price and R. H. Guy, *Crit. Rev. Toxicol.*, 1993, **23**, 171.
20. K. E. Burke, R. G. Burford, G. F. Combs, I. W. French and D. R. Skeffington, *Photodermatol. Photoimmunol. Photomed.*, 1992, **9**, 52.
21. International Program on Chemical Safety, *Environmental Health Criteria, Selenium*, WHO, Geneva, 1987.
22. S. Ranganathan and T. Mukhopadhyay, *Ind. J. Dermatol.*, 2010, **55**, 130.
23. US Patent *US2694669*, *Stabilized Selenium Bisulphide*, US Patent Office, 1954.
24. Hazardous Substances Data Bank, *Selenium Compounds*, US National Library of Medicine, Bethesda, MD, 2011.
25. J. Matson, *J. Soc. Cosmet. Chem.*, 1956, **7**, 459.
26. S. C. Mitchell, R. M. Nickson and R. H. Waring, *Sulphur Reports*, 1993, **13**, 279.
27. Kirk-Othmer, *Encyclopaedia of Chemical Technology*, John Wiley, New York, 1978–1984, 3rd edn, p. 12.
28. S. J. Fairweather-Tait, Y. Bao, M. R. Broadley, R. Collings, D. Ford, J. E. Hesketh and R. Hurst, *Antioxid. Redox Signal.*, 2011, **1/4**, 1337.
29. National Research Council (NRC), *Selenium in Diet, Recommended Dietary Allowances*, National Academy Press, Washington, DC, 1989, 10th edn, p. 217.
30. Office of Dietary Supplements, National Institutes of Health (NIH), *Dietary Supplement Fact Sheet: Selenium*, NIH, Washington, DC, 2009.
31. H. E. Ganther, *Carcinogenesis*, 1999, **20**, 1657.
32. C. C. Reddy and E. J. Massano, *Toxicol. Sci.*, 1983, **3**, 431.
33. K. Schwarz and C. M. Foltz, *J. Amer. Chem. Soc.*, 1957, **79**, 3292.
34. T. Rannem, K. Ladefoged, E. Hylander, J. Hegnhøj and S. Jamum, *Am. J. Clin. Nutr.*, 1992, **56**, 933.
35. L. D. Koller and J. H. Exon, *Can. J. Vet. Res.*, 1986, **50**, 297.
36. S. G. Patching and P. H. Gardiner, *J. Trace Elem. Med. Biol.*, 1999, **13**, 193.
37. L. Patrick, *Altern. Med. Rev.*, 2004, **9**, 239.
38. T. C. Stadtman, *Ann. Rev. Biochem.*, 1990, **59**, 111.
39. N. E. Vinton, K. A. Dahlstrom, C. T. Strobel and M. E. Ament, *J. Pediatr*, 1987, **111**, 711.
40. K. Ge and G. Yang, *Am.J. Clin. Nutr.*, 1993, **57**(Suppl.2), 259S.

41. J. S. McLester, *Nutrition and Diet in Health and Disease*, W.B. Saunders, Philadelphia, PA, 1943, p. 128.
42. J. D. Wallach, *Quantum Med.*, 1988, **1**, 1.
43. P. A. M. Rogers, S. P. Arora, G. A. Fleming, R. A. P. Crinion and J. G. McLaughlin, *Irish Vet. J.*, 1990, **43**, 151.
44. J. R. Glover, *Trans. Assoc. Ind. Med.*, 1954, **4**, 94.
45. M. F. Raisbeck, *Vet. Clin. North Am. Food Anim. Pract.*, 2000, **16**, 465.
46. K. L. Nuttall, *Ann. Clin. Lab. Sci.*, 2006, **36**, 409.
47. MSD, Nutritional myopathy in calves and lambs, in *The Merck Veterinary Manual*, Merck, Sharpe and Dohme, Whitehouse, NJ, 2011.
48. H. E. Ganther, *Biochemistry of Selenium*, Zingaro, Copper, VNR, New York, 1974, p. 546.
49. W. A. House and R. M. Welch, *J. Nutr.*, 1989, **119**, 916.
50. V. Minnich, M. B. Smith, M. J. Brauner and P. W. Majerus, *J Clin Invest.*, 1971, **50**, 507.
51. M. Noda, T. Takano and H. Samurai, *Mutat. Res.*, 1979, **66**, 175.
52. R. F. Whiting, L. Wei and H. F. Stich, *Mutat. Res.*, 1980, **78**, 159.
53. J. H. Ray, *Mutat. Res. Lett.*, 1984, **141**, 49.
54. I. Rahden-Staroń, P Suchocki and H. Czeczot, *Mut.Res.-Genet. Toxicol. Environ.Mutagen.*, 2010, **69**, 44.
55. R. W. Teel and S. R. Kain, *Mutat. Res.*, 1984, **127**, 9.
56. F. Abdullaev, L. Riverón-Negrete, I. Péreez-López, R. Camacho-Carranza and J. J. Esponosa-Aguirre, *Rev. Int. Contam. Ambient*, 2005, **21**(Suppl. 1), 85.
57. F. Geyikoglu and H. Türkez, *Brazilian Arch. Biol. Technol.*, 2006, **49**, 393.
58. G. Batist, A. G. Katki, R. W. Klecker and C. E. Myers, *Cancer Res.*, 1986, **46**, 5482.
59. S. Gabor, M. Ciugudeanu and D. Surcel, *Environ. Res.*, 1985, **37**, 293.
60. M. A. Medinsky, R. G. Cuddihy, J. O. Hill and R. O. McLellan, *Toxicol. Lett.*, 1981, **8**, 289.
61. J. F. Ramos and T. J. Webster, *Int. J. Nanomed.*, 2012, **7**, 3907.
62. J. E. Oldfield, *J. Nutr.*, 1987, **117**, 2002.
63. J. Huff, *J. Natl. Canc. Inst.*, 2004, **96**, 333.
64. J. Marshall, M. Reid and A. Duffield Lillico, *J. Natl. Canc. Inst.*, 2004, **96**, 334.
65. A. A. Nelson, O. G. Fitzhugh and H. O. Calvery., *Cancer Res.*, 1943, **3**, 230.
66. H. A. Schroeder and M. Mitchener, *J. Nutr.*, 1971, **101**, 1531.
67. National Toxicology Program, *Natl. Toxicol. Progr. Tech. Rep. Ser.*, 1980, **197**, 1.
68. National Toxicology Program, *Natl. Toxicol. Progr. Tech. Rep. Ser.*, 1980, **199**, 1.
69. National Toxicology Program, *Natl. Toxicol. Progr. Tech. Rep. Ser.*, 1980, **194**, 1.

70. L. Björkhem-Bergman, U.-B. Torndal, S. Eken, C. Nyström, A. Capiano, E. Huusfeldt, M. Björnstedt and L. C. Eriksson, *Carcinogenesis*, 2005, **26**, 125.
71. M. E. Reid, A. J. Duffield-Lillico, E. Slate, N. Natarajan, B. Turnbull, E. Jacobs, G. F. Combs, D. S. Alberts, L. C. Clark and J. R. Marshall, *Nutr. Cancer*, 2008, **60**, 155.
72. M. W. Russo, S. C. Murray, J. I. Wurzelmann, J. T. Woosley and R. S. Sandler, *Nutr. Cancer*, 1997, **28**, 125.
73. G. F. Combs and W. P. Gray, *Pharmacol. Therap.*, 1998, **79**, 179.
74. C. K. Baskett, V. L. Spate, M. M. Mason, T. A. Nichols, A. Williams, I. M. Dubman, A. Gudino, J. Denison and J. S. Morris, *J. Radioanalyt. Nucl. Chem.*, 2001, **249**, 429.
75. J. C. Fleet, *Nutr. Rev.*, 2009, **55**, 277.
76. R. A. LeBoeuf, B. A. Laishes and W. G. Hoekstra, *Cancer Res.*, 1985, **45**, 5489.
77. W. Lijinski, J. A. Milner, R. M. Kovatch and B. J. Thomas, *Toxicol. Health.*, 1989, **5**, 63.
78. M. V. Marshall, M. S. Arnott, M. M. Jacobs and A. C. Griffin, *Cancer Lett.*, 1979, **7**, 331.
79. H. J. Thompson, L. D. Meeker, P. J. Becci and S. Kokoska, *Cancer Res.*, 1982, **42**, 4954.
80. C. Ip and F. B. Daniel, *Cancer Res.*, 1985, **45**, 61.
81. B. Prokopczyk, J. G. Ross, D. Desai, S. Amin, O. S. Sohn, E. S. Faila and K. El-Bayoumy, *Cancer Lett.*, 2000, **161**, 35.
82. A. M. Fan and K. W. Kizer, *West. J. Med.*, 1990, **153**, 160.
83. B. Davidson and R. S. Fisher, *Kentucky Geological Survey*, 2005, Circ. **10**, Ser. XII.
84. J. C. Fleet, *Nutr. Rev.*, 1997, **55**, 277.
85. K. Wallace, T. Byers, J. S. Morris, B. F. Cole, E. R. Greenberg, J. A. Baron, A. Gudino, V. Spate and M. R. Karagas, *Cancer Epidemiol. Biomarkers Prev.*, 2003, **12**, 464.
86. G. F. Combs, L. C. Clark and B. W. Turnbull, *Med. Klin. (Munich)*, 1997, **92**(Suppl. 3), 42.
87. M. A. Moyad, *Urology*, 2002, **59**(Suppl. 1), 9.
88. A. J. Duffield-Lillico, E. H. Slate, M. E. Reid, B. W. Turnbull, P. A. Wilkins, G. F. Combs, H. K. Park, E. G. Gross, G. F. Graham, M. S. Stratton, J. R. Marshall and L. C. Clark, *J. Natl. Canc. Inst.*, 2003, **95**, 1477.
89. M. E. Reid, A. J. Duffield-Lillico, E. Slate, N. Natarajan, B. Turnbull, E. Jacobs, G. F. Combs, D. S. Alberts, L. C. Clark and J. R. Marshall, *Nutr.Cancer.*, 2008, **60**, 155.
90. A. Hareyan, *NCI Review of Prostatic Cancer Prevention Study Shows no Benefit for Use of Selenium and Vitamin E Supplements*, National Cancer Institute Release, 2008, Cancer Information Service, http// www.cancer.gov.

91. Occupational Safety and Health Administration (OSHA), *Sampling and Analysis of Crystalline Silica – X-ray Diffraction (XRD)*, OHSA, Cyber Conference, 1997.
92. IARC, *IARC Monographs on the Evaluation of Carcinogenic Risks to Human*, IARC, Lyon, 1973, vol. 2, p. 1.
93. IARC, *IARC Monographs on the Evaluation of Carcinogenic Risks to Human*, IARC, Lyon, 1977, vol. 13, p. 1.
94. IARC, *IARC Monographs on the Evaluation of Carcinogenic Risks to Human*, IARC, Lyon, 1987, vol. 42, p. 1.
95. J. C. Barrett, *Environ. Health Perspect.*, 1994, **102**(Suppl. 5), 19.
96. J. C. Barrett, P. W. Lamb and R. W. Wiseman, *Environ. Health Perspect.*, 1989, **81**, 81.
97. M. C. Jaurand, *Ann. N.Y. Acad. Sci.*, 1991, **643**, 258.
98. J. S. Harington, *Ann. N.Y. Acad. Sci.*, 1991, **643**, 465.
99. W. F. Cooke, *Br. Med. J.*, 1924, **2**, 140.
100. Health and Safety Executive (HSE), UK, *Mesothelioma – the Human Face of an Asbestos Epidemic*, HSE, London, 2012.
101. IARC, *IARC Monographs on the Evaluation of Carcinogenic Risks to Human*, IARC, Lyon, 1999, vol. 74, p. 1.
102. Food and Drug Administration (FDA), *Silicone Gel-Filled Breast Implants: Updated Safety Information*, FDA, Silver Spring, MD, 2011.
103. S. A. Bondurant, V. Ernster and R. Herdman, ed. Safety of Silicone Breast Implants, *Institute of Medicine*, National Academy Press, New York, 1999.
104. P. Chadwick, F. Taherinejad, K. Hamberg and M. Waring, *J. Wound Care*, 2009, **18**, 483.
105. A. B. G. Lansdown and A. Williams, *J. Wound Care*, 2007, **16**, 404.
106. K. Schwartz, *Proc. Natl. Acad. Sci.*, 1973, **70**, 1608.
107. R. Jugdaohsingh, S. H. Anderson, K. L. Tucker, D. P. Kiel, R. P. Thompson and J. J. Powell, *Am. J. Clin. Nutr.*, 2002, **75**, 887.
108. R. Jugdaohsingh, K. L. Tucker, N. Quao, L. A. Cupples, D. P. Kiel and J. J. Powell, *J. Bone Miner. Res.*, 2004, **19**, 29.
109. J. J. Powell, S. A. McNaughton, R. Jugdaosingh, S. H. C. Anderson, J. Dear, F. Khoth, L. Mowatt, K. L. Gleason, M. Sykes, R. P. H. Thompson, C. Bolton-Smith and M. J . Hodson, *Br. J. Nutr.*, 2005, **94**, 804.
110. W. Mertz, *Proc. Nutr. Soc.*, 1974, **33**, 307.
111. E. M. Carlisle, *Science*, 1970, **167**, 279.
112. E. M. Carlisle, *Sci. Total Environ.*, 1988, **73**, 9.
113. A. Lassus, *J. Int. Med., Res*, 1993, **4**, 209.
114. J. Najda, A. Gminski, M. Drodz and A. Danch, *Biol. Trace Elem. Res.*, 1992, **34**, 185.
115. C. D. Seaborn and F. H. Nielsen, *Biol. Trace Elem. Res.*, 2002, **89**, 251.
116. E. M. Carlisle, *Nutr. Rev.*, 1982, **40**, 193.
117. G. S. Smith, A. L. Neumann, V. H. Gledhill and C. A. Arzola, *J. Anim. Sci.*, 1973, **36**, 271.

118. K. L. Tucker, *Curr. Pharm. Dis.*, 2003, **9**, 2687.
119. C. D. Seaborn and F. H. Nielsen, *Biol. Trace Elem. Res.*, 2002, **89**, 239.
120. M. R. Calomme and D. A. Van den Berghe, *Biol. Trace Elem. Res.*, 1997, **56**, 153.
121. S. Fregert, *J. Invest. Dermatol.*, 1958, **31**, 95.
122. E. M. Carlisle, *J. Nutr.*, 1976, **106**, 478.
123. E. M. Carlisle, *Fed. Proc.*, 1975, **34**, 927.
124. E. Bissé, T. Epting, A. Bell, G. Lindinger, H. Lang and H. Wieland, *Anal. Biochem.*, 2005, **337**, 130.
125. T. Tamada, *Anal. Sci.*, 2003, **19**, 1291.
126. D. M. Reffitt, R. Jugdaohsingh, R. P. Thompson and J. J. Powell, *J. Inorg. Biochem.*, 1999, **76**, 141.
127. L. Burns, M. Ashwell, J. Berry, C. Bolton-Smith, A. Cassidy, M. Dunnigan, K. T. Khaw, H. McDonald, S. New, A. Prentice, J. Powell, J. Reeve, S. Robins and B. Teucher, *Br. J. Nutr.*, 2003, **89**, 835.
128. J. A. T. Pennington, Silicon in foods and diets, *Food Additives Contamin.*, 1991, **8**, 97.
129. Food and Drug Administration (FDA), *Select Committee on GRAS Substances (SCOGS) Opinion; Sodium Silicate*, FDA, Springfield, VA, 2011, SCOGS-Report, **61**.
130. F. Liebau, In *The Physics and Technology of Amorphous SiO_2*, R. A. Devine, Plenum (NY) 1992.
131. World Health Organization (WHO), *Principle for Evaluating Health Risks in Children Associated with Exposure to Chemicals*, Environmental Health Criteria, 237, WHO, Geneva, 2006.
132. K. Steenland and L. Stayner, *Cancer Causes Control*, 1997, **8**, 491.
133. W. Noll, *Chemistry and Technology of Silicones*, Academic Press, New York, 1968.
134. G. S. Brody, *Breast Implants, Safety and Efficacy of Silicone*, Medscape, CDC Commentary Series, 2009, pp. 1–12.
135. N. Kossovsky and C. J. Freiman, *Arch. Pathol. Lab. Med.*, 1994, **118**, 686.
136. D. Napierska, L. C. J. Thomassen, D. Lison, J. A. Martens and P. H. Hoet, *Particle Fibre Toxicol.*, 2010, **7**, 39.
137. H.-S. Choi, Y.-J. Kim, M. Song, M.-K. Song and J.-C. Ryi, *Toxicol. Environ. Sci.*, 2011, **3**, 7.
138. C. J. Johnson, K. E. Driscoll, J. N. Finkelstein, R. Baggs, M. A. O'Reilly, J. Carter, R. Gelein and G. Oberdörster, *Toxicol. Sci.*, 2000, **56**, 405.
139. X. Shi, V. Castranova, B. Halliwell and V. Vallyathen, *J. Toxicol. Environ Health, B Crit. Rev.*, 1998, **1**, 181.
140. F. Seiler, B. Rehm, S. Rehm and J. Brisch, *Arch. Toxicol.*, 2001, **74**, 716.
141. T. Kajiwara, A. Ogami, H. Yamato, T. Oyabu, Y. Morimoto and I. Tanaka, *J. Occup. Health*, 2007, **49**, 88.
142. R. M. Gilberti, G. N. Joshi and D. A. Knecht, *Am. J. Respir. Cell Mol. Biol.*, 2008, **39**, 619.

143. L. M. Constantini, R. M. Gilberti and D. A. Knecht, *Plos ONE*, 2011, **6**, e14747.
144. P. Cocco, C. H. Rice, J. Q. Chen, M. A. McCawley, J. K. McLaughlin and M. Dosemeci, *Am. J. Ind. Med.*, 2001, **40**, 674.
145. C. Pelucchi, E. Pira, G. Piolatto, M. Coggiola, P. Carta and C. La Vecchia, *Ann. Oncol.*, 2006, **17**, 1039.
146. F. H. Y. Green, R. Althouse and K. C. Weber, *Am. J. Ind. Med.*, 1989, **16**, 605.
147. K. Ulm, H. Waschulzik, H. Ehnes, K. Guldner, B. Thomasson, A. Schwebig and H. Nuss, *Thorax*, 1999, **54**, 347.
148. K. A. Mundt, T. Birk, W. Parsons, E. Borsch-Galetke, K. Siegmund, K. Heavner and K. Guldner, *J. Occup. Environ. Med.*, 2011, **53**, 282.
149. W. Chen and J. Chen, *Occcup. Environ. Med.*, 2002, **59**, 113.
150. C. O'Neill, P. Jordan, T. Bhatt and R. Newman, *CIBA Foundation Symposium*, 1986, **121**, 214.
151. E. Hnizdo, *Br. J. Ind. Med.*, 1990, **47**, 656.
152. T. Brown, *Occup. Med.*, 2009, **59**, 89.
153. T. Partanen, E. Pukkata, K. Kurppa and H. Koskinen, *J. Occup. Med.*, 1994, **36**, 616.
154. S. Honnons and J. M. Porcher, *J. Environ. Pathol. Toxicol. Oncol.*, 2000, **19**, 391.
155. G. Oberdorster, *Inhalation Toxicol.*, 1996, **8**(Suppl.), 7.
156. D. F. Goldsmith, R. P. Ruble and C. O. Klein, *Scand. J. Work Environ. Health*, 1995, **21**(Suppl. 2), 104.
157. P. A. Hessel, J. F. Gamble, J. B. Gee, G. Gibbs, F. H. Green, W. K. Morgan and B. T. Mossman, *J. Occup. Environ. Med.*, 2000, **42**, 704.
158. A. Kolling, S. Rittinghausen, U. Heinrich and F. Pott, *Exptl. Toxicol. Pathol.*, 2008, **60**, 282.
159. R. C. Merrill, *Ind. Eng. Chem.*, 1949, **41**, 337.
160. A. K. Lührs and W. Geurtsen, *Progr. Mol. Subcell. Biol.*, 2009, **47**, 359.
161. IARC, *IARC Monographs on the Evaluation of Carcinogenic Risks to Humans*, IARC, Lyon, 1998, vol. 42, p. 1.
162. National Toxicology Program (NTP), *Natl. Toxicol. Progr. Tech. Rep. Ser.*, 1993, **421**, 1.
163. R. Davies, J. W. Skidmore, D. M. Griffiths and C. B. Moncrieff, *Fd. Chem. Toxicol.*, 1983, **21**, 201.
164. M. Coggiola, D. Bosio, E. Pira, P. G. Piolatto, C. La Vecchia, E. Negri, M. Michelazzi and A. Bacaloni, *Am. J. Ind.Med.*, 2003, **44**, 63.
165. K. Reijula, P. Pääkko, R. Kerttula, O. Taikina-Aho, T. Tuuponen and J. Hassi, *Br. J. Ind. Med.*, 1991, **48**, 140.
166. N. Guha, K. Straif and L. Benbrahim-Tallaa, *IARC Monographs International Agency for Research on Cancer*, WHO, *La Medicina del Lavoro*, 2011, **102**, 310.
167. E. Tátrai and G. Ungváry, *Pol. J. Occup. Med., Environ. Health*, 1993, **6**, 27.

168. V. N. Pylev, L. A. Vasil'eva, S. A. Khrustalev and T. A. Krasnova, *Gig. Sanit.*, 2003, **2**, 53.
169. G. R. Pearson, J. A. Longstaffe, V. M. Lucke, S. Yo and W. J. Henderson, *Vet. Rec.*, 1986, **118**, 450.
170. P. J. van Diest, W. H. Beekman and J. J. Hage, *J. Clin. Pathol.*, 1998, **51**, 493.
171. P. A. Dewan, S. K. Condron, P. N. Morreau, R. W. Byard and J. Terlet, *Arch. Dis. Child.*, 1999, **81**, 71.
172. L. Garrido, B. Pflederer, M. Papisov and J. L. Ackerman, *Magn. Reson. Med.*, 1993, **29**, 839.
173. G. R. D. Evans and B. J. Baldwin, *Plast. Reconstr. Surg.*, 1997, **99**, 1354.
174. J. Weinzweig, P. L. Schnur, J. P. McConnell, J. B. Harris, P. M. Petty, T. P. Moyer and D. Nixon, *Plast. Reconstr. Surg.*, 1998, **101**, 1836.
175. S. S. Teuber, D. A. Reilly, L. Howell, C. Oide and M. E. Gershwin, *J. Rheumatol.*, 1999, **26**, 699.
176. W. B. Greene, D. S. Raso, L. G. Walsh, R. A. Harley and R. M. Silver, *Plast. Reconstr. Surg.*, 1995, **95**, 513.
177. R. M. Silver, E. H. Sahn, J. A. Allen, S. Sahn, W. Greene, J. C. Maize and P. D. Garen, *Arch. Dermatol.*, 1993, **129**, 63.
178. B. Pflederer and L. Garrido, *Magn. Res. Med.*, 1995, **33**, 8.
179. E. Tavazzani, S. Xing, J. E. Waddell, D. Smith and E. L. Boynton, *J. Biomed. Mater. Res. A.*, 2005, **72A**, 161.
180. L. Pimentel, M. Barnadas, D. Vidal, F. Sancho, A. Ramon Fontarnau and A. Alomar, *Dermatology*, 2002, **205**, 162.
181. W. Peters, D. Smith and S. Lugowski, *Ann. Plast. Surg.*, 1999, **43**, 324.
182. L. F. Smith, T. T. Smith, E. Yeary, J. M. McGee and K. Malnar, *J. Okla. State. Med. Assoc.*, 1999, **92**, 126.
183. Y. Tanaka, I. Morishima and K. Kikuchi, *World J. Surg. Oncol.*, 2008, **6**, 33.
184. J. W. Jakub, M. D. Ebert, A. Cantor, M. Gardner, D. S. Reintgen, E. L. Dupont, C. E. Cox and A. R. Shons, *Plast. Reconstr. Surg.*, 2004, **114**, 1737.
185. S. Kamasaki, M. Tsurumaru, T. Kamano, S. Kobayashi, M. Hino and R. Kuwatsuru, *Breast Cancer*, 2000, **7**, 71.
186. S. Sahoo, P. P. Rosen, R. M. Feddersen, D. S. Viswanatha, D. A. Clark and A. Chadburn, *Arch. Pathol. Lab. Med.*, 2003, **127**, e115.
187. M. K. Newman, N. J. Zemmel, A. Z. Bandak and B. J. Kaplan, *J. Plast. Reconstr. Aesthet. Surg.*, 2008, **61**, 822.
188. M. R. Y. Bishara, C. Ross and M. Sur, *Diagn. Pathol.*, 2009, **4**, 11.
189. A. Grubstein, M. Cohen, A. Steinmetz and D. Cohen, *Clin. Imaging*, 2011, **35**, 228.
190. N. J. Nelson, *J. Natl. Canc. Inst.*, 2000, **92**, 1714.
191. L. A. Brinton, P. Toniolo and B. S. Pasternack, *J. Clin. Epidemiol.*, 1995, **48**, 557.
192. L. A. Brinton, *Plast. Reconstr. Surg.*, 2007, **120**(Suppl. 1), 94S.

193. H. Berkel, D. C. Birdsell and H. Jenkins, *New Engl. J. Med.*, 1992, **326**, 1649.
194. K. A. Kern, J. T. Flannery and P. G. Kuehn, *Plast. Reconstr. Surg.*, 1997, **100**, 737.
195. M. J. Duffy and J. E. Woods, *Plast. Reconstr. Surg.*, 1994, **94**, 295.
196. E. P. Winer, K. Fee-Fulkerson, C. C. Fulkerson, G. Georgiade, K. E. Katoe, M. Conaway, C. Brunatti, V. Holmes and B. K. Rimer, *J. Natl. Canc. Inst.*, 1993, **85**, 1407.
197. C. M. McCarthy, A. F. Klassen, S. J. Cano, A. Scott, N. Vanlaeken, P. A. Lennox, A. K. Alderman, B. J. Mehrara, J. J. Disa, P. G. Cordeiro and A. L. Pusic, *Cancer*, 2010, **116**, 5584.
198. G. S. Brody, *Breast Implant, Safety and Efficiency of Silicone*, Medscape, CDC Commentary Ser., Atlanta, GA, 2009.
199. C. W. Su, D. A. Dreyfuss, T. J. Krizek and K. J. Leoni, *Plast. Reconstr. Surg.*, 1995, **96**, 513.
200. W. Liu, Y. Yang and Q. F. Liu, *Adv. Mater. Res.*, 2012, **583**, 95.

Part 2
Xenobiotic Elements of No Nutritional Value

CHAPTER 9

Aluminium and Zirconium

9.1 General Introduction

Aluminium and zirconium have no nutritional role in the human body but trace quantities may be detected in the circulation following ingestion in food and drinking water. Zirconium is discussed alongside aluminium on account of its usage in antiperspirant and deodorant preparations and ability to cause granuloma following topical application or subcutaneous or intramuscular injections in humans and animals.[1–5] Like aluminium, zirconium is widely disseminated in the geosphere, particularly in Australia and India. Both elements are allergenic, but allergic contact granuloma is a well known complication following contact with zirconium compounds used in cosmetics and in treatment of *Rhus* dermatitis (due to contact with the oils of poison oak or poison ivy).[6–8] Neither element nor their inorganic or organic compounds found in the environment or occupational situations are presently listed as human carcinogens, but their usage in antiperspirant and deodorant cosmetics has been linked to breast cancer on the basis that bioactive radicals are absorbed percutaneously and bind with oestrogens, which are well documented human carcinogens.[9–11] Aluminium [Al(III)] and possibly aluminium–zirconium (IV) complexes acting as metallo-oestrogens are claimed to exhibit genotoxic profiles capable of causing DNA damage and epigenetic change, and to be capable of inducing breast cancer.[12–14]

9.2 Aluminium

Aluminium is the third most abundant element in the Earth's crust and deposits are found worldwide in the form of oxides, hydroxides, and numerous complexes with fluorine, silicon, sodium and other metallic elements. Alums are a large group of hydrated double salts, usually consisting of aluminum sulfate, water of hydration, and the sulfate of another element. Important alums are

Issues in Toxicology No. 18
The Carcinogenicity of Metals: Human Risk through Occupational and Environmental Exposure
By Alan B. G. Lansdown

Published by the Royal Society of Chemistry, www.rsc.org

potassium aluminium sulfate, ammonium aluminium sulfate, and sodium aluminium sulfate. Some alums are brightly coloured, as in rubies, emeralds and sapphires which consist mainly of crystalline aluminium oxide. Common ores of aluminium include bauxite ($Al_2O_3 \cdot 2H_2O$), Gibbsite or hydrargillite ($Al_2O_3 \cdot 3H_2O$), Diaspore ($Al_2O_3 \cdot H_2O$), Cryolite ($AlF_3 \cdot 3HF$), Corundum (Al_2O_3), Potash felspar ($K_2O \cdot Al_2O_3 \cdot 6SiO_2$), Potash mica ($K_2O \cdot 3Al_2O_3 \cdot 6SiO_2 \cdot 2H_2O$), and Alunite [$K_2SO_4 \cdot Al_2(SO)_3 \cdot 4Al(OH)_3$]. Distribution of these and more than 250 other aluminium salts is influenced by geophysical activity, volcanic action, water and natural erosion. Atmospheric aluminium particles range from 0.0005 $\mu g\ m^{-3}$ in the Antarctic to 1 $\mu g\ m^{-3}$ in industrial areas. Aluminium salts are widespread in inland and marine waters throughout the world.[15]

Aluminium is a trivalent metal in Group IIIb of the Periodic Table with chemical similarities to gallium, indium and thallium which are more toxic and of questionable carcinogenic risk to humans. It is a highly reactive element and is not found naturally in its elemental form. Toxicologically, aluminium has attracted considerable attention as a potential cause of neuropathy, encephalopathy and Alzheimer's disease symptoms, although this is not substantiated.[16,17] Aluminium is a cause of occupational asthma but sensitisation to aluminium salts appears to be rare.[8] Aluminium salts (chloride, hydroxide, chlorhydrate and aluminium zirconium chlorhydrate) used in cosmetics are associated with follicular irritancy and milaria. Recent concern over the possible relationship between long-term use of under-arm antiperspirants and mammary cancer in women is currently controversial and unresolved.

Aluminium and its compounds are not currently listed as human carcinogens by NTP in RoC 12th Edition,[18] but the IARC identified several reports claiming excesses in leukaemia and cancers of lung, gastrointestinal tract and bladder in men employed in aluminium production. Its major use is in cladding for nuclear reactors and ceramics.[19]

9.2.1 Industrial Use and Human Exposure

Elemental aluminium is a soft, ductile, corrosion resistant lightweight metal with high electrical conductivity. It is widely used for foil and conductor cables, but alloying with other elements is necessary to provide the higher strengths needed for other commercial uses in alloys in aircraft and vehicle construction, household products and building. More than 300 alloys containing aluminium are in industrial use, the main alloying metals being copper, zinc, magnesium, silicon, manganese and lithium. Small additions of chromium, titanium, zirconium, lead, bismuth and nickel are also made and iron is invariably present in small quantities. Alum tanning has been practised in leather industries for more than 2000 years. Medicinal uses include adjuvants in vaccine production, renal dialysis fluids and in antiperspirants in cosmetics. Aluminium salts are widely used in water treatment as coagulants to reduce organic matter.[15]

There are few situations when humans do not come into contact with aluminium products or residues in everyday life or industry. Absorption of bioactive Al(III) aluminium is expected through inhalation of industrial particulates and soil erosion dusts, combustion of fossil fuels, ingestion with water and food, use of intestinal antacids, vaccines and renal dialysis fluids, and through dermal application of antiperspirants and cosmetics. Aluminium residues from cooking utensils, pots, pans, *etc*. migrate into food as consumed. Intestinal absorption of aluminium salts may be less than 1%, or be up to 2.5% but between 50 and 150 mg accumulates in the body, most in the lungs, brain, kidneys, liver and thyroid.[20,21] Percutaneous absorption from most soluble aluminium salts is low.[22–25] Soluble aluminium salts are intensely astringent and irritant, but the Al(III) ion is biologically active and binds strongly to epidermal and hair keratin and several proteins involved in iron, calcium and zinc metabolism. Aluminium–keratin complexing leading to occlusion of sweat ducts may be contributory to the antiperspirant action attributable to aluminium chloride and aluminium chlorhydrate.[26] A study with ^{26}Al-labelled aluminium chlorhydrate as used in antiperspirants showed that only 0.012% of the aluminium was absorbed percutaneously within 6 days, equivalent to about 4 μg from a single application on both armpits.[27] Aluminium absorbed gastrointestinally or through renal dialysis is metabolised throughout the body in the form of albumin or γ-globulin complexes; some is excreted in the urine but some accumulates in bone and soft tissues.[28] The extent to which aluminium is absorbed from inhaled particles is unclear. Chronic inhalation of aluminium dust is a known cause of pulmonary irritancy, sore throat and cough, and cases of fibrosis are reported. A fatal case of pulmonary nodular fibrosis with severe emphysema occurred in a young worker exposed to fine aluminium dust (19 mg m^{-3}); in 27 other cases responses in aluminium workers exposed chronically to dust varied from fatality to no obvious effect.[29,30] The authors noted that the finer the dust the greater the pathological response. Pulmonary damage has not been observed in experimental animals exposed to aluminium dust, suggesting that either insufficient dust was inhaled or that the species used were insensitive to the exposures.[31,32] No deaths were reported following chronic exposure to 2.18–2.45 mg Al m^{-3} as refractory alumina fibre for 86 weeks in groups of 50 male and female Wistar rats.

Topical aluminium preparations have been in use for many years and many millions of people have been exposed to them, but skin cancers have not been reported following chronic dermal exposure to aluminium or any other antiperspirant as far as I know. On the other hand, aluminium ores or compounds penetrating the epidermal barrier are a known cause of subcutaneous granuloma: these have been reported following application of aluminium (or zirconium) compounds to damaged skin, in tattoos, and following intradermal injections of aluminium hydroxide in laboratory animals.[3–5,33,34] Post-vaccination granulomas are a well-known reaction due to aluminium adsorbed vaccines. Histiocytes have been shown to contain aluminium using energy dispersive X-ray microanalysis.

9.2.2 Cytotoxicity and Carcinogenicity

9.2.2.1 In Vitro Studies

Al(III) is a bioactive ion and there is good evidence that it is absorbed by cultured cells and interacts with DNA, and inhibits DNA proliferation and repair at high doses.[35–38] Cyclic voltammetry, ultraviolet (UV) and fluorescence spectroscopy have demonstrated that aluminium binds strongly to chromatin. Studies using calf thymus have demonstrated that binding is principally to phosphate groups of DNA backbones and not to DNA bases. Aluminium induces DNA damage, apoptosis and cell death, mediated through reactive oxygen intermediates in a dose-related manner, in human peripheral lymphocyte cultures and plant root cells.[39,40] Aluminium induces oxygen reactive radicals and oxidative stress in cell cultures, and chronic aluminium administration in rats led to increased expression of p53 protein and cyclin D1 in brain tissue, providing evidence of disturbances of the cell cycle.[41] Lipid peroxidation is a further indication that high doses of aluminium can induce cytotoxic change in cultured cells through oxidative stress.[42] When the cells were pretreated with aluminium, they showed changes consistent with the ability of Al(III) to inhibit DNA repair. Aluminium-related damage in human peripheral lymphocyte cultures is closely related to plasma concentrations and presumably levels of oxidative stress.[43]

No evidence has been seen to show that aluminium is mutagenic in cultured cells or reverse transcription assays in bacteria.[44] Assay of aluminium oxide (Al_2O_3) nanomaterials of 30 or 40 nm diameter failed to evoke mutagenic change in five strains of *Salonella typhimurium* with or without S9 liver homogenate.[45] Extracts of sputum or urine from workers in an aluminium plant in Söderberg in Norway failed to show positive results in the Ames test, but when cultures were exposed to the S9 fraction, urine from workers who smoked was mutagenic.[46,47] Similarly negative findings were obtained in an exploratory study to examine the mutagenicity of aluminium oxide in dental implant ceramics.[48] Administration of aluminium oxide by intraperitoneal injection in mice at high concentration (1 g kg^{-1}) led to minimal levels of mutagenicity.

9.2.2.2 Experimental Animal Studies

Studies in experimental animals have failed to provide evidence of carcinogenicity for aluminium compounds commonly encountered environmentally or in industry.[44] No correlation has been seen between tissue and serum aluminium concentrations and toxicity in rats, although high brain concentrations were reported.[28,49] The brain seems to be particularly susceptible to aluminium, and structural and pathological changes including neurofibrillary degeneration and gliosis were observed in cats injected subdurally with aluminium hydroxide.[50] Animals injected subcutaneously or intraperitoneally developed granulomas.[3–5,51] Acute and long-term toxicity studies in rats, mice and dogs at tolerable doses have failed to reveal clear dose–responses to soluble aluminium salts, and high 50% lethal dose (LD_{50}) levels are recorded at 200–1000 mg kg^{-1} body weight in acute exposures.[52] Implantation of powdered alumina in subcutaneous tissue, muscle or intraperitoneal sites did not give rise to sarcoma as seen with many metals but invoked a transitory inflammatory reaction.

9.2.2.3 Human Studies

Pulmonary fibrosis, pneumoconiosis and pulmonary distress are occupational hazards for workers engaged in bauxite mining and processing of alumina. Clinical studies show that levels of pulmonary damage correlate closely with particle size, fibre length and concentration and the duration of exposure.[53,54] Alveolar macrophages serve a primary protective function in absorbing inhaled aluminium particles; lysosomal enzymes precipitate the metal thereby preventing diffusion through alveolar membranes.[55] Fourteen epidemiological studies, conducted in Norway, Sweden, France, Poland, Australia, and the Quebec area of Canada and involving 35,066 men and women engaged in aluminium-related industries (bauxite mining, refining smelter work, aluminium reduction plants and foundaries), report high levels of respiratory distress and cardiovascular symptoms.[53,56–65] Twelve other cases of acute poisoning have been reported following inhalation of aluminium chloride.[66] Many of these studies indicate increased tumour incidence as an occupational risk in workers with 20 or more years' exposure in the industry, but in view of considerable inconsistency, it is not possible to implicate aluminium residues as a principal cause of any cancer type.[63]

Aluminium-related industries include bauxite mining, aluminium refining, smelting, *etc*., occupations in which workers are exposed to a large range of toxic and potentially carcinogenic materials, notably polyaromatic hydrocarbons (PAH; *e.g.* benzo[a]pyrene), silica, arsenic, cigarette smoking and coal tars. Any of these, singly or together, are a potential cause of increased mortality from lung and urinary bladder cancers, or mesothelioma.[54] Several studies highlight the inherent carcinogenic risks associated with PAH exposure in aluminium metallurgy as a cause of lung cancer. Two such studies, involving a total of 3923 men working in a French and Norwegian reduction plants exposed to PAH in their working lives, showed no increase in lung or bladder cancers;[62,65] Szymczak emphasised that cancer risk was associated with environmental benzo[a]pyrene concentrations but that the risk was similar to or lower than that related to smoking.[67] Other studies implicate PAH exposure as a cause of increased urinary bladder cancer which is "unusually high in aluminium smelter workers.[57,58,63] This was illustrated by reference to potroom workers with an average of 20 years employment in the Soderberg plant in Quebec. The relative risk of exposure to benzo[a]pyrene at a concentration of 0.2mg/m^3 in a working life of 40 years is estimated to increase the risk of bladder cancer in the aluminium industry by a factor of 2.4.[63] However, in this and other studies, the cancer risk was exacerbated more by cigarette smoking and occasional exposure to coal-tar pitch volatiles from coke ovens, rather than through inhalation of aluminium residues released from the Soderberg process.

A study involving mortality records of 5891 men working in a reduction plant in Quebec emphasised that the duration of exposure to coal tars (tar–years) was a major factor in lung and bladder cancer incidence.[60,61] In the authors' view, mortality from lung cancer was approximately doubled in men with 20 or more tar–years compared with those not exposed to tar-related products. It is notable that, in this large study, the authors noted that death certificates recorded a vast number of neoplastic and non-neoplastic conditions in long-term employees in the aluminium smelter including tuberculosis, cardiovascular disease, trauma, leukaemia and cancers of pancreas, genital organs, brain, kidney and gastrointestinal tract.

In a recent study in two Australian pre-bake aluminium smelters, cardiovascular, respiratory or injury-related fatalities were similar to or lower than usual and the incidences of lung and bladder cancers were not occupationally increased.[56] However the authors analysed health records of 4396 employees working in the Soderberg process and listed increased incidence of mesothelioma and cancers of prostatic, stomach and kidney as a cause of death in production and maintenance workers with more than 20 years' exposure. They regarded these tumours as incidental tumours and of no occupational significance! It seems highly likely that these workers were exposed at some time to arsenic or asbestos, or other defined causes of prostatic cancer and mesothelioma.[68,69]

A retrospective analysis of multicentre epidemiological data derived from fatality records over 20 or more years demonstrate that workers in aluminium-related industries in various parts of the world are exposed to a range of toxic materials at some stage in their working lives. In many of the men, cancers were not identified until after 20 or more years of work in the industry, although on occasions increased cancer risk was seen in men with less than 3 months' exposure. Psychoses, neurological problems, accidental traumas and stresses as recorded must play a significant role in the ill-health recorded in several studies,[53,57,58,61,65] but none of the reports seen implicate aluminium as a cause of industrially-related cancers in any organ.

9.2.2.4 Aluminium in Cosmetics as a Cause of Breast Cancer

Breast cancer attributable to long-term usage of antiperspirant cosmetics and the action of aluminium as a metallo-oestrogen is still questionable. Phillippa Darbre reasoned that since aluminium is known to have genotoxic and epigenetic effects in cultured cells, and is claimed to act on oestrogen receptors in isolated mammary cancer cells in the form of a metallo-oestrogen, it has the molecular capacity to impair molecular signalling pathways, leading to mammary cancer.[12–14] Her argument rests on prolonged absorption of biologically active Al(III) ion through breast skin leading to accumulation in mammary tissue, where it interferes with ligand binding and oestrogen-regulated gene expression, but the full range of its genotoxic potential remains to be determined. Interestingly, the author claimed that a range of other metals exhibit oestrogenic properties, but that there is minimal or no evidence to show that antimony, arsenic, barium, cadmium or copper compounds applied topically initiate or promote mammary cancers. Of these, compounds of copper, barium and antimony are not known to be carcinogenic in humans, and chemically they are quite different from aluminium. Minute and uneven levels of aluminium have been detected in mammary tissue (4–437 nmol g^{-1}), but the significance of this in terms of mammary tumour induction is not known.[70,71]

Aluminium has a very low rate of percutaneous absorption in animal and human skin on account of the strong binding capacity of the cation to –SH groups in epidermal keratin and hair.[24,72–74] (Figure 9.1). Aluminium accumulating in mammary tissue through this route is expected to be appreciably lower than that acquired through gastrointestinal absorption or by inhalation. Aluminium uptake from drinking water has more recently estimated to be 0.1 mg l^{-1}, and for food 5 mg per day.[15] In individuals in whom local aluminium levels are higher than normal as a result of tattoos or intradermal injection, local granulomas have resulted without evidence of hyper-oestrogenic activity.[75] This has been demonstrated in intradermal guinea pig tests.[3–5] Rageth emphasised that the disproportionately higher number of cancers seen clinically in the upper quadrant of the breast cannot be substantiated reliably by pathological evidence in biopsies or through interrogation of breast cancer patients.[76] Gikas *et al.* commented that, although certain cosmetic chemicals show a possible link with mammary cancer in animal models and laboratory studies, there is no credible evidence to show that they cause breast cancer in susceptible women, irrespective of their oestrogenic profile.[77]

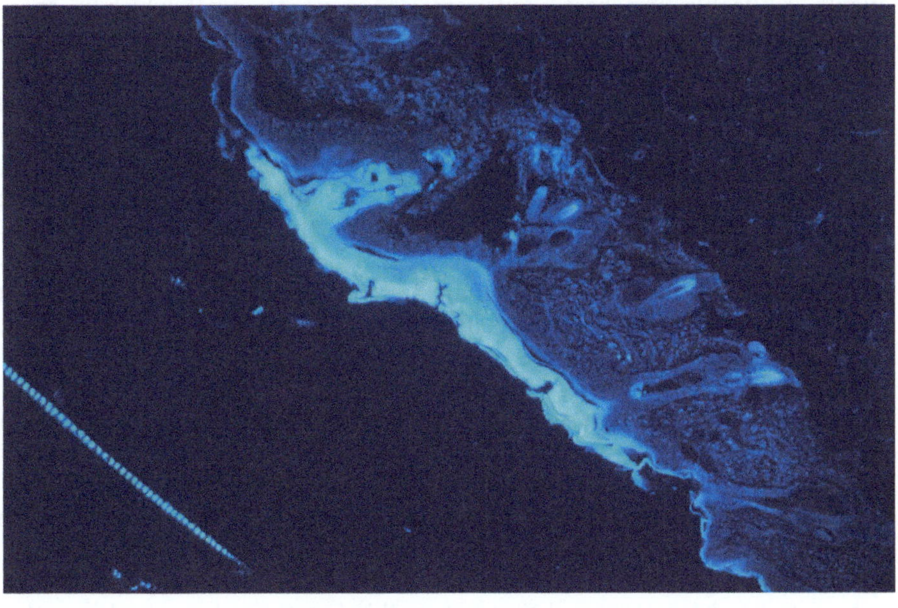

Figure 9.1 Mouse skin treated with aluminium chloride showing aluminium fluorescence in the stratum corneum and dermis.
(Lansdown, *Br. J. Dermatol.*, 1973, 105, Fig. 4).

9.3 Zirconium

Zirconium is best known through its principle ore zircon which is widely used in semi-precious gemstones and in the synthetic hyacinth and zirconite jewellery stones made from zircon and zirconium oxide. Zircon as zirconium silicate is found extensively in beach sands and deposits in Australia, India and Sri Lanka, Brazil, South Africa and the USA. It constitutes about 0.03% of the Earth's crust with estimated world reserves exceeding 1 billion tonnes. Apart from zircon ($ZrO_2 \cdot SiO_2$ or $ZrSiO_4$), baddetite (ZrO_2) is the only other ore of importance. Zirconium is commonly found in deposits with the chemically similar element hafnium, both of which are used in nuclear reactors. Zirconium was developed in the USA as an alternative to aluminium as a cladding for nuclear reactors. Zirconium proved to be a very satisfactory metal from all standpoints but had to be separated from hafnium in the so-called Manhattan Project established by the US Atomic Energy Commission in the 1940s. Zirconium does not absorb neutrons and thus was excellent for cladding reactor fuel for navy reactors. Leakage of radioactive zirconium from these reactors is a human health and environmental problem, addressed by international radiological protection authorities.[78]

Zirconium is a tetravalent metal in Group IVa of the Periodic Table. Zirconium occurs in nature as five stable isotopes: ^{90}Zr is the prevalent form, comprising approximately half of natural zirconium, with the other four stable isotopes, ^{91}Zr, ^{92}Zr ^{94}Zr and ^{96}Zr, comprising 11%, 17%, 17% and 2.8% respectively. ^{95}Zr is the principle radioactive isotope found in spent nuclear fuel and in wastes resulting from reprocessing this fuel. Zirconium is a strong, ductile, malleable and heat- and corrosion-resistant metal which is spontaneously combustible as a finely divided powder, making it a major fire hazard. In pyrotechnics, powdered zirconium is used in the production of flares and tracer bullets. Most inorganic zirconium compounds are insoluble and exhibit low toxicity in humans and animals. Humans may ingest up to 50 μg per day of zirconium but intestinal absorption is low and much of that absorbed is deposited in bone.[79] The zirconium content in the human body is given as 1mg but higher concentrations may be expected in orthopaedic patients with zirconium alloy prostheses, and also in workers exposed to zirconium occupationally.[80,81] Ecologically, zirconium presents minimal risk since it is not noticeably absorbed by land plants and its ability to enter food chains is low.

Three main health issues arise from human exposure to zirconium:

- Exposure to radioactive zirconium compounds used as cladding in nuclear reactors[82,83]
- Zirconium salts in the treatment of *Rhus* dermatitis[7,8,84,85]
- Zirconium as a constituent of antiperspirants and deodorants in cosmetics.[12–14]

Anxiety has been expressed that polymethbyl methacrylates used in orthopaedic surgery and containing zirconium oxide (ZrO_2) as an X-ray contrast medium may contain trace levels of radioactivity, and hence be carcinogenic.

A German study reported that analyses of several zirconium oxide samples did show low levels of radiation but predicted that the cancer risk was probably low because the implants were usually used in patients at least 60 years of age and levels of radioactive emission were very low.[86]

The literature is replete with the use of zirconium salts in the treatment of dermatitis resulting from contact with poison ivy and certain plants of the *Rhus* group.[7,8,84,85] *Rhus* dermatitis is a major healthcare problem in certain areas of the USA and Canada where poison dogwood, sumac and poison ivy are ubiquitous in gardens, swamps and wildlife areas. Hydrous zirconium oxide or ziradryl cream has been shown to inactivate the poison of the *Rhus* plants by binding the hydroxyl groups of urushiol.[87–89] However, dermatological use of zirconium is contraindicated on account of its strong sensitising ability and capacity to induce granulomas following use in *Rhus* dermatitis therapy, as well as in antiperspirants and deodorants.[8] Occasionally, zirconium-related granulomas are of the sarcoid type, similar to those induced by beryllium.[90–92] Zirconium is not classified presently as a human carcinogen, whereas beryllium is classified as a human carcinogen on the sufficiency of clinical evidence in humans.[18,93]

9.3.1 Industrial and Commercial Applications and Human Exposure

Major applications of zirconium include production alloys for use in ceramics, nuclear reactors, superconductors, orthopaedic and dental devices, vacuum tubes, crucibles and catalytic converters, and steel production.[78] In each case, zirconium alloys are lightweight, malleable and show ductile properties, and hardness. Micronised zircon and zircon flour used in ceramic products account for almost half of the worldwide consumption of zircon. Zircon is a common opacifying constituent of glazes applied to ceramic tiles and sanitary ware and is also used as an opacifier in porcelain tiles by incorporation directly into the mixture used for forming the body of the tile. Zirconium and zirconium alloys show resistance to strong acids and alkalis and can be used in chemical processes where other material families are unsuitable. Zirconium alloys are corrosion resistant and biocompatible, and therefore can be used for surgical implants, orthopaedic prostheses and dentistry. Zirconium is known for its use in cladding of nuclear reactors, owing to its ability to "capture" electrons, corrosion resistance and stability at high temperatures, and resistance to neutron damage. Zirconium comprises a primary safety barrier in nuclear reactors. Zirconium compounds are used for waterproofing of textiles, manufacture of pigments and printers' inks, in leather production (with alums), and in the production of flame-resistant materials, ceramics, glass and enamel. Ceramics based on high purity zirconia have become key constituents of microelectronic and semiconductor components used in applications such as filters in communications equipment, acceleration sensors, buzzer elements and ultrasonic devices and photographic flashbulbs. As a stable salt, zirconium silicate (zircon, $ZrSiO_4$) is used also in jewellery, television glass and surgical instruments.

Table 9.1 Industrial and commercial uses of zirconium compounds.

Zr Compound	*Usage*
Acetate	High temperature conductors, water repellents, refractory applications
Hydroxide	Glass pigment
Naphthenate	Ceramics, lubricants, paints, varnishes
Nitrate	Electronics, ceramic glazes, special glasses
Oxychloride	Cosmetics (deodorants, antiperspirants), greases, textiles
Lactate (Na^+ salt)	Cosmetics, chemical syntheses, wound dressings
Potassium fluoride	Welding fuses
Sulfate	Chemical syntheses, lubricants, chemical reagents and catalysts, leather production and tanning
Tetrafluoride	Production of molten salts for nuclear reactors
Oxide	*Rhus* dermatitis therapy
Zirconia	Ceramics, alloy castings, mould linings

Absorption of zirconium from soluble salts into the human body is probably very low. Minimal quantities ingested with diet are absorbed intestinally and various salts [*e.g.* $ZrOCl_2$, ZrO(OH)Cl, zirconium/aluminium chlorhydrate] used in antiperspirant preparations show negligible percutaneous absorption.[23,26] Inhalation of zirconium or zirconium silicate is a potential cause of pneumoconiosis and pulmonary fibrosis but alveolar absorption is unclear. Zirconium dust particles absorbed by pulmonary macrophages are birefringent in polarised light microscopy.[94]

The numerous uses of zirconium compounds are summarised in Table 9.1. In each case advantage is taken of such physico-chemical properties as their solubility, stability in air and water, conductive properties, hardness and colour.

9.3.2 Cytotoxicity, Mutagenicity and Carcinogenicity

9.3.2.1 In Vitro Studies

Minimal evidence is available to show the mutagenicity or genotoxicity of zirconium or its compounds.[3,95,96] Bioassays in bacteria, algae and fish indicate that zirconium exhibits low environmental toxicity.[97] It was not mutagenic in the fluctuation test and failed to induce DNA-related changes in chromotests. Zirconium, as a ceramic material (tetragonal zirconia polycrystal) used increasingly in medical devices and stabilised with yttrium, was not found to be mutagenic or capable of transforming cultured fibroblasts.[95] In a *Salmonella* mutagenicity test with and without S9 mix, zirconyl chloride was not mutagenic.[98] *In vitro*, 10 and 20 $\mu g\ ml^{-1}$ zirconium oxychloride added to cultures of human blood cells from healthy donors led to a significant increase in chromosomal aberrations, breaks and sister-chromatid exchanges as evidence that zirconium is capable of clastogenicity.[99]

The cytotoxicity of aluminium and zirconium used in cosmetics was examined in cultured skin fibroblasts and macrophages.[3] Whereas, at high concentrations, aluminium hydroxide, aluminium chlorhydrate and a zirconium–aluminium–glycine complex were shown to be cytotoxic and haemolytic, zirconium hydroxide was without obvious effect. The study demonstrated a good correlation between the ability of zirconium and aluminium compounds to evoke sensitisation reactions and granuloma in guinea pig tests and their cytotoxic properties in cultured cells.

9.3.2.2 *Experimental Animal Studies*

Experimental studies in animal models have failed to demonstrate the carcinogenicity or mutagenicity of zirconium compounds in chronic studies.[100–102] Rats, mice, rabbits and dogs have been shown to tolerate high concentrations of zirconium salts for lifetime periods of inhalation or administration of compounds in diet without showing evidence of neoplastic change in any tissue. Studies conducted to evaluate the acceptability of zirconium-based prosthetic materials by subcutaneous implantation have yielded negative results.[95,103]

A more accurate assessment of the ability of zirconium to evoke granulomas is seen in experimental studies where topical, intradermal or subcutaneous implantation or injection of sodium zirconium lactate or zirconium tetraisopropoxide [$Zr(OCHCH_3CH_3)_4$] led to induction of profound inflammatory reactions and granulomas in guinea pigs and rabbits (Figure 9.2).[4,5] In contrast, zirconium oxide rods stabilised with yttrium oxide failed to evoke subcutaneous reaction within 24 months following implantation in mice, thereby indicating the safety of the product in orthopaedic implants.[104] In their experiments with rabbits,[103] in which aluminium compounds used in cosmetic preparations were

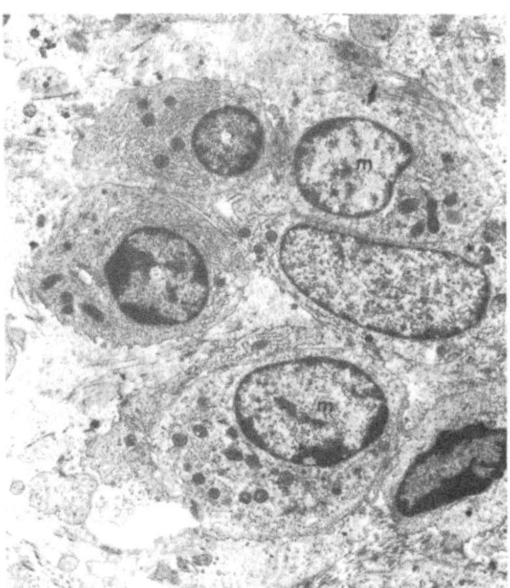

Figure 9.2 Zirconium granuloma.

compared with zirconium tretraisopropoxide, Cronk and Prior had previously shown that 22% aluminium chlorohydroxide produced acute inflammatory reactions and granuloma following repeated subcutaneous injection (but not topical application), whilst the zirconium salt produced a response only when it was injected subcutaneously or intracutaneously with hexachlorophene (a profound skin irritant.[105] Percutaneous application of zirconium tetraisopropoxide with hexachlorophene did not induce reactions following topical application, suggesting that insufficient bioactive ion was absorbed percutaneously.

When zirconia-toughened alumina implants for orthopaedic use or dental devices were implanted in rabbits, tumours did not develop at implantation sites or at other sites in the body and showed good long-term biocompatibility.[96] In chronic toxicity studies in mice, 5 ppm zirconium sulfate in the diet did not evoke an increase in tumours in any organ although the metal was located in bone, heart, lung, kidney, liver and spleen.[106] (Administration of 5 ppm sodium nitrate or potassium antimony tartrate to mice also failed to demonstrate significant carcinogenicity in these lifetime studies.) Only one study has been seen where primary lung tumours were induced by zirconium; this was an experimental study in rabbits exposed to radio-labelled aerosol microsphere particles (10 μm diameter) containing plutonium oxide ($^{238}PuO_2$) and zirconium oxide (ZrO_2).[107,108] α-Emissions in the range 8 to 143 nCi from ^{238}Pu are held responsible for the adenocarcinomas and adenomas reported, but the incidence was only 2% for adenocarcinoma and 9% for benign adenomas.

9.3.2.3 Epidemiogical and Human Case Studies

Human studies focus upon the strong sensitising capacity of zirconium or its compounds and their capacity to induce dermal granulomas.[7,8] Dermal exposure to sodium zirconium lactate and other zirconium-based antiperspirants is a known cause of axillary dermatitis and subcutaneous granuloma.[90,109,110] At least four of these cases related to patients who had used proprietary stick deodorants/antiperspirants containing sodium zirconium lactate in a recommended manner and developed axillary granulomas of a tuberculoid type. One patient showed the equivalent of 0.3 μ g^{-1} zirconium residue in her tissue.

Several older studies show that chronic inhalation or occupational exposures to zirconium dust leads to fibrosis, pneumoconiosis and pulmonary granulomas, but no zirconium-related pulmonary or other respiratory tract carcinomas have been reported.[85,94,111,112] Granulomatous changes have been recorded in the skin when zirconium salts were used to treat *Rhus*-dermatitis.[92,113–116] Sarcoid-type granulomas were seen in a 25-year-old ceramic tile worker who developed a dry cough and dyspnoea after 3.5 years as a sorter and glazier exposed to zirconium silicate and clay materials; this seems to be characteristic of pulmonary hypersensitivity reactions to inhaled zirconium.[112] Open lung biopsy revealed an intense granulomatous reaction associated with zirconium deposits nearly 100-fold higher than background levels. The patient was a nonsmoker and had no history of exposures to other aetiological agents. In other pulmonary biopsies small birefringent particles characteristic of inhaled zirconium were observed around terminal airways and within granulomas. Prospectively, Bartter *et al.* indicated that pulmonary fibrosis and pneumoconiosis progress with an appropriate history of zirconium-related exposure with a latent period of about 15 years before the onset of dyspnoea and radiographic changes, with a honeycomb effect associated with end-stage fibrosis and accumulation of birefingent zirconium particles.[94] It is surprising, therefore, to see a report of 32 hand-finishers who had worked long term (1–17 years) in a metal reactor components plant and who presented with no specific zirconium-related abnormalities on chest X-radiographs or pulmonary function tests, and certainly no signs of neoplastic transformation.[111]

References

1. P. J. LoPresti and G. W. Hambrick, *Arch. Dermatol.*, 1965, **92**, 188.
2. H. G. Skelton, K. J. Smith, F. B. Johnson, C. R. Cooper, W. F. Tyler and G. P. Lupton, *Am. J. Acad. Dermatol.*, 1993, **28**, 874.
3. P. Badenoch-Jones, J. L. Turk and D. Parker, *J. Pathol.*, 1978, **124**, 51.
4. J. L. Turk and D. Parker, *J. Invest. Dermatol.*, 1977, **68**, 336.
5. J. L. Turk, P. Badenoch-Jones and D. Parker, *J. Pathol.*, 1978, **124**, 45.
6. W. B. Shelley and H. J. Hurley, in *Immunological Diseases*, ed. M. Samter, Little Brown & Co., Boston, 1971, p. 2.
7. W. L. Epstein and J. R. Allen, *J. Am. Med. Ass.*, 1964, **190**, 162.
8. A. A. Fisher, *Contact Dermatitis*, Lea Febiger, Philadelphia, 1987.
9. World Health Organization (WHO), *Statement: Carcinogenicity of Combined Hormonal Contraceptives and Menopausal Treatment*, WHO, Geneva, 2007.
10. A. O. Mueck and H. Seeger, *Maturitas*, 2007, **57**, 42.
11. International Agency for Research on Cancer (IARC), *IARC Monographs on the Evaluation of the Carcinogenic Risk of Chemicals to Humans*, 2007, vol. 91.
12. P. D. Darbre, *J. Inorg. Biochem.*, 2005, **99**, 1912.
13. P. D. Darbre, *J. Appl. Toxicol.*, 2006, **26**, 191.
14. P. D. Darbre, *Best Pract. Res. Clin. Endocrinol. Metab.*, 2006, **20**, 121.
15. World Health Organization (WHO), *Aluminium in Drinking Water: Background document for the development of WHO Guidelines for drinking water quality*, WHO, Geneva, 1998.
16. R. L. Bertholf, *Crit. Rev. Clin. Lab. Sci.*, 1987, **25**, 195.
17. P. B. Klepak, *J. Alzheimers Dis.*, 2006, **9**(Suppl. 3), 291.
18. US Department of Health and Human Services, *Report on Carcinogens*, US DHHS, Research Triangle Park, NC, 12th edn, 2011.
19. International Agency for Research on Cancer (IARC), *IARC Monographs on the Evaluation of the Carcinogenic Risk of Chemicals to Humans*, IARC, Lyons, 1987, Suppl. 7, p. 8.
20. R. A. Yokel, *NeuroToxicol.*, 2000, **21**, 813.
21. R. A. Yokel and P. J. McNamara, *Pharmacol. Toxicol.*, 2001, **88**, 159.

22. I. H. Blank, J. L. Jones and E. Gould, *Proc. Sci. Sect. Toilet Goods Assoc.*, 1958, **29**, 1.
23. A. B. G. Lansdown, *Br. J. Dermatol.*, 1975, **89**, 67.
24. M. Wilhelm, J. Passlick, T. Busch, M. Szydlik and F. K. Ohnesorge, *Human Toxicol.*, 1989, **8**, 5.
25. M. Wilhelm, F. K. Ohnesorge, I. Lombeck and D. Hafner, *J. Anal. Toxicol.*, 1989, **13**, 17.
26. A. B. G. Lansdown, *J. Soc. Cosmet. Chem.*, 1973, **24**, 677.
27. R. Flarend, T. Bin, D. Elmore and S. L. Hem, *Fd. Chem. Toxicol.*, 2001, **39**, 163.
28. G. M. Berlyne, R. Yagil, J. Ben Ari, G. Weinberger, E. Knopf and G. M. Danovitch, *Lancet*, 1972, **1**, 564.
29. J. Mitchell, *Br. J. Ind. Med.*, 1959, **16**, 123.
30. J. Mitchell, G. B. Manning, M. Molyneau and R. E. Lane, *Br. J. Ind. Med.*, 1961, **18**, 10.
31. P. Gross, R. A. Harley and R. T. P. deTreville, *Arch Environ Health*, 1973, **26**, 227.
32. G. H. Pigott, B. A. Gaskell and J. Ishmael, *Br. J. Exper. Pathol.*, 1981, **62**, 323.
33. J. T. Prior and G. A. Cronk, *Arch. Dermatol*, 1959, **80**, 105.
34. M. Vogelbruch, B. Nuss, M. Körner, A. Kapp, P. Kiehl and W. Bohm, *Allergy*, 2000, **55**, 883.
35. D. Dyrssen, C. Haraldson, E. Nyberg and M. Wedborg, *J. Inorg. Biochem.*, 1987, **29**, 67.
36. A. R. Llorente, P. Del Castillo and J. C. Stockert, *J. Microscopy*, 1989, **155**, 227.
37. A. C. Johnson and M. Wood, *Appl. Environ. Microbiol.*, 1990, **56**, 3629.
38. R. Y. Zhang, Y. Liu, D. W. Pang, R. X. Cai and Y. P. Qi, *Anal. Sci.*, 2002, **18**, 761.
39. V. M. M Achery and B. B. Panda, *Mutagen.*, 2010, **25**, 201.
40. A. Lankoff, A. Banasik, A. Duma, E. Ochniak, H. Lisowska, T. Kuszewski, S. Góźdź and A. Wojcik, *Toxicol. Lett.*, 2006, **161**, 27.
41. V. Kumar, A. Bal and K. D. Gill, *Toxicol.*, 2009, **264**, 137.
42. R. Anane and E. E. Creppy, *Hum. Exp. Toxicol.*, 2001, **20**, 477.
43. H. Celik, N. Celik, A. Kocyigit and D. Dikilitas, *Clin. Biochem.*, 2012, **45**, 1629.
44. A. Léonard and G. B. Gerber, *Mutat. Res.*, 1988, **196**, 247.
45. A. Balasubramanyam, N. Dailaja, M. Mahboob, M. E. Rahman, S. M. Hassain and P. Grover, *Toxicol. in Vitro*, 2010, **24**, 1871.
46. A. Krøkje, A. Tiltnes, E. Mylius and B. G. Gullvåg, *Mutat. Res.*, 1985, **156**, 147.
47. A. Krøkje, A. Tiltnes, E. Mylius and B. Gullvåg, *Mutat. Res.*, 1988, **204**, 163.
48. O. Zelić, B. Dimitrijević, M. Vasilijevska, A. Dijić and A. Lekić, *J. Clin. Periodontol.*, 1998, **25**, 892.
49. R. Verberckmoes, *Lancet*, 1972, **ii**, 750.

50. R. F. de Estable-Puig, J. F. de Estable-Puig and C. Romero, *Virchows Arch. Abt. B Zellpath*, 1971, **8**, 267.
51. M. Di Silvestre, S. Guizzardi, N. Bettini, G. Gargiulo and R. Savini, *Chir. Organi Mov.*, 1991, **76**, 167.
52. World Health Organization (WHO). *Aluminium International Programme on Chemical Safety, (Environmental Health Criteria)*, WHO, Geneva, 1997, p. 194.
53. M. C. Friesen, L. Fritschi, A. Del Monaco, G. Benke, M. Dennekamp, N. de Klerk, J. L. Hoving, E. MacFarlane and M. R. Sim, *Occup. Environ. Med.*, 2009, **66**, 615.
54. A. Rønneberg and A. Andersen, *Occup. Envion. Med.*, 1995, **52**, 250.
55. P. Galle, J. P. Berry and C. Galle, *Environ. Health. Perspect.*, 1992, **97**, 145.
56. M. R. Sim, A. Del Monaco, J. L. Hoving, E. MacFarlane, D. McKenzie, G. Benke, N. de Klerk and L. Fritschi, *Occup. Environ. Med.*, 2009, **66**, 464.
57. G. Thériault, C. Tremblay, S. Codeier and S. Gingras, *Lancet*, 1984, **i**, 947.
58. G. Thériault, C. G. Tremblay and B. G. Armstrong, *J. Occup. Med.*, 1980, **32**, 869.
59. A. L. Seldén, H. B. Westberg and O. Axelson, *Am. J. Ind. Med.*, 1997, **32**, 467.
60. G. W. Gibbs and I. Horowitz, *J. Occup. Med.*, 1979, **21**, 347.
61. G. W. Gibbs, *J. Occup. Med.*, 1985, **27**, 761.
62. P. Romundstad, T. Haldorsen and A. Andersen, *Occup. Environ. Med.*, 2000, **57**, 495.
63. B. G. Armstrong, C. G. Tremblay, D. Cyr and G. P. Thériauld, *Scand, J. Work Environ. Health*, 1986, **12**, 486.
64. A. Andersen, B. E. Dahlberg, K. Magnus and A. Wannag, *Int. J. Cancer.*, 1982, **29**, 295.
65. J. J. Moulin, T. Clavel, B. Buclez and G. Laffitte-Rigaud, *Int. Arch. Occup. Health*, 2000, **73**, 323.
66. P. Harry, A. Caubet, G. Durand, C. Mathieu, Y. Mauras and P. Allain, *J. Toxicol. Clin Exp.*, 1989, **9**, 359.
67. W. Szymczak, *Med. Pr.*, 1996, **47**, 547.
68. V. D. Martinez, E. A. Vucic, D. D. Becker-Santos, L. Gil and W. I. Lam, *J. Toxicol.*, 2011, ID431287.
69. Health and Safety Executive (HSE), *Mesothelioma*, HSE, London, 2011–2012.
70. K. G. McGrath, *Eur. J. Cancer Prev.*, 2003, **12**, 479.
71. C. Exley, L. M. Charles, L. Barr, C. Martin, A. Polwart and P. D. Darbre, *J. Inorg. Biochem.*, 2007, **101**, 1344.
72. J. J. Hostýnek, R. S. Hinz, C. R. Lorence, M. Price and R. H. Guy, *Crit. Rev. Toxicol.*, 1993, **25**, 171.
73. J. F. Alder, M. C. C. Batoreu, A. D. Pearse and R. Marks, *J. Anal. At. Specrtum*, 1986, **1**, 365.

74. A. B. G. Lansdown, in *Metal Ions in Toxicology: Effects, interactions interdependencies*, ed. A. Sigel, H. Sigel and R. O. Sigel, RSC Publishing, Cambridge, 2011, vol. 8, p. 187.
75. N. McFadden, T. Lyberg and A. Hensten-Pettersen, *J. Am. Acad. Dermatol.*, 1989, **20**, 903.
76. C. J. Rageth, *The Breast*, 2004, **14**, 85.
77. P. D. Gikas, L. Mansfield and K. Mokbel, *Int. J. Fertil. Woman's Med.*, 2004, **49**, 212.
78. International Atomic Energy Authority (IAEA), *Safety Reports Series*, 2007, No 51, IAEA, Vienna.
79. I. Veronese, A. Giussani, M. C. Cantone, T. Maggioni, C. Birattari, F. Groppi, E. Werner, P. Roth and V. Höllriegl, *Appl. Radiat. Isot.*, 2003, **58**, 431.
80. P. Heitland and H. D. Köster, *J. Trace Elem Med. Biol.*, 2006, **20**, 253.
81. S. Nag and R. Banerjee, *ASM Handbook; Materials for Medical Devices*, R. Narayan, ASM International, Madison, WI, 2012, p. 23.
82. B. S. Rodchenkov and A. N. Ivanov, *Fourth International Conference on the Application of Zirconium in the Atomic Industry, Soviet Atomic Energy*, 1978, **45**, p. 1204.
83. G. D. Moan and P. Rudling, *Zirconium in the Nuclear Industry: Thirteenth International Symposium, Issue 1423*, ASTM International, West Conshohocken, PA, 2002.
84. G. R. Baler, *Arch. Dermatol.*, 1965, **91**, 145.
85. P. J. LoPresti and G. W. Hambrick, *Arch. Dermatol.*, 1965, **92**, 188.
86. W. Hopf, C. G. Hopf and B. Glöbel, *Acta Orthop. Belg.*, 1990, **56**, 332.
87. G. A. Cronk and D. E. Naumann, *J. Lab. Clin. Med.*, 1951, **37**, 909.
88. G. A. Cronk, *Arch. Dermat. Syphilol.*, 1952, **66**, 282.
89. A. P. Caspers, *Can. Med. Assoc. J.*, 1957, **76**, 852.
90. L. Rubin, A. H. Slepyan, L. F. Weber and I. Neuhauser, *J. Am. Med. Assoc.*, 1956, **162**, 953.
91. R. M. Williams and G. B. Skipworth, *Am. Med. Ass. Arch. Dermatol.*, 1959, **80**, 273.
92. C. Sheard, F. E. Cormia, S. C. Atkinson and EL. Worthington, *J. Am. Med. Assoc.*, 1957, **164**, 1085.
93. International Association for Research on Cancer (IARC), *IARC Monographs on the Evaluation of Carcinogenic Risks to Humans*, IARC, Lyons, 2012, 100C.
94. T. Bartter, R. S. Irwin, J. L. Abraham, A. Dascal, G. Nash and P. J. Jederlinic, *Arch. Int. Med.*, 1991, **151**, 1197.
95. V. Covacci, N. Bruzzese, G. Maccauro, C. Andreassi, G. A. Ricci, C. Picconi, E. Marmo, W. Burger and A. Cittadini, *Biomaterials*, 1999, **20**, 371.
96. G. Maccauro, A. Cittidini, G. Magnani, S. Sangiorgi, F. Muratori, P. F. Mnicone, P. Rossi Iometti, D. Marotta, A. Chierichini, L. Raffaelli and A. Sgambato, *Int. J. Immunopathol. Pharmacol.*, 2010, **23**, 841.

97. P. Couture, C. Blaise, D. Cluis and C. Bastien, *Water, Air and Soil Pollution*, 1989, **47**, 87.
98. K. Mortelmans, S. Haworth, T. Lawlor, W. Speck, B. Tainer and Zeiger, *Environ. Mol. Mutagen.*, 1986, **8**(Suppl. 7), 1.
99. S. Ghosh, G. Takukder and A. Sharma, *Toxicol. in Vitro*, 1991, **5**, 295.
100. J. W. E. Harrisson, B. Trabin and E. W. Martin, *J. Pharmacol. Exp. Therap.*, 1951, **102**, 179.
101. C. J. Spiegl, M. C. Calkins, J. J. DeVoldre, J. J. Scott, L. T. Steadman and H. E. Stokinger, *Inhalation Toxicology of Zirconium Compounds I. Short-term Studies*, Technical Report, US Atomic Energy Commission, Rochester, NY, 1956.
102. M. Kanisawa and H. A. Schroeder, *Cancer Res.*, 1969, **29**, 892.
103. J. T. Prior and G. A. Cronk, *Am. Med. Ass. Arch. Dermatol.*, 1959, **80**, 447.
104. K. Takamura, K. Hayashi, N. Ishinishi, T. Yamada and Y. Sugioka, *J. Biomed. Mater. Res.*, 1994, **28**, 583.
105. W. S. Gump, *J. Soc. Cosmetic Chemists*, 1969, **20**, 173.
106. M. Kanisawa and H. A. Schroeder, *Cancer Res.*, 1969, **29**, 892.
107. R. G. Thomas and D. M. Smith, *Int. J. Cancer*, 1979, **24**, 594.
108. R. G. Thomas, G. A. Drake, J. E. London, E. C. Anderson, J. R. Prine and D. M. Smith, *Int,. J. Radiat. Biol.*, 1981, **40**, 605.
109. J. T. Prior, H. Rustas and G. A. Cronk, *I. Invest. Dermatol.*, 1957, **29**, 449.
110. L. F. Weber, I. Neuhauser, L. Rubin, A. H. Slepyan and H. Shellow, *J. Am. Med. Ass.*, 1956, **162**, 65.
111. O. C. Hadjimichael and R. E. Brubaker, *J. Occup. Med.*, 1981, **23**, 543.
112. K. K. Liippo, S. L. Anttila, O. Taikina-Aho, E. L. Ruokonen, E. L. Toivenen and T. Tuomi, *Am. Rev. Respir. Dis.*, 1993, **148**, 1089.
113. W. L. Epstein, *Adv. Biol. Skin*, 1971, **11**, 313.
114. N. A. Hall, *J. Am. Pharm. Assoc.*, 1972, **12**, 576.
115. G. A. Cronk, *Arch. Dermat. Syphilol.*, 1952, **66**, 282.
116. M. E. Obermayer, *Arch. Dermatol.*, 1969, **100**, 505.

CHAPTER 10

Cadmium and Mercury

10.1 General Introduction

Cadmium and mercury are bivalent transition metals classified with zinc in Group 12 (formerly IIb) of the Periodic Table.[1] They are highly reactive elements but differ greatly in their biological activities and in their action in the human body. In each case, cellular and molecular interactions have been studied extensively in humans and in experimental animal models. The cations bind strongly to extracellular cellular and intracellular proteins, they strongly induce metallothioneins (MT-I and MT-II), and their metabolic pathways in the human body are well defined. Cadmium [Cd(II)] and mercury [Hg(II)] are xenobiotic ions and acknowledged cytoplasmic poisons.[1–3] Their toxicities are well illustrated by the major environmental and epidemiological disasters in Japan of which Minamata Bay and *itai-itai byo* ("ouch ouch disease") are major examples of environmental contamination through inappropriate management of methyl mercury and cadmium respectively.[4–6] In retrospect, the Minamata Bay incident should be considered a milestone of ecological ignorance in world history with improper management of industrial waste resulting in loss of life and long-lasting harm.

Natural deposits of both elements are widespread in the geosphere. Geochemical deposits are commonly associated with other elements including zinc, arsenic, lead and silica from which the metals are extracted using selective exchange processes or other physico-chemical means.[7] However, natural deposits are not stable but influenced by soil erosion, geochemical activity, water action and environmental change. Environmental problems are exacerbated by the extensive industrial use of both elements and the environmental contamination arising from factory waste, landfill sites, combustion of industrial debris and fossil fuels, and aerobic emissions. Both elements have been in use as pesticides and agricultural products for many years and are now recognised as major environmental hazards.[8–10]

Issues in Toxicology No. 18
The Carcinogenicity of Metals: Human Risk through Occupational and Environmental Exposure
By Alan B. G. Lansdown

Published by the Royal Society of Chemistry, www.rsc.org

10.2 Cadmium

Cadmium is a soft silvery white bivalent metal found widely in the human environment. It exists in only one valence state, Cd(II), and does not readily form organo-metallic compounds of toxicological significance. Metallic cadmium does not occur naturally but is extracted from the oxide, sulfide, sulfate and chloride, often found in deposits with zinc, copper, lead and arsenic. Extraction workers will be exposed to the toxic effects of these elements as an occupational risk. Weathering and erosion of sedimentary rocks over many years has led to vast amounts of cadmium being disseminated into surface soils, washed into inland waterways and the world's oceans to be concentrated in marine life, thereby entering human food chains. Over many generations, combustion of fossil fuels, cadmium in agrochemicals, industrial emissions and factory wastes have resulted in widespread dissemination of cadmium residues in the environment.[11–13] Human health risks from cadmium exposure were recognised largely through environmental contamination resulting from cadmium in fertilisers and fungicides used in tea plantations and rice paddies in Far Eastern countries. Renal toxicity, osteoporosis and increased rates of prostatic and lung cancers were reported. The World Health Organization, International Agency for Research on Cancer and many national authorities issued guidelines for human exposure. Cadmium is recognised as a carcinogen and teratogen in animals and humans.[8,14–16]

Cadmium resembles zinc in its biological action but whereas zinc is a component or co-factor of at least 300 essential enzymes and is present in every living cell in the body,[17] cadmium has no micro-trace elemental value in higher organisms and is cytotoxic, mutagenic, teratogenic and potentially carcinogenic.[14] Low levels of cadmium occur in the human body through ingestion, inhalation and dermal contact but, in a normal individual, inherent protective functions are sufficient to mitigate its toxic effects. Cadmium, like zinc, strongly induces and binds cysteine-rich metallothioneins (MT-I and II) which act as cytoprotectants against several toxic metals.[18–21] It displaces zinc from Zn–MT complexes and zinc-metalloenzymes as an essential part of its cytotoxic mechanism.[21,22] The MTs provide a cytoprotective mechanism by mopping up and complexing free Cd(II) but experimental studies illustrate that when the cytoprotective effects afforded by MTs become saturated, so Cd(II) exerts inflammatory and immunological changes.[20] Further illustration of the cytoprotective role of MTs is provided by experimental studies using animals with deficient MT-gene expression. MT-null type animals are susceptible to cadmium-induced cytotoxicity and carcinogenicity.[18,23–25] Zinc supplements mitigate the toxic influence of cadmium, indicating that cadmium toxicity is largely attributable to impairment of zinc-related metabolic pathways. Examples are provided to show that individuals showing high Cd : Zn plasma concentrations are at a significantly greater risk of prostatic cancer.[26,27] This is particularly true of smokers, who may be at 4.5 times greater risk of developing cancer than normal, and have 20-fold higher blood cadmium level.[28]

A major route of cadmium uptake is through drinking water and ingestion of vegetables and meat, the latter from animals grazing cadmium-rich pastures. Children absorb cadmium more readily than adults and are more susceptible to cadmium-induced renal toxicity.[29] Plants grown on cadmium-rich soils accumulate cadmium residues at high concentrations and seem to be more tolerant of the toxic effects of cadmium than animals.[30,31] Tobacco plants accumulate cadmium and residues persist into the cigarette-making phases and into cigarettes and mainstream smoke. Cigarettes contain up to 1.61 mg kg^{-1} Cd depending on their source and the soil in which the tobacco plants are grown, but between 20 and 50% of inhaled mainstream smoke may be absorbed.[32–38] Pfeifer *et al.* identified marked increases in *p53* gene mutations and a 30% prevalence of base-pair transversions in tobacco-related cancers compared to 12% in non-smoking related lung cancers.[36]

10.2.1 Industrial Use and Human Exposures

Cadmium has many industrial uses including anti-corrosion metal alloys, PVC and plastics manufacture, colorants and pigments (in plastics, ceramics, glasses and paints), nickel : cadmium batteries and anticorrosion electroplating. Cadmium forms stable alloys with copper, tin and other non-ferrous metals, and is used in the production of cadmium-coated aluminium, brass, copper and steel.[39] Smelting of non-ferrous metal ores is estimated to be the largest single source of cadmium release into the marine environment.[13] The use of cadmium in batteries may account for up to 75% of current usage of the metal.

Cadmium sulfide-based yellow, orange, maroon and red pigments have long been used in colouring paints, plastics, ceramics and glass. The colours are stable and heat resistant and can be incorporated into acrylic and latex coatings, polymers used in telephones, gas pipes and fittings, electricity cables, beverage crates and motor vehicle radiator fans. The pigments are usually incorporated in plastics in proportions of 0.01 to 0.75% by weight. Additionally, cadmium has extensive value as a plasticiser and stabiliser in PVC although in recent years this has been dramatically restricted on safety grounds. The European Commission (EC) phased out cadmium from all PVC products in 2010 on the grounds that safer alternatives were available.[40] They have also taken steps to control recycling and disposal of cadmium-containing plastic waste. The electrical manufacturing industry in the USA stressed that whilst alternatives had been found to replace cadmium in electrical devices, many are inferior and that a blanket ban on the use of cadmium-bearing electrical contacts will do more harm than good.[41]

Many occupational health problems have been experienced in Far Eastern countries where cadmium chloride has been used as a fungicide for treating rice fields, tea plantations and vineyards. In Sri Lanka, the organic-rich, acidic soils with 5300 mm annual rainfall with 65% runoff create ideal conditions for cadmium to leach from agricultural fields to contaminate local reservoirs, where cadmium levels of 0.03–0.06 mg l^{-1} were 10–20 times higher than those set by the World Health Organization (0.003 mg l^{-1}).[42,43] Large-scale release of

Table 10.1 Cadmium uptake in rats and mice treated topically with cadmium chloride.[58]

$CdCl_2$ *Concentration (%)*	*Skin (μg g^{-1})*	*Blood (μ l^{-1})*	*Liver (μ g^{-1})*	*Kidney (μ g^{-1})*
Rat				
1.0	6.109	11.65	0.526	0.216
0.1	1.830	1.55	0.061	0.028
0.01	0.248	<0.50	<0.005	<0.005
Control	<0.003	<0.10	<0.005	<0.005
Mouse				
1.0	6.008			
0.1	0.765			
0.01	0.138			
Control	<0.005			

cadmium from superphosphate fertilisers (containing <40 mg kg^{-1} inorganic cadmium) into ground water and food products is a reported cause of renal disease.[11,44–46] In Thailand and other Asian countries, fertilisers applied to paddy fields, wastewater, sludge, and phosphate mine tailings contribute greatly to the human intake of cadmium.[47]

Bioactive cadmium ion is absorbed into the circulation following ingestion, inhalation and minimally following dermal contact.[15] Gastrointestinal uptake may be as high as 40–60% of that ingested but, regardless of its route of uptake, at least 50% of the cadmium absorbed is transported to the kidney with lesser amounts to liver, muscle and bone.[48] The renal proximal tubular epithelium is a prime target for cadmium toxicity, and cadmium accumulates in the kidney.[49] Cadmium excreted in the urine, hair and fingernails provides a useful guide to body burden or patterns of uptake.[50–52] Inhalation is a major route of cadmium absorption through cigarette smoke and in industry where welders exposed to cadmium fumes are at risk of developing fume fever, pneumonitis and impaired lung function.[53–55] A study of 101 men exposed to cadmium fumes and/or smoking for more than one year showed bronchitis, significant reduction in pulmonary function and radiological change consistent with emphysema. Levels of pulmonary damage are a reflection of the duration and intensity of cadmium exposures.[56]

Radiotracer studies using ^{109}Cd have shown percutaneous penetration of cadmium from soil samples.[57] Much of the ion is bound strongly to epidermal keratin with <12% absorbed by plasma. Experimental studies in rodents have shown that much of the cadmium absorbed is transported to the kidneys and liver (Table 10.1).[58]

10.2.2 Genotoxicity, Mutagenicity and Carcinogenesis

Cadmium is a cumulative toxin capable of inducing acute and chronic changes, teratogenicity, reduced fertility and fatality.[15,59,60] Cadmium and its inorganic compounds are recognised in the 12th RoC as human carcinogens on the basis of sufficient epidemiological and experimental evidence.[8,14,61]

10.2.2.1 Cytotoxicity and In Vitro Mutagenicity

Cadmium absorbed by passive diffusion or through phagocytic means binds avidly with erythrocytes and plasma proteins to be transported systemically. Free ion binds strongly to receptors on cell membranes and intracellular proteins to induce irreversible enzymatic changes, mitochondrial and lysosomal dysfunction and disorders in other intracellular target sites.[62–66] Toxic changes include mutagenic and genotoxic damage, and cyto-degeneration.[15,67,68] Cadmium shows a strong affinity for sulfydryl groups and binds strongly to epidermal keratins. It impairs intracellular sulfydryl homeostasis through inducing oxidative stress and increasing lipid peroxidation in key tissues, including neuronal cells and renal tubular cells. In the opinion of some, much remains to be understood regarding the intracellular action of cadmium in target cells.[48,49,69]

Cadmium is strongly genotoxic and evokes DNA damage, transcriptional responses and gene deletion in rodent osteoblasts, ovarian cells, myoblasts and liver cell cultures, apoptosis, impaired proliferation and differentiation, and other cytoplasmic changes in cultured cells.[70,71] It is mutagenic in bacterial and mammalian cells systems but recent evidence suggests that inhibition of DNA repair is a predominant mechanism in cadmium-induced carcinogenesis.[72] Mutagenicity is attributed also to induction of reactive oxygen species, which is enhanced by addition of the chelator diethydithiocarbamate to the culture medium.[73] Cd(II) has been shown to enhance the mutagenicity of two nitrosamines in reverse transcription assay with *Salmonella typhimurium* in a synergistic fashion.[74]

10.2.2.2 Experimental Animal Studies

Experimental animal studies provide a very limited insight into the carcinogenicity of cadmium or its inorganic compounds in humans.[75] Some evidence of carcinogenicity is provided by studies in which metallic cadmium or cadmium compounds (soluble or insoluble) were administered by intramuscular implantation or subcutaneous injection, or inhalation of cadmium chloride in aerosol form. There is minimal evidence that cadmium or its compounds are carcinogenic in lifetime oral/in diet dosing studies even though animals exhibited profound signs of cadmium-related toxicity.[76,77] Rats of the Noble (NBC/Cr) strain given 25, 50, 100 or 200 ppm cadmium chloride in drinking water for up to two years showed hyperplastic lesions of the prostate and testis, and a low dose-related increase in renal and adrenal tumours. No evidence is provided by experimental studies to show that dermal exposure to soluble cadmium compounds is a cause of neoplastic change even though they are profoundly irritating in skin and are a cause of impaired healing.[20,58] Cadmium is absorbed percutaneously and accumulates in kidney and liver.

Single or repeated subcutaneous injections of metallic cadmium, cadmium oxide and cadmium sulfide in rats were a cause of injection site sarcomas (fibrosarcomas with varying degrees of pleomorphism, spindle cells tumours, rhabdomyosarcoma) but their ability to spread locally or to metastasise to other sites varied according to the nature of the injection and the duration of observation.[78–82] Female Syrian hamsters given a single subcutaneous injection of 50 μg kg^{-1} cadmium chloride did not show an increased incidence of injection site or other tumours in a 78-week observation period, but ovarian lesions were reported in sub-acute studies.[83,84] One strain of mice (DBA) given single or repeated cadmium chloride and observed over two years showed dose-related increases in primary lymphoma.[85] In contrast, an NFS strain of mice failed to show increased lymphomas in response to cadmium chloride but developed injection site and hepatocellular tumours not seen in DBA or control animals. Tumour yields in rats and sensitivity to subcutaneous cadmium chloride also varied with strain. Total tumour yields ranged from 10 to 37% and included a high proportion of injection site tumours and varying numbers of interstitial cell tumours of the testis, prostatic tumours and islet cell tumours of the pancreas. Whilst subcutaneous injections of cadmium chloride were shown to induce testicular degeneration, atrophy and soft tissue tumours in some strains of rat, when prostatic tumours (adenomas) did occur, they were not dose related and developed at doses well below those causing prostatic atrophy (*i.e.* 2.5 μmol kg^{-1}).[86] Prostatic hyperplasia increased following single injections of up to 20 μmol kg^{-1}.

Surprisingly, testicular tumours, which show a high spontaneous incidence in F344 rats, were not promoted or otherwise influenced by subcutaneous injections of cadmium, which suppressed the leukaemias and lymphoid lesions which are prevalent in this strain.[79] Repeated subcutaneous injections of 10, 20 or 30 μmol at weekly intervals for 18 weeks resulted in a dose- and strain-related incidence of injection site tumours and enhanced metastatic potential. Primary tumours exhibited high MT activity, whereas MT was not demonstrated histochemically in metastatic lesions. In another study, these researchers demonstrated that Wistar rats given single or repeated subcutaneous injections of cadmium chloride showed a dose-related increase in Leydig cell tumours of the testis which they claimed were dependent upon levels of cadmium-induced testicular degeneration.

Tumour incidence in rats injected subcutaneously with cadmium chloride was influenced by dietary zinc.[69,81] Thus intramuscular injection of zinc, which is well known to mitigate the toxic effects of cadmium through induction of cytoprotective MT, exhibited a reduction in the incidence of injection site and testicular intramuscular tumours in Wistar rats given cadmium chloride, but only when administered by subcutaneous injection. Oral zinc was ineffective. In contrast, dietary zinc deficiency was associated with enhanced appearance of cadmium-related injection site and testicular tumours in rats. Simultaneous injection of calcium or magnesium acetate had no obvious influence on the incidence of injection site or interstitial cell testicular tumours in rats.[87] However, these authors did report that, in their strain of Wistar rats, a single injection of 0.04 mmol kg^{-1} cadmium chloride led to a four-fold increase in islet cell tumours of the pancreas.

One study has been seen in which pulmonary tumours were induced in a W strain of rats exposed to cadmium chloride aerosols for 18 months.[88] After a further 13 months, animals surviving 50 μg m^{-3} cadmium chloride aerosol exhibited a 71.4% incidence of carcinomas whereas those receiving lower doses of 25 or 12.5 μg m^{-3} showed 52.6 and 15.4% tumour incidence, respectively. No other tumours were reported.

10.2.2.3 Human Case and Epidemiological Studies

Cadmium enters the human body through inhalation of fumes, dusts and factory emissions, ingestion of contaminated food and drinking water, and through percutaneous absorption. In the UK in 1978, the Applied Geochemistry Research Group emphasised that high levels of cadmium (11–998 ppm) in the ground in the village of Shipham in Somerset presented health risks to the population.[89] The average cadmium level in the soil in the UK is <1.0 ppm. Population studies of the people of Shipham showed elevated blood cadmium, and some evidence of vascular and renal changes, but health-related changes were inconclusive and the daily cadmium intake was below the estimated threshold for renal damage. A small but statistically significant excess mortality was seen from hypertension and cerebrovascular disease but not cadmium-related cancer.

Inhalation of cadmium in fumes and dust in cadmium-related industries is a cause of acute pneumonitis, emphysema, oedema, increased lung cancer mortalities and multisystem disorders.[53,90–92] Renal dysfunction marked by irreversible pathological change in tubular filtration, proteinuria and enzymatic changes commonly occurs.[93–95] Bone, liver, spleen, prostate and muscle are vulnerable to cadmium according to the route, duration and levels of exposure and the concentration accumulating in the tissue.[96–99] Cadmium interacts with calcium, magnesium and zinc in hydroxyapatite metabolism leading to increased bone resorption, defects in mineralisation and osteoporosis.[100,101] Approximately 13–19% of cadmium inhaled is absorbed by the lungs, equivalent to 0.06 μg per day.[102] Estimates suggest that 50% of particles of ~1 μm diameter depositing in lung tissue are absorbed systemically and then transported to liver, kidney, bone and other tissues.[15] Ambient air cadmium concentrations have generally been estimated to range from 0.1 to 5 ng m^{-3} in rural areas, from 2 to 15 ng m^{-3} in urban areas, and from 15 to 150 ng m^{-3} in industrialised areas.[92] Cadmium is absorbed from animal feed and is distributed to food products and human food chains in rural areas.[103] Up until 1996, the lowest mean airborne level of cadmium causing toxicity in individuals was estimated to be 88 μg m^{-3} over 8.6 years in men and 129 μg m^{-3} over 20 years in women.[104] However, Lamm *et al.* estimated that risks of dying of lung cancer in a cadmium recovery plant in Colorado were more due to exposure to arsenic and cigarette particulates, rather than to cadmium.[105] Concentrations in hair and urine are clinical markers of inhalation of environmental cadmium or its ingestion in contaminated food and drinking water.[15,59,60]

Excellent reviews document inorganic cadmium compounds as causes of cancer of the human lung, prostate and possibly liver, kidney, breast, pancreas and stomach.[8,56,75,79,92] The first clinical evidence that cadmium might be a cause of human cancer was recorded in a Ni–Cd battery manufacturing plant in

1965, when, among 4 of 74 workers exposed to cadmium oxide dust for at least 10 years, one died of lung cancer and three of prostatic cancer.[93] Subsequent occupational health studies conducted in the USA, Sweden and the UK claim that significant increases in respiratory tract and prostatic cancer fatalities, and non-malignant respiratory diseases, in cadmium production workers of at least two years' employment are attributable to inhalation of cadmium.[90,97–99,106–108] The weight of competent epidemiological evidence points to inhalation of cadmium fumes in smelting, extraction and Ni–Cd battery manufacture as a potential cause of respiratory tract cancer and that the risk is proportional to the duration of exposure.[75,109,110] Quantitative assessment of lung cancer risk in a 43-year survey of one smelter showed that fatalities were more frequent in workers with at least six months' employment, and that standardised mortality ratios (SMR) from lung cancer were 149, rising to 272 in the highest cadmium exposure group (20 + years).[90] Hispanic workers were seemingly more sensitive to cadmium-induced lung cancer than white males, but overall the cancer risk from cadmium exposure (100 μg m^{-3}) was estimated to be approximately 50–100 deaths per 1000 workers exposed for 45 years.

Although experimental studies in rodents implicate parenteral cadmium as a possible cause of prostatic and possibly lung cancers, interpretation of human case and epidemiological evidence is problematic and inconclusive. Many studies are reported in the past 40 years but none clearly allow for smoking as a cause of bronchitis and lung cancer, nor do they allow for other toxic factors in the workplace.[111] Nickel residues in Ni–Cd battery manufacture are a good example. One study reporting a 50% increase in unspecified lung cancer mortality among workers with a cumulative exposure to 41–200 μg m^{-3} over 40 years regarded the incidence as being "not statistically significant".[97] In a second study, cancer mortalities amongst in 522 workers employed in Swedish Ni–Cd battery plant for at least one year showed that, whilst lung and prostatic cancer rates appeared to be increased following cadmium exposure and the duration of the latent period, the numbers of cases did not attain statistical significance.[92] Cumulative data from six cohorts showed 28 cases of prostatic cancer and 195 cases of lung cancer, suggesting that high cadmium inhalation is a causative factor, but the concomitant exposure to nickel residues might point to co-carcinogenesis. Other negative observations of a link between cadmium exposure and cancer were reported in a study of 6995 men exposed for more than one year (1942–1970); there was no excess of prostatic cancer mortality and only a marginal but not significant rise in lung cancer, but it is emphasised that only 3% of these men were exposed to high cadmium levels.[112] In this and most other cadmium studies, inhalation was associated with life-threatening bronchitis and fume fever. No allowances were made for the incidence of cigarette smoking amongst workers, long lag phases between presumed initial exposure and evidence of lung malignancy, or risks attributable to other potentially carcinogenic substances such as arsenic, lead and nickel commonly found in the workplace.[92,105,113] Epidemiological studies of 3025 workers in a British Ni–Cd battery industry plant (1946–1984) showed that lung cancer risk was closely allied to working practice and periods of inhaling cadmium oxide dust, cigarette smoking, nickel hydroxide dust and oxyacetylene welding fumes.[99,92] They did not anticipate that inhalation of nickel oxide dust might be contributory to or co-carcinogenic in workers employed in this Ni–Cd battery plant.

Cancers of the prostate and genital tract may be associated with industrial exposures to high levels of cadmium in smelters and in Ni–Cd battery manufacture,[93,114,115] but estimates are inconsistent and whilst some epidemiologists claim that exposure to cadmium oxide dust is implicated this is not adequately supported by statistically significant numbers.[116–118] Kazantzis and Armstrong reported no excess of cancers of the urinogenital tract in a survey of 6995 men exposed to cadmium for more than one year, although cadmium-related renal damage was evident.[117] Epidemiological data showing 28 cases of prostatic cancer mortality in 522 Swedish workers in six cohorts exposed to cadmium dust for more than one year were also insignificant.[92] No adjustments were made for the known fact that cigarette smokers are more at risk of prostate cancer irrespective of the cadmium content of the tobacco used.[120,121] Suspected cadmium-induced malignancies in other tissues including liver, pancreas, breast, kidney and bladder, are not supported by statistically sound evidence.[75,112,119,122,123]

Cigarette smoking is a major source of cadmium absorption and a confounding factor in many occupational health reports showing increased lung cancer mortality.[37,38,91,105,124] Cadmium in cigarettes ranges from 1.56 to 1.96 $\mu g\ g^{-1}$ and in cigars and pipe tobacco 0.93 to 1.86 $\mu g\ g^{-1}$. Chronic inhalation of 0.5–2.0 $\mu g\ g^{-1}$ Cd(II) in cigarette smoke contributes significantly to the body burden and is a major cause of emphysema. Smoking habits and the occupations of parents have been shown also to influence hair cadmium concentrations in children.[125,126] It has to be emphasised with cadmium, as with other metals, that cigarettes and related products contain a wide range of carcinogens including tars, polycyclic hydrocarbons and organic chemicals, all of which are known to be carcinogenic in rodents and which can be expected to contribute singly or cumulatively to human lung and prostatic cancer. In UK alone, about 86% of all lung cancer deaths are attributed to smoking.[127] Recent estimates show that, overall, prostate cancer is the second most common cancer worldwide for males, with an estimated 900,000 new cases diagnosed in 2008.[128] Interestingly, current estimates of prostatic cancer incidence worldwide show lower rates in countries such as China and other Far Eastern countries that currently produce the highest levels of cadmium.

Clinical evidence supporting the claim that environmental cadmium exposure alone is a cause of human lung cancer mortality is seriously questioned on account of the presence of other carcinogenic materials, including arsenic or nickel, in the atmosphere.[99,105,106] Inconsistencies in statistical evaluations, and lack of pathological details, employment information or personal records presented present further complications in the interpretation of some studies. Systematic and critical reviews of cytogenic and human population studies discussed below shows:

1. The correlation between cadmium exposure and "cytogenic endpoints" is poor.

2. Published occupational health studies are insufficient to show a clear link between cadmium exposure and cancer.[70,108]
3. Insufficient allowance is made for the presence of toxic/potentially carcinogenic materials in the workplace.

This view is not held by the IARC, US DHHS and most national or state authorities.[14,16]

10.3 Mercury

Mercury is a silvery white, volatile, liquid metallic element at room temperature. It exists in a monovalent (mercurous) or bivalent (mercuric) state and forms a large number of inorganic and organic compounds all of which are biologically active and cytotoxic.[169] Mercury exists naturally but substantial amounts are extracted from its principal ores including cinnabar (mercuric sulfide), silicate ($HgSiO_4$), Edgarbaileyite (after the distinguished American geologist Dr Edgar Herbert Bailey) ($Hg^{1+}{}_6Si_2O_7$), and Calomel (Hg_2Cl_2). Distribution of inorganic and organic compounds of mercury is greatly influenced by weathering of rocks, volcanic activity and rainfall but environmental dissemination of inorganic mercury and organomercurials is greatly augmented by human activities including mining, industrial processes, combustion of fossil fuels and waste disposal and widespread use in past years of alkyl and phenyl mercury agrochemicals. These anthropogenic activities possibly contribute to 80% of mercury residues released into inland water courses and into marine areas to be concentrated in fish and human food chains.[129–132] Mercury vapour is oxidised to the water-soluble biologically active Hg(II) forms by the action of bacteria and micro-organisms.[133,134] Major international problems have arisen through large-scale use of organomercuric fungicides, seed-dressings and agrochemicals. In the 1950s, profound mercury poisoning resulting from industrial waste disposal over 37 years or more in the Minamata Bay area of Japan resulted in widespread human illness. More than 900 people suffered major neurological damage and fatality through consumption of fish contaminated with methyl mercury residues pumped into sea water.[5,135–137] Methyl mercury poisoning in Iraq was a cause of many fatalities, fetal abnormalities and lasting health problems through consumption of bread made from contaminated grain.[129,136,138,139] Mercury used in gold amalgamation has for many years constitute a major environmental problem in developing countries of South America and elsewhere, where contamination of water, sediments, phytoplankton and zooplankton is taken up by fish and other species.[140–143] High concentrations of mercury are reported in seals, harbour porpoises and whales basking in off-shore waters. Mercury is possibly second only to lead in being the most toxic of all metals.

Mercury has no nutritional value in any mammalian species but blood concentrations of 6 mg may occur as a result of eating contaminated food or inhaling mercury vapour from broken thermometers and instruments. Workers in many diverse professions are at risk of mercury poisoning, particularly those employed in mining and extraction of metal from cinnabar ore, chloralkali industries, electrical, and makers of pressure instruments, fluorescent and mercury discharge lamps and dentistry. Broken thermometers are a cause of contact hypersensitivity and granulomas.[144,145] Dental amalgams may contain 50% mercury and dental personnel exposed daily may show blood levels of mercury of up to 57 μg l^{-1}, or 50–300% more than non-dental professionals.[146–148] Analyses have also shown that mercury leaches out from dental amalgams and that people with dental fillings show higher blood levels of mercury than others, albeit that the health risk is probably low.

Documentation on the toxicology of mercury is extensive in view of the wide range of people exposed to it intentionally or unintentionally in industry or home life. Principal health risks are experienced through ingestion of mercuric residues in food and water, and through inhalation. Health anxieties have also been raised concerning use of the organomercurial Thimerosal, used since the 1930s as a preservative in vaccines.[149,150] Thimerosal [mercury, ethyl(2-mercaptobenzoato-S)-,sodium salt] (Figure 10.1) has been used as a efficacious antibiotic but it contains 50% of mercury by weight, meaning that a child is exposed to 25 μg of Hg per 0.5 ml dose at vaccination.

Mercury is highly allergenic, and symptoms ranging from contact dermatitis or stomatitis to encephalitis, and mortality may even follow exposure to broken thermometers.[144,151] Elemental mercury is oxidised to Hg(II) in the body by the hydrogen peroxidase–catalase pathway and is primarily excreted *via* the kidneys. Mercury, like cadmium, induces and binds metallothioneins (MT-I, MT-II and MT-III) which act a cytoprotectants.[152–154] Hg(II) displaces zinc and copper from MT-complexes and binds other proteins, including proteinase K and carboxypeptidases, with free thiol groups and high cysteine residues.[155–157] Albumin is the major mercury binding protein in human serum and there are preferential mercury-binding sites on the molecule.[152,158] Albumin binds at least 90% of serum mercury, serving to transport it to soft tissues. The kidney is the principal site for deposition of mercury, which is eliminated *via* the urine, faeces, breast milk and hair; all of which have been used beneficially in biomonitoring of the body burden of the metal.

Mercury has a strong affinity for sulfur, and among the amino acids, L-cysteine shows the highest affinity for the cation. The primary mode of toxic action for mercury in living organisms is thought to be interference with enzyme function and protein synthesis by binding to sulfydryl or thiol groups. Marked differences are seen in the uptake, metabolism and toxicity of inorganic and organic mercurials; inorganic mercury binds plasma proteins to a greater extent than organic compounds. Organomercurial compounds are up to 1000

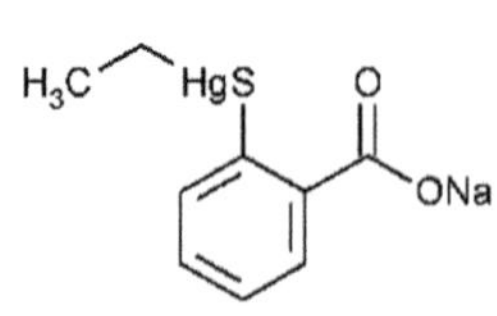

Figure 10.1 Thimerosal.

times more toxic than inorganic forms and children are particularly vulnerable.[136,159] Whereas methyl mercury and related compounds are possibly carcinogenic in humans, metallic mercury and inorganic mercury compounds are not presently classifiable as carcinogenic to humans.[8] After absorption, each type of mercury is specifically metabolised. Thus, elemental mercury absorbed into erythrocytes is oxidised to Hg(II) by a peroxidatic reaction, with the reaction depending upon generation of hydrogen peroxide and the specific activity of catalase.[160] In an albumin complex, this reaches the kidneys, liver, brain and pituitary gland; some accumulates whereas the balance is eliminated *via* the kidneys or bile.[161] Some mercury is exhaled.

10.3.1 Industrial Use and Human Exposure

Mercury has been known to humans for more than 3000 years, when it was extracted by heating the bright mineral cinnabar in air or in the presence of iron.

$$HgS + O2 \rightarrow SO_2 + Hg$$

or

$$HgS + Fe \rightarrow FeS + Hg$$

Principal routes of human exposure include mining and extraction, use of mercury in measuring equipment (thermometers, barometers, *etc.*), electrical appliances, extraction of gold, and in antiseptics, fungicides, seed dressings, battery manufacture, intermediates in chemical syntheses (*i.e.* electrolytic production of chlorine and chloralkalis), as a colorant and preservative in paints and in vaccines (Table 10.2).

Table 10.2 Commercial uses of mercury and its compounds.

Compound	*Commercial use*	*Health risk*
Metallic mercury	Thermometers, barometers, electrical devices, batteries, jewellery, extraction of gold	Broken instruments, inhalation of vapour
Mercury vapour	Light bulbs, neon signs	Inhalation of vapour
Mercadmium Orange (Orange 23) Vermilion (Cinnabar Red 106)	Paint colorant	
Mercuric oxide	Batteries	Inhalation of valour
Ammoniated mercury	Skin lightening soaps and creams	Dermal contact
Mercury fulminate	Explosives	Inhalation, dermal contact
Thimerosal	Vaccine preservative	Injection (especially children)
Mercury amalgams	Dental fillings	Inhalation of vapour

Mercury does occur as an occasional contaminant of natural gas plants. Condensates and analyses of dimethyl mercury, diethyl mercury, diphenyl mercury, methyl mercury chloride, ethyl mercury chloride, phenyl mercury chloride and mercuric chloride at five sites revealed total mercury concentrations of 7–50 ng ml^{-1}, whereas in two commercial heavy gas oil samples the levels were 22.2 ± 0.6 ng ml^{-1} and 2.3 ± 0.1 μg ml^{-1} respectively.[162] Compounds of mercury have a long history as anti-infective agents in human medicine, but the therapeutic value of agents such as calomel (mercurous chloride, Hg_2Cl_2), mercuric chloride ($HgCl_2$), potassium mercuric iodide, yellow mercuric oxide, ammoniated mercury, mercurochrome and various ointments containing finely divided elemental mercury is outweighed by the readiness of the metal to precipitate and concentrate in soft tissues, its profound irritancy following topical, oral or parenteral administration, and its systemic toxicity.[163]

Mercury was introduced into dentistry as amalgams for filling tooth cavities more than 150 years ago to restore decayed or damaged teeth and is still widely practices internationally, although in recent years alternative tooth-coloured materials have been developed for dental restorations and are increasingly used across the EU.[164] In combination with other metals such as silver, gold and tin, the use of dental amalgams has been a source of controversy, with concerns over safety, cost and efficiency, such that newer materials including ceramics, composite materials, glass ionomer cements and a variety of hybrid structures have been introduced progressively. For many of these alternatives, safety issues remain unresolved. With regard to mercury amalgams, analyses have shown that mercury levels in the hair, blood and urine of dental personnel average at least twice those seen in non-dental professions.[166,167] A nationwide survey in the USA indicated that dentistry was the third largest user of mercury (45 tons annually), most of which is disposed of in the environment.[165–169] Mercury is a profound contact allergen, with dentists and dental nurses being at greatest risk through dermal contact and through inhalation of mercury vapour.[144,170–172] Gingival bacteria probably transform metallic mercury in dental amalgams to more toxic organomercury compounds.[173]

Several traditional Chinese and Ayurvedic medicines (*e.g. Zhuifeng Tougu Wan, Fufang Lu Hui Jiaonang* and *SAFI*) are still based on Hg(II) compounds and their toxicity is appreciated.[174–176] Some are known to contain arsenic and lead! Additionally, mercury has been used for hundreds of years for cultural and religious reasons. A number of practices exist today that use mercury, including Santería (an Afro-Hispanic belief system), Palo Mayombé (Caribbean), Candomblé (Afro-Brazilian), Voodoo (Afro-Haitian), Espiritismo (Puerto Rican) and Yoruba Orisha (Afro-Hispanic). Mercury is also used in Hindu practice as a major constituent of Parad, from which religious relics are made. In some cases, mercury is injected subcutaneously, intravenously or intramuscularly to improve athletic prowess or protect users from evil.

Mercury is absorbed into the human body through ingestion, inhalation and through percutaneous absorption. Exposure to mercury vapour constitutes an environmental and occupational hazard in many professions, notably in mining, extraction, battery manufacture and dentistry.[8,167,177] Mercury vapour is readily absorbed through the lungs and diffuses percutaneously to accumulate in the circulation.[161,178] Approximately 80% of mercury and mercury vapour released into the environment results from mining procedures (cinnabar and gold), smelting, fossil fuel combustion and incineration of industrial waste containing discarded batteries, dental amalgams, electrical equipment and damaged thermometers

10.3.2 Cytotoxicity, Mutagenicity and Carcinogenicity

The IARC evaluated carcinogenic risks associated with inhalation of mercury vapour, and exposure to mercury residues in battery plants, gold mining and agrochemicals.[8] They concluded that there was inadequate epidemiological evidence for carcinogenicity in humans, but that sufficient experimental animal evidence was available to classify methyl mercury as a potential human carcinogen. Additionally, the National Toxicology Program (NTP) nominated the organomercuric compound Thiomerosal for further "toxicological testing".[179] Thiomerosal, containing 50% by weight of mercury, was introduced in the 1930s as a preservative for vaccines and other biological products, but its safety in use was questioned in view of the large number of people exposed to it. Thiomerosal is a cytostatic agent and has been shown to influence calcium-dependent proliferation in human oral cancer cells.[180]

Occupational health reports by WHO and the Department of Health showed methyl mercury to be a cause of serious neuromuscular complaints, loss of intelligence, neonatal abnormalities and miscarriage, anaemia and liver disease in an increasing numbers of people.[181–183] They concluded that all mercury compounds are toxic but the small-chain alkyl compounds are the most hazardous.[136,159] Mercury is a cumulative protoplasmic poison with the developing brain, central nervous system, kidney and liver being prime target organs.[139,169] Local tissue damage resulting from inhalation of mercury vapour, contaminated dusts and aerobic particles, dermal exposure, oral ingestion and parenteral administration progresses to granuloma formation with a rapid systemic dissemination of bioactive Hg(II) ion and transfer to soft tissues.[152,184–186]

10.3.2.1 *In Vitro Studies*

The genotoxicity of inorganic or organomercurial compounds in exposed mercury workers is also questionable.[187] Results of mutagenicity and genotoxicity testing have been inconsistent and whereas inorganic and methyl compounds of mercury have been shown to induce DNA single-strand breaks at low concentrations in mammalian cells, bacterial reverse mutagenicity assays have produced negative results.[188,189] In a mutagenicity study in NIH 3T3 cells transfected with a reporter gene, inorganic mercury significantly increased mutation frequency at concentrations of 0.5 μM and this was directly related to intracellular induction of reactive oxygen species and oxidative stress. Induction of metallothionein failed to prevent mutations, which were more pronounced following glutathione depletion.[190]

Further evidence of mercury-induced oxidative stress is provided by a cell viability and gene profile assay in human hepatoma $HepG_2$ cells.[184] Mercury was highly cytotoxic at 0.5 μg ml^{-1} within 48 hours and a clear relationship was demonstrated between cytotoxicity and gene induction in 13 stress-related recombinant cell lines. Using a variety of genetic endpoints, mammalian cell lines, bacteria, moulds and yeasts, mercury compounds failed to induce point mutations, but often exerted clastogenic effects in eukaryotic cells by binding sulfydryl groups and acting as spindle inhibitors. Inorganic mercurials induced the generation of oxygen reactive species. The primary toxic action is through depletion of glutathione and bonding sulfydryl residues on proteins, leading to oxidative stress in mammalian cells.[190]

Some evidence for the cytogenicity of mercury in workers exposed occupationally is provided by a study of 29 workers exposed to mercury fulminate $[Hg(CNO]_2$.[191] This is an unstable compound used in explosives manufacture. The workers exhibited a three-fold increase in urinary mercury excretion (123.2 ± 54.1 μg l^{-1}) and haematological analysis revealed significantly higher chromosomal aberrations, including gaps, breaks and fragmentation and micronucleated lymphocytes. However, the changes correlated neither with the duration of exposure of the workers nor with their urinary mercury excretion.

Thimerosal was shown to induce genotoxic changes in cultured human peripheral lymphocytes.[192] Thus sister-chromatid exchanges, decreased mitotic index and proliferation rates were reported following exposure to Thimerosal containing 49.5% ethyl mercury at concentrations of 0.2 to 0.6 μg ml^{-1}.

10.3.2.2 Experimental Studies in Animal Models

The earliest study, by Druckrey in 1957, showed that rats injected intraperitoneally twice daily with 0.05 ml metallic mercury developed non-metastatic spindle cell sarcomas after 22 months.[193] A more detailed study by NTP showed that male but not female Fischer-344 rats dosed orally with 1.9 or 3.7 mg kg^{-1} mercuric chloride daily for two years showed a statistically significant increase in squamous cell papillomas of the forestomach and thyroid follicular cell carcinomas.[194] They showed also that B6C3F1 mice exposed to 3.7 or 7.4 mg kg^{-1} per day mercuric chloride developed a low incidence of renal tubule tumours but the incidence of tumours was not significantly increased. Both studies showed that mercuric chloride was highly toxic and most animals in higher-dose groups died early. Male ICR-mice fed diets containing methyl mercury at 15 ppm for 78 weeks developed renal tumours after 53 weeks; female mice showed no tumours.[195]

A chronic toxicity study in rats gavaged with maximum tolerated doses of mercuric chloride (1.25–20 mg kg^{-1}) for up to two years showed a clear dose-related trend in renal damage with early and progressive changes in renal marker enzymes and accumulation of mercury in the tissues, but no evidence of renal carcinogencity.[196] Mercury did exacerbate the spontaneous rate of nephrotoxicity in F344 rats, and mortality. The authors reported that direct contact with bioactive mercury induced squamous cell carcinomas in the forestomachs of both sexes, a change not reported in humans or in other experimental studies.

A 90-day study in which 50 piglets were given phenylmercuric chloride daily in the dose range to 0.19–4.56 mg kg^{-1} over 90 days showed that at doses exceeding 2.28 mg kg^{-1} the treatment was toxic to the kidneys and large intestine.[197] The pathological damage of renal necrosis, profound colitis and degenerative changes was not accompanied by preneoplastic change and possibly resembled the subacute toxicity seen in humans following mercury poisoning. Much phenyl and methyl mercury ingested is voided in the faeces.

Experimental studies provide further evidence for the neurotropic action of mercury. Thus mice exposed to mercury vapour on alternate days for two weeks (0.1 mg m^{-3}) and then exposed to 4.1 mg m^{-3} for a further 11 weeks showed a marked increase in MT-I, MT-II and MT-III indicating neuroaccumulation of mercury.[154] In contrast MT-null mice were unresponsive. Granules of mercury were observed in nerve cells and glial cells in both strains. Marmosets exposed to methyl mercury showed that mercury-induced pathology includes oedema in the white matter of the brain and selective cortical degeneration along deep cerebral fissures.[198]

10.3.2.3 Human Case and Epidemiological Studies

The IARC and 12th RoC do not presently classify mercury or inorganic mercury compounds as human carcinogens.[8,14] Whilst there is a lack of conclusive epidemiological evidence for the carcinogenicity of inorganic or organomercurials in humans,[138,191,199] there is considerable evidence to show that mercury compounds are extremely toxic and that death through neuropathy, renal failure or hepatic damage will occur before overt carcinogenic change.[200,201] Characteristic symptoms of mercury poisoning illustrated by the large-scale epidemiological studies in Japan and other countries include impaired neurological development, neurodegeneration, cerebellar ataxia, blindness, anaemia and blood dyscrasias, abnormal liver and renal function tests, tremors and convulsions, paralysis and mortality.[129,131,136,137,159,186,202–204] The brain and central nervous system are most at risk following chronic exposure to mercury residues in dental personnel and through food and environmental contact. Mercury residues have been detected in human brain tumours but it is unclear whether mercury should be considered as a causative factor.[205] The neurotoxic action of mercury is mitigated by antioxidants such as selenium and vitamin E.[206] Numerous experimental studies in rodents, dogs, pigs and marmosets have substantiated the susceptibility of neurological tissues to bioactive mercury ion but, additionally, they have shown the vulnerability of kidney and liver to mercury toxicity.[196,207–210]

Boffetta *et al.* conducted an epidemiological study of cancer rates in a large European cohort of mercury miners. In 6784 men and 265 women working in mines in Spain, Slovenia, Italy and Ukraine, there was no excess of cancer mortality amongst men although lung cancer mortality was increased in Slovenian and Ukrainian miners.[211] Increases in liver cancer were present in miners and millers in Italy and Slovenia, but whilst there was a trend with estimated exposure, it was not related to duration in employment. No occupation-related increased mortality was seen with other types of cancer including brain and renal tumours. Three women died of ovarian cancer. Whilst this study might provide some evidence of a link between mercury exposure and lung cancer, regional variations in incidence and the strong possibility of concurrent exposure to radon and silica in mining prohibit a valid conclusion.

Doubts remain over the possible relationship between mercury exposure and liver cancer. Exposure to mercury was assessed in a mine in Almadén (Spain) in which workers engaged in drilling, furnace operation or filling bottles showed blood and urine Hg concentrations of 374 $\mu g\ l^{-1}$ and 2194 $\mu g\ l^{-1}$ respectively.[204] Air levels of mercury ranged from 2.26 to 3.37 mg m^{-3} initially but were reduced to 0.32–0.83 mg m^{-3} following introduction of revised working practices. Although environmental concentrations of mercury were high in this study, no neoplasia or other pathology was reported.

The IARC predicted that methyl mercury is "possibly carcinogenic to humans" on the basis of limited evidence.[8] Thus epidemiological studies in the Minamata Bay region of Japan showed "excess mortality from liver and oesophageal cancer in areas of highest exposure". Occupational health studies conducted in Scandinavia claiming an increase in brain and renal tumours in relation to mercury exposures are inconclusive. More comprehensive literature surveys provide some evidence that occupational mercury exposure may be a cause of human cancer, but the consistency and statistical viability of the information is questioned. Thus, Cragle *et al.* reported an increased incidence of lung, brain, and kidney cancers within a cohort of 2133 men exposed to mercury vapour in 1953 to 1963 with a follow up to 1978, but concluded that in view of the high incidence of tumours in trades unrelated to mercury exposure, the cancer incidence is "a factor of lifestyle or exposure to other contaminants" rather than to mercury alone".[177] Two Scandinavian cohort studies are reported. The first in Norway showed no excesses of brain or kidney tumours in 674 men exposed to mercury vapour for 1 or more years in a chloralkali plant.[212] The second, in a Swedish factory where lung cancer was increased, showed that workers had been exposed to asbestos in their environments.[213]

Abandoned mercury mines are frequently a source of anxiety for the health of people living in local communities. A survey of 127 adults and children living within a 10 km radius of such a mine in the Philippines with an earlier production of 140 000 kg of mercury annually and releasing about 1 million tons of ore into the Honda Bay area (>0.002 ng ml^{-1}), showed an association with increased miscarriages, muscular problems, paralysis, tremors and anaemia, but no increased incidence of cancer of any type.[214] Blood mercury levels in 12 individuals exceeded 20 ppb but wide variations were reported in mercury contamination in local sea water in Honda Bay and in soil and sea food.

Surveys among dentists showing high blood mercury levels through long-term exposure to mercury vapour have also yielded conflicting results. Whilst no excess overall cancer mortality or brain cancer was reported among dentists in the US Armed Forces,[215] a census of 9332 dental nurses and dentists in Sweden showed excesses of brain cancer. Gliomas were estimated to be twice as common as expected, and a standardised mortality ratio of 1.2–1.3 was seen in patients with glioblastoma.[216] The standardised morbidity ratio for glioma was 1.8, but with a wide confidence interval. To test the hypothesis that dental amalgam fillings are a causal factor in tumours of the central nervous system, Rodvall *et al.* matched 333 patients with glioma, meningioma or acoustic neuroma by age, gender and location with 343 controls.[217] There was no association between the number of amalgam fillings and tumours of the central nervous system and the authors concluded that there was no evidence that amalgam fillings are a cause of central nervous system tumours. Other suspected adverse effects in dentists through exposure to mercury in dental amalgams, including Alzheimer's disease, multiple sclerosis and dementia, have not been substantiated by competent organisations.[218]

No evidence is seen to show that thimerosal is carcinogenic in humans or animals, although its use in vaccines has been reduced as a precaution. *In vitro* studies indicate that thimerosal can influence the movement and viability of human oral cancer cells by impairing cytosolic free calcium concentrations, and it is genotoxic in cultured human lymphocytes with or without metabolic activation.[180,192,219]

References

1. W. B. Jensen, *J. Chem. Edu.*, 2003, **80**, 952.
2. H. A. Schroeder and A. P. Nason, *J. Invest. Dermatol.*, 1969, **53**, 71.
3. H. A. Schroeder, A. P. Nason, I. H. Tipton and J. J. Balassa, *J. Chron. Dis.*, 1967, **20**, 179.
4. K. Tsuchiya, *Keio J. Med.*, 1969, **18**, 181.
5. J. Watts, *Lancet*, 1997, **350**, 422.
6. T. Ishikawa and Y. Ikegaki, *J. Water Pollution Contr. Fed.*, 1980, **52**, 1013.
7. M. Qureshi, R. Kumar and H. S. Rathore, *Anal. Chem.*, 1972, **44**, 1081.
8. International Agency for Research on Cancer (IARC), *IARC Monographs on the Evaluation of Carcinogenic Risks to Humans,* IARC, Lyon, 1993, vol. 58.
9. J. Järup, M. Berglund, C. G. Elinder, G. Nordberg and M. Vahter, *Scand. J. Work, Environ. Health*, 1998, **24**(Suppl.), 1.
10. J. Weinberg, *An NGO Introduction to Mercury Pollution,* IPEN, International POP Elimination Network, 2010.
11. H. Muntau and R. Baudo, *IARC, Sci. Publ.*, 1992, **118**, 133.
12. G. F. Nordberg, *Biometals*, 2004, **17**, 485.
13. J. O. Nriagu and J. M. Pacnya, *Nature*, 1988, **333**, 134.
14. US Department of Health and Human Services (US DHHS), National Toxicology Program, *Report on Carcinogens,* US DHHS, Washington, DC, 12th edition 2011.

15. Occupational Safety and Health Administration (OSHA), *Cadmium: Sect 5, Health Effects,* OSHA, US Department of Labor, Washington, DC, 1993.
16. IARC, *IARC Monographs on the Evaluation of Carcinogenic Risks to Humans*, IARC, Lyon, 1999, vol. 11.
17. A. B. G. Lansdown, U. Mirastschijski, N. Stubbs, E. Scanlon and M. S. Ågren, *Wound Rep. Regen.*, 2007, **15**, 2.
18. C. D. Klaassen, J. Liu and S. Choudhuri, *Ann. Rev. Pharmacol. Toxicol.*, 1999, **39**, 267.
19. A. B. G. Lansdown, *Wound Rep. Regen.*, 2002, **10**, 130.
20. A. B. G. Lansdown, B. Sampson and A. Rowe, *Int. J. Exp. Pathol.*, 2001, **82**, 35.
21. I. H. Bremner and J. K. Campbell, *Environ. Health Perspect.*, 1978, **25**, 125.
22. M. M. Brzöska and A. Moniuszko-Jaconiuk, *Fd. Chem. Toxicol.*, 2001, **39**, 267.
23. Y. Liu, J. Liu, S. M. Habeebu, M. P. Waalkes and C. D. Klaassen, *Toxicol., Sci.*, 2000, **57**, 167.
24. W. Kennette, O. M. Collins, R. K. Zalups and J. Koropatnick, *Toxicol. Sci.*, 2005, **88**, 602.
25. I. Sabolić, D. Breljak, M. Škarica and C. Herak-Kramberger, *Biometals*, 2010, **23**, 897.
26. J. I. Anetor, F. Ajose, G. O. Anetor, A. A. Iyanda, O. O. Babalola and F. A. Adeniyi, *Nigerian J. Physiol. Sci.*, 2008, **23**, 41.
27. C. A. Krone and L. C. Harms, *J. Natl. Cancer Inst.*, 2003, **95**, 1556.
28. M. F. Leitzmann and E. L. Giovannuci, *J. Natl. Cancer Inst.*, 2003, **95**, 1556.
29. N. Mead, *Environ. Health Perspect.*, 2010, **118**, A528.
30. J. Deckert, *Biometals*, 2005, **18**, 475.
31. S. A. Hasan, Q. Fariduddin, B. Ali, S. Hayat and A. Ahmad, *J. Environ. Biol.*, 2009, **39**, 165.
32. L. Yue, *Biomed. Environ. Sci.*, 1992, **53**, 1992.
33. S. S. Hecht, *J. Natl. Canc. Inst.*, 1999, **91**, 1194.
34. S. S. Hecht, *Natl. Rev. Cancer*, 2003, **3**, 733.
35. H. Oliviera, E. A. N. Fernandez, M. A. Bacchi, G. A. Sarriés and F. S. Tagliaferro, *J. Radioanal. Nucl. Chem.*, 2000, **244**, 299.
36. G. P. Pfeifer, M. F. Denissoenko, M. Olivier, N. Tretyakova, S. S. Hecht and P. Hainaut, *Oncogene*, 2002, **21**, 7435.
37. D. J. Hendrick, *Thorax*, 2004, **59**, 184.
38. E. E. Menden, V. J. Elia, L. W. Michael and H. G. Petering, *Env. Sci. Technol.*, 1972, **6**, 830.
39. P. H. Kuck, *Cadmium*, US Legislation and Governmental Legislation, OHSA, Washington, DC, 1993.
40. Royal Society of Chemistry, *Chemistry World,* 2011, **8**, May.
41. NEMA, *NEMA White Paper: Cadmium in electrical contacts,* ed. N. A. Czarnecki and VA Rosslyn, 2008.

42. A. Batra, *Science and Environmental Online "Down to Earth",* Society for Environmental Communications, New Delhi, India, 2011.
43. World Health Organization (WHO), *Cadmium in Drinking Water,* WHO, Geneva, 2011.
44. R. Srikant, V. Rao and N. Vengupal, *Environ. Monit. Assess.*, 1964, **33**, 71.
45. R. M. Smith, R. M. Leach, L. D. Muller, L. C. Griel and D. E. Baker, *J. Anim. Sci.*, 1991, **69**, 4096.
46. A. Gothberg, M. Greger, K. Holm and B. E. Bengtsson, *J. Environ. Qual.*, 2004, **33**, 1247.
47. N. L. Schouw and J. C. Tjell, *J. Transdiscipl. Environ. Studies*, 2003, **2**, 1.
48. J. Godt, F. Scheidig, C. Grosse-Siestrup, V. Esche, P. Brandenburg, A. Reich and D. A. Groneberg, *J. Occup. Med. Toxicol.*, 2006, **1**, 22.
49. A. Bernard, *Indian J. Med. Res.*, 2008, **128**, 557.
50. M. Wilhelm, F. K. Ohnesorge, I. Lombeck and D. Hafner, *J. Anal. Toxicol.*, 1989, **3**, 17.
51. M. Wilhelm, D. Hafner, I. Lombeck and F. K. Ohnesorge, *Sci. Total Environ.*, 1991, **103**, 199.
52. M. Wilhelm, I. Lombeck and F. K. Ohnesorge, *Sci. Total Environ*, 1994, **141**, 275.
53. A. G. Davidson, A. J. Newman-Taylor, J. Darbyshire, D. R. Chettle, C. J. G. Guthrie, D. O'Malley, H. T. Mason, P. M. Fayers, K. M. Venables, C. A. C. Pickering, D. Franklin, M. C. Scott, H. Holden, A. L. Wright and D. Gompertz, *Lancet*, 1988, **331**, 663.
54. J. S. Johnson and K. H. Kilburn, *Am. J. Ind. Med.*, 1983, **4**, 533.
55. S. Barnhart and L. Rosenstock, 1984, **86**, 789.
56. B. G. Armstrong and G. Kazantzis, *Br. J. Ind. Med.*, 1985, **42**, 540.
57. R. C. Wester, H. I. Maibach, L. Sedik, J. Mellendres, S. Di Zio and M. Wade, *Fundam. Appl. Toxicol.*, 1992, **19**, 1.
58. A. B. G. Lansdown and B. Sampson, *Lab. Anim. Sci.*, 1996, **46**, 549.
59. World Health Organization (WHO), *Cadmium: Evaluation and risk assessment,* WHO, Geneva, 2013.
60. Agency for Toxic Substances and Disease Registry (ATSDR), *Public Statement for Cadmium,* ATSDR, Atlanta, GA, 2011.
61. Industrial Industries Advisory Council, *Cadmium and Genito-urinary Cancer,* AAAC, Position Paper, 2009, **24**, 1.
62. M. P. Waalkes, S. B. Cherenoff and C. D. Klaassen, *Biochem. J.*, 1984, **220**, 819.
63. M. E. Figueiredo-Pereira, S. Yakushima and G. Cohen, *J. Biol. Chem.*, 1998, **273**, 1.
64. N. A. Wolff, M. Abouhamed, P. J. Veroust and P. Thévenod, *J. Pharmacol. Exp. Therap.*, 2008, **318**, 782.
65. G. Gobe and D. Crane, *Toxicol. Lett.*, 2010, **198**, 49.
66. B. A. Fowler and G. F. Nordberg, *Toxicol. Appl. Toxicol.*, 1978, **46**, 609.
67. M. J. Thun, C. G. Elinder and L. Friberg, *Am. J. Ind. Med.*, 1991, **20**, 629.

68. S. G. Han, V. Castranova and V. Vallyathan, *J. Toxicol. Environ. Health*, 2007, **70**, 852.
69. M. P. Waalkes., R. Kovatch and S. Rehm, *Toxicol. Appl. Pharmacol.*, 1991, **108**, 448.
70. V. Verougstraete, D. Lison and P. Hotz, *Mutat. Res.*, 2002, **511**, 15.
71. T. Schwerdtie, F. Ebert, C. Thuy, C. Richter, L. H. F. Mullenders and A. Hartwig, *Chem. Res. Toxicol.*, 2010, **23**, 432.
72. M. Filipic, I. Fatur and M. Vudrag, *Hum. Exp. Toxicol.*, 2006, **25**, 67.
73. D. A. Pagano and E. Zeiger, *Environ. Mol. Mutagen.*, 1992, **19**, 139.
74. R. Mandel and H. J. P. Ryser, *Mutat. Res.Genet. Toxicol.*, 1984, **136**, 9.
75. J. Huff, *Int. J. Occup. Environ. Health.*, 2007, **13**, 202.
76. E. Loser, *Cancer Lett.*, 1980, **9**, 191.
77. M. P. Waalkes, M. R. Anver and B. A. Diwan, *J. Toxicol.*, 1999, **58**, 199.
78. G. Kazantzis and W. J. Hanbury, *Br. J. Cancer*, 1966, **20**, 190.
79. M. P. Waalkes, S. Rehm and H. G. Cherian, *Toxicol. Sci.*, 2000, **54**, 110.
80. J. C. Heath and M. R. Daniel, *Br. J. Cancer*, 1964, **18**, 124.
81. M. P. Waalkes, S. Rehm, C. W. Riggs, R. M. Bare, D. E. Devor, L. A. Poirier, M. L. Wenk and J. R. Henneman, *Cancer Res.*, 1989, **49**, 4282.
82. M. P. Waalkes, S. Rehm, B. Sass, N. Konishi and J. M. Ward, *Environ. Res.*, 1991, **55**, 40.
83. S. Rehm and M. P. Waalkes, *Toxicol. Sci.*, 1988, **10**, 635.
84. M. P. Waalkes and S. Rehm, *Toxicology*, 1998, **126**, 173.
85. M. P. Waalkes and S. Rehm, *Fundam. Appl. Toxicol.*, 1994, **23**, 21.
86. M. P. Waalkes, S. Rehm, C. W. Riggs, R. M. Bare, D. E. Devor, L. A. Poirier, M. L. Wenk, J. R. Henneman and M. S. Balaschak, *Cancer Res.*, 1988, **48**, 4656.
87. L. A. Poirier, K. S. Kasprzak, K. L. Hoover and M. L. Wenk, *Cancer Res*, 1983, **43**, 4575.
88. S. Takenaka, H. Oldiges, H. König, D Hochrainer and D. Oberdörster, *J. Natl. Cancer Inst.*, 1983, **70**, 367.
89. R. Philipp, *J. Roy. Soc. Med.*, 1985, **78**, 328.
90. L. Stayner, R. Smith, M. Thun, T. Schnorr and R. Lemen, *Ann. Epidemiol.*, 1992, **2**, 177.
91. G. Oberdörster, *Scand. J. Work. Environ. Health*, 1986, **12**, 523.
92. C.-G. Elinder, T. Kjellström, C. Hogstedt, K. Anderson and G. Spång, *Br. J. Ind. Ed.*, 1985, **42**, 651.
93. C. L. Potts, *Ann. Occup. Hyg.*, 1965, **8**, 55.
94. H. J. Mason, A. J. Davidson, A. L. Wright, G. J. G. Guthrie, P. M. Fayers, K. M. Venables, N. J. Smith, D. R. Chettle, D. M. Franklin, M. C. Scott, H. Holden, D. Gompertz and A. J. Newman-Taylor, *Br. J. Ind. Med.*, 1988, **45**, 793.
95. P. W. Meuller, *Clin. Chem.*, 1993, **39**, 743.
96. M. Piscator, *Environ. Health Perspect.*, 1981, **40**, 107.
97. M. J. Thun, T. M. Schnorr, A. B. Smith, W. E. Halperin and R. A. Lemen, *J. Natl. Cancer Inst.*, 1985, **74**, 325.

98. T. Sorahan and R. J. Lancashire, *Occup. Environ. Med.*, 1997, **54**, 194.
99. T. Sorahan, *Br. J. Ind. Med.*, 1987, **44**, 803.
100. A. Åkesson, P. Bjellerup, T. Lundh, J. L. Nerbrand, G. Samsioe, S. Skerving and M. Vahter, *Environ. Health Perspect.*, 2006, **11**, 830.
101. M. M. Brzóska, J. Roalska, M. Galazyn-Sidorczuk, M. Jurczuk, A. Roszczenko, E. Kulikowska-Karpińska and J. Moniuszko-Jakoniuk, *Toxicology*, 2007, **237**, 89.
102. Central Directorate on Environmental Pollution, *Cadmium in the Environment and Its Significance to Man,* HMSO, London, 1980.
103. R. P. Sharma, J. C. Street, M. P. Verma and J. L. Shupe, *Environ. Health Perspect.*, 1979, **28**, 59.
104. World Health Organization (WHO), *Environmental Health Criteria 134 - Cadmium*, International Programme on Chemical Safety (IPCS), Monograph, WHO, Geneva, 1992.
105. S. H. Lamm, M. Parkinson, M. Anderson and W. Taylor, *Ann. Epidemiol.*, 1992, **2**, 195.
106. J. M. Peters, D. Thomas, H. Falk, G. Oberdörster and T. J. Smith, *Environ. Health Perspect.*, 1986, **70**, 71.
107. G. Kazantzis, G. Lamm and K. R. Sullivan, *Scand. J. Work Environ. Health*, 1988, **14**, 220.
108. T. Sorahan and J. A. Waterhouse, *Br. J. Ind. Med.*, 1983, **40**, 293.
109. G. Kazantzis, R. G. Blanks and K. R. Sullivan, *IARC Sci Publ.*, 1992, **118**, 435.
110. R. A. Lemen, J. S. Lee, J. K. Wagoner and H. P Blejer, *Ann. N.Y. Acad. Sci.*, 1978, **271**, 273.
111. A. E. Sahmoun, L. D. Case, S. A. Jackson and G. C. Schwartz, *Cancer Invest.*, 2005, **23**, 256.
112. G. Kazantzis and B. G. Armstrong, *Environ. Health Perspect.*, 1984, **54**, 193.
113. D. E. Keil, J. Berger-Richie and G. A. McMillan, *Lab. Med.*, 2011, **42**, 735.
114. M. D. Kipling and J. A. H. Waterhouse, *Lancet*, 1967, **i**, 730.
115. G. Bako, E. S. Smith, J. Hanson and R. Dewar, *Can. J. Publ. Health.*, 1982, **73**, 92.
116. G. F. Nordberg, *Scand. J. Work Environ. Health*, 1993, **19**(Suppl. 1), 104.
117. Y.-C. Chen, Y. S. Pu, H.-C. Wu, T. T. Wu, M. K. Lai, C. Y. Yang and F.-C. Sung, *BioMed. C. Cancer*, 2009, **9**, 429.
118. R. Doll, *Ann. Occup. Hyg.*, 1984, **28**, 291.
119. G. Kazantzis and B. G. Armstrong, *Scand. J. Work Environ. Health*, 1982, **8**(Suppl.1), 157.
120. Z. Gong, I. Agalliu, D. W. Lin, J. L. Stanford and A. R. Kristal, *Cancer Causes Control*, 2008, **19**, 25.
121. J. L. Walter, Y. Park, A. Hollenback, A. Schatzkin and D. Albanes, *Cancer Epidemiol. Biomarkers Prev.*, 2009, **18**, 2427.
122. G. G. Schwartz and I. M. Reis, *Ann. Occup. Hyg.*, 1965, **8**, 55.
123. B. Julin, A. Wolk, L. Bergkvist, M. Bottai and A. Åkesseon, *Cancer Res*, 2012, **15**, 1459.

124. G. Scherer and H. Barkemeyer, *Ecotoxicol. Envir. Safety*, 1983, **7**, 71.
125. M. A. Bosque, J. L. Dominigo, J. M. Llobet and J. Corbella, *Biol. Trace Elem. Res., i*, 1991, **28**, 147.
126. T. A. Őzden, G. Gökcav, H. V. Ertem, Ő. D. Süoălu, A. Kilic, S. Sökücü and G. Saner, *Clin. Biochem.*, 2007, **40**, 52.
127. Cancer UK Research, *Smoking Statistics,* London, 2012.
128. Cancer UK Research, *Prostatic Cancer Statistics,* London, 2013.
129. F. Bakir, S. F. Damluji, L. Amin-Zaki, M. Murtadha, A. Khalidi, N. Y. Al-Rawi, S. Tikriti, H. I. Dhahir, T. W. Clarkson, J. C. Smith and R. A. Doherty, *Science*, 1993, **181**, 230.
130. M. M. Velga, J. A. Meech and N. Onate, *Nature*, 1994, **368**, 816.
131. M. Nobuo, 6th International Conference on Mercury a Global Pollutant (ICMGP), Minamata, Japan, 2001.
132. J. McCurry, *Lancet*, 2006, **367**, 99.
133. T. Smith, K. Pitts, J. A. McGarvey and A. O Summers, *Appl. Environ. Microbiol.*, 1998, **64**, 1328.
134. E. Monsson, *Mercury,* in *Encyclopaedia of Earth,* ed. J. Cutler, Environmental Information Coalition, National Council for Science and the Environment, Cleveland Washington, 2007.
135. M. Harada, *Crit. Rev. Toxicol.*, 1995, **25**, 1.
136. K. Eto, *Toxicol. Pathol.*, 1997, **25**, 614.
137. S. Ekino, M. Susa, T. Ninomiya, K. Imamura and T. Kitamura, *J. Neurol. Sci.*, 2007, **262**, 131.
138. A. W. Al Mufti, J. F. Copplestone, G. Kazantzis, R. M. Mahmoud and M. A. Majid, *Bull. WHO*, 1976, **53**(Suppl.), 23.
139. M. R. Greenwood, *J. Appl. Toxicol.*, 1985, **5**, 148.
140. R. Eisler, *Rev. Environ. Contam. Toxicol.*, 2004, **181**, 139.
141. M. P. Simmons, A. P. Johnson and M. C. French, *Vet. Rec.*, 1993, **132**, 291.
142. J. Marrugo-Negrete, L. N. Benitez and J. Oivero-Verbel, *Arch. Environ. Contam. Toxicol.*, 2008, **55**, 305.
143. D. E. Gaskin, K. I. Stonefield, P. Suda and R. Frank, *Arch. Environ. Contam. Toxicol.*, 1979, **8**, 733.
144. A. Fisher, *Contact Dermatitis,* Lea and Febiger, Philadelphia, 1987.
145. K. Gebauer and B. Connor, *Austr. J. Dermatol.*, 1991, **32**, 129.
146. I. Akesson, A. Schultz, R. Attewell, S. Skerfving and P. O. Glantz, *Arch. Environ. Health*, 1991, **46**, 102.
147. B. Möller-Madsen, J. C. Hansen and J. Kragstrup, *Scand. J. Dent. Res.*, 1988, **96**, 56.
148. K. Ritchie and E. MacDonald, *Occup. Environ. Med.*, 1995, **52**, 813.
149. Food and Drug Administration (FDA), *Thimerosal in vaccines,* FDA, Silver Spring, MD, 2012.
150. H. Larson, *New Sci.*, 2013, **339**, 1.
151. M. Sato, *Nippon Hifuka Gakkai Zasshi*, 1989, **99**, 15.
152. S. Lau and B. Sarkar, *J. Toxicol. Environ. Health*, 1979, **5**, 907.

153. M. Aschner, T. Syversen, D. O. Souza and J. B. Rocha, *Exp. Biol. Med.*, 2006, **231**, 1468.
154. A. Tasutake, M. Sawada, A. Shimada, M. Satoh and C. Tohyama, *Arch. Toxicol.*, 2004, **78**, 48.
155. S. Bagger, B. R. Byberg and K. Breddane, *J. Inorg. Biochem.*, 1991, **42**, 97.
156. A. E. Funk, F. A. Day and F. O. Brady, *Comp. Biochem. Physiol. C*, 1987, **86**, 1.
157. T. M. DeSilva, G. Veglia, F. Porcelli, A. M. Pranter and S. J. Opella, *Biopolymers*, 2002, **64**, 189.
158. W. L. Hughes, *Ann. N. Y. Acad. Sci.*, 1957, **65**, 454.
159. L. Trasande, P. J. Landrigan and C. Schechter, *Environ. Health Perspect.*, 2005, **115**, 590.
160. S. Halbach and T. W. Clarkson, *Biochem. Biophys Acta*, 1978, **523**, 522.
161. S. Gladys, B. van Meerbeek, G. Vanherle and P. Lambrechts, *Ned. Tijdsc hr. Tandheelkd.*, 1993, **100**, 179.
162. A. Shafawi, L. Ebdon, M. Foulkes, P. Stockwell and W. Corns, *Analyst*, 1999, **124**, 18.
163. T. Sollemann, *A Manual of Pharmacology*, Saunders, Philadelphia, 1942.
164. Scientific Committee on Emerging and Newly Identified Health Risks (SCENIHR) SCHER, *Tooth Filling Materials, Dental Amalgams and Alternative Materials, Secretariat@ec.europa.eu*, 2008.
165. C. Saengsirinavin and P. Pringsulaka, *J. Dent. Ass. Thai.*, 1988, **38**, 170.
166. R. F. Herber and A. A. Wilbowo, *Comm. Dent. Oral Epidemiol.*, 1988, **16**, 153.
167. N. H. Rowe, K. S. Sidhu, L. Chadzynski and R. F. Babcock, *J. Mich. Dent. Ass.*, 1996, **78**, 32.
168. I. Skare and A. Engqvist, *Arch. Environ. Health*, 1994, **49**, 384.
169. N. Langford and R. Ferner, *J. Hum. Hypertens.*, 1999, **13**, 651.
170. M. Wekkeli, G. Hippmann, A. R. Rosenkrantz, R. Jarisch and M. Gotz, *Contact Dermat.*, 1990, **22**, 295.
171. L. Kanerva, M. Komulainen, T. Estlander and R. Jolanki, *Contact Dermat.*, 1993, **28**, 26.
172. W. Aberer, *Wien. Klin. Wochenschr.*, 1996, **108**, 98.
173. S. B. Chang, C. Siew and S. E. Gruninger, *J. Dental Res.*, 1992, **71**, 66.
174. J. Liu, J.-Z. Shi, L.-M. Yu, R. A. Goyer and M. P. Waalkes, *Exp. Biol. Med.*, 2008, **233**, 810.
175. Medicines and Healthcare Regulatory Association (MHRA), *Press Release: Toxic poisoning alert for online Chinese medicine,* MHRA, London, 2013.
176. X. Yuan, H. L. Kohl and W. K. Chui, *Phytochem. Anal.*, 2009, **20**, 293.
177. D. L. Cragle, D. R. Hollis, J. F. Qualters, W. G. Tankersley and S. A. Fry, *J. Occup. Med.*, 1984, **26**, 817.
178. J. B. Hursh, T. W. Clarkson, E. F. Miles and L. A. Goldsmith, *Arch. Environ. Health*, 1989, **44**, 120.

179. National Toxicology Program, (NTP) *Announcement of and Request for Public Comments on NTP Recommendations,* Federal Register, 2002, **67**, 40329.
180. L. N. Kuo, C. J. Huang, Y. C. Fang, C. C. Huang, J. L. Wang, K. L. Lin, S. T. Chu, H. T. Chang, J. M. Chien, H. H. Su, C. C. Chi, W. C. Chen, J. Y. Tsai, W. C. Liao, L. L. Tseng and C. R. Jan, *Hum. Exp. Toxicol.*, 2009, **28**, 301.
181. World Health Organization (WHO), *Concise International Chemical Assessment Document,* WHO, Geneva, 2003, **50**, p. 1.
182. Institute of Medicine (IOM), *Thimerosal-containing Vaccines and Neurodevelopmental Disorders,* National Academy Press, Washington, DC, 2001.
183. A. W. McMahon, J. K. Iskander, P. Haber, M. M. Braun and R. Ball, *Vaccine*, 2008, **26**, 427.
184. D. J. Sutton, P. B. Tchounwou, N. Ninashvili and E. Shen, *Int. J. Mol. Sci.*, 2002, **3**, 965.
185. D. T. Netscher, J. A. Friedland and R. M. Guzewicz, *Ann. Plast. Surg.*, 1991, **26**, 592.
186. S. Asano, K. Eto, E. Kurisaki, H. Gunji, K. Hiraiwa, M. Sato, H. Sato, M. Hasuike, N. Hagiwara and H. Wakasa, *Pathol. Internat.*, 2000, **50**, 169.
187. A. Léonard, P. Jacquert and R. R. Lauwerys, *Mutat. Res.*, 1983, **114**, 1.
188. F. Schurtz, M. Sabeter-Vilar and J. Fink-Gremmels, *Mutagenesis*, 2000, **15**, 525.
189. S. De Flora, C. Bennicelli and M. Bagnasco, *Mutat. Res.*, 1994, **317**, 57.
190. M. Valko, H. Morris and M. T. Cronin, *Curr. Med. Chem.*, 2005, **12**, 1161.
191. W. A. Anwar and M. S. Gabal, *Mutagenesis*, 1991, **6**, 189.
192. D. Eke and A. Çeilik, *Toxicol. in Vitro*, 2008, **22**, 927.
193. H. Druckrey, H. Hamperl and H. Schmahl, *Z. Krebsforsch.*, 1957, **61**, 511.
194. National Toxicology Program (NTP), *NIH Publication,* No. 91-3139, US Dept. Health and Human Services, Washington, DC, 1991.
195. K. Mitsumori, K. Maita, T. Saito, S. Tsuda and Y. Shirasu, *Cancer Lett.*, 1981, **12**, 305.
196. M. P. Dieter, G. A. Boorman, C. W. Jameson, S. L. Eustis and L. C. Uriah, *J. Toxicol. Environ. Health*, 1992, **36**, 319.
197. L. Tryphonas and N. O. Nielsen, *Can. J. Comp. Med.*, 1970, **34**, 181.
198. K. Eto, A. Yasutake, T. Kuwana, Y. Korogi, M. Akima, T. Shimozeki, H. Tokunaga and Y. Kaneko, *Toxicol. Pathol.*, 2001, **29**, 565.
199. P. Boffetta, E. Merier and H. Vainio, *Scand. J. Work Environ. Health*, 1993, **19**, 1.
200. N. K. Mottett, C. M. Shaw and T. M. Burbacher, *Environ. Health Perspect.*, 1965, **63**, 133.
201. B. Weiss, *Toxicol. Sci.*, 2007, **97**, 223.
202. S. B. Elhassani, *J. Toxicol. Clin Toxicol.*, 1982, **19**, 875.
203. P. B. Tchounwou, W. K. Avensu, N. Ninishvili and D. Sutton, *Environ. Toxicol.*, 2003, **18**, 149.

204. M. G. Gómez, J. D. C. Klink, P. Boffetta, S. Español, G. Sällsten and J. G. Quintan, *Occup. Environ. Med.*, 2007, **64**, 389.
205. A. Simning and E. van Wijngaarden, *Occup. Environ. Med.*, 2007, **64**, 432.
206. H. E. Ganther, *Ann. N.Y. Acad. Sci.*, 1980, **355**, 212.
207. N. Mukai and T. Nishida, *Prog. Neuropathol.*, 1979, **4**, 319.
208. P. J. Kostyniak and A. J. Soiefer, *J. Appl. Toxicol.*, 1984, **4**, 206.
209. A. F. Castoldi, N. Onishchenko, C. Johansson, T. Coccini, E. Roda, M. Vahter, S. Ceccatelli and L. Manzo, *Regul. Toxicol. Pharmacol.*, 2008, **51**, 215.
210. A. Yasutake, M. Marumoto and M. Yoshida, *J. Toxicol. Sci.*, 2010, **35**, 767.
211. P. Boffetta, M. Garcia-Gómez, V. Pompe-Kirn, T. Bellander, M. Bulbulyan, J. D. Cabellero, F. Ceccarelli, D. Colin, T. Dizdarevic, S. Español, A Kabal, N. Petrova, G. Sällsten and E. Merler, *Cancer Causes Control.*, 1998, **9**, 591.
212. D. Ellingsen, A. Andersen, H. P. Nordhagen, J. Efskind and H. Kjuus, *Br. J. Ind. Med.*, 1993, **50**, 875.
213. L. Barregård, G. Sällsten and B. Järvholm, *Br. J. Ind. Med.*, 1990, **7**, 99.
214. N. P. C. Maramba, J. P. Reyes, A. T. Fransisco-Rivera, L. Crisanta, R. Panganiban, C. Doiquino, N. Dando, R. Timbang, H. Akagi, M. T. Castillo, C. Quitoriano, M. Afuang, A. Matsuyama, T. Eguchi and Y. Fuchigami, *J. Environ. Med.*, 2006, **81**, 135.
215. Z. Hrubec, A. E. Blair, E. Rogot and J. Vaught, *Mortality Risk by Occupation among US Veterans of Known Smoking Status 1954-1980* (NIH Publication No 92-3407), US National Institutes of Health, Bethesda, MD, 1992.
216. A. Ahlbom, S. Norell, Y. Rodvall and M. Nylander, *Br. Med. J.*, 1986, **292**, 62.
217. Y. Rodvall, A. Ahlbom, G. Pershagen and B. Spännare, *Oral Oncol.*, 1998, **34**, 265.
218. M. J. Wahl, *Quintess. Int.*, 2001, **32**, 696.
219. S. Argots, T. G. Asgari, J. Schulz, M. Bünger, Müller and E. Hallier, *Arch. Toxicol.*, 2003, **77**, 50.

CHAPTER 11

Lead

11.1 Introduction

Lead is a common contaminant of the human environment. It serves no nutritional value in humans or animals and is probably the most toxic of all metallic elements.[1,2] It has been used by humans for more than 5000 years for water conduits and storage, coinage and cooking utensils but in more recent times has acquired a wider value in paints, glass and ceramics, metal alloys, batteries, cosmetics and hair colorants, ammunition, protection against radiation, and weights for fishing nets. The present focus on lead contamination in urban environments and in the proximity to major road systems relates to the use of tetraethyl lead [$Pb(C_2H_5)_4$] as an anti-knock agent in petrol and diesel fuel.[3] Tetraethyl lead was widely used (*ca.* 1925 to 1990) to increase the octane rating of gasoline (petrol). On combustion the molecule cleaves to release carbon dioxide, water and lead (Pb):

$$Pb(CH_3CH_2)_4 + 13O_2 \rightarrow 8CO_2 + 10H_2O + Pb$$

Exhaust emissions from vehicles, discard and combustion of lead waste from factories and urban communities near active and disused lead mines, and plants recycling lead waste are a cause of widespread and long lasting lead accumulation in ecosystems. Non-degradable lead residues in water courses, soil and plants contaminate wild and domestic animals and enter human food chains.[4] Plant and wildlife living in the proximity of large motorways accumulate lead. Drinking water drawn from lead-rich wells, or conducted through lead piping and stored in lead-lined containers, has been a common source of human contamination over many years and is still a problem in many city areas. Low concentrations of lead are present in the bodies of most people living in urban communities.

Issues in Toxicology No. 18
The Carcinogenicity of Metals: Human Risk through Occupational and Environmental Exposure
By Alan B. G. Lansdown

Published by the Royal Society of Chemistry, www.rsc.org

Lead is a soft, malleable, heavy metallic element in Group IVb of the Periodic Table. It exhibits two oxidation states, Pb(II) and Pb(IV), but Pb(IV) is less stable and is oxidised to Pb(II). Four stable isotopes of lead exist naturally (^{204}Pb, ^{206}Pb, ^{207}Pb, ^{208}Pb) but 57 other isotopes have been identified with half-lives ranging from >300 nsec to many years. Lead forms a large number of inorganic and organic compounds, most of which are highly toxic to humans and other animals. Lead is a biologically active metal which dissolves slowly in water and in cold acids. It does not burn or readily oxidise in air. Chief lead-bearing ores include galena (PbS, 86.6% Pb), cerussite ($PbCO_3$) and anglesite ($PbSO_4$). Natural deposits of lead and lead ores are commonly associated with zinc, copper, silver, gold, arsenic, antimony and cadmium, to which miners will be exposed in the extraction and smelting processes. Deposits containing lead are found undersea, in superficial rock formations and in volcanic areas where its distribution is influenced by geochemical action, weathering, hydrothermal water currents and human activities including mining, smelting, industrial waste disposal, slags and combustion of lead-containing waste, petroleum fumes and fossil fuels. Solubility of lead in conduits is influenced by the acidity of the water, temperature, pressure and duration of exposure.[5] Inland waterways, coastal and oceanic water contain lead residues according to the proximity of lead mining areas, geochemical deposits, and contaminated water flowing from waste sites and industrial plant.

Environmental problems resulting from lead contamination were the subject of a conference organised in 1996 by the Institute of Mining and Metallurgy, when principle global emissions were estimated to be:

- Gasoline combustion: 128,889 tonnes
- Fossil fuel consumption: 9913 tonnes
- Non-ferrous metal production: 54,595 tonnes
- Waste incineration: 3304 tonnes
- Cement production: 1880 tonnes
- Iron and steel production: 10,066 tonnes.

Of the 208,247 tonnes released globally, estimated emissions by the UK alone were 7590 tonnes. In recent times, considerable effort has been made nationally and internationally to reduce environmental lead contamination.[6] In 1996, the European Council had visions of reducing lead in drinking water to 10 $\mu g\ ml^{-1}$ with maximum tolerable weekly intake estimated to be 3 mg for adults (430 μg per day) and 25 $\mu g\ kg^{-1}$ body weight in children. This is achievable but children are well known to be at risk of developing neurological problems and learning difficulties as a consequence of lead exposure in food and drinking water, paints and colorants. The World Health Organization reviewed problems arising from lead in drinking water in 2011 and accepted that most lead in drinking water arises from plumbing in buildings and that removal of lead plumbing and fittings should be a priority.[7] This is costly and impractical in the short term, but all reasonable measures are imposed in many countries to reduce dependence on lead paint as an anticorrosion material and lead in petrol.

Lead and lead compounds are reasonably anticipated to be human carcinogens, based upon limited evidence of carcinogenicity from human epidemiological studies and upon the sufficiency of evidence provided from studies in experimental animals.[2] Since the carcinogenic risk was first associated with exposures to lead phosphate and lead tetra-acetate in 1981, more than 4000 scientific communications have been issued and a vast amount of clinical, occupational and experimental data published, such that the carcinogenic risk is now extended now to cover all lead compounds. As in the validation of the carcinogenicity of other metals, the diagnostic value of case and epidemiological data on lead is limited by the inadequacy of information relating to exposure patterns, presumed durations of exposure, lifestyle factors including smoking, and exposure to other putative carcinogens including arsenic and cadmium present in the environment.[8,9] Radon is an acknowledged hazard in lead mines as it is in iron, and gold mining.[10] Lead is a late product of the decay of radon in mining, and radioactive ^{210}Pb absorbed by miners is deposited in bone, liver and kidney. Stable and radioactive isotopes (^{210}Pb) of lead also occur in cigarettes and cigarette smoke and are carcinogenic in the lungs and respiratory tract in long-term smokers.[11–14] Induction of lung cancer is complex and may result from interactions among radioactive emissions, lead and numerous other carcinogens commonly present in cigarette smoke or related products.[15]

11.2 Occupational, Industrial and Environmental Exposures

Lead is ubiquitous in the human environment, in food and drinking water, particles and fumes in the air, and in many occupational and industrial environments. Few people these days are not in contact with lead in some aspect of everyday life. The various ways in which populations are or have been exposed to lead industrially or occupationally are summarised in Table 11.1.

Uses of lead have changed greatly since early times. In the Roman era, the sweet taste of lead acetate was used to flavour and preserve wine and fruit. The thought that lead-induced infertility was in part responsible for the demise of the Roman Empire may be anecdotal, but it is well known that lead exposure is a cause of reproductive failure and targets the male reproductive system.[16,17] These days, accurate figures for lead absorption in different parts of the world are difficult to assess and wide variations are reported according to levels of industrialisation, traffic volumes and automobile exhaust fumes, and local legislation for lead control. A recent survey involving 1593 men in population studies in Montreal showed that the lifetime prevalence of exposure was 3–4% for organic lead, 19–40% for inorganic lead from engine emissions and 16–17% for inorganic lead from other sources.[18] No associations were observed when comparing ever-to-never exposed subjects in pooled analyses [organic lead, odds ratio (OR) = 1.39, 95% confidence interval (CI) 0.77 to 2.52; inorganic lead from engine emissions: OR = 0.89, 95% CI 0.72 to 1.09; inorganic lead

Table 11.1 Occupational and industrial exposure to lead and compounds.

Lead usage	*Compound*
Chemical syntheses	Metallic lead and inorganic lead compounds
Battery manufacture	Metallic lead
Alloys, soldering, canning	Metallic lead
Gold cyanidation	Lead nitrate
Lead coatings and linings	Metallic lead
Paints and anticorrosive preparations	Lead oxides (red lead, black lead), lead chromates, lead carbonate, *etc.*
Pharmaceutical (earlier times as astringents and anticancer agents)	Lead acetate or subacetate (lead water), colloidal lead
Cosmetics: hair colorants, eye liner, lipstick	Kohl (lead sulfide), lead oxides, metallic lead, lead acetate
Petrol and diesel fuels	Lead acetates, tetraethyl lead
Glass and ceramics	Lead oxides
Building industries: roofing, guttering, water tanks, drainage pipes	Metallic lead
Water industries: piping, storage tanks, conduits	Metallic lead
Recreational: fishing net weights, gun shot	Metallic lead
Protection against radioactive emissions, X-ray (lead aprons)	Metallic lead
Jewellery (in earlier times)	Metallic lead
Food and drink flavouring (in earlier times)	Lead acetate
Printing industry	Metallic lead
Military: ammunition and armouries	Metallic lead
Mining, refining, smelter work	Metallic lead and lead ores
Agrochemicals (in vineyards), pesticides	Lead arsenate ($PbHAsO_4$), lead phosphate

from other sources: OR = 0.99, 95% CI 0.76 to 1.29]. Lead exposure in this Canadian study was not associated with lung cancer despite substantial cumulative exposure.

Considerable exposure to lead, lead dust and fumes occurs in mining, refining and smelting, but occupational surveys show that workers involved in production of alloys and steels, batteries and motor industries are also at risk. A survey of radiator repair workers in Thailand showed that lead in hair ranged from 42 to 999 mg kg^{-1} and did not correlate well with duration of exposure.[19] Ill-heath was not reported in all 163 workers evaluated.

Lead has been used in cosmetic products in Europe at least since prehistoric times and lead-based powders have been identified in ancient tombs. Kohl or surma is an ancient eye cosmetic and may contain up to 80% lead. It is traditionally made by grinding galena (lead sulfide) and other ingredients and is still used in South Asia, the Middle East and parts of northern Africa to darken the eyelids and as mascara for the eyelashes. Although lead acetate is not a typical water soluble dye, lead is absorbed from commercially available hair preparations, and concentrations in hair exposed to a 2% solution rose to 466 ppm with blood concentrations of 50 μg 100 ml^{-1}.[20]

11.3 Absorption and Metabolism

Lead absorption in humans occurs largely through ingestion, of food and drinking water, and inhalation. Percutaneous absorption of inorganic lead is low, through binding of Pb(II) to epidermal keratin, but amphiphilic and lipid-soluble organic lead compounds including tetrabutyl-butyl-, tetraethyl- and tetramethyl lead in gasoline are rapidly absorbed following dermal contact.[21] Body lead burdens attributable to inhalation of lead in factory dusts or vehicle exhaust emissions and consumption of contaminated food and water are commonly monitored by analysis of blood, hair, toenails, sweat and urine.[22] Approximately 60% of inhaled lead is absorbed into the body. Adults absorb up to 20% of all ingested inorganic lead, but uptake and metabolism is greatly influenced by a person's age and dietary sufficiency, gut bacteria, and competing metal ions in the intestine.[23–25] Approximately 60% of an oral dose of 100 μg of ^{203}Pb was retained in 85 fasting subjects, with no difference noted between males and females. Most inhaled lead enters the circulation and can result in a 30–50% increase in blood lead concentrations. Lead acetate is absorbed into the body 1.5 times more rapidly than any other lead compound but organo-lead compounds are appreciably more toxic than inorganic lead compounds.[26] Children are particularly at risk of absorbing lead residues and considerable efforts have been made in recent years to protect their environment from lead pollution.[27] This includes prohibition of lead in domestic paints and plastic colorants.

Commonly, blood, urine and hair lead levels are monitored as an indication of lead exposure, but because lead is a cumulative poison and may be retained in tissues for considerable periods, these are poor guides as to total body concentrations or patterns of uptake. Lead has a biological half-life in human blood of approximately 36 days.[25] Lead-binding proteins are a series of low molecular weight proteins, analogous to metallothionein, which segregate lead in a non-toxic form in several organs (kidney, brain, lung, liver, erythrocytes). Surveys of the literature show also that lead poisoning is frequently asymptomatic or is accompanied by subtle changes that are not readily appreciated.[21,28] Bluish lead lines on gums and discoloration of teeth are tell-tale signs of chronic lead poisoning and, in the case of a 61-year-old lead smelter, these changes were associated with blood lead of 6.3 μmol l^{-1}.[29] The element is not strongly bound in these tissues and is progressively released into the circulation; in children this can lead to encephalopathies and mental impairment. Bone is the major site for accumulation of lead in the human body and clinical observations show that it is mobilised from this reservoir to be a long-term risk for tissues such as brain, kidney and liver.[30–33] Up to 95% of body lead is contained within actively mineralising regions of bones and teeth where osteoclasts serve a principal role in its metabolism and mobilisation.[34,35] The half-life of lead in mid-shaft regions of major long bones may be as long as 30 years, but much of this is readily exchangeable.[31] Osteoporosis is a feature of lead toxicity but bone tumours have not been reported.

Lead in drinking water and blood levels, particularly in women and children, have been monitored in the USA at least from the 1970s.[3] Mean concentrations in women aged 20–59 years decreased from 1.8 μg dl^{-1} in 1988–1994 to 1.2 μg dl^{-1} in 1999–2002; similar levels were seen in children. In its 2012 survey, the US Centers for Disease Control and Prevention noted that, whilst significant efforts had been made to control lead in the human environment, children are still being exposed and pregnant women are still vulnerable. Neonates and young children are particularly vulnerable to the neurotoxic effects of dietary lead exposure and ingestion of lead paints.[27,36] Lead absorption in pregnancy presents a serious risk of abortion, congenital abnormality and neonatal distress.[37]

Metallic lead, lead oxides and lead acetates used in paints, varnishes and hair dyes are poorly absorbed percutaneously except where the protective function of the stratum corneum is disturbed or denatured by lipid solvents.[38,39] At least 99% of Pb(II) ion absorbed binds sulphydryl moieties on metallothionein-like proteins in human erythrocytes.[40,41]

11.4 Cytotoxicity, Mutagenicity and Carcinogenicity

11.4.1 *In Vitro* Studies

Experimental studies in cultured mammalian cell lines have provided some information relevant to the potential carcinogenic mechanisms of lead, which is shown to be a protoplasmic poison and cytotoxic to most eukaryotic cells through its actions on mitochondria, lysosomes, membrane structure and function and chromosomal integrity.[42] The Pb(II) ion interacts strongly with and displaces sodium, potassium, zinc, copper, iron, manganese, calcium and magnesium ions from essential enzymes involved in cellular respiration, membrane transport, RNA and DNA synthesis and transcription, mitosis, and biosynthetic pathways.[43–46] Cell degeneration, apoptosis, clastogenicity, chromosomal damage and cell transformation are features of lead toxicity. Other cellular actions include an inhibition of catalase and superoxide dismutase (SOD), suppression of polymorphonuclear leucocyte chemotaxis and Langerhans cell function (immunosuppression), impairment of haem synthesis and oxidative stress.[47–49] *In vitro* studies and analysis of peripheral polymorphonuclear leucocytes from lead-exposed workers showed an inhibition of the mitochondrial enzymes ferrochelatase and coporphyrin oxidase, accumulation of zinc protoporphyrin in erythrocytes, and increased urinary coporphyrin excretion.

Erythrocytes, leucocytes, fibroblasts, renal tubular epithelium, hepatocytes, osteoclasts, pulmonary macrophages, peripheral blood neutrophils, and glial cells of the brain are highly vulnerable to lead-induced toxicity.[50,51] Lead particles and sparingly soluble compounds such as lead sulfide are phagocytosed and evoke mutagenic changes, impaired DNA repair, chromosomal damage, micronucleus formation, aberrant metaphase and sister-chromatid exchanges in peripheral blood lymphocytes in lead workers.[52,53]

The mutagenicity and clastogenicity of lead compounds in cell cultures and bacterial tests seem to vary according to their solubility, ionisation and availability of free ion to bind to DNA. Thus, lead acetate was shown to be mutagenic in Chinese hamster V79 cells and in the *gpt*-locis of a G12 cell line, whereas lead nitrate was inactive at any concentration in this screen.[54] Lead acetate caused "nicks" in chromosomal DNA and at non-toxic concentrations inhibited DNA repair. Lead bromide, as a combustion product of lead in gasoline products, and lead chromate were mutagenic in the Ames *Salmonella typhimurium* reversion test with at least one strain of the bacterium, whereas lead chloride was negative.[55–57] The anion has a major bearing on the toxicity of the lead compound. Lead chromate is confirmed as a mutagen in a battery of microbial tests including *E. coli, Saccharomyces cerevisiae* and *Salmonella* assays with or without addition of the rat liver S9 homogenate. However, supplementary tests have confirmed that this mutagenicity is attributable principally to the chromate anion and not Pb(II). In a similar way, the toxicity of arsenate is largely contributory to the changes seen in response to lead arsenate. Trivalent arsenicals are highly cytotoxic and induce reactive oxygen species in cell cultures.

In vitro studies point to direct DNA damage and mutagenicity leading to deficiencies in transcription and DNA synthesis and repair as plausible mechanisms of lead-induced carcinogenicity.[33] Certain compounds such as lead acetate can generate reactive oxygen species and cause oxidative damage, whereas the insoluble salts, lead chloride, bromide and iodide, exhibit weak or negligible mutagenicity in bacterial tests including the Ames *Salmonella typhimurium* reverse transcription assay. Apostoli *et al.* demonstrated that occupational exposure to lead had no effect on urinary mutagenicity in *Salmonella* or *E. coli* mutagenicity with or without the S9 metabolic activation.[58]

11.4.2 Experimental Animal Studies

A large number of experimental studies in animal models have been conducted in the past 50 years on the toxicity and physiological action of lead. They can be grouped under:

a. General toxicology and biochemical mechanisms of pathological action
b. Carcinogenicity
c. Reproductive toxicity, teratogenicity and transplacental carcinogenicity.

In many studies, metallic lead or soluble lead compounds have been administered at high or maximal tolerated doses in order to obtain a dose-related effect. As in the case of subcutaneous or intraperitoneal dosing studies, routes of administration fail to represent potential routes of human exposure and circumvent or overwhelm normal physiological and inherent protective mechanisms, including metallothioneins and other metal-binding proteins. Limitations in the extrapolation of experimental data to humans are imposed by the relatively narrow range of rat and mouse strains used. Despite these deficiencies, studies in animal models have confirmed the neurotoxicity and nephrotoxicity of lead and its capacity to disturb iron and copper metabolism in haemopoiesis and calcium, zinc and magnesium in epithelial cell differentiation.[42,59–65] Experimental evidence is available also to show that lead intoxication is associated with osteological abnormality and defects in mineralisation through inhibition of alkaline phosphatase and other key enzymes involved in calcium metabolism.[66,67]

Experimental studies in rodents provide strong evidence that certain lead compounds should be considered carcinogenic following oral or in-diet administration, and through parenteral administration.[68,69] However, the studies reported show considerable inconsistency both in the patterns of tumours induced and susceptibility according to the age, strain and duration of dosing and experimental detail. Published studies have examined a diverse range of lead compounds including chromate, halides, arsenate, acetate, phosphate and tetraethyl lead and shown increased tumour yields over lifetime studies. (Arsenates and chromates are listed multi-site carcinogens in rodents). Several studies allude to the capacity of lead to cause nephrocarcinogenicity, but it may also be a cause of tumours of the adrenal, lungs, thyroid, testis and mammary glands.[70–77] There is little evidence from rodent studies that lead or lead compounds administered by any route are a cause of brain or central nervous system tumours as seen in some human studies.

Early studies showed that administration or 0.1 or 1.0% basic lead acetate or lead phosphate in the diet led to a high incidence of renal tumours in rats, although viral infections and residual radioactivity were possible complications.[70,71] The nephrocarcinogenicity of 1.0% dietary lead acetate was later confirmed in studies with Wistar rats where 15 of 16 animals surviving 320 days developed profound nephritis and multiple adenomas or adenocarcinomas.[72] Similar observations were made in rats of the CB strain following repeated subcutaneous or intraperitoneal injections of lead phosphate over 34 weeks, but other tumours were not treatment related.[73] Experiments in which powdered lead or lead chromate solutions were administered orally or by intramuscular injection confirm the carcinogenicity of lead chromate but provide little evidence of risk following injection of powdered lead.[74] Lead is virtually insoluble in water but marginally soluble in serum (0.58 g l^{-1} at 25 °C). Lead chromate is also sparingly soluble (0.06 g l^{-1}) but is positive in bacterial mutation assays *in vitro* following chemical solubilisation. Lead chromate (<40 mg kg^{-1}) injected repeatedly led to injection site sarcomas in 64% of animals and a low incidence of renal tumours in male rats. In contrast, female mice showed low carcinogenicity following administration of tolerable levels of lead chromate, the tumours identified being lymphomas and lung carcinomas. The carcinogenicity of lead chromate in rodents is largely attributable to the action of the hexavalent chromate anion rather than the Pb(II) cation.

Mice were sensitive to lead-related toxicity but male animals tolerated lead chromate better than females. Poirier *et al.* demonstrated that repeated intraperitoneal injection of soluble lead subacetate at 24 doses equivalent to 1, 3, 10 or 30 times the maximal tolerated dose (0.04 mmol kg^{-1}) induced multiple lung tumours in an A strain within 30 weeks.[75] This tumorigenicity was nullified by concurrent administration of calcium or magnesium acetate, suggesting interaction between the metallic cations. Interestingly, administration of calcium acetate alone led to increased murine lung tumours. Elsewhere, Sprague–Dawley rats receiving 1% lead acetate in their diet for up to 18 months showed a massive accumulation of lead in kidneys and liver, and 45% developed renal tumours within 68 weeks, but no other tumours.[76] Addition of calcium acetate in this study enhanced renal carcinogencity but reduced renal accumulation of lead. The few toxicological studies conducted in hamsters and cynomolgus monkeys provide minimal advance in understanding lead-related carcinogenesis. Studies in which hamsters were given 0.1 or 0.5% lead subacetate for up to two years failed to confirm the nephrocarcinogenicity of the salt.

Experimental studies in mice have shown that genetic status is important. They emphasise that lead, like many other metals, induces and binds the cysteine-rich proteins metallothioneins (MT) I and II, and that the MT–Pb(II) complex plays a central protective role against the toxicity and carcinogenicity of free lead ion.[78,79] This is illustrated in experimental studies with genetically engineered MT-null mice. No carcinogenic action was observed in *wild type* animals exposed to high concentrations (2000 or 4000 ppm) of lead acetate in drinking water, whereas MT-*null* animals developed renal carcinoma. Renal carcinogenicity was also demonstrated in animals exposed during gestation, through lactation and observed throughout life. MT-*null* animals were subject to teratomas, cystic renal tumours and bladder tumours.

Mice did not show nephrocarcinogenic responses to subcutaneous injections of the organo-lead compound tetraethyl lead (TEL), but female animals developed an increased incidence of lymphomas.[69,80] Epstein and Mantel demonstrated that a total dose of 1.2 mg administered to neonatal mice was highly toxic; 12% of female animals injected with 0.6 mg of TEL and surviving for 36 weeks developed malignant lymphomas between the 36th and 51st weeks (control: 0 of 48).[80] Lymphomas developed in 1 of 26 males in this dose group (compared to 1 in 26 of the control group). The significance of this result is marginal because Swiss mice are susceptible to this form of malignancy.

The potent teratogenicity of lead compounds in humans and laboratory animal models is not disputed.[37,81,82] Skeletal abnormalities and impaired osteogenesis are recorded but more recent concern relates to the possibility that Pb(II) passing transplacentally is a cause of carcinogenicity in fetuses and offspring of mothers exposed in mid to late gestation.[77,79,83] Transplacental lead is shown to disturb trace metal ion balances in fetuses and is a cause of abnormal skin development and changes associated with aberrant glycosaminoglycan synthesis, possibly through covalent binding.[84] In a limited study, mice injected with the high oral dose of 10 mg kg^{-1} lead acetate in mid-pregnancy showed pathological changes in the developing kidneys of fetuses pre-term, but no tumours. Recent work emphasises the susceptibility of MT-*null* mice to transplacental carcinogenicity.[77,79] Thus, pregnant dams given 2000 or 4000 ppm lead acetate in drinking water from mid-pregnancy through parturition into lactation produced offspring that developed testicular tumours and teratomas. Offspring of MT-sufficient *wild type* animals did not show increased tumour yield. Some offspring developed pre-neoplastic hyperplastic lesions in the renal epithelium and bladder. Transplacental lead has yet to be shown to cause lung or neurological tumours.

11.4.3 Human Case and Epidemiological Studies

Lead and lead compounds are presently assessed as "anticipated to be human carcinogens" on the basis of limited evidence of carcinogenicity from studies in humans and sufficient evidence of carcinogenicity in experimental animals.[2] Assessment of the true carcinogenic effect of lead and lead compounds in humans is complex and putative mechanisms of action are difficult to evaluate from at least 24 epidemiological studies embracing more than 90,000 patients working in a variety of lead-related industries.[19,85–88] First indications that environmental lead might be a cause of lung cancer were published in 1956 when Jecklin noticed an apparent correlation between inhalation of lead dust powder in an urban environment and increased lung cancer.[89] He drew attention to the detrimental effects of airborne lead from exhaust fumes from gasoline engines.[90,91] Since that time, epidemiological and case studies have been published on tumour incidence in workers engaged in glass manufacture, printing and publishing, battery factories, mining and smelter works, and undefined lead-related industries, each with widely differing patterns of lead exposure. Many of these reports involve retrospective analyses of *post-mortem* records or questionnaires but have implicated environmental or occupational exposure to metallic lead or lead compounds in cancers of the kidney, gall bladder, stomach and colorectal tissues and brain.[92–95] Greater awareness of lead-related toxicity has led to marked improvements in working practice with a corresponding diminution of lead-related cancers in many developed countries. In the USA, for example, lead concentrations in the air arising through combustion of automobile fuels have declined greatly in recent years, in line with human blood concentrations (Figure 11.1).[96] Lead levels in soil samples taken in Newcastle, UK in 1993 ranged from 25 to 2500 ppm and these were comparable to or lower than those seen in the USA and Australia in 1995.[96,98] In 1993, the National Health and Medical Research Council in the UK set a national goal for blood lead level of 10 $\mu g\ dl^{-1}$.[98] In humans as much as 60% of lead inhaled is absorbed in the circulation.

11.4.3.1 Lung Tumours

In view of Jecklin's early observations and awareness of the risk of lead inhalation from gasoline emissions,[89–91] considerable attention is given to the risk of lung cancer. Human and experimental studies show that, at concentrations below the acceptable air standard for occupational exposure, lead (as oxides, PbO and Pb_3O_4) can disrupt pulmonary macrophages and evoke an immune response.[99–101] Changes include mitochondrial damage, alterations in the nuclear membrane and endoplasmic reticulum associated with phagocytosis, and

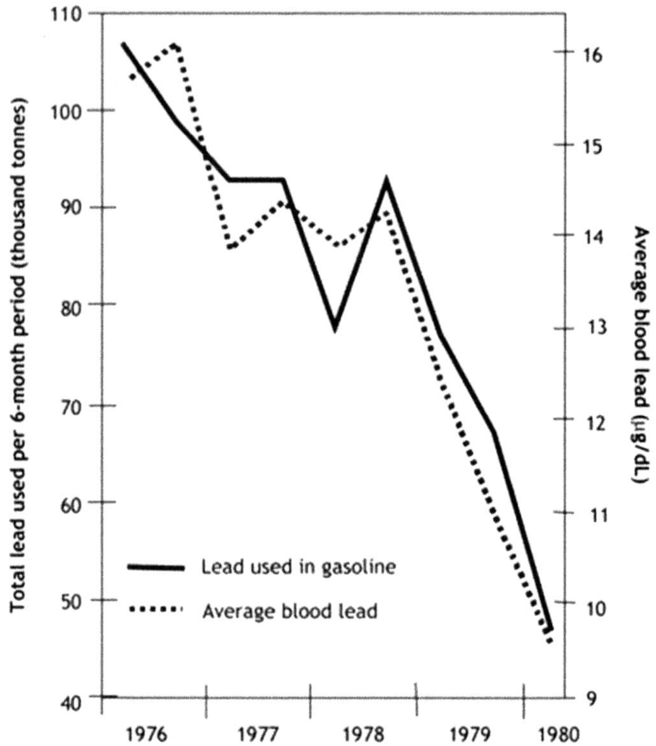

Figure 11.1 Lead concentrations in petrol and children's blood (USA) 1975–1980.[96,97]

precipiatation of lead complexes within the nuclear heterochromatin and cytoplasm. (Macrophages were reduced in rats and rabbits inhaling lead oxides.) Zelickoff has further shown that inhaled lead (PbO) induced neutrophil infiltration and in pulmonary lavage increased lactic dehydrogenase as a marker of early pulmonary damage, raised lysozyme concentrations indicating enhanced pulmonary permeability, and increased hydrogen peroxide and superoxide anion radical in a time-dependent manner.[98]

Cancers of the lung and stomach are listed as occupational risks through chronic high-level industrial exposure to lead dust or soluble inorganic lead compounds.[2,68,88] However, in a large comprehensive study conducted in Quebec, Rousseau *et al.* examined the prevalence of 11 different primary cancer types including lung cancer in 3730 patients and failed to find a statistically significant correlation between any form of lead inhalation, including organic lead (tetraethyl and tetramethyl lead), inorganic lead and lead in gasoline emissions, and lung cancer.[102] The lifetime prevalence of occupational exposure to any lead compounds in the entire study population was 47%, with relative exposure patterns including 39% exposed to lead in gasoline fumes, 4% lead fumes (unspecified), 4% from lubricating oils, greases and vehicle gasoline, 3% from lead pigments (chromates and oxides), and <2% from lead carbonate, aviation fuel and lead dusts. The study recognised that smoking was a significant cause of lung, oesophagus and bladder cancers in the various industries evaluated. Rousseau and others did emphasise that stomach cancer risk was a major problem in lead workers and that a possible risk of rectal cancer and lung cancers existed in people exposed to organic lead compounds.[102] However, their statistics showed no association between ever-exposures to inorganic lead and to lead in gasoline emissions with any of the cancer types analysed. Organic lead compounds have been progressively phased out in Canada from the mid-1970s to the early 1990s, a trend that is evident in many other parts of the world.

Lung cancer is not shown to be an occupational hazard among workers engaged in making lead chromate pigments.[103] Studies involving 1152 men engaged in three English factories from the 1930s or 1940s until 1981 showed that medium or high levels of exposure were a cause of increased mortality, but inhalation of lead residues at concentrations sufficient to cause to lead poisoning was not a cause of lung cancer. On the other hand, heavy exposure to zinc chromate was a cancer risk. In the State of New Jersey (USA) in 2007, lead chromate was still labelled "handle with extreme care" as a carcinogen.[104] This is possibly due to the results of earlier mixed-race studies in 1940–1969, which showed that white males with more than 10 years of employment showed a statistically significant relative risk of 1.6 for lung cancer. Other cancers reported amongst white males included stomach and pancreatic cancers, but those risks were attributed more to chromate exposure than the lead cation. Workers exposed to chromates showed a 1.9 fold increased risk of developing cancer.[105]

Assessments of occupational risks of cancer in lead mining and smelter work have regularly been confounded by the presence of arsenic in the workplace. One such study, involving 3979 workers employed in non-ferrous smelters for at least one year between 1928 and 1979, showed a elevated risk of lung cancer, but detailed analyses showed that the risk was largely attributable to excessive cigarette smoking and arsenic exposure, and not to lead.[106,107] Cigarette butts and filters contain lead, arsenic and cadmium and an array of carcinogenic hydrocarbons.[12–15] More recent studies in Italian lead smelters have failed to substantiate a lung cancer risk for lead exposure, although the hazards of silica-related pneumoconiosis and respiratory tract diseases were prevalent.[108] The mortality study of 1388 labourers employed in 1950–1992 showed that rates of lung, stomach and other cancers were actually lower than expected. Lung cancer excesses have been reported in populations living in the vicinity of abandoned lead mining and smelting in central USA.[109] The lung cancer risk in white men living in this area persisted for more than 28 years but geological analyses showed that they were exposed to mine-tailings containing lead, zinc, cadmium, germanium and many other toxic chemicals. Radium (^{226}Ra) was present in untreated ground water.

Risks of lung cancer through inhalation of lead in automobile exhaust emissions have not been confirmed. One competent study conducted in Montreal in 1979–1986 and 1996–2002, and listing 1593 men with confirmed lung cancer, found no correlation between cancer risk and exposure to lead from engine exhaust fumes or cumulative exposure to inorganic lead from other sources.[18] The authors made allowances for socio-economic factors, smoking and occupational histories, and pooled analyses of lifetime exposures to organic and inorganic lead from exhaust emissions and lead from other sources showed no increased risk of lung cancer. Similarly negative correlations were made when evaluating two databases in New Jersey State Center where 3165 workers were followed for 10 years and contributed a total of 30,401 person–years.[110] The analyses showed a deficit of lung, kidney and brain cancers, whereas cancers of the stomach, breast, larynx, intrahepatic bile duct and chronic myeloid leukaemia were non-significantly increased. However, only 83 cancer cases were recorded and further analyses with larger cohorts are required to verify the observations and to look into possible reasons for higher than expected increases in prostatic cancer.

Anttila *et al.* studied 20,700 workers who had been biologically monitored for their blood lead concentrations by the Finnish Institute of Occupational Health during 1973–1983.[85] The internal comparison within the cohort showed a 1.4-fold increase in the overall cancer incidence and a 1.8-fold increase in the incidence of lung cancer among those who had ever had a blood lead level of 1.0 mmol/l^{-1}. In the case–referent analysis, an increased odds ratio was found for lung cancer for concomitant exposure to lead and engine exhaust.

11.4.3.2 Stomach Cancer

Several epidemiological studies have drawn a correlation between lead exposure and gastrointestinal cancers, particularly in the region of the cardia.[87,92,93,102,108,111–114] Rousseau observed from her survey in 2007 that the risk is specifically linked to working around organic lead – mainly in jobs where personnel are exposed to leaded gasoline in its liquid form and to tasks where workers are liable to breathe in large amounts of gasoline fumes.[102] Exposure to organic lead in the workplace, especially from leaded gasoline, could increase stomach cancer risk but it is emphasised that engine mechanics are exposed to a range of toxic polyaromatic hydrocarbons, alkanes and organic solvents. Lead is absorbed readily from the gastrointestinal tract, possibly by passive diffusion aided by vitamin D and proteins relevant to the calcium and iron uptake pathways. In a study of 4576 patients with confirmed primary incident cancer in Montreal, Rousseau reported that of 3730 interviewed, stomach cancer was associated with exposure to organic lead. Odds ratios for the association between ever-exposure to lead and population and cancer patient controls were 3.0 (95% CI 1.2,7.3) and 2.0 (95% CI 1.1,3.8) respectively. Cancers of the oesophagus and rectum were also higher than in control groups.[102] No such observations were made in respect of inorganic lead exposures or exposures to lead in gasoline emissions.[102]

Cocco *et al.* evaluated the occupational risks of stomach cancers using death certificates in 24 US states.[93,111,114] This multistate study performed in 1984–1992, which evaluated 1056 men of >20 years of age engaged in paper mills, newspaper and printing industries, water supplies and irrigation and general industry, claimed to show a 30% rise in cancers in the cardia of the stomach in workers with high exposures to lead. However, it is emphasised that, as with many other epidemiological studies, the statistical evaluations were confounded by co-exposure to asbestos, smoking, undefined dusts and other environmental pollutants.

Lead exposure in battery and smelter workers is a potential cause of stomach cancer.[87,115,116] In studies of 6818 men working in battery plants and smelting, high death rates were attributed to chronic vascular and renal diseases and unidentified "external" causes, but cancers of the stomach, liver and lungs were higher than expected from national averages. Whereas stomach and lung cancers were higher in battery workers, these lesions were not in seen in a cohort of lead production workers. Lifestyle factors and smoking could not be excluded as possible contributory factors. In that study at least, the authors did not report increased mortality from cancers of the kidney, bladder, liver, trachea, central nervous system, or lymphatic and haematopoietic systems, as reported elsewhere.[93,117–119]

An analysis of work histories, blood lead levels and mortality data on 3832 lead workers with "considerable lead exposures" in a Swedish copper smelter in 1950–1981 also showed an excess of stomach and lung cancers and non-malignant cardiovascular disease compared with the general population.[112] Workers exposed to the highest lead levels exhibited a low but not significant risk of lung cancer. It is unclear from the results to what extent mortality was attributable to cancer or lead-related ischaemic heart disease or cerebrovascular conditions. Fu and Boffetta also claimed an association between lead exposure and stomach, lung and other cancers in a meta-analysis of studies published up until 1995 but emphasised that wider extrapolation of observations was limited by lack of information on smoking and dietary habits, exposure to other contaminants and lifestyle factors.[92]

An attempt was made in an occupational health study at the DuPont factory (1956–1987) to identify the carcinogenic risks to staff through occupational exposure to tetraethyl lead.[120] Manufacture of tetraethyl lead involves exposure to both organic and inorganic lead compounds. Whilst the evaluation showed a four-fold risk of colorectal cancer at high exposure levels, the epidemiologists were unable to determine whether the principle risk lay with the organic or inorganic moieties. The risk of colorectal cancer was more pronounced after a 10-year period of latency. It is noteworthy that colorectal cancers attributable to multiple causes were very common in the USA at the time.[120] Rousseau had shown earlier that stomach cancer was associated with organic lead, in her meta-analysis involving many lead-related occupations.[102]

In a small study in a Swedish art glassworks, the incidence of cancers of the lung, colon, pharynx and prostate was higher than expected from national averages but stomach cancer did not feature in the analyses.[121]

11.4.3.3 Cancers of the Brain and Central Nervous System

Lead is a potent neurotoxin following inhalation or ingestion and the physiological, cellular and molecular effects are well researched.[32,51] Chronic inhalation of lead residues and dusts is claimed to be a cause of human brain cancers, including gliomas and meningiomas.[93–95,118,119] The National Longitudinal Mortality Study of the USA (1979–1989) evaluated case histories and job-exposure matrices of 28,108 brain cancer patients and hospital-based controls and provided unequivocal statistical evidence that high-level lead exposure is a cause of brain cancer mortality with a hazard ratio of 2.3 (95% CI 1.3–4.2).[95] This risk level is much higher than an earlier estimates showing a 1.06 risk (95% CI 0.8–1.4) as calculated by Steenland and Boffetta.[94] However, the chance of developing glioma/meningioma following lead working was considered to be low except where cigarette smoking was implicated. This is illustrated in a large study of 27,060 cases of brain cancer in African-American men employed in lead-related industries.[93] Smoking was positively correlated with increased tumour risk in women but not men in 183 cases of meningioma reported in six hospitals in northeast China.[118]

Children are particularly at risk of developing encephalopathy and pathological changes in their central nervous systems following inhalation of high levels of airborne lead and exposure to recycled lead waste from batteries.[122,123] One study in the vicinity of a lead-acid battery recycling plant in Senegal reported that 18 children died and 81 were subject to severe poisoning following massive lead exposures in their homes (up to 14,000 mg kg^{-1}) and outdoors (up to 320,000 mg kg^{-1}) as a result of used lead-acid battery recycling. Ingestion and inhalation of dust heavily contaminated with lead resulted in blood lead levels of 39–614 μg dl^{-1}.[123] It is important to note that the children died from fatal encephalopathy rather than brain cancers. The true incidence of brain tumours developing following lead intoxication is not known but in a study of two institutionalised children with mental retardation and neuropsychological dysfunction following lead exposure, raised urinary lead concentrations were associated with increased astrocyte proliferation and neuronal toxicity.[122]

11.4.3.4 Renal and Other Cancers

Chronic renal disease manifest by nephritis, sclerosis, tubular atrophy and functional damage is a major concern in lead-related industries.[60,124–127] A multi-centre epidemiological survey in German metal-related industries involving 935 patients showed that women were approximately twice as susceptible to nephrocarcinoma through lead exposure in solder fumes and smelting than men.[117] Also, long-term exposures to lead in the chemical, rubber and printing industries can accentuate the risk of renal cell carcinoma, but workers are exposed to a range of other toxic chemicals including solvents, accelerants, benzene, paints, mineral oils, hydrocarbons and even asbestos. Several other worker analyses show that kidney carcinomas are higher in lead-related industries although numbers are small and the significance marginal.[92,94,102] These analyses also show a marginal rise in bladder cancers but the risk cannot, on present knowledge, be related to lead exposure in any profession. Fu and Boffetta claimed that the relative risk of developing bladder cancer in lead-related industries was 1.41 (95% CI 1.16–1.71).[92] This figure was based upon five studies and whilst a higher rate of bladder and kidney tumours was seen in lead smelters with increasing durations of exposure, with a latent period of >20 years, the true risk is difficult to predict on the basis of lead exposure alone.

The increased susceptibility of the kidney to lead-related toxicity and carcinogenicity is confirmed by experimental studies in rodents.[71–73] On the basis of experimental evidence, it seems that cells of the renal tubular epithelium show a specific sensitivity to lead-related mitogenicity.[83,128] Further, as Silbergeld suggests, allowing for interspecies differences (rodent-specific proteins, differences in cellular responsiveness, *etc.*) plausible mechanisms of lead-induced renal carcinogenicity should include direct DNA damage, clastogenicity, and inhibition of DNA synthesis or repair.[33] Certain lead compounds are known to generate reactive oxygen species and cause oxidative damage to DNA and there is abundant evidence that lead will displace zinc, calcium, magnesium and iron in several enzymes acting as transcriptional regulators, leading to altered gene expression.

11.4.3.5 Perinatal and Childhood Carcinogenicity

A comprehensive review of studies involving many lead-related occupations acknowledges the transplacental toxicity and teratogenicity of lead, but found no "statistically valid evidence" that any specific parental exposure to lead was a cause of childhood cancer.[92,129] Evidence provided by a small number of cases suggesting that paternal exposure to lead in printing and painting professions is associated with childhood Wilm's tumour of the kidney remains to be verified.[130,131] Wilm's nephroblastoma is a highly malignant embryonal condition of the kidney and one of the commonest abdominal malignancies of childhood.[132,133] In the USA, approximately 500 children of less than 20 years of age are diagnosed with Wilm's tumor annually.[124] Kantor reviewed 149 children from the Connecticut Tumour Registry during the period 1935 to 1973 and reported a positive correlation between development of this rare tumour and paternal occupation, compared with controls.[130] Although lead exposure was implicated, the study failed to consider the possibility of exposures to other environmental hazards or familial factors. In a more circumspect study, risks attributable to hydrocarbon exposure were considered separately; birth records of 143 children with Wilm's tumour showed that when fathers were exposed to lead in the paint industry, infants were six times more likely to develop renal cancer.[131] Savitz and Chen conducted a large review of childhood cancers in relation to parental occupation and whilst higher risks of leukaemia and brain cancers could be linked to certain professions, the survey found it impossible to associate the tumours with lead, hydrocarbons or any other specific chemical.[134] Paternal work and exposure to carcinogens in electrochemical industries, petrochemical production, printing and paper manufacture and mechanical engineering require more detailed evaluation as a cause of childhood cancer. Maternal occupation and exposure to lead is a further possible contributory factor to tumours in children, but this is virtually unexplored.

11.5 Conclusions

Research in the past 50 years has provided strong clues to the carcinogenic risk posed by exposure to lead in the workplace and home environment. However, there are many inconsistencies and whilst experimental studies provide essential information on the sub-cellular, genetic and physiological basis of lead-induced malignant changes, interspecies differences in response are confounding. Much of the evidence on the capacity of organic or inorganic lead compounds to induce tumours in human patients and meta-analyses relies upon on retrospective analyses of cancer mortality records and follow-up studies of workers dying in lead-related industries. Several studies allude to long periods of latency, possibly extending for >20 years, between commencing work in the lead industry and frank manifestation of tumours.

Lead is a profound neurotoxin. It impairs intellectual development and can evoke cellular and genetic change predisposing to brain tumours.[95,102,119,122,135–137] However, the actual numbers of brain tumours confirmed in the studies evaluated are limited and it has not been possible in many cases to exclude smoking, exposure to other toxicants or aspects of lifestyle and genetic predisposition as contributory factors. Considerable inconsistency has been observed in the tumour type recorded, and the expression "brain cancer" is used to include meningioma, astrocytoma, glioblastoma multiforme and gliomas. Oxidative stress and genetic polymorphisms underlie development of lead-related brain tumours including glioma and glioblastoma multiforme.[137] Abnormal metaphases with chromatid and chromosomal aberrations occur in healthy people exposed to lead in the manufacture of electrical storage batteries.[143] High levels of lead from vehicle exhaust emissions in the atmosphere of inner cities are no doubt a contributory factor in the high levels of brain tumours seen in some countries but other detrimental factors including radiation, hydrocarbon pollution and other contaminants cannot be excluded.[138] The implementation of the National Petrol-Lead Phase out Program in Taiwan provides some evidence for lead emissions as a cause of brain cancer (Figure 11.2). A significant increase was found from low to high lead in the time-related incidence of brain tumours.

Principle carcinogenic risks from lead exposure indicated by published work are those of respiratory and stomach cancers.[92,94,102] At least five major surveys failed to confirm any association between occupational lead exposures and increased mortality from brain, kidney or other cancers; statistical evaluations showed that cancer mortality was lower than expected.[86,87,112,139] Inconsistencies are attributable to the presence of carcinogens such as arsenic, chromates, silica, hydrocarbons and cigarette smoke in the working environment.[140–142] In one study, 437 lead smelter workers with >3 years' employment exhibited an increased lung cancer incidence, but the lead concentrates contained 5.6% silicates and traces of cadmium, bismuth, tin, antimony and arsenic.[112] The highest incidence of lung cancer was seen in roasters and those exposed to greater arsenic contamination.[94] Anttila *et al.* surveyed records of 20,700 Finnish workers and reported a 1.4-fold increase in the overall cancer

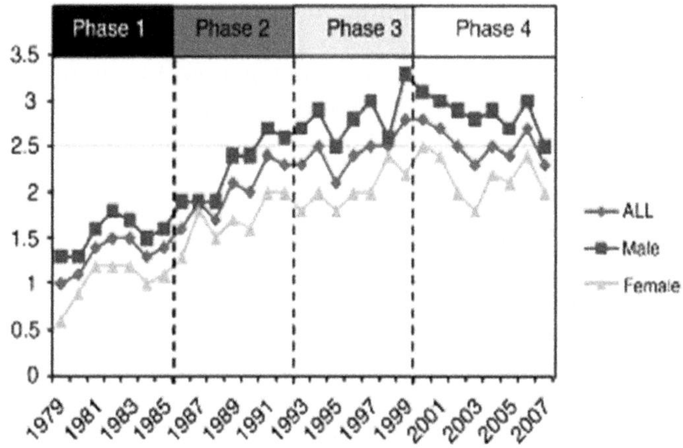

Figure 11.2 A phased long-term trend of brain cancer age-standardised incidence rates (per 100 000 people) in Taiwan from 1979 to 2007.[135]

incidence and a 1.8-fold rise in lung cancer cases but co-exposure to lead and hydrocarbons in exhaust fumes contributed to the increased lung cancer risk, and blood lead levels did not correlate well with the incidence of lung cancer.[85]

Children are at particular risk, from lead-related neuropathy, impaired brain development and depressed IQ.[144,145] Lead is a well known cause of anaemia through its capacity to impair iron metabolism and haem synthesis by inhibiting enzymes such as ferrocheletase and δ-aminolaevulinic acid synthetase.[146,147] Additionally, analyses of urinary mutagenicity and genetic aberrations in peripheral lymphocytes in lead workers have revealed mutagenicity, genotoxicity and chromosomal damage in the absence of tumours.[58,148–151] Increased overall mortality does occur in lead-related professions,[152–155] and water drawn through leaded pipes is a cause of increased childhood fatality.[156] Published epidemiological and case studies show that lead-induced cancer does occur, albeit that the incidence of brain, lung, gastrointestinal and renal tumours varies greatly according to working practice, duration of exposure, age, genotype, race and lifestyle, and the presence of other toxic contaminants in the environment. No evidence is seen to confirm the results of limited animal studies that lead, in any form, is a transplacental carcinogen.

References

1. D. M. Settle and C. C. Patterson, *Science*, 1980, **207**, 1167.
2. US Department of Health and Human Services (US DHHS), *Report on Carcinogens: Background Document for Lead and Lead Compounds*, US DHHS, Research Triangle Park, NC, 2003.
3. M. J. Brown and S. Margolis, *Lead in Drinking Water and Human Blood Levels in the United States*, Centers for Disease Control and Prevention (CDC), *CDC, Supplements*, 2012, **61**, 1.

4. A. R. Flegal and D. R. Smith, *Rev. Environ. Contam. Toxicol.*, 1995, **143**, 1.
5. I. Thornton, *Metals in the Global Environment*, International Council on Metals and the Environment, Ontario, Canada, 1995.
6. Institute of Mining and Metallurgy, *Lead into the Future*, Proceedings of the Conferenc, Buxton, UK, 1996.
7. World Health Organization (WHO) *Lead in Drinking Water*, WHO, Geneva, 2011, WHO/SDE/WSH/03.04/09/Rev/1.
8. M. C. Navarro, C. Pérez-Sirvent, M. J. Martinez-Sánchez, J. Vidal and J. Marimón, *Chemosphere*, 2006, **63**, 484.
9. L. Williams, *Sci. Total Environ.*, 1990, **93**, 1.
10. F. A. Fry, J. L. Smith-Briggs and M. C. O'Riordan, *Br. J. Ind. Med.*, 1983, **40**, 58.
11. E. S. Ferri and H. Christiansen, *Publ. Health Rep.*, 1967, **82**, 828.
12. A. M. Massadeh, F. Q. Alai and Q. M. Jaradat, *Environ. Monito. Assess.*, 2005, **104**, 163.
13. A. Sakoda, K. Fulao, A. Kawabe, T. Kataoka, K. Hashimoto and K. Yamaoka, *Radiat. Prot. Dosimetry*, 2012, **150**, 109.
14. H. S. Karagueuzian, C. White, J. Sayre and A. Norman, *Nicotine Tobacco Res.*, 2011, *doi: 10.1093/ntr/ntr145.*
15. S. S. Hecht, *J. Natl. Cancer Inst.*, 1999, **91**, 1194.
16. M. Joffre, *Br. Med. J.*, 2003, **68**, 47.
17. S. C. Sikka, *Endocrine and Hormonal Toxicology*, P. W. Harvey, K. C. Rush and A. Cockburn, *Wiley and Sons*, Chichester, 1999, p. 91.
18. W. Wynant, J. Siemiatycki, M.-E. Parent and M.-C. Rousseau, *Occup. Environ. Med.*, 2012, **10**, 1136.
19. S. Alam, J. Hammill and K. Karta, *Srinagarind Med. J.*, 1996, **11**, 101.
20. F. N. Marzulli, P. M. Watlington and H. I. Maibach, *Curr. Pobl. Dermatol.*, 1978, **7**, 196.
21. W. C. Bress and J. H. Bidanset, *Vet. Human Toxicol.*, 1991, **33**, 212.
22. M. Wilhelm, D. Hafner, I. Lombeck and F. K. Ohnnesorge, *Sci. Total Environ.*, 1991, **103**, 199.
23. H. A. Ragan, *Sci. Total Environ.*, 1983, **28**, 317.
24. P. R. Flanagan, M. J. Chamberlain and L. S. Valberg, *Am. J. Clin. Nutr.*, 1982, **36**, 823.
25. Agency for Toxic Substances and Disease Registry (ATSDR), *Case Studies in Environmental Medicine*, ATSDR, Atlanta, GA, 2007.
26. World Health Organization, International Programme on Chemical Safety (IPCS), *Inorganic Lead. Environmental Health Criteria*, WHO, Geneva, 1995, p. 165.
27. C. R. Angle and M. S. McIntire, *J. Toxicol. Environ. Health*, 1979, **5**, 855.
28. S. J. DeMichele, *Comp. Biochem. Physiol.*, 1984, **78**, 410.
29. S. Nogué and A. Culla, *N. Eng. J. Med*, 2006, **364**, e21.
30. B. T. Emmerson and B. R. Thiele, *Med. J. Austr.*, 1960, **1**, 243.
31. J. G. Pounds and J. F. Rosen, *Toxicol. Appl. Pharmacol.*, 1986, **83**, 531.
32. E. Tiffany-Castiglioni, *Neurotoxicology*, 1993, **14**, 513.

33. E. K. Silbergeld, M. P. Waalkes and J. M. Rice, *Am. J. Ind. Med.*, 2000, **38**, 316.
34. R. Eisenstein and S. Kawanoue, *Am. J. Pathol.*, 1975, **80**, 309.
35. P. S. I. Barry, *Br. J. Ind. Med.*, 1975, **32**, 119.
36. Centers for Disease Control and Prevention (CDCP), *Advisory Committee Recommendations: Lead in Children's Products*, CDCP, Atlanta, GA, 2009.
37. Swedish Council of Environmental Information (EIS), *Workshop on the Utility of an Evaluated Data Bank in Teratology and its Use for a State-of-the-Arts Report on Adverse Effects of Lead on Reproduction and Development*, Nynäshamn, Sweden, 1977.
38. A. J. Cohen and F. J. C. Roe, *Food Chem. Toxicol.*, 1991, **29**, 485.
39. F. L. Filon, M. Boeniger, G. Adami, P. Spinellli and A. Damian, *J. Occup. Epidemiol. Med.*, 2006, **48**, 692.
40. T. B. Griffin, F. Coulston, H. Willis and J. C. Russell, *J. Clin. Invest.*, 1975, **2**, 152.
41. H. J. Church, J. P. Day, R. A. Braithwaite and S. S. Brown, *J. Inorgan. Biochem.*, 1993, **49**, 55.
42. R. Goyer, C. D. Klaassen and M. P. Waalkes, *Metal Toxicology*, Academic Press, San Diego, p. 995.
43. D. D. Ulmer and B. L. Vallee, *Trace Substances in Environmental Health*, D. D. Hempshill, Columbia University, Missouri, 1969, vol. II, p. 7.
44. A. Dubreuil, E. Hollande, G. Bouley and Cl. Boudene, *Toxicology*, 1979, **13**, 24.
45. S. Hajem, T. Moreau, P. Hannaert, J. Lellouch, G. Huel, G. Hellier, G. Orssaud, J. R. Claude, B. Jaguet and B. Festy, *Environ. Res.*, 1990, **53**, 105.
46. J. P. Wise, J. C. Leonard and S. R. Patierno, *Mutat. Res.*, 1992, **278**, 69.
47. R. A. Goyer and B. C. Rhynr, *Int. Rev. Exp. Pathol.*, 1973, **12**, 1.
48. M. Valentino, M. Govern, I. Marchiseppe and I. Viscona, *Arch. Toxicol.*, 1991, **65**, 685.
49. E. Rossi, S. Taketani and P. Garcia-Webb, *Biomed. Chromatogr.*, 1993, **7**, 1.
50. J. T. Zelikoff, E. Parsons and R. B. Schliesinger, *Environ. Res.*, 1993, **62**, 20.
51. J. Hsiang and E. Diaz, *Curr. Neurobiol.*, 2011, **2**, 35.
52. M. L. Queiroz, M. Almeida, M. I. Gallao and N. F. Hoehr, *Pharmacol. Toxicol.*, 1993, **72**, 73.
53. J. Mäki-Paakkanen, M. Sorsa and H. Vainio, *Hereditas*, 1981, **94**, 269.
54. N. K. Roy and T. G. Rossman, *Mutat. Res.*, 1992, **298**, 97.
55. A. O. Maslat and H. J. Haas, *J. Trace Elem. Electrolytes Health Dis.*, 1989, **3**, 187.
56. E. R. Nestmann, T. I. Matula, G. R. Douglas, K. C. Bora and D. J. Kowbel, *Mutat. Res.*, 1979, **66**, 357.
57. P. Grover, P. V. Rekhadevi, K. Danadevi, S. B. Vuyyuri, M. Mahboob and M. F. Rahman, *Int. J. Hyg. Environ. Health*, 2010, **213**, 99.

58. P. Apostoli, R. Leone, S. Porru, M. E. Fracasso and L. Alessio, *Mutat. Res.*, 1989, **222**, 245.
59. D. S. Klauder and H. G. Petering, *J. Nutr.*, 1977, **107**, 1779.
60. D. J. White, *Br. J. Exp. Pathol.*, 1977, **58**, 101.
61. H. M. Mykkanen, J. W. T. Dickerson and M. C. Lancaster, *Toxicol. Appl. Pharmacol.*, 1979, **51**, 447.
62. B. A. Fowler, C. A. Kimmel, J. S. Woods, E. E. McConnell and L. D. Grant, *Toxicol. Appl. Pharmacol.*, 1980, **56**, 59.
63. J. P. Bressler and G. W. Goldstein, *Biochem. Toxicol.*, 1991, **41**, 479.
64. Y. M. Hwang, I. N. Sunwoo, I. H. Chung and B. Jung, *J. Korean Med. Sci.*, 1989, **4**, 185.
65. F. Monton and F. Coria, *Neuropathol. App. Neurobiol.*, 1991, **17**, 231.
66. H. E. Gruber, Y. Ding, A. A. Stasky, M. Meyer, M. R. Pandian, D. Pandian, N. D. Vaziri, J. Grigsby and H. C. Gonick, *Miner. Electrolyte Metab.*, 1999, **25**, 143.
67. J. D. Hamilton and E. O'Flaherty, *Toxicol. Sci.*, 1970, **26**, 265.
68. International Agency for Research on Cancer (IARC), *IARC Monographs on the Evaluation of Carcinogenic Risk to Humans*, IARC, Lyons, France, 1987, Suppl. 7.
69. IARC, *Monographs on the Evaluation of Carcinogenic Risks to Humans*, WHO/IARC, Lyons, France, 2006, vol. 87.
70. H. V. Zollinger, *Virchows Arch. Pat. Anat. Physiol.*, 1953, **323**, 694.
71. G. J. Van Esch, H. Van Genderen and H. H. Vink, *Br. J. Cancer*, 1962, **16**, 289.
72. E. Boyland, C. E. Dukes, P. L. Grover and B. C. Mitchley, *Br. J. Cancer*, 1962, **16**, 283.
73. F. J. C. Roe, E. Boyland, C. E. Dukes and B. C. V. Michley, *Br. J. Cancer*, 1965, **19**, 860.
74. A. Furst, M. Schlauder and D. P. Sasmore, *Cancer Res.*, 1976, **36**, 1779.
75. L. A. Poirier, J. C. Theiss, L. J. Arnold and M. B. Shimkin, *Cancer Res.*, 1984, **44**, 1520.
76. K. S. Kasprzyak, K. L. Hoover and L. A. Poirier, *Carcinogenesis*, 1985, **6**, 279.
77. E. J. Tokar, B. A. Diwan and M. P. Waalkes, *Toxicology*, 2010, **276**, 5.
78. A. B. G. Lansdown, *Wound Rep. Regen.*, 2002, **10**, 130.
79. M. P. Waalkes, J. Liu, R. A. Goyer and B. A. Diwan, *Cancer Res.*, 2004, **64**, 7766.
80. S. S. Epstein and N. Mantel, *Experientia*, 1968, **24**, 580.
81. V. H. Ferm and S. J. Carpenter, *Exp. Molec. Pathol.*, 1967, **7**, 208.
82. R. Jabeen, M. Tahir and S. Wagas, *J. Ayub. Med. Coll. Abbottbad*, 2010, **22**, 76.
83. M. P. Waalkes, B. D. Diwan, J. M. Ward, D. E. Devor and R. A. Goyer, *Cancer Res.*, 1995, **55**, 5265.
84. S. Dey, J. Arun, M. Das, C. R. Bhatterjee and P. S. Dkhar, *Cytobios*, 2001, **106**(Supp. 2), 245.

85. A. Anttila, P. Heikkilä, E. Pukkala, E. Nykyri, T. Kauppinen, S. Hernberg and K. Hemminki, *Scand. J. Work Environ. Health*, 1995, **21**, 460.
86. I. Dingwall-Fordyce and R. E. Lane, *Br. J. Ind. Med.*, 1963, **20**, 313.
87. O. Wong and F. Harris, *Am. J. Ind. Med.*, 2000, **38**, 255.
88. US DHHS, *Report on Carcinogens, Background Document for Lead and Lead Compounds*, US DHHS, Research Triangle, Park, NC, 2003.
89. L. Jecklin, *Schweiz Med. Wochenschr.*, 1956, **86**, 891.
90. L. J. Jecklin, *Audio, Trans. IRE Prof.*, 1956, **14**, 626.
91. L. J. Jecklin, *Schweiz. Med. Wochenschr.*, 1955, **85**, 685.
92. H. Fu and P. Boffetta, *Occup. Environ. Med.*, 1995, **52**, 73.
93. P. Cocco, M. Dosemeci and E. F. Heineman, *J. Occup. Environ. Med.*, 1998, **40**, 937.
94. K. Steenland and P. Boffetta, *Am. J. Ind. Med.*, 2000, **38**, 295.
95. E. van Wijngaarden and M. Dosemeci, *Int. J. Cancer*, 2006, **119**, 1136.
96. L. Jarup, *Br. Med. J.*, 2003, **68**, 167.
97. National Academy of Sciences/National Research Council, *Measuring Lead Exposure in Infants, Children, and Other Sensitive Populations*, National Academy Press, Washington, DC, 1993.
98. P. Devey and L. Jingda, *Austr. J. Publ. Health*, 1995, **19**, 198.
99. J. T. Zelickoff, E. Parsons and R. B. Schlesinger, *Environ. Res.*, 1993, **62**, 207.
100. C. R. de Vries, P. Ingram, S. R. Walker, R. W. Linton, W. F. Gutnecht and J. D. Shelburne, *Lab. Invest.*, 1983, **48**, 35.
101. E. J. Kaminski, C. A. Fischer, G. L. Kennedy and J. C. Calandra, *Br. J. Exp. Pathol.*, 1977, **58**, 9.
102. M. C. Rousseau, M.-E. Parent, L. Naden, B. Latreille and J. Siemiatycki, *Am J. Epidemiol.*, 2007, **166**, 1005.
103. J. M. Davies, *Br. J. Ind. Med.*, 1984, **41**, 158.
104. A. Sheffet, I. Thind, A. M. Miller and D. B. Lourie, *Arch. Environ. Health*, 1982, **37**, 44.
105. A. M. Urbano, C. F. D. Rodrigues and M. C. Alpoim, *Gene Ther. Mol. Biol.*, 2008, **12**, 219.
106. V. Englyst, N.-G. Lundström, L. Gerhardsson, L. Pylander and G. Nordberg, *Sci. Total Environ.*, 2001, **273**, 77.
107. N.-G. Lundström, V. Englyst, L. Gerhardsson, J. Taiyi and G. Nordberg, *J. Occup. Environ. Med.*, 2006, **48**, 376.
108. P. Cocco, F Hua, P. Boffetta, P. Carta, C. Flore, V. Flore, A. Onnis, G. F. Picchari and D. Colin, *Scand. J. Work Environ. Health*, 1997, **23**, 15.
109. J. S. Neuberger and J. G. Holowell, *Sci. Total Environ.*, 1982, **25**, 287.
110. T. V. Lam, P. Agovino, X. Niu and L. Roché, *Sci. Total Envion.*, 2007, **372**, 455.
111. P. Cocco, M. H. Ward and M. Dosemeci, *J. Occup. Environ. Health*, 1998, **40**, 855.
112. L. Gerhardsson, N.-G. Lundström, G. Nordberg and S. Wall, *Br. J. Ind. Med.*, 1986, **43**, 707.

113. D. Fanning, *Arch. Environ. Health*, 1988, **43**, 247.
114. P. Cocco, M. H. Ward and M. Dosemeçi, *J. Occup. Environ. Health*, 1999, **56**, 781.
115. W. C. Cooper, *Ann. N.Y. Acad. Sci.*, 1976, **271**, 250.
116. W. C. Cooper, O. Wong and L. Kheifets, *Scand, J. Work. Environ. Health*, 1985, **111**, 331.
117. B. Pesch, J. Haerting, U. Ranft, A. Klimpel and B. Oelshlägel, *Int. J. Epidemiol*, 2000, **29**, 1014.
118. J. Hu, J. Little, T. Xu, X. Zhao, L. Guo, X. Jia, G. Huang, D. Bi and R. Liu, *Int. J. Cancer*, 1999, **83**, 299.
119. P. Rajaraman, P. A. Stewart, J. M. Samet, B. S. Schwartz, M. S. Linet, S. H. Zahm, N. Rothman, M. Yeager, H. A. Fine, P. M. Black, J. Loeffler, W. R. Shapiro, R. G. Selker and P. D. Inskip, *Cancer Epidemiol. Biomarkers Prev.*, 2006, **15**, 2514.
120. W. E. Fayerweather, M. E. Karns, I. A. Nuwayhid and T J. Nelson, *Am. J. Ind. Med.*, 1997, **31**, 28.
121. G. Wingren and V. Englander, *Int. Arch. Occup. Environ. Health*, 1990, **62**, 253.
122. H. A. Schreier, N. Sherry and E. Shaughnessy, *Ann. Neurol.*, 1977, **1**, 599.
123. P. Haeffliger, M. Mattieu-Nolf, S. Lociciro, C. Ndiaye, M. Coly, A. Diouf, A. L. Faye, A. Sow, J. Tempowski, J. Pronczuk, A. P. Filipe Jr, A. P. Bertolini and M. Neira, *Environ. Health Perspect.*, 2009, **117**, 1535.
124. A. Navas-Acien, M. Tellez-Plaza, E. Guallar, P. l. Muntner, E. Silbergeld, B. Jaar and V. Weaver, *Am. J. Epidemiol.*, 2009, **170**, 1156.
125. E. B. Ekong, B. G. Jaar and V. M. Weaver, *Kidney Internat.*, 2006, **70**, 2074.
126. E. L. Baker, P. J. Landrigan, A. G. Barbour, D. H. Cox, D. S. Folland, R. N. Ligo and J. Throckmorton, *Br. J. Ind. Med.*, 1979, **36**, 314.
127. R. Lilis, *Am. J. Ind. Med.*, 1981, **2**, 293.
128. E. J. Calabrese and L. A. Baldwin, *Drug Metab. Rev.*, 1992, **24**, 409.
129. J. S. Colt and A. Blair, *Environ. Health Perspect.*, 1998, **106**(Suppl. 3), 909.
130. A. F. Kantor, M. G. Curnen, J. W. Meigs and J. T. Flannery, *J. Epidemiol. Comm. Health*, 1979, **33**, 253.
131. J. R. Wilkins and T. H. Sinks, *J. Epidemiol. Comm. Health*, 1984, **38**, 7.
132. C. R. Sharpe and E. L. Franco, *Epidemiol. Rev.*, 1995, **17**, 415.
133. National Cancer Institute (NCI), *Wilms Tumour and Other Childhood Kidney Tumours: Treatment (PDQ®)*, NCI, US DHHS, Washington, DC, 2012.
134. D. A. Savitz and J. Chen, *Environ. Health Perspect.*, 1990, **88**, 32.
135. T. I. Lidsky and J. S. Schneider, *Brain*, 2003, **126**, 5.
136. S. Cairney, P. Maruff, C. B. Burns, J. Currie and B. J. Currie, *J. Neurol. Neurosurg. Psychiatry*, 2004, **75**, 472.
137. P. Bhatti, P. A. Stewart, A. Hutchison, N. Rothman, M. S. Linet, P. D. Inskip and P. Rajarman, *Cancer Epidemiol. Biomark. Prev.*, 2009, **18**, 1841.

138. W. T. Wu, Y. J. Lin, S. H. Liou, C. Y. Yang, K. F. Cheng, P. J. Tsai and T. N. Wu, *Environ. Int.*, 2012, **40**, 97.
139. W. C. Cooper, *Environ. Health Pespect.*, 1988, **78**, 61.
140. M. H. Sweeney, J. J. Beaumont, R. J. Waxweiler and W. E. Halperin, *Arch. Environ. Health*, 1986, **1**, 23.
141. H. Vainio, *Scand. J. Work Environ. Health*, 1997, **23**, 1.
142. K. Tollestrup, J. R. Daling and J. Allard, *Arch. Environ. Health*, 1995, **50**, 221.
143. A. Forni, A. Sciame, P. S. Bertazzi and L. Alessio, *Arch. Environ. Health*, 1980, **35**, 139.
144. H. Needleman, *Ann. Rev. Med.*, 2004, **55**, 209.
145. B. Maruff, C. D. Burns, P. Tyler, B. J. Currie and J. Currie, *Brain*, 1998, **121**, 1903.
146. S. Yadev and V. Sharma, *Int. J. Pharma Bio-Sci.*, 2011, **2**, P-70.
147. T. Sakai, *Ind. Health*, 2000, **38**, 127.
148. N. B. Jain, F. Laden, U. Guller, A. Shankar, S. Kazani and E. Garshick, *Am. J. Epidemiol.*, 2005, **161**, 968.
149. P. J. Landrigan and A. C. Todd, *West Med. J.*, 1994, **161**, 153.
150. A. Pasha Shaik, S. Sankar, S. C. Reddy, P. G. Das and K. Jamil, *Drug Chem. Toxicol.*, 2006, **29**, 111.
151. M. Stoia, S. Oancea and D. C. Obreja, *Rom. J. Leg. Med.*, 2009, **4**, 287.
152. S. G. Seelvan, P. L. Landrigan, F. B. Stern and J. H. Jones, *Am. J. Epidemiol.*, 1985, **122**, 673.
153. A. Spivey, *Environ. Health Perspect.*, 2007, **115**, A30.
154. D. Michaels, S. R. Zoloth and F. B. Stern, *Int. J. Epidemiol.*, 1991, **20**, 978.
155. W. C. Cooper and W. R. Gaffey, *J. Occup. Med.*, 1975, **17**, 100.
156. National Bureau of Economic Research, *Lead Pipes and Childhood Mortality*, ed. K. Clay, W. Troesken and M. Haines, NBER Working Paper Series, Cambridge, MA, 2006.

CHAPTER 12

Tungsten (Wolfram) and Hard Metals

12.1 Introduction

Tungsten is a hard metal with the highest tensile strength and melting point of all metals. The name is derived from the Swedish words "*tung*" meaning heavy and "*sten*" for stone. The history of tungsten goes back to the 17th century when miners in the Erz Mountains of Saxony-Bohemia noticed that certain ores disturbed the reduction of cassiterite (tin dioxide, SnO_2, the primary ore of tin found in hydrothermal veins and pegmatites associated with granite) and induced slagging. These early tin miners noticed that the ores contained less tin and gave the unrecognised ore German nicknames such as *wolfert* or *wolfrahm* (wolf froth), implying that it "devoured tin like wolves". In modern chemistry, tungsten takes the chemical symbol "W" from the German rather than the Swedish name. Tungsten does not occur naturally in its metallic form, but is found mainly as tungstates in combination with iron and manganese. Scheelite ($CaWO_4$, calcium tungstate) and wolframite ($Fe,MnWO_4$, iron–manganese tungstate), ferberite ($FeWO_4$, iron tungstate) and huebnerite ($MnWO_4$, manganese tungstate) are the principal ores but at least 20 others exist in some parts of the world. Tungsten ores constitute a minor part of the Earth's crust, possibly <1.3 ppm. Major deposits are found in China, Canada, Russia and Australia but a significant amount of tungsten used in industry is recovered through recycling of scrap tungsten products. In the USA, recycled tungsten accounts for nearly one-third of the tungsten required annually. Major production of tungsten concentrates occurs in Austria, Bolivia, Canada, China, Portugal, and Russia.

Tungsten is classified as a transitional element in Group VI (Period 6) of the Periodic Table, with chromium, molybdenum and uranium. It is a silvery white,

Issues in Toxicology No. 18
The Carcinogenicity of Metals: Human Risk through Occupational and Environmental Exposure
By Alan B. G. Lansdown

Published by the Royal Society of Chemistry, www.rsc.org

brittle and stable metallic element at room temperature, but at high temperatures it becomes ductile and can be sawn, spun, drawn and forged. It has excellent corrosion resistance and is minimally attacked by most inorganic acids. Thirty-five isotopes are now known, with five stable forms occurring naturally. Tungsten does not react with air, oxygen or water at room temperature, although it oxidises at higher temperature to yield the trioxide (WO_3). Tungsten residues and dusts are released into the atmosphere through erosion of rocks and natural deposits, wind action and human activity. The latter include mining and smelting, discard and combustion of industrial and domestic waste, engineering and manufacture of electrical components and chemical syntheses. Manufacture and use of the so-called "hard metals" for cutting, drilling and grinding have become the subject of toxicological concern in recent years and exposure to cobalt–tungsten carbide powders as a cause of interstitial lung disease and possible lung cancer is subject to intense scrutiny.[3,4]

Tungsten is found in the human body at low levels but has no recognised trace metal value, despite some biochemical studies suggesting that it might function as a cofactor in a small number of enzymes, in a similar fashion to molybdenum.[5] It interacts with and can displace molybdenum from essential enzymes as part of its toxicity profile.[1,2] It occurs to a small extent in most food chains although levels are much higher in the areas of mining and waste deposits where residues accumulate in plants and food animals. Main routes of tungsten uptake are through ingestion with food and drinking water, and through inhalation. Some residues will be washed into well water and inland waterways but the greatest human risk lies in inhalation of tungsten in dusts and fumes. Human exposure to tungsten residues is conveniently monitored by blood and urine concentrations. These are normally low but can rise dramatically in workers engaged in tungsten-related industry or those living in the region of tungsten mines and smelting works.

The toxicology of tungsten and most of its compounds in humans was for many years regarded as being low under expected conditions of exposure.[6] Tungsten powder increases the acidity of soil and is toxic to bacteria, some plants and lower animals. Tungstates entering water courses are moderately toxic to fish.[7] However, increasing concern is shown by regulatory authorities including the International Agency for Research on Cancer (IARC) and the US National Toxicology Program on toxic and potential carcinogenic risks posed by exposure to tungsten carbide in manufacturing industry and in tools and powders used in grinding, cutting and drilling works.[4,8] Tungsten monocarbide (WMC; $W{\equiv}C^{-}$) is a dense, light bluish-grey, metallic material prepared by heating powdered tungsten with carbon black in the presence of hydrogen at 1400–1600 °C.[9]

WMC is an important member of the class of inorganic compounds of carbon (carbides), used alone or with 6–20% of other metals to impart hardness to cast iron, cutting edges of saws and drills, and penetrating cores of armour-piercing projectiles. As cemented carbides, hard metal mixtures comprising one or more of the finely divided carbides of tungsten, titanium, tantalum and vanadium are embedded in a matrix of cobalt or nickel by sintering. Principal uses are in cutting tools, where for many applications they have replaced conventional high-speed steels. Nemery and Abraham, in an effort to establish the relationship between inhalation of hard metal dusts and development of the so-called "hard metal lung disease", emphasised that hard metals should not be confused with heavy metals and are produced by compacting powdered WMC, cobalt and certain other metals into a polycrystalline material in a sintering process to produce sintered or cemented carbides.[3] They have a hardness equivalent to that of diamond and have great value in machine parts and drills that require high heat resistance. In 2009, the National Toxicology Program of the USA invited recommendations on the listing status for cobalt–tungsten carbide and hard metals for inclusion in the 12th Report on Carcinogens on the basis that they may pose a carcinogenic risk to workers exposed occupationally.[10] At about this time, evidence in the hard metal industries suggested a possible carcinogenic effect of cobalt in the presence of WMC, although the findings were inconclusive.[11–15]

12.2 Industrial Use and Human Exposure

Tungsten has been identified as a hard metal since the 18th century when tungsten alloys were used in the production of gun barrels, armaments and ammunition. An early report detailed the case of a French artilleryman who drank wine from his freshly fired gun barrel as part of an induction ritual. The barrel was of a new type made of tungsten alloy. Tungsten reacts slowly with certain acids and presumably sufficient ion eluted into the wine to be absorbed intestinally. The soldier developed signs of acute poisoning, and sustained seizures and renal damage were signs of tungsten poisoning.[16] This suggests that clearing up firing ranges where tungsten has been used in ammunition will constitute a health risk. Tungsten in shrapnel from tank shells is recognised now as a health hazard for military personnel, although the risks of intramuscular sarcomas indicated by experimental studies in rats remain to be substantiated.[17,18] The so-called "green training ammunition" introduced by the US army contains tungsten and nylon or tungsten and tin alloys, but so far no carcinogenic risk has been identified. Today, high-density tungsten alloys are still proving their worth in ordnance applications around the world.[19]

The unique physico-chemical properties of tungsten, its alloys and inorganic compounds have provided a wide range of applications in industrial and commercial environments, notably in metallurgy generally, production of tough heat-resistant alloys, electronics, engineering, aviation equipment, chemical syntheses and medicine (Table 12.1).[20] Few of these have been adequately investigated toxicologically and patterns of tungsten absorption by the human body are poorly understood. Available information indicates that main routes of uptake are through ingestion of soluble tungstates or through inhalation. Some of that absorbed accumulates in bone, lung, liver, muscle, kidney and spleen.

Tungsten and tungsten alloys have high tensile strength and are widely used in electrical and engineering industries, X-ray targets in medical engineering,

Table 12.1 Industrial and commercial uses of tungsten.

Compound	*Physico-chemical properties*	*Use*
Metallic tungsten	High melting point, sp. gravity 19.3, ductile, low vapour pressure	**Electronics**: electric lamp filaments, television tubes, electrical contacts, high performance switches, wires, coils, incandescent lamps, electrodes, integrated circuits, heating elements for electrical furnaces, electrical contacts in automobile distributors **Medical engineering**: X-ray targets, interventional radiology **Engineering**: glass to metal seals
Tungsten disulfide	Dark grey crystalline solid with high melting point	**Engineering**: dry film lubricant
Tungsten monocarbide (WMC)	Exceedingly hard	**Hard metals**: cutting, drilling, grinding tools, abrasive powders, cemented carbides, heavy duty machinery heat resistant parts, mining tools, military equipment, ammunition, armaments, high speed steels
Tungsten oxide (WO_3)	Amorphous, transparent and crystalline powder	**Pigments**: paints and ceramics, bronzes **Fire proofing**: fabrics Chemical catalysts, petroleum industry
Tungsten alloys	Strong and durable, heat resistant, light weight	**Engineering**: aviation materials **Sports equipment**: golf clubs
Tungstates		Tanning
Calcium and magnesium tungstates		Fluorescent lighting
Tungsten phosphate	A new class of quasi-low-dimensional materials which exhibit charge density wave instabilities	Glass manufacture, bronzes

and hard-metal industries including high speed steels, military products and aviation components. Hard metal is a most important usage of tungsten, its main constituent being WMC with hardness comparable to that of diamond. WMC particles provide hard and abrasion-resistant constituents which are bonded with tough, comparatively soft, metallic cobalt (Co). The structure of two-phase WMC–Co hard metals is conventionally defined by key parameters, including:

- Grain size of the WMC phase
- Co binder (grain size/shape)
- Volume (or wt) fraction and composition of the Co binder phase.

Increased levels of tungsten in blood, urine and faeces, following industrial or environmental exposure, provide tangible evidence that the element is absorbed into the human body.[21–23] Absorption patterns are not known but, in view of molecular similarities with molybdenum, uptake of tungstates or W(IV) ion is assumed to be through the gastrointestinal mucosa. Workers engaged in grinding operations and exposed to tungsten dusts exhibited urinary levels 900-fold higher than in the general population. In a clinical situation, patients implanted with tungsten coils to treat embolism showed elevated tungsten in blood, urine and hair as an indication of uptake.[24,25] Minimum toxic levels of tungsten in the human body have not been reported. The element has no nutritional value as suggested in some reports. A population survey[26] conducted in the USA in 2003 showed average urine concentration to be 0.083 μg l^{-1}.

For tungsten and insoluble tungsten compounds in air, the National Institute for Occupational Safety and Health (NIOSH)[26] recommended an exposure limit of 5 mg m^{-3} (average over a 10-hour period) and a short-term (15 minutes) exposure limit of 10 mg m^{-3}. The Occupational Safety and Health Administration (OSHA) set limits for tungsten of 5 mg m^{-3} (insoluble compounds) and 1 mg m^{-3} (soluble compounds) for construction and shipyard industries. Environmental concentrations estimated by the European Community were set at a maximum of 1% and it was estimated that use of tungsten from discharged ammunition containing 51–95% would result in 6.5–13.5 mg W kg^{-1} in soil.[27] Environmental tungsten is readily taken up by wild animals, notably waterfowl, occasionally with toxic repercussions.[7]

An important ecotoxicological study on tungsten was set up in 2003 to evaluate environmental pollution in Fallon, Churchill County, Nevada in the close vicinity of tungsten mines and where children experienced statistically significant increases in acute lymphocytic and myelocytic leukaemias.[28] Waters of the Carson River showed tungsten levels of 1.5–23 μg l^{-1}, although in well water concentrations were as high as 300 μg l^{-1}. Correlations between human exposure and urine tungsten levels are illustrated in Table 12.2, which indicates

Table 12.2 Urinary tungsten levels in adults and children living in areas of Nevada with high environmental tungsten contamination.[a]

	Geometric Mean Tungsten (μg/L) (95% confidence interval)			
Location	*Adults*	*Children*	*Total*	*Tap water*
Lovelock	0.38 (0.33–0.45)	0.62 (0.50–0.76)	0.48 (0.34–0.68)	0.11 (0.07–0.19)
Pahrump	0.40 (0.38–0.53)	0.56 (0.48–0.66)	0.51 (0.37–0.69)	0.04 (0.02–0.06)
Yerington	1.04 (0.84–1.30)	1.18 (1.00–1.39)	1.11 (0.97–1.27)	3.32 (1.82–6.04)
Churchill County	0.81 (0.56–1.16)	2.31 (1.66–3.22)	1.19 (0.89–1.59)	4.66 (2.98–7.30)
National Average	Age >20 years 0.07 (0.07–0.08)	Age 6–11 years 0.15 (0.12–0.18) Age 12–19 years 0.10 (0.09–0.12)	0.08 (0.07–0.09)	Not available

[a]Statistics provided by the 2nd National Report on Human Exposure to Environmental Chemicals (2003).

clearly that children are appreciably more vulnerable than adults. The table also shows that mean urinary levels of tungsten were at or above the 95% percentile of the level established by the reference population (0.48 μg l^{-1}). Children living in Churchill County showed a 2.8-fold higher tungsten excretion than adults, which is a cause for concern in relation to leukaemia.[29,30]

12.3 Toxicology and Carcinogenicity of Tungsten and Related Compounds

Very little is published on human exposure to metallic tungsten or inorganic tungstates, but the inherent risks of lung diseases following inhalation of hard metal complexes, alloys and composites containing tungsten are appreciated.[3,12,14,15] The mechanisms of uptake, metabolism and protein binding of tungsten are minimally understood. Biochemical similarities between tungsten and molybdenum in certain bacteria, and the ability of tungsten to replace molybdenum in certain enzyme systems, suggest that uptake and metabolism of the two elements will be similar in humans even though no information has been seen to substantiate this view.[31–33]

Van der Voet *et al.* concluded that exposure to tungsten and tungsten compounds presented a small toxic risk to humans, although some long-term hazard might be associated with internalisation of shrapnel and certain tungsten-containing alloys.[19] Isolated cases are reported in military medicine where shrapnel containing tungsten alloys were thought to present carcinogenic risk but these are unconfirmed in human injury victims.[27] A statistically sound experimental study showed that weapons grade tungsten alloy shrapnel implanted intramuscularly in rats induced aggressive rhabdomyosarcomas with metastases following within 4–5 months.[18] Nickel implants also induced tumours in this study as a positive control, whereas metallic tantalum served as negative control. In contrast, when biodegradable tungsten coils were implanted into the subclavian artery in rabbits, no local or systemic toxicity or carcinogenic changes were observed within 4 months, yet serum tungsten levels were increased.[34] No evidence has been seen to implicate tungsten or WMC as a tumour "initiator" in humans or experimental animals. However, an experimental study in Sprague–Dawley rats treated the experimental carcinogen *N*-nitroso-*N*-methyl urea (MNU) provided evidence that dietary administration of metallic tungsten "promoted" mammary carcinogenesis.[35] The mechanism is not known but low levels of molybdenum were inhibitory to tumour formation. This further illustrates the interaction between the two metals.

Local increases in childhood myelocytic and lymphocytic leukaemias observed in the region of tungsten mines in Nevada present perhaps the strongest evidence that tungsten may be a human carcinogen.[29,30] Tungsten levels in the trees of the town of Fallon in 1994 were seen to increase whilst those in neighbouring towns remained the same. From 1997, 17 cases of leukaemia were diagnosed in children living in the Fallon area for some time prior to diagnosis. However, the reports also emphasise the fact that cobalt concentrations in the area had been high for at least 15 years, thereby making a definite aetiological diagnosis impossible. No information has been seen in experimental animals of the ability of tungsten or its inorganic compounds to evoke haematological change. Further research is clearly required.

Inhalation of dust containing WMC and hard metals is implicated as a cause of pneumoconiosis, and the so-called "hard metal lung disease" in workers.[3,36] There is little evidence at the moment that inhalation of tungsten dust in humans is a cause of respiratory disease but an acute study in mice did show that calcium tungstate (250 μg in saline) when instilled intratracheally was a cause of interstitial pneumonitis and alveolar leucocyte infiltrations. These changes occurred at the periphery of bronchiolar walls within 1–21 days of treatment but no pre-carcinogenic changes were reported.[37] This study emphasised the protective role of alveolar macrophages in sequestering tungstate particles and mitigating local toxicity.

Great caution is advised in identifying tungsten *per se* as a toxic agent in human case studies or experimentally in the presence of cobalt, nickel and certain other elements used in alloys and hard metals. Pulmonary fibrosis and interstitial pneumonitis as key features of hard metal disease are more consistent with the effects of inhaled cobalt in the vast majority of reported cases.[3] In a similar way, dermatitis, occasionally reported in hard metal workers as an occupational risk, bears characteristic symptoms of cobalt dermatitis. No evidence has been seen to implicate tungsten as a cause of allergy or contact dermatitis.

In vitro studies provide some evidence that metallic tungsten and WMC are cytotoxic and clastogenic in certain cell lines but no evidence of bacterial mutagenicity for tungsten or WMC has been seen. *Salmonella typhimurium* and *E. coli* strains exposed to UV light emitted by halogen tungsten lamps exhibited significant genotoxic change and dose-related DNA damage.[38] Tungsten nanoparticles were cytotoxic to cultured mammalian lung, skin and colonic cell lines and this effect was accentuated in the presence of Co(II).[39] In micronucleus tests, Comet assays showed that WMC alone did not induce a dose-dependent DNA breakage effect, but was active in the presence of Co(II).[40] WMC with Co(II) induced greater levels of DNA damage in cultured human leucocytes and production of hydroxyl radicals than cobalt alone, showing that the clastogenic effect of cobalt was enhanced by the presence of WMC. Other evidence of interaction between tungsten and cobalt, nickel or iron is provided by experiments demonstrating the neoplastic transformation of human osteoblasts.[41] Miller *et al.* claimed that hard metal mixtures of tungsten, cobalt, nickel and tungsten, or nickel and iron, which mimic alloys used in military applications, evoke malignant transformation in immortalised human osteoblasts. Their experiments point to a transformation involving direct DNA damage, formation of micronuclei and chromosomal breaks. Mixtures of tungsten, nickel and cobalt (WNiCo) and tungsten, nickel and iron (WNiFe) led to 8.9- and 9.5-fold increases in cell transformation respectively, whereas nickel sulfide alone induced DNA damage or chromosomal aberration leading to a 7.7-fold increase in cell transformation.

The genotoxicity of Co–WMC mixtures was further investigated in cultured human lymphocytes and in rat type II pneumocytes exposed to a single intratracheal administration of Co–WMC dust.[42–44] In these experiments, cellular apoptosis, with DNA breaks, chromosomal damage and mutations exhibited a dose-dependent and time-related trend. Alkaline Comet assay of cells obtained after bronchiolar lavage and peripheral blood mononuclear cells showed a reduction in protein content and epidermal cell count with increased lactic dehydrogenase activity. In contrast, Ding *et al.* compared the ability of nano- and fine-sized Co–WMC particles to form free radicals and to activate transcription factors and mitogen-activated protein kinase in mouse epidermal cell lines.[45] The nanoparticles generated higher levels of oxygen radicals and greater oxidative stress, as shown by decreased glutathione, and faster cell growth and proliferation in the JB6 P(+) cell line.

12.4 Epidemiological Evidence for the Carcinogenicity of Hard Metals

The carcinogenicity of hard metal complexes and alloys used in cutting tools, grinding and sharpening instruments, and medical prostheses is exceedingly complex. Wide variations exist in the composition of metals used, their applications and patterns of human exposure. Hard metal complexes such as cobalt–tungsten carbide (Co–WMC) are preferred for cutting tools, and grinding and sharpening instruments, but these frequently contain small unspecified amounts of other carbides for which negligible toxicity data are available. No epidemiological studies are published to substantiate a carcinogenic risk through topical exposure or by parenteral implantation of hard metal composites, but there is increasing concern in the USA and Europe that metal ion release, leading to high serum levels of chromium, cobalt and other ions released from metal-on-metal implants, presents a potential risk of ill-health.[46,47] In their monograph on cobalt in hard metals, the IARC discussed the merits of five epidemiological studies conducted in the early 1990s in France (four) and Sweden (one), including men and women involved in different aspects of the industry.[48]

In the Swedish study, Hogstedt and Alexandersson reported on 3163 male workers, each with at least 1 year of occupational exposure at hard metal manufacturing plants in Sweden during 1940–1982 and followed from 1951 to 1982.[49] Exposures included a number of other substances used in the production of hard metal, such as tungsten carbide. The lung cancer risk was 1.34 (95% CI = 0.77–2.13) whereas the all-cause mortality was slightly less than unity. Among workers with more than 10 years of employment and more than 20 years since first exposure, a significant excess of lung cancer mortality was observed (SMR = 2.78, 95% CI = 1.11–5.72). Smoking habits among hard metal workers were reported to be similar to those of the male Swedish population. The job specifications and exposure patterns of the workers in the cohort are not known from the hospital or general practitioner records presented. The four French studies reported by Moulin and Wild[12–15] used a job exposure matrix to assess occupational exposures to cobalt and tungsten carbide in 10 factories producing hard metals from 1945 to 1994. In their initial report, they observed no increased mortality overall. However, the incidence of 10 lung cancer cases amongst 709 workers with at least one year of employment was regarded as significant, but unrelated to the duration of employment or time since starting work.[11] Cobalt was used as a "binder" in the hard metals. In the more definitive study, Moulin *et al.* conducted an industry-wide evaluation of cancers and health records of 5777 men and 1682 women employed for at least three months, but employment records were incomplete for three of the ten factories evaluated and some works were not traceable at the conclusion of the study. Standard mortality rates for men and women were lower than expected for all respiratory and cardiovascular diseases but higher than predicted for total malignant neoplasms (247 *vs.* 231.91). Of 247 workers showing malignant neoplasms, 66 of these tumours involved the lungs and pleura, 24 extended to the buccal cavity and pharynx and 19 were oesophageal. Trends in cancer mortality were seen in relation to the duration of work in the industry, allowing for adjustment for the smoking habits. When exposures over the last 10 years were ignored, the authors claimed "a twofold lung cancer risk amongst workers exposed simultaneously to cobalt and WMC". Adjustments for exposure to other potential carcinogens in the working environment did not alter the predictions. Wild *et al.* also reported increased risks of lung cancer mortality in a cohort of hard metal workers who had worked in an industrial site for at least three months.[14,15] They identified a job-related trend with an excess mortality rate significantly higher than expected in workers engaged in hard metal production before sintering and in maintenance workers. Tumour incidences were low but, in the opinion of the clinicians, the raised mortality rates could not be attributed to smoking alone.

12.5 Conclusions

This review has failed to find convincing evidence that tungsten or tungstates are carcinogenic to humans through occupational or environmental exposure. The carcinogenicity of tungsten carbide and cobalt containing hard metal complexes is not unequivocally established, although increased incidence of lung tumours attributed to inhalation of factory dusts in grinding, milling and production work points to an occupational hazard.[11–15] Risks of childhood leukaemia in areas of high environmental tungsten residues will need further examination.[29,30] The carcinogenicity of hard metals is supported by several competent *in vitro* studies indicating the ability of particles to induce genotoxic, clastogenic and epigenetic changes in several different human cell lines. It is notable that in virtually every situation where WMC or hard metal complexes are implicated as causes of carcinogenicity, cobalt is present. The 12th Report on Carcinogens unequivocally lists cobalt sulfate as "reasonably anticipated to be a human carcinogen" from results of experimental animal studies even though available human case and epidemiological substantiation was inadequate.[10]

The conclusion of the US Department of Health and National Toxicology Program that cobalt–tungsten carbide-containing powders and hard metals should be listed as reasonably anticipated human carcinogens is cautioned on the adequacy of published studies.[10] The prediction is based upon the work of two groups of workers, Hogstedt and Alexandersson in Sweden[49] and Wild, Moulin and their co-workers in 10 factories in the French hard metal industries.[12,14,15] The actual number of lung cancer cases in both studies is low and further meta-analyses are required, possibly involving other countries. In the French work a 3.3% cancer mortality rate reported in the 7459 workers assessed is low, and lack of accurate details on worker and site records reduces the statistical power of the observations. A conclusion has to be reached as to whether the risks apply to "hard metal" dusts, being an admixture of carbides of several metals, or are attributable primarily to cobalt exposure.

If cobalt–hard metal dusts are proven clinically and occupationally carcinogenic to workers through chronic inhalation of dust, then *in vitro* studies provide a basis for understanding their mechanisms of action.[41–45] These studies in cultured mammalian cell lines show unequivocally that the dusts are capable of inducing DNA-strand breakages, positive micronucleus tests, genotoxicity and mutagenicity. Further, hard metal and cemented carbide mixtures have been shown to promote formation of oxygen radicals and to activate transcription factors and signalling pathways.[45] The oxygen stress induced in cultured cells was greater with nanocrystalline particles than with fine granules, indicating a surface area-dependent effect.

It is concluded that inhalation of hard metal dusts may constitute an occupational risk of lung cancer, but human evidence available at the moment is insufficient and experimental studies in animal models are unhelpful. If tungsten carbide containing hard metal dust is carcinogenic, further multicentre meta-analyses are required, supported by experimental evidence of mechanisms of action.

References

1. F. M. Girio, J. C. Roseiro and A. I. Silva, *Curr. Microbiol.*, 1998, **36**, 337.
2. R. A. Rothery, B. Stein, M. Solomonson, M. L. Kirk and J. H. Weiner, *Proc. Natl. Acad. Sci.*, 2012, **11**, 14773.
3. B. Nemery and J. L. Abraham, *Am. J. Respir. Crit. Care*, 2007, **176**, 2.
4. National Toxicology Program (NTP), *Report on Carcinogens*, US Department of Health and Human Services (US DHHS), Research Triangle Park, NC, 2009; *Fed. Reg.*, 2009, **74**, 6886.
5. G. B. Seiffert, G. M. Ullmann, A. Messerschmidt, B. Schink and O. Einsle, *Proc. Natl. Acad. Sci.*, 2007, **104**, 3073.
6. R. W. Leggett, *Sci. Total Environ.*, 1997, **206**, 147.
7. N. Strigul, A. Koutsospyros and C. Christodoulatos, *Ecotoxicol. Environ. Safety*, 2010, **73**, 164.

8. M.-C. Rousseau, K. Straif and J. Siemiatycki, IARC Update, *Environ. Health Perspect.*, 2006, **113**, A580.
9. F. Wang and T. C. Steimle, *J. Chem. Phys.*, 2012, **136**(Issue 4), 04431.
10. US Department of Health and Human Services (US DHHS), *Report on Carcinogens,* US DHHS, National Toxicology Program, Research Triangle Park, NC, 2011.
11. G. Lasfargues, P. Wild, J. J. Moulin, B. Hammon, B. Rosmorduc, C. Rondeau du Noyer, M. Lavendier and J. Moline, *Am. J. Ind. Med.*, 1994, **26**, 585.
12. J. J. Moulin, P. Wild, S. Romazini, G. Lasfargues, A. Peltier, C. Bozec, P. Deguerry, F. Pellet and A. Perdrix, *Am. J. Epidemiol.*, 1998, **148**, 241.
13. J. J. Moulin, S. Romazini, G. Lasfargues, A. Peltier, C. Bozec, F. Pellet, P. Wild and A. Perdrix, *Rev. Epidemiol. Sante Publique*, 1997, **45**, 41.
14. P. Wild, A. Perdrix, S. Romazini, J. J. Moulin and F. Peltier, *Occup. Environ. Med.*, 2000, **57**, 568.
15. The MAK Collection for Occupational Health and Safety, *Hard Metal. MAK Value Documentation, 2007,* Wiley-VCH Verlag, Gmbh, 2012, p. 218.
16. P. Marquet, B. François and P. G. Vignon Lechâtre, *Lancet*, 1996, **348**, 1070.
17. N. Seppa, *Sci. News*, 2005, **167**, 188.
18. J. F. Kalinich, C. A. Emond, T. K. Dalton, S. R. Mog, G. D. Coleman, J. E. Kordell, A. C. Miller and D. E. McClain, *Environ. Health Perspect.*, 2005, **113**, 729.
19. G. B. van der Voet, T. I. Todorov, J. A. Centeno, W. Jonas, J. Ives and F. G. Mullick, *Mil. Med.*, 2007, **172**, 1002.
20. M. G. Andrews, ed., *Tungsten: The story of an indispensible metal,* The Tungsten Institute, Washington, DC, 1955.
21. D. C. Paschal, B. G. Ting, J. C. Morrow, J. L. Pirkle, R. J. Jackson and E. J. Sampson, *Environ. Res.*, 1998, **76**, 53.
22. T. Kraus, P. Schramel, K. H. Schaller, P. Zobelein, A. Weber and J. Angerer, *Occup. Envir. Med.*, 2001, **58**, 631.
23. Centers for Disease Control and Prevention (CDC), *Fourth National Report on Human Exposure to Environmental Chemicals, Updated table*, CDC, Atlanta, GA, 2012.
24. M. Bachthaler, M. Lenhart, C. Paetzel, S. Feuerbach, J. Link and C Manke, *Cathet. Cardiovasc. Intervent.*, 2004, **62**, 380.
25. G. Nicolaou, R. Pietra, E. Sabbioni, G. Mosconi, G. Cassina and P. Seghizzi, *J. Trace Elem. Electrolytes Res.*, 1987, **2**, 73.
26. Agency for Toxic Substances and Disease Registry (ATSDR), *Toxicological Profile for Tungsten (Update),* US Department of Public Health and Human Services, Public Health Service ATSDR, Atlanta, GA, 2005.
27. V. G. Thomas, M. J. Roberts and P. T. C. Harrison, *Ecotoxicol. Environ. Safety*, 2009, **72**, 1031.
28. Centers for Disease Control and Prevention (CDC), National Center for Environmental Health, *Churchill Clusters: Churchill County (Fallon) Nevada Exposure Assessment*, CDC, Atlanta, GA, 2003.

29. P. R. Sheppard, R. J. Speakman, G. Ridenour and M. L. Witten, *Environ. Health Perspect.*, 2007, **115**, 715.
30. P. R. Sheppard, R. J. Speakman, G. Ridenour and M. L. Witten, *Environ. Health Perspect.*, 2008, **116**, A196.
31. D. G. Barceloux, *J. Toxicol.-Clin. Toxicol.*, 1999, **37**, 231.
32. R. R. Mendel, *J. Royal Soc. Chem, Dalton Transact.*, 2003, **2005**, 3404.
33. L. E. Bevers, P. L. Hagedoom, G. C. Krijger and W. R. Hagen, *J. Bacteriol.*, 2006, **188**, 6498.
34. M. Peuster, C. Fink, P. Wohlsein, M. Bruegmann, A. Günther, V. Kaese, M. Niemeyer, H. Haferkamp and C. Schrakenburg, *Biomaterials*, 2003, **24**, 393.
35. H. J. Wei, X. M. Luo and X. P. Yang, *Zhonghua Zhong Liu*, 1987, **9**, 204.
36. Centers for Disease Control and Prevention (CDC), *National Biomonitoring Program,* CDC, Atlanta GA, 2012.
37. M. N. Peão, A. P. Aguas, C. M. de Sá and N. R. Grande, *Lung*, 1993, **171**, 187.
38. A. Camoirano, C. Bennicelli, M. Bagnesco and S. De Flora, *Mutat. Res.*, 1999, **441**, 21.
39. S. Bastian, W. Busch, D. Kühnel, A. Springer, T. Meissner, R. Holke, S. Sholtz, M. Iwe, W. Pompe, M. Gelinsky, A. Pontthoff, V. Richter, C. Ikonomidou and K. Schirner, *Environ. Health Perspect.*, 2009, **117**, 530.
40. F. Van Goethem, D. Lison and M. Kirsch-Volders, *Mutat. Res.*, 1997, **392**, 31.
41. A. C. Miller, S. Mog, L. McKinney, L. Luo, J. Allen, J. Xu and N. Page, *Carcinogenesis*, 2001, **22**, 115.
42. M. De Boeck, P. Hoet, N. Lombaert, B. Nemery, M. Kirsch-Volders and D. Lison, *Carcinogenesis*, 2003, **24**, 1793.
43. M. De Boeck, M. Kirsch-Volders and D. Lison, *Mutat.Res.,* 2003, **533**,135; corrigendum, *Mutat. Res.,* 2004, **548**, 218.
44. N. Lombaert, M. De Boeck, I. Decordier, E. Cundari, D. Lison and M. Kirsch-Volders, *Toxicol. Lett.*, 2004, **154**, 23.
45. M. Ding, E. R. Kisin, J. Zhao, L. Bowman, Y. Lu, B. Jiang, S. Leonard, V. Vallyathan, V. Castranova, A. R. Murray, B. Fadeel and A. A. Shvedova, *Toxicol. Appl. Pharmacol.*, 2009, **241**, 260.
46. IARC, *IARC Monographs on the Evaluation of Carcinogenic Risks to Humans*, IARC, Lyons, France, 1999, vol. 74.
47. J. J. Jacobs, J. L. Gilbert and R. M. Urban, *J. Bone Joint Surg.*, 1998, **89**, 268.
48. International Agency for Research on Cancer (IARC), *Monographs on the Evaluation of Carcinogenic Risks to Humans,* IARC, Lyons, France, 2006, vol. 86, p. 35.
49. C. Hogstedt and R. Alexandersson, *Arbete Hälsa*, 1990, **21**, 1.

CHAPTER 13

Precious Metals: Silver, Gold and Platinum-related Metals

13.1 Introduction

Silver, gold and the platinoid metals (iridium, osmium, palladium, platinum, rhodium, and ruthenium) have no known trace metal value in the human body although minute traces may be present in serum or soft tissues through ingestion of contaminated food and drink, industrial exposures or through dermal contact.[1,2] They are transitional metals and form a potentially large number of inorganic and organic compounds. The eight metals have important applications in jewellery, trophies and other precious metal artefacts, each imparting specific properties of colour, hardness and lustre. Ruthenium gives jewellery a "high-tech" jet-black or gun-metal grey look and can be plated onto a wide variety of metals. It is very hard and highly resistant to scratches. Iridium is an extremely rare precious metal with a more luxurious patina than gold, silver or platinum. The so-called "white" gold is a gold-based alloy with sufficient alloying constitution to give a bright white coloration. It contains up to 10% palladium but may also contain platinum, zinc, nickel, copper and silver. Rhodium plating is applied to all white gold jewellery to provide a shiny white finish.

The metals fall into two discrete groups, the monovalent elements silver and gold classified with copper in Group Ib of the Periodic Table, and the platinoid elements which are placed with iron, cobalt and nickel in Group VIII. Silver and gold have been known to humans for several millennia and feature in the icons and funerary legacies of ancient civilisations, whereas the platinum-related elements were discovered in more recent times. Platinum, discovered in gold mines in Columbia in 1736, was named after the Spanish "platina" for little silver. Chemically silver and platinum have little in common although they

Issues in Toxicology No. 18
The Carcinogenicity of Metals: Human Risk through Occupational and Environmental Exposure
By Alan B. G. Lansdown

Published by the Royal Society of Chemistry, www.rsc.org

are used together in alloys. Although these eight metals have established identities in industrial and applications, jewellery is a major route of human exposure and from which a large literature has accumulated on their allergenicity and their ability to induce contact hypersensitivity.[3] Gold, silver, platinum and palladium are allergenic in humans but contact dermatitis attributable to rings and jewellery is frequently caused by the nickel content.

Silver and gold are amongst the least toxic metals in the Periodic Table and the stable isotopes are not associated with carcinogenic risks to humans. Very little toxicological information has been seen on levels of exposure to soluble platinum compounds in the general environment and there is limited evidence that exposure is a cause of adverse health effects. Silver, gold, platinum and palladium induce and bind the cytoprotective metallothioneins (MT-I and MT-II) in soft tissues.[4–7] However, the 12th edition of the Report on Carcinogens lists the organic complex, *cis*-platin, as reasonably anticipated to be a human carcinogen on the basis of limited experimental studies in animal models.[8] Other platinoid elements are not presently known to be carcinogenic although much further research is necessary to justify this view. *Cis*-platin (*cis*-diamminedichloroplatinum) and chemically similar agents were first prepared in 1845 and, since their approval in 1979, they have been shown as highly effective anticancer therapies for ovarian, testicular, lung and bladder cancers, melanomas, myeloma and lymphomas.[9] Their clinical value as antimitotic agents, however, is severely limited on account of major toxic side effects and intrinsic or acquired drug resistance.[10,11] Gold preparations such as Auranofin®, tetrahedral Au(I) compounds and square-planar Au(III) compounds have also been claimed to be highly effective anti-cancer therapies.[12–15]

Cis-platin
[*cis*-diamminedichloroplatinum(II)]

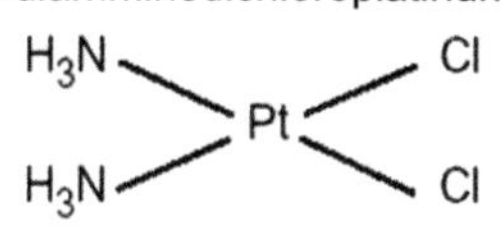

13.2 Silver and Gold

Silver and gold are found naturally in their elemental state in many parts of the world but chief mining areas are in South Africa, North and South America, Australia and Siberia. The commonest silver-bearing minerals recovered by flotation are native silver, argentiferous galena, argentite (Ag_2S), tetrahedrite ($Cu,Fe,Zn,Ag1_2Sb_4S1_3$), freibergite (tetrahedrite with up to 30% Ag), pyragyrite (Ag_3SbS_3), proustite (Ag_3AsS_3) and ceragyrite (AgCl). Often these minerals occur in mixed ores with base metals such as copper and lead sulfides. Native gold commonly occurs as tiny particles contained within sulfide minerals such as pyrite. Iron pyrite is commonly associated with gold, but it also serves as a reducing agent. Gold is found also in complexes with chalcopyrite, galena and arsenopyrite (FeAsS), and stibnite (Sb_2S_3), and as such will be separated from the toxic elements arsenic, antimony and lead in the extraction process. Gold–silver ores are mined in the lode mines of Nevada and in New Zealand. Chemical compounds of gold are rare in nature and consist mostly of tellurides: calaverite $AuTe_2$, krennerite $(Ag,Au)Te_2$, sylvanite $AuAgTe_4$, petzite Ag_3AuTe_2, muthmannite (Ag,Au)Te, montbrayite Au_2Te_3 and others.

Silver and gold are transitional metals, with silver displaying three valences [Ag(I), Ag(II), and Ag(III)] and gold two [Au(I) and Au(III)].[16] Gold is a less reactive metal than silver but is the most malleable and ductile of all metals known. Both elements are excellent conductors of thermal energy and electricity. Silver exhibits two stable isotopes in similar proportions, ^{107}Ag and ^{109}Ag, and numerous synthesised radioactive isotopes with half-lives ranging from microseconds to a few hours. Gold has one natural isotope, ^{197}Ag, but many synthesised radioactive nuclides. The gold isotope ^{198}Au is used for treating cancer and other conditions. Most compounds of Ag(II) and Ag(III) are unstable and have minimal industrial or toxicological significance. The Au(III) salts of gold are more active than Au(I) compounds and of greater toxicological significance. Whereas metallic silver has a low solubility in moisture, and ionises, gold is resistant to oxidation and does not readily ionise. In each case, the ions are highly reactive in the presence of biological materials and show an affinity for ligands on proteins, cell surfaces and organic debris. Both elements are sequestered lysosomally and may be deposited as inert precipitates in soft tissues. Silver and gold ions are bound in the hydroxyapatite complex in bone following absorption into the systemic circulation after oral administration, inhalation (silver), parenteral injection, or percutaneous absorption.[16] Chrysiasis and argyria are the principal adverse effects of gold and silver, respectively. Neither condition is health threatening, but profound and long-term deposition of inert dark coloured deposits of gold or silver sulfide or selenide in dermal tissues leading to severe disfigurement can be psychologically disturbing and highly embarrassing for affected people.[17] Experience in working practices in industries in which workers are exposed to silver, silver oxide and silver nitrate dusts, aerosol droplets or dermal exposures over many years, leading to argyria and argyrosis (silver in the cornea or conjunctiva), has led to greatly improved environmental controls in many countries.

Silver and gold share several common medical and biological properties. Although silver is well known for its antimicrobial properties, gold is mildly active and at one time gold cyanide was researched as a possible therapy for tuberculosis and syphilis.[18–22] It was found to be severely toxic. Silver and gold alloys are employed in stents, dentistry and prostheses, although this use has declined in recent years as more cost-effective treatments have become available. Gold has been used in dentistry and in dental devices for at least 2000 years and, these days, gold implants, dental fillings, wires, screws, foil, and prostheses represent a major medical use of the metal. In each case, gold is preferred over other metals on account of its durability, malleability and resistance to corrosion. Gold wire and gold prostheses are used in tympanoplasty, ossiculoplasty and other surgery of the ear. Gold therapies are well known for their efficacy in the treatment of rheumatoid arthritis, but this is also

Table 13.1 Industrial and commercial uses of silver and gold.

Silver		*Gold*	
Medicinal	*Industrial*	*Medicinal*	*Industrial*
Antimicrobial: Wound care, Catheters, Dental implants	Electronics/ electrodes, Fuel cells	Antimicrobial (formerly) therapy for tuberculosis and venereal disease	Electronics and computers, Gold thread in needlework
Heart valves, Stents	Chemical syntheses	Anti-arthritic therapy	Coinage, medals
Hygiene textiles	Coinage, medals	Dental implants	Protection against solar irradiation (astronauts)
	Cups and trophies	Orthopaedics, stents	Cups, trophies and icons
	Jewellery	X-ray contrast medium	Jewellery
Water purification	Photography	Acupuncture needles	Alloys
X-radiography	Production of alloys, silver plating	Surgery, wires, screws	Photographic toning
Acupuncture needles	Silver thread		Paints in decorative work, gilding and fine arts

contraindicated on account of their ability to evoke contact sensitisation.[3] Principal industrial and commercial uses of silver and gold are shown in Table 13.1.

Some silver and gold leaches from medical devices into the circulation and may deposit in soft tissues, notably the dermis and cornea, to give long-lasting discolorations in the form of argyria and chrysiasis respectively. Although these discolorations resulting from intracellular precipitation of insoluble salts are socially undesirable and a possible cause of psychological problems, they are rare these days, non-toxic and not associated with progressive degenerative disorders or neoplastic change.[17,23] This is illustrated by a case of localised argyria with chrysiasis in a 41-year-old Japanese woman attributable to implanted acupuncture needles.[24] Electron microscopy and X-ray microanalysis demonstrated extracellular deposits of silver selenide and sulfide, whereas free gold was found intracellularly.

Silver and gold are absorbed into the circulation following inhalation, ingestion in diet and drinking water, parenteral injection, implantation of medical devices (including stents, dental inserts, acupuncture needles, indwelling catheters and prostheses) and through percutaneous absorption. Uptake through the latter route is low except in cases of dermal contact with skin wounds and abrasions. Both elements dissolve to a small extent in surface skin moisture to release bioactive ions (Ag^+, Au^+ or Au^{3+}) but these ions complex strongly with –SH groups in epidermal keratin and with serum albumin, macroglobulins and skin surface debris. Minimal toxic levels of silver and gold in the human circulation are not known.

Brief mention is necessary of the biological significance, tissue reactivity and toxicity of nanoparticulate forms of silver and gold in view of their increasing use in antibacterial devices, medical and photodynamic therapies, electronics, probes, diagnostics and analytical chemistry. Nanotechnology is becoming a highly specialised discipline with far-reaching implications in environmental and medical science.[25] Nanoparticles display different physico-chemical properties from foil or analytical grade powders, and as such exhibit different or possibly more efficacious value in therapeutics and diagnostics. Particle size will be of toxicological and environmental significance. Comparative studies have shown that nanocrystalline silver with higher solubility in water exhibits a six-fold or higher log reduction in *Pseudomonas aeruginosa* in culture.[26]

13.2.1 Silver

Silver is a white lustrous transitional metallic element found widely in the human environment. It is the 63rd most abundant metal in the Earth's crust and average concentrations are estimated to be 0.5 ppb in water and 10 ppb in soil. Metallic silver is inert in the presence of human tissues but ionises in the presence of moisture, body fluids and secretions to release the biologically active Ag(I) ion which shows a strong affinity for sulfydryl groups and other anionic ligands of proteins, cell membranes and tissue debris.[4] Ionisation of metallic silver is proportional to the surface area of particles, such that ionic release from nanocrystalline particles of 20 nm may be more than 100-fold higher than that from silver as foil or analytical grade powders.[26]

Considerable use is made nowadays of the antimicrobial properties of silver and its efficacy in controlling pathogenic bacteria, fungi and protozoal infections associated with skin wounds and burns, indwelling catheters, implants and prostheses, water filters (in the Katadyne process), textiles and in a range of domestic products.[27,28] Many preparations based upon silver nitrate, silver sulfadiazine, nanoparticulate silver, silver "glasses" and other inorganic salts have been introduced in recent years with justified claims for antibiological efficacy, clinical suitability and safety in use.[29–31] Recent advances in the biotechnology of medical devices and the ability to impregnate or coat alginates, polyurethane, silicones and textile fibres with ionisable silver compounds now provide clinicians with efficacious means of overcoming infections which have proved costly in terms of hospital care and patient stress as well as being a major cause of fatalities.[31–34] In each case, the antimicrobial value of the silver-containing preparations correlates directly with the amount of free silver ion released into the medium and its capacity for absorption by genetically silver-sensitive organisms.[35,36]

Other than its varied medical applications, extensive human contact with silver and silver alloys occurs through photography and X-radiography, jewellery, silverware and coinage; batteries, bearings, brazing and soldering, chemical catalysis, electronics, mirrors and coatings, solar energy protection and water purification. The Silver Institute estimated that demand for silver in these applications has increased such that, in 2007, 456.5 million ounces were employed with 128 million ounces used in the photographic sector alone, 163.4 million ounces in jewellery and 58.8 million ounces in the silverware markets. High demands and cost have led to increased effort to recover silver (and gold) from waste in photography and radiography, and recycling of silver from alloys used in jewellery, and silverware.[37–39] In the extraction processes, which involve crushing, leaching and separation of the heavy metals, metal scrap is exposed to cyanidation, thiosulfation, acidothiourea and strong acids, hydrometallurgy and pyrometallurgical processes, all of which present human health hazards. The toxicology of metal recovery processes is expected to be a complex issue which clearly needs detail study and regulation.

13.2.1.1 Cytotoxicity, Mutagenicity and Carcinogenicity

Metallic silver is inert in the presence of living tissues but in the presence of moisture, body fluids and excretions, it ionises readily to release biologically active Ag(I).[26] This interacts strongly with protein receptors on cell surfaces. Intracellular uptake by mammalian cells probably occurs, as in sensitive bacteria, by a genetically modulated active process involving receptor proteins or by a pinocytic process. Silver ion exhibits minimal toxicity in humans and other animals on account of its strong propensity to precipitate in an inert form as silver sulfide, silver selenide, or silver–protein complexes in the body or in excretory products. Argyria is the main contraindication for silver exposure. Silver nitrate, which was once widely used as an antimicrobial agent, is corrosive to the skin on account of release of nitric acid in the presence of body tissues. Silver sulfadiazine [SSD; Flamazine, silver 2-(4-aminobenzenesulfonamido)-pyrimidine], introduced in 1968 as a broad-spectrum antibiotic, is not corrosive or irritant and exhibits few side effects.[30,31] Presently, there is little experimental or clinical evidence predicting that silver and its inorganic or organic compounds to which humans will be exposed occupationally, clinically or environmentally are a cause of cancer in any organ. Silver arsphenamine, which was introduced 100 years ago as an intravenous therapy for venereal disease, presents a greater health risk.[18] It is a toxic drug which oxidises slowly following injection to release arsenic, which as discussed later is carcinogenic. Arsphenamines were in use until the 1940s but have been largely superseded in modern pharmacopoeias by safer and more efficacious therapies.

Experimental *in vitro* studies have failed to demonstrate the mutagenicity of silver compounds used medicinally. They provide evidence that silver nitrate and SSD, at least, are cytotoxic to fibroblasts and keratinocytes in monolayer cultures, but not mutagenic in Ames reverse transcription assays with *Salmonella typhimurium*.[40,41] At concentrations 100–700 times more dilute than normally used in clinical practice, silver nitrate was cytotoxic to human fibroblasts in a time- and dose-related manner for all exposure periods tested.[42] Poon and Burd used monolayer cultures of human keratinocytes and fibroblasts to demonstrate the antimitotic effects and influence on cellular behaviour of Ag(I) released from silver nitrate using a three-dimensional planar model.[43] They considered that this model was useful in assessing the cellular effects of silver under conditions simulating *in vivo* exposure in wound care. Other experiments have demonstrated that silver nanoparticles inhibited mitosis in human lung fibroblast cultures or human glioblastoma cell cultures and that the changes were consistent with mitotic arrest in the G_2/M phase of the cell cycle, possibly reflecting impairment in DNA repair.[44] Transmission electron microscopy showed that the silver nanoparticles were located intracellularly within mitochondria and nuclei, indicating their direct involvement in mitochondrial toxicity and DNA damage. Intracellular changes have been associated with depression in expression of growth factors, including platelet-derived growth factor (PDGF), epidermal growth factor (EGF) and basic fibroblast growth factor (b-FGF), that modulate cell motivation, migration and functional maturation. Zapata-Sirvant and Hansbrough claimed that silver sulfadiazine impaired the oxidative burst in cultured human leucocytes in flow cytometric assays at levels at or below clinical concentrations, and that silver impairs the intracellular flux of calcium ion in lymphocytes.[45] Whilst this may disturb immunoresponsiveness and impair metal ion balances in tissue repair, it is unlikely to have a significant effect in terms of neoplastic transformation.[46] Cells in tissue culture show variations in response to silver, and experiments have demonstrated that fibroblasts are more sensitive than keratinocytes and that sensitivity is related to the age and presumably the genotype of the donor.[42,47]

Experimental studies in animal models suggest that silver exhibits a low order of toxicity and that LD_{50} estimates are high. Silver nitrate is profoundly toxic and potentially fatal following parenteral injection, but in the experiments observed, the concentrations administered were unrealistically high and far in excess of those expected in humans, even in extreme cases of exposure.[48–51] In an acute inhalation study in which rats were exposed to micro-fine silver particles for 6 h (133 mg m^{-3}), silver particles accumulated in the nasal tract, lung-associated lymph nodes and lungs, but declined rapidly such that by 7 days only 4% of the initial lung burden remained.[52] As with other metals, pulmonary macrophages provided a primary protective function in the uptake and mobilisation of silver but toxic changes were not reported. On the other hand, lung function and respiratory tract physiology were impaired in subchronic and long-term toxicity studies in rodents exposed to nanoparticles of metallic silver for 90 days.[53] Despite a rapid clearance of silver nanoparticles (18 nm diameter) from the respiratory tract, dose-related inflammatory changes and histopathological responses of thickened alveolar walls and small granulomatous lesions were reported. These changes were associated with reductions in tidal volume and physiological changes. In summary, experimental studies in laboratory animals fail to provide evidence that metallic silver, silver nitrate or colloidal silver are carcinogenic. Local sarcomas have been reported following repeated subcutaneous injections or implantation of metallic silver subcutaneously but predictions based on this model are unreliable since a variety of known non-carcinogens induce tumours by this route.[54–56]

Guidelines drawn up by the US Department of Health and Social Service in 1978 specified that exposure to particulate silver at concentrations of

0.5 mg m^{-3} or less requires workers to be equipped with high-efficiency face filters, and that at higher silver concentrations face filter quality should reflect the gravity of risk, ambient environmental concentrations, and the size of the micro-particles.[57] In more recent times, safety recommendations have been introduced to deal with health risks associated with nanoparticulate silver, silver levels in drinking water, and silver in medical devices.[58] Their conclusion that, on the basis of present epidemiological and pharmacokinetic knowledge, a total lifetime oral intake of about 10 g of silver can be considered as the human no-effect level is still valid. The level of silver ingested with drinking water contributes a negligible part of this intake. A survey of the clinical literature published in the past 100 years and relating to a wide range of occupational, environmental and clinical uses of silver, silver nitrate and soluble silver compounds has failed to provide evidence of carcinogenicity, even in extreme circumstances.[27,28,59,60] On rare occasions where patients were exposed to silver in bizarre circumstances, such as the use of silver nitrate in inducing abortion, patients have endured severe discomfort, but not silver-related fatality or carcinogenicity. However, although cases are not reported, it is anticipated that human contact with radioactive isotopes of silver may lead to carcinogenic changes as is noted below in cases of humans exposed to residual radioactive emissions of gold in jewellery.

13.2.2 Gold

Gold has been known to mankind for more than 4000 years and is recognised as an object of prestige, religious significance and value. Gold comprises no more than 0.004 ppm of the Earth's crust but has a worldwide distribution with major deposits in South Africa, Siberia, China, South and North America, India and Australia. Gold was amongst the first metals to be mined although it occurs naturally in its metallic form admixed with quartz and pyrite in alluvial deposits in rivers and streams. Much history relates to the famous gold rushes and discovery of gold in California and the Klondike region of Canada in the 19th century. Gold has a yellow colour in its natural state but as a finely divided powder it may appear blackish, ruby or purple. Gold is a soft, malleable and ductile metallic element that is stable in air and unreactive in the presence of most reagents. It is an excellent conductor of heat and electricity and of value in electrical contacts and in computer technology. In its ionized form, Au(III) is held to be the most bioactive of all metals. Gold exhibits two oxidation states, Au(I) and Au(III), and a single natural isotope, ^{197}Au. A total of 54 radioactive isotopes have been produced with half-lives ranging from 17 microseconds to 186 days. Decay modes for these radionuclides include proton emission, α-decay, electron capture, β-minus decay and isomeric transition (Jefferson Lab). Gold chloride ($AuCl_3$) is the commonest of gold salts and is chemically more stable than the monovalent chloride.

Gold salts in the environment are readily reduced to metallic gold, which is not readily absorbed into the human body. Human exposure occurs through inhalation of gold dust in mining areas, dermal contact in manufacturing industries and in the domestic environment, and minimally through food and drinking water and use of gold therapies (chrysotherapies) in human medicine and dentistry. Although reports have been seen of gold being taken up by fish from streams near gold mines, there is little evidence that significant amounts of gold enter human food chains.[2] Gold dust is expected as an atmospheric contaminant in the vicinity of mining areas but, in view of their high density, gold particles are expected to settle rapidly and present minimal risk of inhalation. On the other hand, micro-fine particles of 20 nm or less remain in the atmosphere for longer and are more likely to be inhaled by humans and animals. Chronic inhalation of gold dust is expected to be the major cause for concern, particularly if gold can be shown to exhibit mutagenic and carcinogenic properties.

The principal source of toxicological information for gold derives from occupational health studies in mining, extraction and refining processes, manufacturing industries and occupational exposures in gold plating, jewellery manufacture and photography. Contact with gold in jewellery, coinage, ornaments and treasured items is a rare source of health problems. Gold and gold salts are not readily absorbed into the human body on account of the unreactive nature of the metal, its lack of oxidation in air and inability to ionise in moisture or body fluids.[61] In gold mining, greatest concern relates to human exposure to contamination of the ores with arsenic, silicates, lead and antimony, as well as exposures to mercury, and cyanides used in the extraction process.[62] Additionally, gold has a long use in human medicine as a therapy for rheumatoid arthritis, in dentistry and as an antibiotic in treating patients with syphilis and tuberculosis. Gold salts including cyanide and gold potassium cyanide were researched by Robert Koch and others from about 1890 and claimed to have efficacy in anti-infective preparations in animal models, but despite some rewarding tests with sanocrystin (a double thiosulfate of gold and sodium), its toxicity was prohibitive.[63,64] Gold therapies including aurothiomalate, aurothioglucose, gold thiosulfate and gold thiopropanosulfate are injectable therapies for arthritis.[65–67] Gold has been used in dentistry for at least 200 years and these days dental fillings, wires, screws, foil and prostheses represent a major medical application of the metal. Other uses include tympanoplasty, ossiculoplasty and other ear surgery.

A major health problem relates to gold as a potent allergen and cause of contact hypersensitivity. Many cases of dermal irritancy are recorded in contact with jewellery, and occupational exposures in gold platers, photographers and glass gilders.[3,68–71] Langerhans cells are activated through dermal contact as part of this hypersensitisation process.

13.2.2.1. Cytotoxicity, Mutagenicity and Carcinogenicity

Gold represents an enigma in the evaluation of metals for carcinogenicity. Gold and related compounds are not currently classified as being carcinogenic to humans by the US Department of Health and Human Services (US DHHS) or the International Agency for Research on Cancer (IARC), yet there are several well documented references to the cancer risk faced by gold miners.[72–76] In each case, the reports show higher levels of cancer in gold miners and those engaged in extraction of gold from native ores, as well as populations living close to working or redundant gold mines, but the true causation is unclear. In addition to gold residues, individuals were exposed to antimony, arsenic, silica, mercury, bismuth, nickel, radon, silver, palladium, copper, zinc and cyanides found in the geosphere with mined gold or used in extraction processes. Thus, Kusiak examined *post-mortem* records of 54,128 men who worked in Ontario gold mines between 1955 and 1986 and reported an excess of lung carcinoma in 13,603 miners with at least 10 years' exposure, but noted that the host rock contained arsenic and radon.[72] A meta-analysis of the health risks of underground gold miners in Australia, the Americas and Africa identified increased risk of stomach, liver and respiratory tract cancers, but emphasised that alcohol consumption, cigarette smoking, human immunodeficiency virus (HIV), malaria, dengue fever, local viral infections and high mercury levels precluded valid appreciation of the carcinogenic risks attributable to gold mining.[73,74] Several of these elements are known carcinogens or may act as co-carcinogens.[1,73,77–79] Subchronic or long-term exposures to gold dust in mining and refining industries might be expected to lead to respiratory diseases and pulmonary damage but gold residues are absorbed by lung macrophages and alveolar cells, seemingly without cancer risk; far greater risks are associated with inhalation of silica (quartz dust), mercury vapour, radon and cyanide commonly reported in the work environment.[62,80,81] Gold dust inhaled by miners and refinery workers has a long half-life in the lung, but silicosis and asbestosis are the main causes of ill-health and lung cancer.[82] Similar problems are encountered in platinum and diamond mining.

Radon emissions are a problem for gold miners. Radon is classified as a known human carcinogen and a higher incidence of cancer is reported in numerous epidemiological studies where people are exposed to the gas in gold and uranium mines.[8,83,84] Lung cancer is a recurrent problem and mortality in underground workers may be increased 4–6-fold depending upon the population exposed. In a study of 54,128 Canadian gold miners in Ontario, lung cancer was reported in 13,603 workers who had commenced work in the mines before 1946. However, the true causation of the cancers is complicated because the bed rock in the mines contained 0.1% arsenic, and inhalation of arsenic dust has to be included as a carcinogenic risk. Similar problems were encountered in a French mine, where increased lung cancer in miners and smelter workers was attributed to inhalation of a combination of gold dust, arsenic (soluble and insoluble forms), radon and silica.[85] Above-ground workers are also at risk of radon exposure. A map of radioactive material in the region of dumps of gold mine tailings in South Africa showed radioactivity at 200–300 Bq kg^{-1} due to radon and uranium residues; this diffuses into the surrounding air when the dumps are disturbed.[86] Emissions of radon gas are a problem in many parts of the world and it is even recorded that indoor emissions are claimed to be a potential cause of renal cancer, leukaemias, malignant melanoma and other cancers.[87]

Chronic or long-term parenteral administration of gold compounds to relieve symptoms of rheumatoid arthritis and related conditions (chrysotherapy) is not carcinogenic although cases of pulmonary fibrosis are reported.[58,88] Other changes related to gold therapies include liver and renal changes, reflecting the role of these organs in gold metabolism and excretion, haematological and immunological change.[65,89,90] Aplastic anaemia, granulocytopenia and thrombocytopenia are reported. Gold, like many metals absorbed into the circulation, is carried in the blood in the form of protein complexes and in erythrocytes. Metallothionein, albumin, fibrinogen, γ-globulin, histone and insulin have been shown to bind gold ions.

Rare incidences of skin cancer are reported in people wearing gold jewellery with residual radioactive contamination.[91–95] (Gold has 54 known isotopes with half-lives ranging from microseconds to a few days). The level of carcinogenic change seen was proportional to the duration of exposure. In a survey of 135 people with gold-related skin cancer, the incidence was 11-fold higher than that predicted for men and 45 times higher for women. More specific information on the carcinogenicity of radioactive gold is provided by case studies. Skin cancers, ranging from squamous cell carcinoma, granuloma annulare, lymphoplasia to lymphocytoma cutis, have been diagnosed in the close vicinity of contaminated rings, earrings, *etc.* They include the case of a woman who developed a basal cell carcinoma on her finger following exposure to 240 rad per week radiation from a gold ring for 37 years and an estimated radiation dose of 4.6×10^4 rad.[91] Other cases of skin tumours are reported including lymphocytoma cutis and benign lymphoplasia following exposure to radiation in gold earrings.[93,94] Granuloma annulare is reported as a complication of injectable gold therapies for juvenile rheumatoid arthritis.[92,96] In one case concerning a 17-year-old girl, multiple dermal papules developed following weekly injections of 25 mg sodium aurothiomalate. On biopsy, the papules were characterised by some necrobiosis of dermal collagen outlined by prominent patchy zones of chronic inflammation with lymphocytes and histiocytes as the dominant cell types.

Experimental evidence for the carcinogenicity of gold is limited to two chronic studies in which rats were injected subcutaneously with multiple injections of sodium aurothiomalate or fine gold powder.[97,98] Dose levels of up to 192 mg kg^{-1} were high and intolerable, but in surviving animals 13% developed injection site sarcomas and severe renal lesions and adenomas. This observation has not been confirmed clinically in human patients.

In vitro studies present a mixed picture of the mutagenicity and genotoxicity of gold. Isolated cytogenicity studies in human cell lines provide evidence for cytotoxicity but no evidence of mutagenicity. Thus gold nanoparticles absorbed endocytically by alveolar type II cell lines and NCIH4441 cells in culture were taken up lysosomally and were cytotoxic in MTT assays.[99] Experimental studies in cultured rabbit eye lenses showed that gold accumulated on surface sites but intracellular penetration was not established.[100] This accumulate was toxic and led to a change in intracellular electrolytes, including loss of sodium ions and increased accumulation of potassium.

13.3 Platinum and Platinoid Elements

Platinum and related elements are classified in Group VIII of the Periodic Table and share a characteristic similarity in their lustrous silvery-white appearance and physico-chemical properties. They are rare elements and constitute a minor proportion of the Earth's crust. They commonly occur together in natural alluvial deposits and in the presence of gold, silver, copper, iron, nickel and arsenic. The platinum group metals (PGM) exhibit tensile strength, ductility and durability, and stability in the presence of heat, chemical action and electrical currents to provide alloys, coatings, electrodes and components for electrical devices and contacts, furnace and crucible linings, medical prostheses, dental implants and jewellery. Platinum and palladium absorb gaseous hydrogen and oxygen in the form of a sponge. The PGM have a capacity to form co-ordination complexes with carbon, nitrogen, phosphorus, oxygen and sulphur, and at least 24 co-ordination complexes of platinum, rhodium, iridium and ruthenium have been investigated as possible anticancer agents. *Cis*-platin at least has been identified as one of the most potent antitumour agents known but its toxicity and possible carcinogenicity may outweigh its therapeutic value.[8,9,101–103] Extensive use is made of the catalytic properties of PGM in chemical reactions including the oxidation of organic vapours in automobile exhausts, petroleum reforming, and in the oxidation of ammonia in production of nitric oxide for fertilizers, explosives and nitric acid. Platinum-supported catalysts are used in refining crude oil, production of high octane gasoline and aromatic compounds in the petroleum industry. Palladium, rhodium and platinum have value as catalysts in petrol engines in catalytic converters to treat automobile exhaust emissions, which are a source of pollution and human exposure in inner city areas (Pt levels 8–87 ng m^{-3}).[1,104–106] Since 1979, the US automobile industry has been the largest consumer of PGM. Other routes of exposure include mining and recovery, and refining the metals from alluvial deposits, manufacture and use of alloys in industry, jewellery and medical devices. Surgical and dental use of alloys may involve long-term exposures and risk of metal residues leaching into the systemic circulation but health risks have not been identified.

Platinum and palladium concentrations in dental technicians were shown to be several fold higher than in road construction workers and school children exposed to automobile exhaust fumes. Average airborne platinum concentrations in Rome, Goteborg and Madrid ranged from 4.1 to 117.7 pg m^{-3} according to the city area and the size of particles, but at these concentrations minimal risk is anticipated.[106] This view is based upon inhalation rates being 8–10% and an environmental level many times lower than the presumed toxicity threshold of 100 ng m^{-3}. In an industrial setting, airborne Pt levels ranging from 2.39 to 4.83 μg m^{-3} have been reported in platinum production, recovery and recycling of catalytic converters.[107] Urine, blood and hair levels, measured in 106 workers, of 1.86 μg l^{-1}, 0.38 μg l^{-1} and 2.26 μg kg^{-1} are an indication of the amount of platinum absorbed and patterns of excretion. No adverse effects were recorded. Platinum particles seem to circulate in the air and even administrative staff not exposed directly to platinum exhibited tissue levels 2–20-fold higher than in people working outside the factory. In a study of 34 workers exposed in four-hour shifts in a refinery, urinary platinum reached 1000 times that seen in non-exposed personnel and correlated well with exposure levels.[108]

Dermal contact with metallic platinum in jewellery and platinum-containing metal alloys in dentistry, prostheses and artefacts seems to present minimal toxic risk although inhalation of platinum catalysts emitted in vehicle exhausts and workplace fumes can be a cause of asthma and respiratory distress.[107,109,110] The PGM are strong sensitisers in predisposed people and dermatitis and asthma are recorded in patients exposed to jewellery, sprays, refining workshop catalysts, and even atmospheric contamination.[3,111–115] Certain platinum compounds, notably the chloroplatinates, are a profound cause of dermatitis and the so-called "platinosis" syndrome is reported in platinum refiners.[116] At least platinum and palladium are known to induce and bind metallothioneins, and Kripp emphasised that these proteins constitute the most potent cytosolic sink for platinum drugs.[6,117] He emphasised that besides the strong Pt(II)–sulfur bond involving 20 cysteine residues of the metallothionein molecule, incorporation of Pt(II) into the complex contributes effectively to the scavenging action of *cis*-platin. Some platinum is stored in the body, possibly in the form of a Pt–MT complex. An experimental study in rats showed that inhaled radiolabelled ^{191}Pt was localised rapidly in the respiratory tract and gastrointestinal tract but most was eliminated within 24 hours, although lung clearance showed a half-time of <8 days.[118] High concentrations of ^{191}Pt were seen in bone and kidney.

Elements of the platinum group are not presently classified as human carcinogens.[8,119] However, the co-ordination complex *cis*-platin, which was identified and patented as an efficacious anticancer drug for use in head, neck, testicular, ovarian and lung carcinomas in 1965, is now widely researched as a potential carcinogen, mainly on the basis of experimental studies.[120] The *trans*-platin isomer has no such effect.[12,120] More than 20 co-ordination complexes of platinum have now been implicated as anticarcinogenic agents; of these carboplatin and oxaliplatin have gained international marketing approval, and nedaplatin, iobaplatin and heptaplatin have received approval in some countries.[15,121] A newer hydrophobic Pt(IV) complex is claimed to be an efficacious anticancer agent without the side effects and acquired resistance which limit the choice of other platinum-based therapies.[122] All are cytostatic and act through direct interaction with DNA to form adducts and intra-strand cross-links which activate several signal transduction pathways including those involving ATP-kinases, p53 and p73 leading to apoptosis.[9,13,108,123] All inhibited DNA synthesis, induced DNA repair, replication and DNA base-pair changes in hamster cells, as reported as an integral part of their apoptotic action. However, in a study of Pt(II) and Pt(IV) complexes, none induced DNA strand breaks or DNA cross-linkages.[12] Although therapeutically appealing, selection of platinum-related drugs as anticancer agents is restricted by their toxicity, increased drug inactivation, acquired cellular resistance and efficacy against different types of solid tumour.[12,121,124,125]

13.3.1 Cytotoxicity, Mutagenicity and Carcinogenicity

At the moment, there is little good evidence to implicate metallic platinum or inorganic platinum compounds as human carcinogens. Numerous environmental and occupational surveys are reported but most record levels of PGM in blood, urine or hair as evidence of exposure rather than toxicological or neoplastic responses. Lack of positive carcinogenicity may be a reflection of the low levels of biologically active ions in the circulation or tissues, or be an indication that the ions are not genotoxic or mutagenic. A single study, showing cancer incidence in precious-metal miners in South Africa, recorded 149 cancers (all types) compared to 385 expected from national averages but "significant deficits of cancer among PGM workers" in 12 sites.[126] The most significant cancers amongst gold miners involved the respiratory tract, liver, oesophagus and buccal cavity, all of which have been diagnosed in some PGM-exposed workers. The authors failed to point out that gold and precious metal miners in South Africa are exposed to radon, arsenic and silica as carcinogenic materials in their work programme.[72,76,85]

The co-ordination compound *cis*-diamminedichloroplatinum (*cis*-platin) [containing 67.24% Pt(IV)] and certain analogues were claimed to act as tumour initiators and tumour promoters in rats and mice and are identified as "reasonably anticipated to be human carcinogens".[8,120] *Cis*-platin does not degrade in the body to release the Pt(IV) ion but reacts spontaneously in solution with glutathione or cysteine to form *cis*platin–glutathione or *cis*platin–cysteine conjugates.[123] Subsequent cleavage of this *cis*platin–glutathione conjugate to a *cis*platin–cysteinyl–glycine conjugate by gamma-glutaryl transferase (GGT) is the first step in the metabolism of the *cis*platin–glutathione conjugate. Cysteine-*S*-conjugate beta-lyase catalysis is considered to be the final step in the pathway converting the *cis*platin–cysteine conjugate to a reactive thiol which is nephrotoxic in rodent kidney.[127–131] The major metabolite of *cis*-platin is trichloroammineplatinate, and pH is a prominent factor affecting cis-platin stability. The pathway proposed by Zhang[130] is shown in Figure 13.1. Inhibition of γ-glutamyl transpeptidase or pyridoxal 5′-phosphate (PLP)-dependent enzymes blocks the nephrotoxicity.

Lacking appropriate human confirmation, experimental studies in cultured cell lines, genetic reversion studies in bacteria, and chronic experimental animal tests reveal strong sufficient evidence that *cis*-platin, at least, is mutagenic, genotoxic and carcinogenic through a direct and transplacental action.[132] *Cis*-platin is a toxic drug acting on the kidney, inner ear and haemopoietic tissues, and presumably actively metabolic and mitosing tissues, but it is mutagenic, cytotoxic and antimitotic.[121,125,133–136] Abundant evidence is available from bacterial tests to illustrate the mutagenicity of *cis*-platin and certain other platinum salts in conventional *Salmonella* tests and other bacterial assays. Early studies confirmed the mutagenicity of *cis*-platin in *Escherichia coli*,

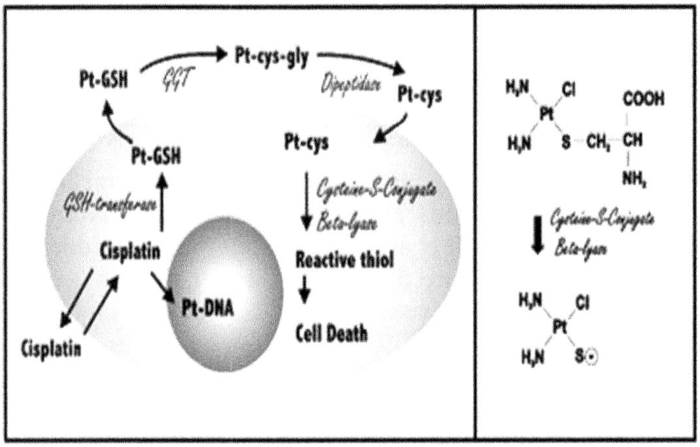

Figure 13.1 Metabolism of *cis*-platin to a nephrotoxin. The proposed pathway of cisplatin activation is based on the metabolism of the glutathione conjugates of the halogenated alkenes to nephrotoxins. The location of intracellular and extracellular reactions that bioactivate *cis*-platin to a nephrotoxin are shown (left). Cisplatin enters cells and binds to DNA, which is toxic to dividing cells but not to quiescent cells such as the proximal tubules in the kidney.[130]

Neurospora sphaeroblast assays and in the *Salmonella* microsome mutagenicity test: cell cultures indicate DNA damage, base-pair substitutions and frame-shift mutations, whereas *trans*-isomers were negative in both tests.[137–142] Genotoxic stress triggers formation of DNA adducts, and transcriptional activation of genes regulating DNA repair, cell survival and viability. Comparative tests substantiate the mutagenicity and genotoxicity of *cis*-platin, carboplatin, *trans*-platin and platinum chlorides ($PtCl_4$ and K_2PtCl_4) in the micronucleus cytokinesis block test and SOS-bacterial chromotest,[143] whereas platinum chloride ($PtCl_2$) and dipotassium platinum (K_2PtCl_6), hexachloride and five palladium salts were negative.[144] Variations have also been observed in the responses of different *Salmonella* strains to mutagenic Pt(II) complexes. Mutagenicity in response to 11 Pt(II) and Pt(IV) compounds in multiple strains of *Salmonella* in the Ames test related specifically to the *cis* configuration and to the presence of the "R" virulence factor", indicating that SOS-repair processes are necessary for initial DNA damage and mutation.[140] (The SOS system is a regulon; that is, it controls expression of several genes distributed throughout the genome simultaneously). Experiments in Chinese hamster cells have further confirmed the ability of cisplatin to evoke clastogenic change; dose-dependent chromosomal aberrations, sister-chromatid exchanges, 6-thioguanine-resistant mutations and morphological transformations were reported.[145] This is in contrast to results of later laboratory tests which demonstrated a lack of correlation between blood levels of platinum in nurses handling cytostatic drugs and incidence of sister-chromatid exchanges or micronuclei in studies spanning two years.[146]

At least 14 experimental studies are published showing that *cis*-platin is capable of inducing carcinogenic changes in mice or rats following intraperitoneal or subcutaneous injection. However, they are inconsistent and show interspecies and inter-strain variation in response to high and often toxic concentrations of the drug over long periods.[132] They may be sufficient to indicate that *cis*-platin is both a tumour inducer and promoter and can induce carcinogenic change in fetuses and neonates following transplacental transfer. In the subcutaneous sarcoma model, F344 rats given repeated injections of *cis*-platin, or the analogues *cis*-dichloro-bis (cyclopentylamine) platinum and *cis*-dichlorobis(pyrrolidine) platinum, over 82 weeks developed metastasising sarcomas at injection sites.[141] In contrast, intraperitoneal injection of *cis*-platin at 1.0 mg kg^{-1} three times weekly for 455 days in BD-XI rats led to 66% mortality.[147] Thirteen animals (33.3%) developed neoplasia: 12 leukaemias and one renal sarcoma. The rats were dehydrated before treatment but no tumours were seen in control animals, and the tumour rate in surviving animals is not known. The ability of *cis*-platin to act as a tumour promoter was demonstrated in the two-stage skin carcinogenesis model in mice where papillomas induced by experimental carcinogens became more numerous and showed malignant change following treatment with platinum or its analogues.[141,148–150] Mice given a single intraperitoneal injection of cisplatin followed by the experimental promoter, 12-*O*-tetradecanoyl-phorbol-13-acetate (TPA), an extract of croton oil, developed dermal papillomas. The tumour yield and malignancy were significantly increased when TPA was given beforehand. Interestingly, animals treated with *cis*-platin without croton oil did not develop overt skin tumours. Even so, *cis*-platin is capable of inducing genotoxic changes and persistent *cis*-platin–DNA adducts have been demonstrated in biopsies of skin, liver and kidney in mice.[150] Analysis of the so-called Harvey *ras*-gene in *cis*-platin treated mouse skin using the polymerase chain reaction for DNA sequencing has shown that codons 12, 13, 61 and 117 were not affected and that neoplastic transformation probably involves transformation of other as yet unidentified genes.[148] Other studies have shown that, whilst A/Jax mice are useful in demonstrating the primary carcinogenic action of *cis*-platin in the murine lung, point mutations in the *K-ras* gene, notably in codon 61, differ from those seen in neoplastic changes induced by other primary lung carcinogens such as benz[a]pyrene and urethane!.[151–153]

Four studies in F344 rats and SENCAR mice conducted at the National Cancer Institute in Bethesda have demonstrated that *cis*-platin acts as a transplacental carcinogen and induces carcinogenic changes in fetuses or neonates.[154–157] The offspring of pregnant mice given a high single dose (7.5 mg kg^{-1}) in late gestation showed a high incidence of skin papillomas, but only when they had been treated postnatally with TPA as a promoter. Nevertheless, offspring exposed transplacentally to *cis*-platin alone did show a high incidence of lymphoid and lung tumours, and renal lesions. Experiments in pregnant F344 rats showed that sub-terminal administration of *cis*-platin resulted in a low incidence of renal tumours in male offspring. The incidence was greatly increased (10.5–45.5%) by administration of high levels of sodium barbiturate. This study demonstrated also that *cis*-platin is a transplacental hepatocarcinogen and may be responsible for a low incidence (3%) of central nervous system tumours, but these tumours were not apparently sensitive to promotion by sodium barbiturate. Evidence that *cis*-platin acts by a direct and tissue-specific mechanism is provided by demonstration of DNA adducts in maternal and fetal tissues.[156] Levels were higher in maternal kidney, liver and lung than in fetal tissues; but higher adducts in fetal brain suggest that the mutagen accumulates more specifically at this site. Tumour induction in offspring involved specific changes in *ras*-mutations in codons 12 or 13 of exon I related to intra-strand cross-linking.[157] In each case, *cis*-platin induced genetic changes in fetal tissues that differed in magnitude and location from those seen in maternal tissues.

13.3.2 Carcinogenicity and Mechanism of Action

No evidence is seen to implicate industrial, environmental or social contact with platinum or related metals as a cause of human cancer. The elements do enter the human body, mainly through inhalation of contaminated dust, but whilst this may be a cause of respiratory distress, lasting damage is not recorded, presumably because insufficient amounts of biologically reactive ion accumulate in the body and residues are readily excreted. However, *cis*-coordination complexes, notably *cis*-platin (*cis*-diamminedichloroplatinum) introduced as a potent anticancer agent should be considered as being potentially carcinogenic. The agent forms conjugates with glutathione and is metabolised to a cysteinyl–glycine conjugate, to a cysteine conjugate, and finally to a reactive thiol without release of Pt(II) ion.[127–130] It exhibits and directs antimitotic action in tumour tissues through binding of DNA, inducing apoptosis and impairing transcriptional pathways; it is efficacious against several types of solid tumour.[158] Whilst no clear relationship has yet been demonstrated between parenteral administration of platinum-related drugs and human cancer, strong evidence is provided from experimental work that *cis*-platin induces DNA adducts, specific mutational changes in target cells and acts as a direct carcinogen. The cytoprotective role of metallothioneins is acknowledged from experimental animal studies. There is an urgent need for long-term critical and circumspective evaluation of *cis*-platin in clinical practice to substantiate the magnitude of the cancer risk.

References

1. J. Begerow, U. Sensen, G. A. Wiesmüller and L. Dunemann, *Zentralbl. Hyg. Umweltmed.*, 1999, **202**, 411.
2. J. Wittsiepe, M. Schrey, M. Wilhelm, J. Begerow and L. Dunemann., *J. Trace Elem. Med. Biol.*, 2003, **17**, 117.
3. A. Fisher, *Contact Dermatitis*, Lea and Febiger, Philadelphia, MD, 1987.
4. Q. Zhang, W. Zhong, B. Xing, W. Tang and V. Chen, *J. Inorg. Biochem.*, 1998, **72**, 195.
5. A. B. G. Lansdown, *Wound Rep. Regen.*, 2002, **10**, 130.
6. M. Knipp, *Curr. Med. Chem.*, 2009, **16**, 522.

7. S. Takahashi, *J. Hematol. Oncol.*, 2012, **5**, 41.
8. US Department of Health and Human Services (US DHHS), *Report on Carcinogens*, National Toxicology Program, 11th edn, 2011.
9. Z. H. Siddik, *Oncogene*, 2003, **22**, 7265.
10. K. V. Floros, H. Thomadaki, G. Lallas, N. Katsaros, M. Talieri and A. Scorilas, *Ann. N.Y. Acad. Sci.*, 2003, **1010**, 153.
11. J. Wang, D. He, Q. Zhang, Y. Han, S. Jin and F. Qi, *Cancer Biother. Radiopharmacol.*, 2009, **24**, 675.
12. A. C. Plooy and P. H. Lohman, *Toxicology*, 1980, **17**, 169.
13. J. Reedijk, *Pure Appl. Chem.*, 1987, **59**, 181.
14. E. R. Tiekink, *Inflammopharmacology*, 2008, **16**, 138.
15. N. J. Wheate, S. Walker, G. E. Craig and R. Oun, *Dalton Trans.*, 2010, **39**, 8113.
16. A. B. G. Lansdown, *Patty's Toxicology*, E. Bingham and B. Cohrssen, Wiley, NJ, 2012, vol.1, p. 75.
17. B. A. Bouts, *New Engl. J. Med.*, 1999, **340**, 1554.
18. T. Sollmann, *A Manual of Pharmacology*, Saunders, Philadelphia, 1942.
19. A. B. G. Lansdown, A. Williams, S. Chandler and S. Benfield, *J. Wound Care*, 2005, **14**, 131.
20. A. B. G. Lansdown, *Adv. Pharmacol. Sci.*, 2010, **910**, 686.
21. W. Y. Chen, J. Y. Lin, W. J. Chen, L. Luo, E. Wei-Guang Diau and Y. C. Chen, *Nanomedicine*, 2010, **5**, 755.
22. J. F. Hernández-Sierra, F. Ruiz, D. C. Pena, F. Martinez-Gutiérrez, A. E. Martinez, J. Ade Guillén, H. Tapia-Pérez and G. M. Castañón, *Nanomdicine*, 2008, **4**, 237.
23. S. S. Bleehan, D. J. Gould, C. L. Harrington, T. E. Durrant, D. N. Slater and J. C. Underwood, *Br. J. Dermatol.*, 1981, **104**, 19.
24. H. Suzuki, S. Baba, S. Uchigasaki and M. Murase, *J. Am. Med. Assoc.*, 1993, **29**, 833.
25. P. J. A. Borm, D. Robbins, S. Haubold, T. Kuhl, H. Busch, K. Fissan, R. Donaldson, V. Schins, W. Stone, J. Kreling, J. Lademann, J. Knutmann, D. Warheit and E. Oberdorster, *Part. Fibre Toxicol.*, 2008, **3**, 11.
26. R. E. Burrell, *Ostomy. Wound Manag.*, 2003, **49**(Suppl, 19).
27. A. B. G. Lansdown, *Adv. Pharmacol. Sci.*, 2010, **2010**, 910686.
28. A. B. G. Lansdown, *Silver in Healthcare: its antimicrobial efficacy and safety in use*, Royal Society of Chemistry, London, 2010.
29. A. B. G. Lansdown, *Wounds*, 2006, **2**, 51.
30. A. B. G. Lansdown, *Curr. Probl. Dermatol.*, 2006, **33**, 17.
31. A. B. G. Lansdown, *Trends Wound Care*, 2004, **3**, 87.
32. T. Elliott, *Intens. Care Med.*, 2000, **26**, 5045.
33. R. G. Sibbald, J. Contreras-Ruiz, P. Coutts, M. Fierheller, A. Rothman and K. Woo, *Adv. Skin Wound Care*, 2007, **20**, 549.
34. F. Furno, K. S. Morley, B. Wong, B. L. Sharp, P. L. Arnold, S. M. Howdle, R. Bayston, P. D. Brown, P. G. Winship and H. G. Reid, *J. Antimicrobial. Chemotherap.*, 2004, **54**, 1019.

35. A. Gupta, M. Maynes and S. Silver, *Appl.. Environ. Microbiol.*, 1998, **64**, 5042.
36. S. Silver, *FEMS Microbiol. Rev.*, 2003, **27**, 341.
37. J. Ficeriova, P. Baláz, E. Dutková and E. Gock, *Open Engin. J.*, 2008, **2**, 6.
38. B. K. Biswas, K. Inoue, K. Ohto, H. Harada, H. Kawakita and A. Hoshino, *Proc. Int. Conf. Environ. Asp. Bangladesh*, 2010, **10**, 207.
39. U. U. Jadhav and H. Hocheng, *J. Achieve. Mat. Manufact. Engin.*, 2012, **54**, 159.
40. A. Burd., C. H. Kwok, S. C. Hung, H. S Chan, H. Gu, W. K. Lam and L. Huang, *Wound Repair Regen.*, 2007, **15**, 94.
41. E. C. McCoy and H. S. Rosenkranz, *Chemotherapy*, 1978, **24**, 87.
42. E. Hidalgo, R. Bartolomé, C. Barroso, A. Moreno and C. Domínguez, *Skin Pharmacol. Appl. Skin Physiol.*, 1998, **11**, 140.
43. V. K. M. Poon and A. Burd, *Burn*, 2004, **30**, 140.
44. P. V. AshaRani, G. L. K. Mun, M. P. Hande and S. Valiyaveettil, *Nano*, 2009, **3**, 279.
45. R. L. Zapata-Sirvant and J. F. Hansbrough, *J. Burn Care Rehabil.*, 1993, **14**, 132.
46. E. C. Smoot, J. O. Kucan, A. Roth, N. Mody and N. Debs, *J. Reconstruct. Surg.*, 1991, **87**, 917.
47. R. L. McCauley, H. A. Linares, V. Pelligrini, D. N. Herndon, M. C. Robson and J. P. Heggers, *J. Surg. Res.*, 1989, **46**, 267.
48. B. Venugopal and T. D. Luckey, *Chemical Toxicology of Metals and Metalloids*, Academic Press, NY, 1978, p. 32.
49. P. Bienvenu, C. Nofre and A. Cier, *Comp. Rend.*, 1963, **256**, 1043.
50. J. Wahlberg, *Arch. Environ. Health*, 1965, **11**, 201.
51. R. J. Lewis, *Sax's Dangerous Properties of Industrial Materials*, Van Nostrand Reinhold, 9th edn, NY, 1999, p. 2934.
52. S. Takenaka, E. Karg, C. Roth, H. Schulz, A. Ziesenis, U. Heinzmann, P. Schramel and J. Heyder, *Environ. Health Persp.*, 2001, **109**(Suppl.4), 547.
53. J. H. Sung, J. H. Ji, J. D. Parke, K. S. Song, H. R. Ryu, J. U. Yoon, K. S. Jeon, J. Jeong, B. S. Han and Y. H. Chung, *Inhalat. Toxicol.*, 2008, **20**, 57.
54. H. Nothdurft, *Strahlentherapie*, 1956, **100**, 192.
55. D. Schmähl and D. Steinhoff, *Z. f. Krebsforsch*, 1960, **63**, 586.
56. A. Furst and M. C. Schlauder, *J. Environ. Pathol. Toxicol.*, 1977, **1**, 51.
57. P. L. Drake and K. J. Hazelwood, *Ann. Occup. Hyg.*, 2005, **49**, 575.
58. World Health Organization, *Guidelines for Drinking Water Quality*, WHO, Geneva, vol. 2, 2nd edn, 1996.
59. R. Faust, *Oak Ridge Reservation, Environmental Restoration Program*, 1992.
60. Integrated Risk Information Service (IRIS), *Silver (Cas No. 7440-22-4)*, www.epa.gov/IRIS/subst/0099.htm, 2010.

61. C. H. Banks, *Chemistry, Foundations and Applications*, J. J. Lagowski, Thompson Gale, New York, 2004, p. 176.
62. F. Balkau, *International Council on Metals in the Environment*, 2000, **8**, 4.
63. R. Koch, *Dtsche. Med. Wochenschr.*, 1890, **16**, 756.
64. L. M. de Witt, *J. Infect. Dis*, 1918, **23**, 426.
65. G. Jones and P. M. Brooks, *Br. J. Rheumatol.*, 1996, **35**, 1154.
66. B. M. Sutton, *Overview and Current Status of Gold-containing Anti-arthritic Drugs*, in *Metals in Medicine*, ed. J. C. Dabroviak, John Wiley, Chichester, UK, 1983, p. 355.
67. E. Alessio and S. J. Berners-Price, *Gold-Based Therapeutic Agents: A New Perspective in Bioinorganic Medicinal Chemistry*, Wiley-VCH Verlag, Weinheim, Germany, 2011.
68. A. Bowyer, *Acta Derm. Venereol., (Stockh.)*, 1967, **47**, 9.
69. K. E. Malten and J. Mali, *Allerg. Asthma*, 1966, **12**, 31.
70. D. Caspi, M. Tishler and M. Yaron, *Ann. Rheum. Dis.*, 1989, **48**, 730.
71. L. Kanerva, *Arch. Dermatol.*, 1989, **281**, 321.
72. R. A. Kusiak, J. Springer, A. C. Richie and J. Muller, *Br. J. Dermatol.*, 1991, **48**, 808.
73. R. Eisler, *Environ. Geochem. Health*, 2003, **25**, 325.
74. A. M. Donoghue, *Occup. Med.*, 2004, **54**, 283.
75. B. K. Armstrong, J. C. McNaulty, L. J. Levitt, K. A. Williams and M. S. Hobbs, *Br. J. Ind. Med.*, 1979, **36**, 199.
76. N. E. Lightfoot, M. A. Pacey and S. Darling, *Chron. Dis. Can.*, 2010, **29**(Suppl. 2), 101.
77. H. Kabir and C. Bigli, *J. Occup. Med.*, 1993, **35**, 1203.
78. A. D. Oxman, D. C. Muir, S. R. Shannon, S. R. Stock, E. Hnizdo and H. J. Lange, *Am. Rev. Respir. Dis.*, 1993, **148**, 38.
79. R. A. House, *Lung, Tracheal and Bronchial Cancer in Nickel, Uranium and Gold Miners*, Discussion paper prepared for The Workplace Safety and Insurance Appeals Tribunal, Toronto, Canada, May 2007.
80. D. Brune, G. Nordberg and P. O. Wester, *Sci. Total Environ.*, 1980, **16**, 13.
81. R. Eisler and S. N. Wiemeyer, *Rev. Environ. Contam. Toxicol.*, 2004, **183**, 21.
82. G. Nelson, *Global Health Action*, 2013, **6**, 19520.
83. International Agency for Research on Cancer (IARC), *IARC Monographs on the Evaluation of Carcinogenic Risk of Chemicals to Humans*, IARC, Lyons, France, 1988, vol. 43, p. 173.
84. Agency for Toxic Substances and Disease Registry (ATSDR), *Toxicological Profile for Radon*, ATSDR, Atlanta, GA, 1990, **90**/23, 1.
85. L. Simonato, J. J. Moulin, B. Javelaud, G. Ferro, P. Wild, R. Winkelmann and R. Saracci, *Am. J Ind. Med.*, 1994, **25**, 625.
86. R. Lindsay, R. J. de Meijer, R. T. Newman, T. G. K. Motlhabane and P. Maleks, *Instr. Meth. Phys. Res.*, 2004, **B213**, 775.
87. O. Axelson, *Environ. Health. Perspect.*, 1995, **103**(Suppl. 2), 37.
88. M. Fernandez Casares, H. A. Casas and H. Leczycki, *Medicina (B. Aires)*, 1991, **51**, 59.

89. H. J. L. Marriott and H. R. Peters, *Edinb. Med. J*, 1949, **56**, 381.
90. M. D. Smith, A. Smith, M. Fehr, M. J. Ahern and P. J. Roberts-Thomson, *Ann. Rheum. Dis.*, 1988, **47**, 15.
91. R. A. Miller and J. E. Aldrich, *J. Am. Acad. Dermatol.*, 1990, **23**(Part 2), 360.
92. N. Martin, I. Belinchon, C. Fuente, A. Vélez and E. Sánchez-Yus, *Arch. Dermatol.*, 1990, **126**, 1370.
93. K. Iwatsuku, M. Yamada, M. Takigawa, K. Inoue and K. Matsumoto, *J. Am. Acad. Dermatol.*, 1987, **16**, 83.
94. Y. Kobayashi, H. Nanko, J. Nakamura and M. Mizoguchi, *J. Am. Acad. Dermatol.*, 1992, **27**, 457.
95. M. S. Baptiste, R. Rothenberg, P. C. Nasca, J. T. Janerich, C. D. Stutzman, K. Rimawi and W. O'Brien, *J. Am. Acad.Dermatol.*, 1984, **10**, 1019.
96. R. S. Rothwell and E. H. Scott, *Arch. Dermatol.*, 1980, **116**, 863.
97. B. J. Payne, *Vet. Pathol.*, 1978, **15**(Suppl. 5), 39.
98. B. J. Payne and L. Z. Saunders, *Vet. Pathol.*, 1978, **15**(Suppl. 5), 51.
99. C. Uboldi, G. Bonacchi, G. Lorenzi, M. I. Hermanns, C. Pohl, G. Baldi, R. E. Unger and C. J. Kirkpatrick, *Part. Fibre Toxicol.*, 2009, **6**, 18.
100. B. R. Grubb and P. J. Bentley, *Exp. Eye Res.*, 1988, **46**, 637.
101. G. Sava, T. Girali, G. Mestroni and G. Zassinovich, *Chem. Biol. Interact.*, 1983, **45**, 1.
102. T. C. Johnstone, J. J. Wilson and S. J. Lippard, *Inorg. Chem.*, 2013, in press.
103. V. Kvardova, R. Hrstka, D. Walerych, P. Muller, E. Matoulkova, V. Hruskova, D. Stelclova, P. Sova and B. Vojtesek, *Mol. Cancer*, 2010, **9**, 147.
104. S. Zimmerman and B. Sures, *Environ. Sci. Pollut. Res. Int.*, 2004, **11**, 194.
105. C. Colombo, A. J. Monhemius and J. A. Plant, *Ecotoxicol. Environ. Sci.*, 2008, **71**, 722.
106. B. Gómez, M. A. Palacios, M. Gómez, J. L. Sanchez, G. Morrison, S. Rauch, C. McLoud, R. Ma, S. Caroli, A. Alimonti, E. Petricci, B. Bocca, P. Schramel, M. Zischka, C. Petterson and U. Wass, *Sci. Total Environ.*, 2002, **299**, 1.
107. F. Petrucci, N. Violante, O. Senofonte, A. Cristaudo, M. Di Gregorio, G. Forte and A. Alimonti, *Occup. Environ. Med.*, 2005, **62**, 27.
108. R. Schierl, H. G. Fries, C. van de Weyer and G. Frumann, *Occup. Environ. Med.*, 1998, **55**, 138.
109. R. Merget, G. Schultz-Werninghaus, T. Muthorst, W. Friedrich and J. Meier-Sydow, *Clin. Allergy*, 1988, **18**, 569.
110. D. B. Baker, P. H. Gann, S. M. Brooks, J. Gallagher and I. L. Bernstein, *Am. J. Ind. Med.*, 1990, **18**, 653.
111. C. Sheard, *Arch. Dermatol.*, 1955, **71**, 351.
112. A. V. Roshchin, V. G. Veslov and A. I. Panova, *J. Hyg. Epidemiol. Microbiol. Immunol.*, 1984, **28**, 17.
113. G. M. Levine, *Br. J. Dermatol.*, 1971, **75**, 590.

114. G. M. Levine and C. D. Calnan, *Clin. Allerg.*, 1971, **1**, 75.
115. P. J. Linnett, *Occup. Med. (London)*, 2005, **55**, 595.
116. A. E. Roberts, *Am. Med. Assoc. Arch. Occup. Med.*, 1951, **4**, 549.
117. M. Knipp, A. V. Karotki, S. Chesnov, G. Natile, P. J. Sadler, V. Brabec and M. Vašák, *J. Med. Chem.*, 2007, **50**, 4075.
118. W. Moore, M. Malanchuk, W. Crocker, D. Hysell, A. Cohen and J. F. Stara., *Environ. Health Perspect.*, 1975, **12**, 35.
119. World Health Organization (WHO), *Platinum. Environmental Health Criteria*, WHO, Geneva, 1991, No. 125.
120. I. Kostova, *Recent Patents on Anti-cancer Drug Discovery*, 2006, **1**, 1.
121. J. T. Hartmann and H. P. Lipp, *Expert Opin. Pharmacotherap.*, 2003, **4**, 889.
122. A. V. Roschin, V. G. Veselov and A. I. Panova, *J. Hyg. Epidemiol., Microbiol. Immunol.*, 1984, **28**, 17.
123. R. C. Todd and S. J. Lippard, *Metallomics*, 2009, **1**, 280.
124. R. S. Goldstein and G. H. Mayor, *Life Sci.*, 1983, **32**, 685.
125. C. Marzano, F. Bettio, F Baccichetti, A. Trevisan, L. Giovagnini and D. Fregona, *Chem. Biol. Interact.*, 2004, **148**, 37.
126. N. D. McGlashan, J. S. Harrington and E. Z. Chelkowska, *Eur. J. Oncol.*, 2003, **8**, 19.
127. D. M. Townsend, M. Deng, L. Zhang, M. G. Lapus and M. H. Hanigan., *J. Am. Soc. Nephrol.*, 2003, **14**, 1.
128. P. Mistry, C. Lee and D. C. McBrien, *Cancer Chemotherap. Pharmacol.*, 1989, **24**, 73.
129. P. A. Zhang, M. Koberda, J. L. Hines, C. Knight, R. Sriram, N. V. Raghavan and B. E. Rabinow, *Am. J. Hosp. Pharmacol.*, 1991, **48**, 1500.
130. L. Zhang and M. H. Hanigan, *J. Pharmacol. Exp.Therap.*, 2003, **306**, 988.
131. M. W. Anders and W. Dekant, *Ann. Rev. Pharmacol. Toxicol.*, 1998, **38**, 501.
132. M. H. Greene, *J. Natl. Cancer Inst.*, 1992, **84**, 306.
133. D. Wierda and T. L. Pazdernik, *J. Pharmacol. Exp.Therap.*, 1979, **211**, 531.
134. S. R. Kempf and S. Ivankovic, *Cancer Treat. Prev.*, 1987, **14**, 365.
135. S. Fields, J. Koeller, R. L. Topper, G. Guritz and D. Von Hoff, *J. Natl. Cancer Inst.*, 1990, **82**, 1649.
136. V. G. Schweitzer, *Laryngoscope*, 1993, **103**, 1.
137. D. J. Beck and R. R. Brubaker, *Mutat. Res.*, 1975, **27**, 181.
138. S. Pope and J. H. Parish, *Carcinogenesis*, 1984, **5**, 1105.
139. W. F. Benedict, M. S. Baker, L. Haroun, E. Choi and B. N. Ames, *Cancer Res.*, 1977, **37**, 2209.
140. I. E. Mattern, L. Cocchiarella, C. G. van Kralingen and P. H. Lohman, *Mutat. Res.*, 1982, **95**, 79.
141. W. R. Leopold, E. C. Miller and J. A. Miller, *Cancer Res.*, 1979, **39**, 913.

142. D. A. Dickinson, G. R. Warnes, G. Quievryn, J. Messer, A. Zhitkovich, E. Rubitski and J. Albrecht, *Mutat. Res.*, 2004, **549**, 29.
143. P. Quillardet and M. Hofnung, *Mutat. Res.*, 1985, **147**, 65.
144. T. Gebel, H. Lantzsch, K. Plessow and H. Dunkelberg, *Mutat. Res.*, 1997, **389**, 183.
145. D. Turnbull, N. C. Popescu, J. A. DiPaolo and B. C. Myhr, *Mutat. Res.*, 1979, **66**, 267.
146. A. Pilger, I Köhler, H. Stettner, R. M. Mader, B. Rizovski, R. Terkola, E. Diem, E. Franz-Hainzi, C. Konnaris, E. Valic and H. W. Rüdiger, *Int. Arch.Environ. Health*, 2000, **73**, 442.
147. S. R. Kempf and A. S. Ivankovic, *J. Cancer Res. Clin. Oncol.*, 1986, **111**, 133.
148. M. A. Nelson, B. W. Futscher, M. R. Loew and G. T. Bowden, *Cancer Lett.*, 1992, **65**, 27.
149. K. M. Barnhart and G. T. Bowden, *Cancer Lett.*, 1985, **29**, 101.
150. H. Hennings, R. A. Shores, M. C. Poirier, E. Reed, R. E. Tarone and S. H. Yuspa, *J. Natl. Cancer Inst.*, 1990, **82**, 836.
151. S. A. Belinsky, S. A. Stephanski and M. W. Anderson, *Cancer Res.*, 1993, **53**, 410.
152. A. Hisamoto, E. Kondo, K. Kiura, T. Okada, S. Hosokawa, J. Mimoto, M. Takigawa, M. Tabata and M. Tanimoto, *Lung Cancer*, 2007, **58**, 15.
153. J. Mimoto, K. Kiura, K. Matsuo, T. Yoshino, I. Takata, H. Ueoka, M. Kataoka and M. Harada, *Carcinogenesis*, 2000, **21**, 915.
154. B. A. Diwan, L. M. Anderson, J. M. Ward, J. R. Henneman and J. M. Rice, *Toxicol. Appl. Pharmacol.*, 1995, **132**, 115.
155. A. J. Giurgiovich, B. A. Diwan, K. B. Lee, L. M. Anderson, J. M. Rice and M. C. Poirier, *Carcinogenesis*, 1996, **17**, 1665.
156. E. F. Munoz, D. A. Diwan, R. J. Calvert, C. M. Weghorst, J. Anderson, J. M. Rice and G. S. Buzzard, *Carcinogenesis*, 1996, **17**, 2741.
157. C. Bischin, A. Lupan, V. Taciuc and R. Silaghi-Dumitrescu, *Mini-Rev. Med. Chem.*, 2011, **11**, 21.
158. A.-M. Florea and D. Busselberg, *Cancers*, 2011, **3**, 1351.

CHAPTER 14

Beryllium

14.1 Introduction

Beryllium is a rare alkaline earth metal found in rocks, coal, soil, and volcanic dust. It is a bivalent metallic element classified in Group IIa of the Periodic Table. Although similar in its atomic configuration to other elements of this group, including magnesium, calcium and strontium, it exhibits certain unique chemical properties including a tendency to form inorganic compounds and complexes by covalent rather than ionic bonding. This is attributed to the fact that beryllium has a relatively high nuclear charge coupled with a small atomic radius.[1] Many of its compounds are insoluble in water and biologically unreactive. Beryllium is a hard and brittle metal with a particularly high melting point of 1278 °C and boiling point of 2500 °C, light weight and exceptional strength in relation to its weight. Estimates suggest that beryllium is two-thirds lighter than aluminium and 40% more rigid than steel, making it particularly attractive for use in the aeronautics industry. Other physico-chemical properties making beryllium suitable in engineering industries include its non-magnetic capacity, tolerance to high temperature, thermal and electrical conductivity, transparency to X-rays and ability to form alloys. At concentrations as low as 2%, these hard and extremely durable alloys of beryllium with aluminium, nickel, copper and even uranium provide for a vast range of uses in engineering in the electronics and nuclear industries. Beryllium has a large number of other uses in industry, but the element is notable for its profound toxicity and immunoreactivity, ability to induce granulomas and as a cause of chronic beryllium disease (CBD).[2–4]

Beryllium is rarely found in its metallic form but more than 50 ores have been identified. They contribute to no more than 6 ppm of the Earth's crust. These ores are not found widely in the environment and possibly no more than 80,000 tones exist, mainly in deposits in China, Khazastan, Portugal, Madagascar and the USA. The chief beryllium ores include beryl (10–12% BeO), phenakite

Issues in Toxicology No. 18
The Carcinogenicity of Metals: Human Risk through Occupational and Environmental Exposure
By Alan B. G. Lansdown

Published by the Royal Society of Chemistry, www.rsc.org

(42–45%), bertrandite (40–4%), helbertrandite (32–35%), chrysoberyl (18–20%), helvite and genthelvite (10–12%), bavenite (6–7% BeO), euclase (16–20%), margante (1–3%) and leucophanite (10–12%). Beryl and emeralds (the green variety of beryl) were known to ancient civilisations and have been used as gemstones for many years. Emerald is regarded as a valuable gemstone on account of its rich green colour, combined with durability and rarity. Many ores such as beryl ($3BeO,Al_2O_3,6SiO_2$) and bertrandite ($4BeO.2SiO_2.H_2O$) are silicates from which most metal is extracted. Beryllium oxide (BeO), commonly referred to as beryllia, is found naturally and possesses a unique combination of thermal, dielectric and mechanical properties which are highly desirable in electronic applications.

The general population is exposed normally to low levels of beryllium residues in air, food and water in the course of their daily lives. Median aerobic concentrations of beryllium in the USA were given as 0.03 ng m^{-3} but as high as 0.2 ng m^{-3} in urban and inner city areas. In each case levels were appreciably lower than those implicated in chronic beryllium diseases.[5] People living or working near sites where beryllium is mined, processed, machined and converted into metal alloys may be exposed to higher levels but, as discussed below, occupational risks are most frequently encountered by workers exposed chronically in beryllium industries. Health risks are identified also in recycling industrial waste, which in some countries constitutes a major source of the metal. In the USA for example, as much as 10% of the annual consumption of beryllium was provided by reclamation of the metal from industrial scrap. In Europe, recycling facilities are available to separate valuable metals such as steel, copper, aluminium and beryllium from electronic and other products. In view of the inherent risks of chronic beryllium disease, or beryllosis, special provision is made these days to protect workers from the toxic effects of beryllium emissions and risks of lung diseases.

Environmentally, beryllium presents a minimal hazard to plants or animals. Although levels found in water and soil are probably <6 ppm, the metal does not accumulate to any great extent in aquatic species nor is it taken up from sediment by bottom-feeding molluscs. Most plants are capable of absorbing beryllium to a small extent from the soil, but few are regarded as beryllium accumulators. The general population is exposed to beryllium in food and drinking water, but very low concentrations can be expected to enter the human body in view of the insolubility of the compounds ingested. There is no evidence that beryllium is a trace nutrient in the human body nor is it a known constituent of human food chains. It is conceivable that a small amount of beryllium may be absorbed from soluble compounds such as chloride, fluoride and hydroxide in food or drinking water, but it is expected that much of this is eliminated in the faeces as insoluble phosphate. The US Public Health Service in 1943 established that gastrointestinal absorption of beryllium is about 0.006% but this has not been ratified.[3] Limited experimental evidence has shown that, when animals were fed 5% beryllium metal or beryllium oxide in their diet, intestinal absorption was minimal and growth rates or viability were unaffected. Dietary 2.5% beryllium carbonate fed to pigs led to an inhibition of phosphate absorption presumably through the precipitation and excretion of insoluble beryllium phosphate.[6]

The International Agency for Research on Cancer (IARC) have identified beryllium concentrations in cigarettes of 0.005 μg per cigarette, but emphasised considerable variation according to the origin of the tobacco and the analytical technique applied.[7] Steinmaus and Balmes evaluated risks of beryllium and inhalation and asbestos exposure as a cause of lung disease with reference to a 73-year-old technician who had been an ardent smoker for most of his life.[8] He had worked in various capacities in pouring and measuring beryllium oxide, growing beryllium crystals and cleaning up after experiments for at least 15 years. They estimated that his risk of developing lung cancer from asbestos was less than through cigarette smoking alone, but that the hazards presented by beryllium exposure were comparable to those attributable to cigarette smoking.

The lungs and respiratory tract are regarded by most epidemiological evidence as the prime target organ for beryllium toxicity and carcinogenicity. Early studies, by Hardy and Tabershaw and Van Ordstrand in the 1940s, identified dermatoses and delayed chemical pneumonitis in beryllium workers involved in the manufacture of fluorescent light bulbs and following long-term work in refineries.[9,10] In more recent times, more than 30 epidemiological and case studies have evaluated the risk of lung cancer in workers exposed to beryllium in extraction and processing of ores, alloy manufacture, precious metal refineries, sawing and milling, the nuclear industry (where the metal is used as a neutron moderator) and the aerospace industry (where it is used in the manufacture of high-performance brakes). Additionally, clinical and experimental evidence has been provided to show that beryllium targets subcutaneous tissues, skin and bone.[11–15] The US DHHS and IARC have concluded that beryllium and its compounds should be classified as Class 1 carcinogens on the basis of sufficient human and supportive experimental evidence in several species of animal.[16,17]

14.2 Industrial Uses and Human Exposure

The unique physico-chemical properties of beryllium and its stability and resistance to corrosion make it an ideal choice for many industrial, investigative and medicinal functions. Beryllium oxide ceramics have a thermal conductivity second only to that of diamond amongst electrically insulating materials, dissipating nearly 300 watts per millikelvin (W mK^{-1}) at room temperature. The principal industrial uses of beryllium relate to its value in hard metal alloys with copper, aluminium and nickel, its thermal and electrical conductivity and suitability for fluorescent light bulbs, X-ray tubes and nuclear reactors, and manufacture of specialist equipment requiring durability and resistance to corrosion. The principal uses of beryllium and its compounds are summarised in Table 14.1. It is notable that the primary mirror for the James Watt Space Telescope is made of beryllium, on account of its high strength to weight ratio, ability to retain its shape across a range of temperatures, capacity to conduct electricity and heat, and its non-magnetic property. Other notable applications

Table 14.1 Industrial application of beryllium and its compounds.

Chemical form	*Physico-chemical properties*	*Uses*
Metal	Hard, lightweight metal, stability over wide temperature range, transparency to X-rays Neutron reflector and moderator	Aerospace industries, space shuttle brakes, defence applications, satellite mirrors and telescopes, gyroscopes, nuclear reactors and triggering device for nuclear weapons, protective cladding,
Beryllium oxide (BeO)	Thermal and electrical conductor	Automobiles, aeronautics, electronics, computers, lasers, electrical insulation, electronic circuits, microwave components, fluorescent lamps
Alloys: a. Be–Copper	Be–Cu alloy (<4% Be), hardness and strength, electrical conductivity, elasticity, corrosion and fatigue resistance, tolerates wear, non-sparking, temperature tolerance	Bearings, cams, gears, springs and products subject to hard wear and vibration, moulds and castings, golf clubs, electrical contacts, switches relays, computers, radar and satellite parts, telecommunications
b. Be–Aluminium	Be–Al alloy (20–60% Be), high technology applications, light weight alloy, hardness and strength	Satellite technology, meteorology devices, packaging for nuclear equipment, high resolution medical and industrial X-ray equipment, high performance motor cars
c. Be–Nickel	Be–Ni alloy (0.275–7% Be) high tensile strength, hardness, age hardening characteristics, light weight	High performance drills, watch balance wheels, aeroplane brakes, golf clubs
d. Be–Uranium	Be releases neutrons when bombarded with α-irradiation	Triggering devices for nuclear weapons, rocket technology and fuel systems

include parts for supersonic aircraft and in the Space Shuttle programme. The *World Nuclear News* recently published a statement that "Nuclear fuel containing beryllium oxide and uranium oxide is longer lasting, more efficient and safer than conventional nuclear fuels".[18]

Only bertrandite and beryl are commercially viable sources of beryllium and more than 95% of the world's supply is obtained by mining minerals containing between 0.3 and 1.5% beryllium. The Beryllium Science and Technology Association of the USA list bertrandite ore, which is mined in Utah where it occurs close to the surface and can be extracted in an environmentally friendly method. A small fraction of beryllium is obtained from beryl as a by-product of small scale emerald gemstone mining operations in Brazil, Argentina and other countries in South America. That material contains between 3.0 and 5.0% beryllium, but it is much harder than bertrandite, and the beryllium is more difficult to extract and refine. In the extraction process, beryl ore is melted in industrial furnaces, solidified and crushed, then treated with sulfuric acid to produce a water-soluble sulfate. Bertrandite ore is crushed, made into slurry and also treated with sulfuric acid to form a sulfate. Beryllium sulfate solutions are processed to produce extremely pure beryllium hydroxide [$Be(OH)_2$]. Beryllium hydroxide is the common input material for copper beryllium alloys, beryllia ceramics and pure beryllium metal manufacturing. Beryllium oxide, with many applications in automobiles, electronics and insulation, is prepared by a calcining or roasting process in the presence of a calcium additive.

14.3 Uptake and Metabolism

Human exposure to beryllium and its compounds in everyday life is expected to be low and of small toxicological consequence, other than in areas of mining, processing and manufacturing industries. However, occupational dermal contact with beryllium is a cause of concern in view of the profound ability of the metal to induce contact sensitivity and granuloma formation.[2] Fisher emphasised that soluble salts of beryllium such as the fluoride, chloride, nitrate and sulfate are readily hydrolysed and release free acids which are corrosive and irritant to the skin and eyes. Beryllium compounds of low solubility may dissolve sufficiently in sweat and surface skin moisture to be absorbed percutaneously and evoke delayed contact sensitisation.[19] Penetration of beryllium granules through skin wounds is a cause of ulceration, and granulomas. This is illustrated by people who have developed granulomas of an allergic type and foreign body reactions requiring urgent surgical excision as a result of beryllium phosphor from broken fluorescent tubes penetrating cuts or abrasions.[3,11] Other skin lesions have been reported where workers have accidentally rubbed beryllium residues into skin wounds whilst using grinding tools or working in beryllium extraction and processing or alloy production. Biopsies of characteristic beryllium granulomas reveal fibrous nodules of epithelioid cells containing birefringent beryllium granules within epithelioid cells.[20] Professions most at risk are listed in Table 14.2.

Table 14.2 Occupations at risk of beryllium-related dermal granuloma (from Williams, 1988).

Occupation	*Contact*
Metal workers	Pure beryllium, alloys, scrap metal, metal disposal
Ceramics manufacturers	Crucibles, ceramics, cermats, jet engine blades, rocket covers, brake pads
Electronics	Transistors, heat sinks, X-ray windows
Atomic energy industries	Rocket fuels, heat shields
Laboratory workers	Miscellaneous beryllium compounds
Beryllium extraction from ores	Metallic beryllium, beryllium oxide, hydroxide
Fluorescent lamp manufacturers	Beryllium phosphor

Systemic beryllium disease is a recognised complication of workers in beryllium-related industries and is accompanied by multiple granulomas involving the lymph nodes and lungs. Dermal contact with contaminated workclothes and instruments represents a plausible route of beryllium exposure, although present views are that percutaneous absorption of beryllium through intact skin is low.[21,22] No evidence has been seen to show that beryllium is a cause of skin tumours in humans or experimental animals.

Inhalation should be considered the main route of beryllium uptake and cause of beryllium-related diseases. Workers inhaling beryllium dusts in mining, processing, recycling and beryllium-related industries are liable to develop the major health problems of berylliosis and chronic beryllium disease (CBD). CBD is characterised by massive pulmonary inflammation with multiple granulomas and intra-alveolar septal fibrosis.[3,23–26] This granulomatous pulmonary condition was originally described as "a delayed chemical pneumonitis", implying an underlying immunological mechanism with hypersensitivity reactions to beryllium through inhalation or dermal contact.[27] Gell and Coombs recorded that CBD resembled sarcoid conditions but was distinguishable through lower populations of epithelioid cells and the presence of beryllium in the tissues.[19] Laser microprobe mass spectrometry analysis of tissue sections presents a major advance in diagnosis of CBD.[20,26] Beryllium is inhaled through cigarette smoke along with many other carcinogens. The Beryllium Case Registry collated lung cancers in patients previously diagnosed with acute chemical pneumonitis and bronchitis secondary to short-term beryllium exposure and in their opinion the excess risk of lung cancer could not be explained on the basis of smoking alone.[28]

Patterns of beryllium uptake following inhalation in industry will vary according to worker job profile, duration of employment and levels of environmental contamination. Beryllium absorbed into the body is metabolised probably as a ferritin complex, and accumulates in bone, lung, the haematopoietic system and liver. In a review of published studies up to 2001, Kolanz *et al.* considered that, despite variations in monitoring methods, available data do not show a consistent dose–response relationship between airborne concentrations of beryllium and the incidence of CBD.[25] Monitors for beryllium exposure or total beryllium mass are inappropriate in predicting risk of CBD or the mass of respirable particles. Measurement of chemical form and particle surface chemistry may be more reliable. A recent pilot study reported that daily inhaled dose of beryllium was 0.022μg kg^{-1} per day in workers exposed to airborne particles and that uptake was related to particle size.[22]

Experimental animal studies have demonstrated that beryllium particles are phagocytosed by pulmonary macrophages and elicit a hypersensitivity immune response.[29] Pulmonary deposition and clearance have been demonstrated in rats exposed to beryllium sulfate.[30] At aerosol concentrations sufficient to invoke pulmonary carcinogenesis (34.25 μg m^{-3}) uptake increased to a plateau after 36 weeks when an equilibrium was established between uptake and clearance, possibly involving pulmonary and tracheobronchial lymph nodes.

Ferritin and apoferritin have been demonstrated in rat liver homogenates to be major beryllium binding proteins.[31] Much of the Be(II) ion absorbed appeared to bind to the iron core of the molecule. Beryllium also binds other intracellular proteins including nuclear non-histone phosphoproteins, various phosphate compounds and some enzymes of possible toxicological significance. However, beryllium does not induce or bind metallothioneins, unlike certain other divalent metals, on account of its inability to complex with mercaptide moieties.[31,32] Tissues accumulating the highest levels of beryllium in the mammalian body include liver and spleen, which also exhibit high levels of ferritin. Ferritin and apoferritin may function as cytoprotectants against the toxic effects of beryllium.

14.4 Genotoxicity, Mutagenicity and Carcinogenicity

14.4.1 Human Studies

The carcinogenicity of beryllium has been recognised for many years although several of the studies published before 1987 are considered to be statistically and scientifically inadequate. Occupational exposure to beryllium and its derivatives occurs in mining, refining and extraction, aerospace industries, alloy manufacture and in the nuclear industry. Saracci pointed out that results of epidemiological surveys conducted up to 1991 indicate an excess of lung cancers of the order of 50% after the onset of exposures.[50] The 12th RoC lists beryllium and beryllium compounds as known human carcinogens on the adequacy of evidence provided by recent statistically sound epidemiological studies.[16] These include statistical evaluations of cancer mortalities recorded in the US Beryllium Case Registry covering the period 1940–1969 in beryllium processing plants in Ohio and Pennsylvania with a follow-up to 1988.[33–40] Additional epidemiological evidence points to risks of cancer experienced by workers exposed to beryllium in the production of atomic energy,[41–43] as well as those employed in beryllium mining and processing. The statistics substantiate inhaled beryllium as a cause of increased lung cancer mortality, and show that lung cancer patients had shorter tenure and lower lifetime cumulative beryllium exposures than the population generally. There is a great deal of evidence from large-scale surveys such as that conducted at the plants in Ohio, that – 20-year periods of latency between work in the industry and the appearance of cancers are to be expected.[38] Steinmaus and Balmes emphasised the contributory role played by environmental contaminants such as silica and asbestos dust to CBD and lung cancer incidence, with reference to a science technician who developed a poorly differentiated pulmonary carcinoma following work in a variety of occupations involving exposures to beryllium compounds.[8]

A recent epidemiological study, of 9199 workers in 7 plants exposed to beryllium during 1940–2005, showed a significantly higher rate of mortality attributable to CBD and lung cancer than seen in the US population generally, but emphasised that occupational beryllium exposure is associated with a wide spectrum of illness including cor pulmonale, chronic obstructive pulmonary disease, and cancers of the urinary tract and nervous system, attributable and related to work in the industry.[44,45] Previous human studies have not recorded beryllium as a multi-system carcinogen, suggesting that workers in this study were exposed to other unspecified carcinogens.

Epidemiological evidence for the pulmonary carcinogenicity of beryllium is disputed. Thus, a meta-analysis of cancers of the lungs, bronchus and trachea in large cohorts at Reading, two plants in Lorain, Ohio and nearly all beryllium plants in the USA covering ~50 years concluded that "standard mortality rates were low and insufficient evidence was presented to show a dose–response relationship to beryllium exposure".[25,46,47] Epidemiologists have criticised the air sampling methods, assessments of inhalation exposure, insufficient detail of occupational environments and unspecified uncertainties that did not warrant the conclusion that beryllium exposure was a cause of CBD or lung cancer. Long latent periods between beryllium exposure and appearance of signs of CBD represent further complications in interpreting epidemiological studies.[23,26] Statistical evaluations have mostly failed to give full details of tumour incidence or the pathological nature of the lesions (primary or secondary tumours, malignancies) in relation to occupational beryllium exposure.[35,46] Nevertheless, statements are made that "occupational exposure to beryllium is the most plausible explanation for increased risk of lung cancer" and it has been concluded that, if there is an association between beryllium exposure and lung cancer, it exists only "at the mega-levels of exposure present in the two oldest plants up to the 1950s".[48] An evaluation of respiratory tract cancers in a cohort of 3055 workers in Ohio showed 27 cases of bronchogenic cancer, 5 epidermoid, 9 anaplastic small cell, 6 bronchiogenic adenocarcinoma, 2 carcinoma of alveolar–bronchiolar type and 3 large cell undifferentiated carcinomas.

Several epidemiological studies claim that cigarette smoking did not influence the results of evaluations of the carcinogenicity of beryllium or its compounds, but it is emphasised that some brands of cigarettes contain minute amounts of beryllium (0.005 μg.Be per cigarette) as well as another 55 potentially carcinogenic substances including cadmium, arsenic, nickel, chromium and polycyclic hydrocarbons.[7,49] Long-term inhalation of these materials in cigarette smoke must contribute to the cancer mortality seen in heavy smokers. Although some studies allude to a lack of evidence implicating cigarette smoking in beryllium-related lung cancer,[8,47,49] MacMahon considered that its importance in beryllium workers was undervalued and stressed that cigarette smoking was a more likely cause of lung cancer mortality complicated by severe pulmonary emphysema and ischemic heart disease than beryllium-related toxicity.[36] This has not been verified.

14.4.2 Experimental Animal Studies

At least 30 experimental studies in rats, mice, rabbits, dogs and monkeys have confirmed the carcinogenicity of beryllium and certain inorganic salts by intratracheal instillation, or by oral, inhalational or intravenous administration.[50] Tumour incidence and pulmonary pathology vary according to the species involved, the duration and route of administration and genetic profile.[4,51,52] Tumour locations also differ between animals and humans. Thus, whereas inhalation or intratracheal instillation of beryllium or an inorganic salt has regularly been shown to induce pulmonary adenocarcinoma, as in humans, carcinomas of the central nervous system and urinary tract as reported an epidemiological study of 9199 workers followed in Ohio beryllium processing plants during 1940–2005 have not been observed experimentally.[44,45] On the other hand, highly aggressive metastatic and transplantable fibrosarcomas, osteosarcomas and chondrosarcomas, seen in early studies in rabbits given multiple intravenous injections of beryllium oxide (BeO), zinc beryllium silicate, or beryllium sulfate ($BeSO_4$) for 39 or more weeks, have not been reported in any human studies.[12–15,53] Whilst the experimental osteosarcomas are histologically and behaviourally similar to human bone tumours, no evidence is seen to show that beryllium is a cause of human osteosarcoma.

Experimental bone tumours seen in rabbits injected with beryllium salts seem to originate in medullary bone and metastasise to liver, lymph nodes and lung. Tapp actually introduced zinc beryllium silicate particles (20 mg) into the marrow cavity to induce sarcoma formation.[15] In that study, tumours developed 12–15 months after the single injection. Beryllium interacts with calcium and locates in bone and cartilage, and in rabbits given intravenous suspensions beryllium oxide (13.0–11.6 mg kg^{-1} for up to 26 weeks), osteosarcomas developed within 16 months. Beryllium particles were phagocytosed by macrophages and transported to lymph nodes, spleen and liver where they invoked foreign body reactions.[12,14] Beryllium-induced bone tumours in rabbits can be transplanted to the inner chamber of the eye, where sustained growth and differentiation was observed in some but not all grafts. Metastatic tumours in the lung were also transplantable. Beryllium-laden macrophages were associated with a diminution of white pulp and lymphocytes in spleen, but neoplastic changes did not develop in these tissues within 120 weeks of dosing. Beryllium-induced bone tumours have not been reported in guinea pigs or rats subject to chronic beryllium exposure but experimental studies in these species do substantiate the risk of lung cancer following chronic beryllium exposure.[52]

Finch *et al.* conducted experiments in beagle dogs exposed to calcined beryllium oxide and showed that the animals developed CBD and specific immune changes but found that CBD did not progress in the same fashion as in humans.[52] Squirrel monkeys exposed to 35 μg m^{-3} beryllium sulfate also developed CBD and of the 12 surviving the study (<4070 h), 8 developed multiple pulmonary carcinomas.[54] Pulmonary granulomas and lymphocytic proliferation have been reported in two strains of mice inhaling beryllium metal, but rats seemed to be the preferred species for demonstrating the pulmonary carcinogenicity of beryllium. At least seven different strains of rat have been exposed to beryllium by aerosol inhalation or by intratracheal instillation. Mean lung burdens as low as 17 μg were sufficient to induce 50% incidence of tumours in F344 rats surviving ≥1 year, but the tumours were not associated with K-*ras* or *p53* gene mutations.[52] In mice, lung beryllium burdens consistent with tumour induction were equivalent to 300 μg and responses were strictly strain dependent. Reeves compared the variability among strains of rat to inhaled beryllium sulfate (6–54.7 μg m^{-3}) and reported progression through CBD to frank neoplasia in Sprague–Dawley, Sherman and Charles River strains. The Charles River strain seemed to be most responsive strain, with adenomas, adenocarcinomas and squamous tumours developing after a latent phase of nine months from initial exposure.[30,55,56] In single inhalation-dose studies, F344 rats developed pulmonary adenocarcinomas following exposure to 500, 410, 830 or 980 mg m^{-3} metallic beryllium in aerosol spray for up to 48 minutes.[57] Tumours developed in 64% of the animals after a 14-month latent phase. Be(II) ion accumulated initially in pulmonary tissue and was transported by macrophages and lymphocytes to local lymph nodes and other tissues.[30,57] Female animals were more efficient in clearing pulmonary beryllium deposits.

The mechanism of induction of lung tumours by beryllium metal, Be(II), is unclear. Experimental studies in rodents, monkeys, dogs and rabbits indicate that impairment in DNA synthesis, mutation or other Be–DNA interaction is not involved.[52,57] They suggest that Be(II)-induced neoplastic change develops through CBD and/or transformation of target cells.[58,59]

14.4.3 *In Vitro* Studies

Mutagenicity studies in bacterial systems and in mammalian cell cultures present conflicting information.[4,60] Bacterial tests including the Ames *Salmonella typhimurium* reversion test have mostly failed to provide evidence of mutagenicity, yet cell cultures have shown that certain soluble salts such as beryllium chloride and nitrate induce sister-chromatid exchanges and have a capacity to damage DNA.[61] In broad ranging mutagenicity tests conducted in accordance with the Organisation for Economical Cooperation and Development (OECD) guidelines, metallic beryllium showed a low solubility in simulated lung fluid but was moderately soluble in lysosomal fluids.[62] Beryllium failed to show evidence of mutagenicity in bacterial reversion tests, cell gene mutation assays, chromosome aberration tests, unscheduled DNA synthesis or cell transformation tests using Syrian hamster embryo cells. Further, beryllium metal failed to induce DNA repair in cells damaged by external stimuli. These studies predict that beryllium is not mutagenic or clastogenic, and also demonstrate its low toxicity in acute oral toxicity and skin and eye irritancy tests.

Lack of mutagenicity is further illustrated by cytogenetic analysis of lung tumours induced in F344 rats.[57] DNA isolates of 24 adenocarcinomas showed no mutations in K-*ras* codons 12, 13 or 16, but 2 of 12 tumours analysed exhibited GGT–GTT transversions. Beryllium failed to induce mutagenicity in the marker gene *p53* or rearrangements in the *c-raf*-1 proto-oncogene. In all, this comprehensive test schedule provides minimal evidence that metallic beryllium induces neoplastic change by direct action on gene function but it may act through an epigenetic pathway.

Although *in vivo* tests in animal models and clinical trials indicate that beryllium is a pulmonary carcinogen and is capable of inducing and promoting neoplastic change, experimental studies in cultured lung fibroblasts show that the Be(II) ion may block replicating cells in the G_0–G_1 pre-S phase of the cell cycle at low concentrations.[63] Whilst the mechanism remains to be investigated further, it is suggested that this involves upregulation of the *p53* gene and synthesis of the cyclin-dependent kinase inhibitor CDKNIA (p21,*Waf-1.Cip1*). It is conceivable that abnormalities in the cell cycle underlie the susceptibility of certain individuals to CBD.

Evidence that beryllium metal or Be(II) ions exhibit direct genotoxicity is equivocal, but there is limited evidence to show that Be(II) binds to and inhibits a range of key metabolic enzymes including alkaline phosphatase, phosphoglucomutase, K^+,Na^+-dependent ATPase, phosphoprotein phosphatase, hexokinase, glucosphosphatase and ribonuclease.[31,64] On the other hand, soluble beryllium salts failed to inhibit respiratory enzymes in rat hepatocyte cultures and actually promoted succinic hydroxide and ATP (in the presence of calcium).[65] Witschi examined the influence of beryllium on enzymatic changes in rats and discussed the possibility that beryllium inhibits enzyme induction at the transcriptional level.[66] Experiments in rats following partial hepatectomy showed that beryllium inhibited DNA-synthesizing enzymes but failed to influence enzymes such as aspartate transcarbamoylase, carbamoyl phosphate synthetase, uridine kinase and glucose 6-phosphatase, which are not involved in liver regeneration. No evidence was seen that that beryllium specifically inhibited thymidine kinase or DNA polymerase. Whilst the IARC and the 12th Report on Carcinogens claim that ionic beryllium caused genetic damage in mammalian cells possibly through nucleic acid binding and infidelity in DNA replication,[16,17] this is not entirely substantiated by many *in vitro* experiments published.[63,67,68] It is clear that the Be(II) ion is capable of inducing many subtle changes at sub-toxic levels including upregulation of cytokines, interferon and immune changes as part of a complex epigenetic profile.

14.5 Conclusions

Beryllium is a toxic metal and chronic inhalation of dust in industrial environments is a cause of respiratory disease characterised by massive inflammatory change, multiple granulomas and systemic disease with immunological change (*i.e.* CBD).[10,39,69,70] In chronic states the condition resembles sarcoid.[19] Beryllium is a cause of delayed hypersensitivity and cytochemical changes manifest by inhibition of key metabolising enzymes, alterations in the cell cycle and release of cytokines and growth factors. These and other epigenetic changes can be expected to impair gene expression and lead to signalling defects, resulting in neoplastic change. Much is still speculative and the mechanism of action of beryllium in inducing lung and/or other cancers requires a lot more study.

Studies in experimental animal models confirm the carcinogenicity of beryllium and provide some guide as to the sequence of events leading to CBD and carcinoma formation, but marked interspecies differences in tissue sensitivity to neoplasia leave questions as to their suitability in predicting human risk.[40] The dosing schedules and routes of administration of beryllium and its compounds do not accurately represent patterns of human exposure and, with the exception of lung cancers, tumours develop at sites not seen in beryllium workers. However, the lung tumours induced in animals are histopathologically similar to human tumours and are metastatic.[8,48]

At least 17 epidemiological studies have been evaluated and while some claim to show a relationship between beryllium exposure and lung cancer, others that have been reanalysed show that standardised mortality rates are lower than expected and not statistically significant. Thus Levy *et al.*, for the National Institute of Occupational Safety and Health, found that the data presented by large scale studies in USA (Ohio) were not compatible with the view that beryllium exposure is a likely cause of lung cancer.[46,47] Other reviewers are of the opinion that the weight of evidence presented by published studies suggests that beryllium at high level is a probable cause of lung cancer, but that at exposure levels experienced in modern industries evidence of risk is inadequate or at best marginal.[37,40] Throughout, concerns have been expressed concerning the validity and quality of environmental contamination, the sufficiency of worker records of employment history, lack of detail on airborne beryllium exposures (particle size and concentration), duration of service, *etc.*, and allowances for cigarette smoking. It is noteworthy that lung cancer incidence attributed to beryllium exposure in one plant in the USA was estimated in an area of high environmental pollution and where local lung cancer rates were far in excess of national averages.[47] A further big problem recognised in occupational studies in beryllium industries is the 10–20-year period of latency between presumed first exposure and identification of tumours of the lung or other tissues.[25,40,44,45]

The contribution of cigarette smoking to beryllium-related lung cancer is equivocal.[47] Cigarette smoking is a recurrent problem in estimating the health risk of workers in any industrial environment and, as in the beryllium industry, allowances are made in calculating statistical significance.[8,49] Authors of major studies have claimed that lung cancer risk in beryllium plants is higher than would be expected in smokers generally and have applied "adjustment factors" in their analyses. It is conceivable that any or all of the 50 or more constituents of cigarette smoke promote or otherwise influence preneoplastic changes attributable to beryllium, or they act with it in the form of co-carcinogens. This aspect will require further study. Pending further clarification and more detailed case-by-case evaluation, beryllium should be classified as a carcinogenic risk to workers exposed for long periods in industry.

References

1. D. A. Everest, *The Chemistry of Beryllium*, Elsevier, New York, 1964.
2. A. A. Fisher, *Contact Dermatitis*, Lea and Febiger, Philadelphia, 1987.
3. D. C. Deubner, Y. W. Lowney, D. Paustnbach and J. Warmerdam, *Appl. Occup. Environ. Hyg.*, 2001, **16**, 568.
4. L. S. Welch, *Patty's Toxicology*, E. Bingham and B. Cohrrsen, John Wiley & Co, Hoboken, NJ, 2012, vol. 1, 113.

5. Agency for Toxic Substances and Disease Registry (ATSDR), *Toxicological Profile for Beryllium*, ATSDR, US DHHS, Washington, DC, 2002.
6. J. H. Moore and C. Tyler, *Brit. J. Nutr.*, 1955, **9**, 389.
7. C. J. Smith, S. D. Livingston and D. J. Doolittle, *Fd. Chem. Toxicol.*, 1997, **35**, 1107.
8. C. Steinmaus and J. R. Balmes, *Environ. Health Perspect.*, 2000, **108**, 1003.
9. H. L. Hardy and L. R. Tabershaw, *J. Ind. Hyg. Toxicol.*, 1946, **28**, 197.
10. H. S. Van Ordstrand, R. Hughes, J. M. DeNardi and M. G. Carmody, *J. Am. Med. Ass.*, 1945, **129**, 1084.
11. J. E. Flynn, *Ann. Surg.*, 1952, **137**, 265.
12. J. M. Barnes, F. A. Denz and H. A. Sissons, *Br. J. Cancer*, 1950, **4**, 622.
13. I. Fodor, *Acta Morphol. Acad. Sci. Hung.*, 1977, **25**, 99.
14. G. M. Higgins, B. M. Levy and B. L. Yollick, *J. Bone Joint Surg.*, 1966, **46**, 789.
15. E. Tapp, *Br. J. Cancer*, 1966, **20**, 778.
16. US Department of Health and Human Services, *12th Report on Carcinogens*, US DHHS, Research Triangle Park, NC, 2011.
17. IARC, Monographs on the Evaluation of Carcinogenic Risks to Humans, IARC, Lyons, France, vol. 58, 1993.
18. World Nuclear News, Beryllium-Uranium Fuel Research Shows Promise, WNN, London, 2013.
19. P. G. H. Gell and R. R. A. Coombs, *Clinical Aspects of Immunology*, Blackwell Scientific, Oxford, 2nd edn, 1967.
20. W. J. Williams, *Postgrad. Med. J.*, 1988, **64**, 511.
21. G. A. Day, A. Dufresne, A. B. Stefaniak, C. R. Schuler, M. L. Stanton, W. E. Miller, M. S. Kent, D. C. Deubner, K. Kreiss and M. D. Hoover, *Ann. Occup. Hyg.*, 2006, **61**, 67.
22. C. Emond, J. P. Robin, R. Breton, S. Philippe and J. Zayed, *J. Toxicol. Environ. Health*, 2007, **70**, 529.
23. P. A. Theodos and R. T. Cathcart, *J. Am. Med. Assoc.*, 1955, **158**, 1428.
24. Y. Kusaka, *Sangyo Igaku*, 1993, **35**, 75.
25. M. E. Kolenz, A. K. Madl, M. A. Kelsh, M. S. Kent, R. M. Kalmes and D. J. Paustenbach, *Appl. Occup. Environ. Hyg.*, 2002, **16**, 593.
26. J. Müller-Quernheim, K. I. Gaerde, A. Prasse and G. Zissel, *Pneumologie*, 2007, **61**, 109.
27. S. S. Tinkle, J. M. Antoni, B. A. Rich, J. R. Roberts, R. Salmen, K. DePree and E. J. Adkins, *Environ. Health Perspect.*, 2003, **111**, 1202.
28. P. F. Infante, J. K. Wagoner and N. L. Sprince, *Environ. Res.*, 1980, **21**, 35.
29. A. F. Eidson, A. Taya, G. L. Finch, M. D. Hoover and C. Cook, *J. Toxicol. Environ. Health*, 1991, **34**, 433.
30. A. L. Reeeves and A. J. Vorwald, *Cancer Res.*, 1967, **27**, 446.
31. D. J. Price and J. G. Joshi, *J. Biol. Chem.*, 1983, **258**, 10873.
32. J. K. Piotrowaki and J. A. Szymanska, *J. Toxicol. Environ. Health*, 1976, **1**, 991.
33. J. K. Wagoner, P. F. Infante and D. L. Bayliss, *Environ. Res.*, 1980, **21**, 15.

34. T. F. Mancuso, *Environ. Res.*, 1980, **21**, 48.
35. E. Ward, A. Okun, A. Ruder, M. Fingerhut and K. Steenland, *Am. J. Ind. Med.*, 1992, **22**, 885.
36. B. McMahon, *J. Occup. Med.*, 1994, **36**(15), 25.
37. P. D. McGavran, A. S. Rood and J. E. Till, *Environ. Health Perspect.*, 1999, **107**, 731.
38. M. T. Sanderson, E. M. Ward, K. Steenland and M. R. Peterson, *Am. J. Ind. Med.*, 2001, **39**, 133.
39. K. Steenland and E. Ward, *J. Natl. Cancer Inst.*, 1991, **83**, 1380.
40. D. M. Hollins, M. A. McKinley, C. Williams, A. Wiman, D. Filos, P. S. Chapman and A. K. Madl, *Crit. Rev. Toxicol.*, 200, **39**(Suppl. 1), 1.
41. D. P. Loomis and S. H. Woolf, *Am. J. Ind. Med.*, 1996, **29**, 131.
42. S. Wing, C. M. Shy, J. L. Wood, S. Wolf, D. L. Cragle, W. Tankersley and E. Frome, *Am. J. Ind. Med.*, 1993, **23**, 265.
43. C. Rooney, V. Berel, N. Maconochie, P. Fraser and G. Davies, *Br. Med. J.*, 1993, **307**, 1391.
44. M. K. Schubauer-Berigan, J. R. Couch, M. R. Peterson, T. Carreón, Y. Jin and J. A. Deddens, *Occup. Environ. Med.*, 2010, doi.10.11.1136/uem, 2010 056515.
45. M. K. Schubauer-Berigan, J. A. Deddens, J. R. Couch and M. R. Peterson, *Occup. Environ. Med*, 2010, doi. 10.11.36/oem. 2010, 056515.
46. P. S. Levy, H. D. Roth and D. C. Deubner, *J. Occup. Environ. Med.*, 2007, **49**, 96.
47. P. S. Levy, H. D. Roth, P. M. T. Hwang and T. E. Powers, *Inhal. Toxicol.*, 2002, **14**, 1003.
48. A. B. Smith and Y. Suzuki, *Environ. Res.*, 1980, **21**, 10.
49. S. S. Hecht, *J. Natl. Cancer Inst.*, 1999, **91**, 1194.
50. R. Saracci, *J. Natl. Cancr Inst.*, 1991, **83**, 1362.
51. D. H. Groth, *Environ. Res.*, 1980, **21**, 66.
52. G. L. Finch, M. D. Hoover, F. F. Hahn, K. J. Nikula, S. A. Belinsky, P. J. Haley and W. C. Griffith, *Environ. Health Perspect.*, 1996, **104**(Suppl. 973).
53. L. V. Gardner and H. F. Heslington, *Fed. Proc.*, 1946, **5**, 221.
54. A. J. Vorwald, *Use of Non-Human Primates in Drug Evaluation*, ed. H. Vagtborg, Univ. Texas Press, Austin, TX, 1968, p. 222.
55. A. J. Vorwald and A. L. Reeves, *Arch. Ind. Health*, 1959, **19**, 190.
56. A. L. Reeves, D. Deitch and A. J. Vorwald, *Cancer Res.*, 1967, **27**, 439.
57. C. Nickell-Brady, F. F. Hahn, G. L. Finch and S. A. Belinsky, *Carcinogenesis*, 1994, **15**, 257.
58. A. L. Reeves, *Int. J. Toxicol.*, 1989, **8**, 1307.
59. K. A. Goel, V. P. Agrawal and V. Garg, *Bull. Environ. Contam. Toxicol.*, 1980, **24**, 59.
60. T. Gordon and D. Bowser, *Mutat. Res.*, 2003, **533**, 99.
61. K. Kuroda, G. Endo, A. Okamoto, Y. S. Yoo and S. Horiguchi, *Mutat. Res.*, 1991, **264**, 163.
62. C. Strupp, *Ann. Occup. Hyg.*, 2011, **55**, 30.

63. N. M. Lehnert, R. K. Gary, B. L. Marrone and B. E. Lehnert, *Toxicol.*, 2001, **160**, 119.
64. M. Thomas and W. N. Aldrich, *Biochem. J.*, 1966, **98**, 94.
65. F. W. Klemperer, J. M. Miller and J. C. Hill, *J. Biol. Chem.*, 1949, **180**, 281.
66. H. P. Witschi, *Biochem. J.*, 1970, **120**, 623.
67. S. Silva, K. Ganguly, T. M. Fresquez, G. Gupta, T. M. McCleskey and A. Chaudhery, *J. Occup. Emviron. Hyg.*, 2009, **6**, 775.
68. J. Y. Lee, O. Alochina, B. King, L. Taylor, M. Elloso, P. Scott and M. D. Rossman, *Infect. Immun.*, 2000, **68**, 4032.
69. J. M. Aronachik, *Radiol. Clin. N. Am.*, 1992, **30**, 1209.
70. K. C. Meyer, *Chest*, 1994, **106**, 942.

CHAPTER 15

Gallium, Indium and Thallium

15.1 Introduction

These three metallic elements are minor elements in the Earth's crust and are not knowingly encountered by most people in everyday life. They are classified with aluminium in Group IIIa of the Periodic system and resemble aluminium in some aspects of their chemistry. Although the elements were isolated and identified in the mid-1800s, only recently have their physico-chemical properties been fully appreciated and their value in industry recognised. Despite differences in their overt form, gallium, indium and to a lesser extent thallium, and certain of their inorganic compounds, show excellent electrical conductivity and suitability for semiconductors used in solid-sate electronic devices, liquid crystal display (LCD) units, computers, lasers, light emitting diodes and solar panels. Their diversity of applications becomes better understood through detailed knowledge of the industry and the specific value of the metals.[1–4]

Gallium and indium were regarded as being relatively non-toxic and without human health risk until concerns were raised in the rapidly expanding semiconductor industry following reports of increased pulmonary disorders, reproductive failures and cancer fatalities.[5–9] In 1985, a report was published evaluating the health risks faced by a group of 1807 workers in the semiconductor industry and at the time fatalities from all causes were close to those expected, but risks of melanoma were suspected.[10] Semiconductor compounds particularly identified as presenting health risks include gallium arsenide, indium phosphide, complexes of indium and tin oxides (ITO), thallium, thallium oxide, thallium carbonate and antimony. Fowler and others emphasised that, in the 1990s, there was a relatively small toxicology base for gallium, indium and thallium but that there was evidence that such effects as alterations in gene products, impairments in metabolising enzymes, induction of oxidative stresses and possible mutagenicity predicted that these compounds may possess carcinogenic activity.[9] Gallium arsenide has not been confirmed as a carcinogen in

Issues in Toxicology No. 18
The Carcinogenicity of Metals: Human Risk through Occupational and Environmental Exposure
By Alan B. G. Lansdown

Published by the Royal Society of Chemistry, www.rsc.org

experimental studies in rodents by Research by the National Toxicology Program (NTP),[11] although persuasive experimental evidence is available to suggest that a carcinogenic risk may accompany exposure to indium phosphide or ITO.[12,13] The carcinogenicity of arsenic and antimony is discussed in Chapter 17. Chen emphasised that the safety of semiconductor materials is of paramount importance in a country such as Taiwan where much of the economic development is based upon the semiconductor industry.[14] Taiwan is claimed to be the largest producer of CD-ROMs, light emitting diodes (LED) and integrating circuit products, and 30,000 people are exposed to gallium, indium arsenic and thallium in the semiconductor industry.

The IARC evaluated, in detail, the specific health risks posed to workers exposed to gallium arsenide and indium phosphide and concluded that whilst the former should be classified as a Group 1 human carcinogen, there was inadequate evidence for the carcinogenicity of indium phosphide in exposed workers.[15] Thallium in contrast is an extremely toxic metal and a serious cause of neurological damage but, under guidelines issued by the US Environmental Protection Agency, there was inadequate information to assess its carcinogenic potential, and there were no adequate studies in animals to evaluate the risks associated with thallium or its compounds in animals.[16,17]

15.2 Gallium

Gallium is a silvery white soft metallic element with chemistry closely resembling that of aluminium. It exists in a near liquid state at room temperature and exhibits a very low melting point of 29.78 °C and high boiling point of 2250 °C. As such, gallium exhibits the largest liquid temperature range of all metals and shows a strong tendency to supercool below its freezing point. It expands greatly on heating, making it more suitable than mercury for high temperature thermometers. An unusual property of gallium arsenide, used in semiconductors and electrical appliances, is its ability to convert electricity into coherent light.

Gallium is found in the form of two stable isotopes, ^{69}Ga and ^{71}Ga, with natural abundance of 60.108 and 39.892% respectively. ^{69}Ga is used for production of the radioisotope ^{68}Germanium which is used for so-called Ge-68/Ga-68 generators. Nine other radioactive isotopes of gallium exist but these decay in a short time to form isotopes of zinc or germanium. Gallium occurs naturally as a trace contaminant in ores of other metals such as bauxite (aluminium ores), sphaerelite, quartz, germanite ($Cu_{13}Fe_2Ge_2S_1$), zinc ores, and coal. Flue dusts produced by combustion of coal may contain up to 1.5% gallium. Bauxite is the most common source. Gallium can be extracted from mineral ores by conventional industrial chemical means and most of the world output (?270 tonnes) from China, Germany, Kazakhstan and Ukraine is used in the semiconductor industry.

Although gallium has no known physiological function in the human body, certain of its characteristics enable it to interact with cellular processes and biologically important proteins, especially those of iron metabolism. This has led to the development of certain gallium compounds as diagnostic and therapeutic agents in medicine especially in the areas of metabolic bone disease, cancer and infectious diseases.[18] Thus, radiolabelled gallium citrate or nitrate is used in gallium scans in diagnostic medicine to visualise bone or soft tissues and in the diagnosis of cancers. Gallium maltolate tris(3-hydroxy-2-methyl-4H-pyran-4-onato)gallium is a soluble salt that is being investigated as a stable form of gallium for oral delivery. Gallium as Ga(III), released from soluble salts such as gallium citrate used in radio-diagnostics, is excreted slowly from the body. Gallium arsenide exhibits a similar three-dimensional (3-D) lattice structure to indium phosphide and is virtually insoluble, and it is unknown whether the molecule cleaves in the body to release As(III), a known carcinogen

Older texts quote gallium and its compounds as being of low toxicity in humans. More recent experimental studies have implicated gallium arsenide as an inhibitor of δ-amino-laevulinic acid dehydratase through displacement of zinc, thereby impairing haem synthesis.[19] Bernstein *et al.* have shown that cancer cells absorb gallium more than normal cells and that, through inhibition of iron-related haem synthesis, gallium may act as an antiproliferative and anticancer agent.[20] This is supported by *in vitro* studies which showed that gallium nitrate and gallium maltolate inhibited proliferation in a human hepatocellular carcinoma cell line in a dose-dependent fashion. Gallium is a potent inhibitor of protein synthesis, particularly the enzyme γ-aminolaevulinate dehydratase in the haem pathway. It binds serum transferrins, reduces pain and improves the quality of life of patients undergoing ^{67}Ga-citrate scans for bone tumour diagnosis. Through its probable action in displacing iron and zinc, gallium maltonate therapy reduced serum bilirubin and inflammation-related liver enzymes and ultimately led to regression of primary liver tumours.

In vitro assays have shown that gallium arsenide is neither mutagenic in the *Salmonella typhimurium* reversion test with or without S9 metabolic enzyme extracts, nor able to evoke DNA damage or changes in a Syrian hamster embryo cell transformation test in a dose-responsive manner.[21–23] It failed to induce micronuclei in peripheral blood of mice exposed by inhalation for 14 weeks. Other biochemical changes effected by parenteral administration of gallium arsenide include upregulation of chemokines, cytokines and mRNA expression, all representing activation of cellular activity and antigen expression.[24] It selectively upregulates inflammatory cytokines at exposure sites and in macrophages may alter gene and cytokine expression as a prelude to pathological damage.[8,19,24,25] In cultured murine macrophages, gallium arsenide was shown to augment expression of cathepsins. These and other cytochemical changes are indicative of the ability of gallium arsenide to modulate inflammatory responses and pathological damage in tissues.

Few experimental and clinical studies are published relating to the toxicology of gallium and its compounds. Human epidemiological studies have generally evaluated the cancer risks faced by workers in the semiconductor production industry, rather than possible carcinogenic changes attributable to individual components such as gallium arsenide or indium phosphide.[5,6,10,26] A major study conducted by the NTP failed to provide substantive experimental evidence for the carcinogenicity of gallium arsenide through subacute to chronic inhalation.[11] Thus, F344 rats and B6C3F1 mice were exposed to inhalates of gallium arsenide ranging from 1 to 150 mg m^{-3} for 6 hours daily 5 days per week for 14 days, 14 weeks or 2 years. The animals showed a dose-related pulmonary response and haematological changes reflecting the ability of gallium to accumulate in the lungs and to interact with and displace iron and zinc from key enzymes. Animals exhibited a dose-related reduction in body and organ weights. In 14-day studies, 150 mg m^{-3} gallium arsenide exposure led to squamous metaplasia of the larynx in both species, with male animals being more susceptible than female animals. Mice were sensitive to subacute exposures as low as 10 mg m^{-3}. Chronic (2-year) studies focused more on low level gallium exposures (1 mg m^{-3}), but even here body weights were significantly reduced and the incidence of pulmonary pathology increased (females > males). Rats inhaling gallium arsenide for 2 years showed a higher number of adrenal medullary tumours and greater prevalence of mononuclear leukaemia, compared with the "historical control range", whereas no neoplastic pathology was reported in mice. Hamsters were not susceptible to carcinogenicity following repeated intratracheal instillation of 3.3% gallium arsenide for 15 weeks.[27]

The toxicity and carcinogenicity of gallium-related compounds used in semiconductors is debatable in view of the lack of human case or epidemiological evidence.[28] Many compounds, including gallium arsenide and aluminium gallium arsenide, are of low solubility and presumably not absorbed into the body in sufficient amounts to cause injury. Experimental studies in rodents show that pulmonary macrophages phagocytose particles as a protective mechanism. Other changes including squamous metaplasia and inflammatory changes in the respiratory tract are attributed to long-term irritancy as seen with non-carcinogenic dusts. The average whole-body retention after 7 days was 65%, with 26% being excreted in the urine and 9% in the stools (Soreq Radiopharmaceuticals, Israel).

At the moment no clinical or experimental evidence has been presented to show that gallium or its common compounds are carcinogenic. Where pathological changes were reported in rodent assays, levels of gallium arsenide in the inhalate were many times the expected level likely to be experienced in the semiconductor or other industry involving human exposure to gallium dust. Consistent with the American Conference of Governmental Industrial Hygienists, a permitted exposure level for gallium arsenide of 0.3 μg m^{-3} is recommended to protect against pulmonary inflammation and other adverse effects.[29] In contrast, the current Occupational Safety and Health Administration (OHSA) give a permissible exposure limit for arsenic at 1 μg m^{-3} as a time-weighted average over an 8-hour work shift.[30] The recommendations for gallium computed from the results of limited experimental experience gained from rodent studies are urgently in need of ratification.

15.3 Indium

Indium is a soft silvery white metal that gives a brilliant intense blue colour (like indigo) in its emission spectrum. It is a rare element and probably composes no more than 0.05 ppm of the Earth's crust. Deposits are found in association with zinc, copper, iron, lead and other ores. Canada is the chief exporter and in 2007 produced about 65 tonnes. In chemical properties, indium resembles gallium in many ways; it remains in a liquid state over a wide range of temperatures and has a high boiling point (2353 °C), has an ability to form alloys and is corrosion resistant. When heated above its melting point, it burns with a violet flame to the yellow sesquioxide (In_2O_3). For many years, indium was regarded as a rare and uninteresting metal but recent research has shown it to have properties conducive to important uses in low melting point alloys for bearing castings, coatings for corrosion resistance, photoconductors and a wide range of electrical applications including transistors, rectifiers and semiconductors. Liquid indium clings to or wets glass and similar surfaces.

Indium forms a number of inorganic and organic compounds mainly with the trivalent oxidation state, but some compounds derived from the In(I) and In(II) oxidation states include halides, selenide, oxide, sulfide and telluride. Organo-indium compounds are useful in the synthesis of materials by the metal-organic chemical vapour phase deposition process involving thermal decomposition.[31] Mixtures of an organo-indium compound and a reagent such as phosphine are transformed in this process to produce the structurally strong compound indium phosphide used in semiconductors and solid-state optical devices. Indium phosphide, like gallium arsenide, is a highly stable compound with a complex cubic lattice-like structure with atoms in each corner. This is illustrated in Figure 15.1 where indium atoms alternate with phosphorous atoms to form a very stable 3-D structure.

Indium has one stable isotope ^{127}In, but six radioactive isotopes of which ^{115}In has a decay period estimated to be 4.4×10^{14} years. Indium is considered to

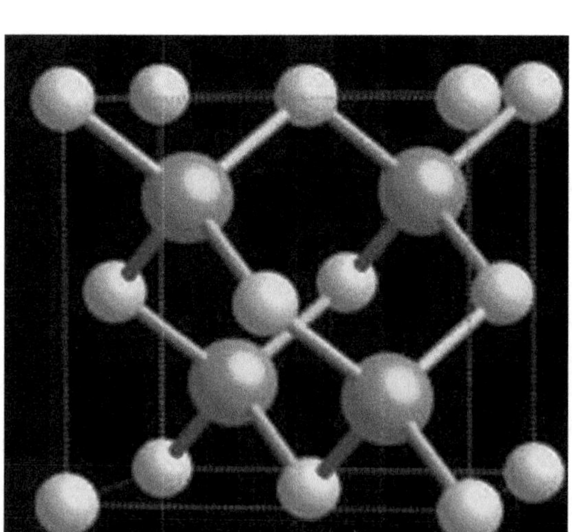

Figure 15.1 Indium phosphide (InP), 3-D lattice structure.

be of low toxicity. It is not recognised as a trace metal in the human body but has been reported at low concentrations of 0.01 and 0.07 μg g^{-1} in tissues and 0.0045 μg g^{-1} in hair in unexposed populations.[32,33] For present purposes it is necessary to concentrate upon the special properties of indium phosphide (InP) which has been classified by the IARC as a class 2 *possible* human carcinogen on the basis of experimental studies in animals.[15] Indium phosphide does not occur naturally but indium metal is found in the air (43 ng m^{-3}), sea water (20 μg l^{-1}), rainwater (0.59 μg l^{-1}) and in food (<10 μg g^{-1}). Higher concentrations have been determined in seafood from contaminated water near metal smelters (*ca.* 15 mg kg^{-1}).[34]

Human occupational exposure is principally through inhalation of dust in grinding and sawing operations, although more epidemiological information seems to be available concerning occupational health in indium mining and processing workers and in production of ITO. Indium salts are unlikely to penetrate the skin, on account of their insolubility, and no toxic effects have been reported following dermal contact or ingestion.[35] Fowler *et al.* emphasised that, other than blood and urine analyses, there were no useful markers of indium uptake.[9] Analysis is by spark source mass spectrometry using electrical and photoplate detection.[30,36] Fisher does not list indium or its inorganic compounds as contact sensitising agents although there is limited evidence that the element may be a sensitisor in rabbit skin.[37] The metabolism of indium in the human body is not known but detection limits in blood, serum and urine measured by graphite-furnace atomic absorption spectrometry have been calculated at 0.7, 0.4 and 0.4 μg l^{-1} respectively.[30]

Studies in indium and ITO production show that inhalation is the principal route of uptake by the human body. Mean whole blood and urine concentrations reported in 107 semiconductor workers exposed to insoluble indium compounds of 16.8 and 2.45 μg l^{-1}, respectively, compared to 0.14 and 0.02 μg l^{-1} in unexposed workers, were largely attributed to pulmonary uptake.[38] In three patients with profound interstitial pneumonitis, blood and urine levels were greatly elevated (*i.e.* 40 and 127 μg l^{-1}).[39] Nakano and colleagues showed a clear causal relationship between indium inhalation and pulmonary function in 12 factories and one research facility from 2003 to 2006.[40] Serum indium concentrations for current workers, past employees and non-exposed workers in this Japanese study were shown to be 8.35, 9.63 and 0.56 μg l^{-1} respectively, indicating that inhaled indium is not readily eliminated from the lungs and may present an ongoing risk, if indeed indium can be shown to be carcinogenic. The Japanese Society for Occupational Health proposed a safe level of serum indium for occupational exposures of 3 μg l^{-1} based upon biological monitoring of workers in indium related industry.[38,41] High levels of indium were observed in 93 workers in an ITO production facility.[42] Geometric mean of indium in serum in exposed workers was 4.55 μg l^{-1} (max. 116.9 μg l^{-1}) compared to 0.25 μg l^{-1} in non-exposed workers.

Indium phosphide is an irritant to the eyes, skin and respiratory tract and a cause of dental erosions and bone and joint pain.[43] Canadian Environmental Guidelines indicate that indium phosphide is a profound cause of lung disease through exposure in processing, LCD manufacture and in semiconductor industries.[34] Symptoms of alveolar proteinosis, cholesterol clefts, macrophage induction and indium-related interstitial changes affect workers inhaling indium phosphide (InP), indium arsenide (InAs) or indium oxide (In_2O_3).[44–49] However the minimal level of inhalation of indium liable to cause lung disease is not known.

Clinical evidence that indium inhalation alone is a cause of human cancer is limited. Lung diseases were reported in indium process workers, liquid crystal manufacturers, ITO production and indium recycling plants, but no incidence of pulmonary or other cancers has been specifically linked to inhalation of metallic indium, indium phosphide or other indium compound encountered industrially or environmentally.[8,40,42,45–51] Similar observations have been observed in several studies in rats, mice or Syrian hamsters given repeated intratracheal installates of indium arsenide or indium phosphide and observed for up to two years.[52–54] However, at least four experimental studies seen, including one conducted by the NTP, have shown that administration of indium phosphide by aerosol (0.01–0.3 mg m^{-3}) for 6 hours per day, 5days a week for up to two years will induce alveolar/bronchiolar adenomas and carcinomas, and that inhalation of ITO is carcinogenic in rats and mice in chronic studies.[12,53,55,56] Tanaka *et al.* also demonstrated that intratracheal administration of indium arsenide with a total concentration of arsenic of 7.5 mg once weekly for 15 weeks induced a 25% incidence of pulmonary malignancies.[28] This last study is not consistent with earlier work which showed that 4 mg kg^{-1} indium arsenide or 3 mg kg^{-1} indium phosphide containing a mean of 2.4 mg kg^{-1} indium failed to induce tumours within two years following a twice-weekly intratracheal dosing regimen for eight weeks.[54]

In rodents, pulmonary absorption of indium is low following intratracheal administration and the granules are only sparingly soluble in lung fluids, although some were phagocytosed by macrophages.[52,57] Gottschling *et al.* proposed that, on the basis of inducible nitric oxide synthetase production, cyclo-oxygenase, and glutathione-*s*-transferase in neoplastic tissues, oxidative stress underlies indium-induced cancer.[55] Fowler indicated also that apoptotic change, impaired protein synthesis, stress-related protein responses and variations in protein expression may be contributory.[9]

The carcinogenicity of the ITO complex is not proven. The ITO mixture comprising 90% indium oxide (In_2O_3) and 10% tin oxide (SnO_2) is used in the manufacture of LCD and high electrical conductivity devices and whilst workers inhaling the dust in production and recycling plants may develop respiratory distress, epidemiological evidence fails to substantiate the carcinogenic risk.[42,50,51] Experimental studies have demonstrated that rats inhaling 0.01 mg m^{-3} ITO repeatedly for 104 weeks did develop broncho-alveolar adenomas and carcinomas.[56] *In vitro* assays have demonstrated that ITO is both cytotoxic and genotoxic in macrophages (but not cultured alveolar epithelial cells) through its capacity to generate reactive oxygen species. Stannous oxide alone was minimally active in either assay.[51]

15.4 Thallium

Thallium is a highly toxic metal, originally identified by Sir William Crookes in 1861 while analysing rocks containing tellurium. It exhibits a brilliant green wavelength in spectroscopic analysis, hence the name *Thallos or* "green twig". It is widely dispersed in the human environment and associated with potassium- and rubidium-containing minerals. Estimates suggest that deposits in the Earth's crust range from 0.1 to 1 mg kg^{-1} with thallium-bearing rocks such as lorandite ($TlAsS_3$), orabite ($TlAs_2SbS_5$) and crooksite [$(TlCuAg)_2Se$] being found in Greece, Sweden, Texas and Brazil and as a minor constituent of iron, copper and selenide ores. Lorandite ore has also been mined in gold ores in Nevada where deposits are reputed to contain 60% thallium.[58] It is also found combined with halogens as yellow or colourless minerals.

Thallium is a lustrous bluish-white soft metal which was isolated originally as a by-product from smelting other metals. It is disseminated into the environment through combustion of coal and geochemical action, rain and snow. Thallium is stable in air and low levels of residue are found at low concentrations in water and soil where they persist for long periods without being broken down.[59] In metallic form, it is insoluble in water but dissolves slowly in acids, forming thallous salts and releasing hydrogen gas. An appealing feature for industrial application is the ability of thallium to alloy with mercury and other elements.

Thallium is a highly bioactive element and in its univalent state it exerts its toxicity largely through its interaction with and substitution of sodium and potassium in nerve conduction, renal and hepatic function. Thallium is a Group IIIb metal with atomic configuration similar to aluminium and the semiconductor elements discussed above. Thallium dust may burn but does not readily ignite, yet on heating it releases toxic fumes. Fluorine reacts with thallium vigorously to produce incandescent emissions. It tarnishes slowly in air with the formation of trioxide (Tl_2O_3), or Tl_2O when heated; in moist air the hydroxide (TlOH) is formed. Thallium forms inorganic compounds as Tl(I) and Tl(III) and is particularly active in the presence of halide elements. It is a ductile metal and capable of being drawn, or cold-rolled for use in electronic devices, switches and occasional use in semiconductors. It has limited value in the production of special low melting point glasses and medical procedures. The profound toxicity of thallium salts has been used in production of rodenticides and pesticides in the form of syrups, jelly or pastes. Ant traps, baits and rat poisons contained 0.5–1.0% thallium, but these were prohibited in the USA at least by 1963, although they may still be in use in other countries.

Thallium has no nutritional value in mammals but is consumed through food chains. It is absorbed by plants, fish and shellfish, and human exposure is mostly through eating contaminated food or through inhaling dust or industrial fumes. Low concentrations of thallium are detected in some brands of cigarette, though uptake through this route is minimal in comparison with cadmium, arsenic, lead and other toxic factors identified in mainstream smoke. The average content of thallium in the human body measured in a US population was approximately 0.1 mg with blood concentrations of 3 μg l^{-1}, with reference values being calculated as 0.15–0.63 μg l^{-1} in blood and 0.02–0.34 μg l^{-1} in serum.[60–62] Thallium has a short half-life in the human body and intake is conveniently measured in urine, fingernails and hair.[63,64] The hair concentrates thallium specifically and provides a useful non-invasive guide to levels of poisoning or environmental contamination to complement blood and urine analyses. Hair values for 12 cases of thallium poisoning were shown to be proportional to urinary concentrations.[65] Fingernails from non-exposed persons showed a three-fold higher level of thallium than hair and averaged 0.051 ppm. Skin and hair have been shown to be target tissues in cases of thallium poisoning.[66]

Low levels can be expected in the workplace environment in manufacture of electronic devices, the semiconductor industry and in the manufacture of special glasses, but until recently the metal had not been widely evaluated for human or animal toxicity.[58,67,68] Available evidence shows that thallium is absorbed into the body through inhalation, ingestion and dermal contact and that its toxic action is largely attributable to its interaction with and displacement of sodium and potassium.[69,70] Electrolyte imbalances arise in cell membranes, mitochondria and physiological systems through Tl(I) displacing K(I) in the activation of Na^+,K^+-ATPase in plasma cell membranes. The ions Tl(I) and K(I) have similar ionic radii. Inorganic thallium has been shown to uncouple oxidative phosphorylation, cause swelling of isolated mitochondria, and lead to oxidative stress through impairment of GSH metabolism and the glutathione-dependent antioxidant defence system.[71] Impaired liver and renal function, cardiac contraction and neuromuscular conduction are indices of thallium toxicity, but alopecia, respiratory distress and altered liver function are reported.[59]

Thallium is a toxic metal in the human body and workers exposed chronically are at risk of developing neurological, respiratory and physiological effects, leading to paralysis due to K^+ and Na^+ disturbances and possible fatality.[70,72] The Environmental Protection Agency, with reference to their *Guidelines for Carcinogen Risk Assessment*, emphasised that "there are no studies available to determine the carcinogenic potential of thallium in animals and no adequately conducted studies in humans;[16,73] *in vitro* studies provide inconsistent evidence of the genotoxic potential of thallium compounds".[63,74,75] However, biochemical and *in vitro* evidence of disturbances in mitochondrial function, metabolic and cell cycle abnormality, and oxidative stress, as seen with certain other metal carcinogens, suggests that thallium should not be overlooked as a possible carcinogen under some circumstances of exposure.[73]

Data available from two epidemiological studies are severely limited not only by the low numbers of workers evaluated but by the lack of positive observations. The two occupational studies evaluated included a total of 214 workers, 86 of whom were exposed in a magnesium seawater battery factory and 128 in cement manufacturing plants.[63,74] Neither study provided statistically significant evidence of tumours or lesions in any organ, and insufficient data was available on environmental monitoring from which to derive a useful control limit. The threshold limit of 0.01 mg m^{-3} set by the American Conference of Governmental Industrial Hygienists, based largely on analogy with other toxic heavy metals, could not be substantiated.[76] Several case studies are reported in which thallium poisoning resulted from suicides or rodenticide consumption but these fail to provide useful information concerning the possible long-term risks associated with thallium consumption. Cavanagh *et al.* calculated that the lowest dose likely to cause overt symptoms was 0.31 g, but the symptoms were transitory and not fatal.[77] Heyl and Barlow examined five young men with thallium poisoning in South Africa.[66] All displayed mild to severe neurological problems and weakness, loss of appetite, abdominal cramps and severe dermatological symptoms with pustular dermatitis and alopecia. Urinary and plasma levels exceeded 500 mg dl^{-1}. The authors drew attention to the presumed lethal dose of 1 g thallium sulfate, and evidence of non-fatal thallotoxicosis in children at 5 mg kg^{-1}. Although thallium concentrations were not measured in skin, this study emphasised that the dermis should be regarded as a target tissue. Repetto and Del Peso list six occupational health studies involving exposure to thallium in pesticides, rodenticides, contaminated food and undefined occupations.[58] Each emphasises the preponderance of neurological disorders, but inconsistencies in the types and durations of exposure make conclusions difficult. Interestingly, none of the data seen refer to risks faced by workers exposed chronically to thallium in mining and extraction processes, which can be expected to involve exposure to arsenic, antimony and other potentially carcinogenic materials.

Available toxicity studies in animals provide little evidence to support the view that thallium is carcinogenic. Acute and subacute toxicity studies in rodents or dogs with thallium oxide or sulfate have failed to identify target organ toxicity or evidence of even pre-neoplastic change.[78–80] Rats given 0.8 or 1.6 mg kg^{-1} sub-lethal doses of thallium showed that the metal accumulates in the brain and evokes a dose-related lipid peroxidation in the corpus striatum, cerebellum and other parts of the tissue but with no clear differential distribution.

Thallium nitrate failed to provide evidence of mutagenicity in five strains of *Salmonella typhimurium* Ames bacterial reversion tests, or in two strains of *Bacillus subtillus*.[81,82] Negative observations were made in mitogenic gene conversion and reverse mutation screens in the yeast *Saccharomyces cerevisiae*.[83,84] Limited mutagenicity tests in mammalian cell lines (sister-chromatid exchanges, single-strand DNA-breaks, depressed DNA synthesis) provide the only tangible evidence of a carcinogenic risk with this element.[85,86] Thallium acetate reduced the viability of Chinese hamster ovary cells in culture and depressed ATP and protein synthesis.

In the absence of statistically sound carcinogenicity tests in animals, clinical case studies or epidemiological evidence, the carcinogenicity of thallium is unproven.

15.5 Semiconductors

Four major epidemiological studies have been conducted in an attempt to understand the cause(s) of increased cancer fatalities in the semiconductor industry.[5,6,10,87] McEvenny's study was undertaken in response to a Health and Safety Executive initiative to investigate concerns regarding the cancer risk faced by current and former workers at the Greenock Plant of National Semiconductor (UH) Ltd.[5] Cancer registry analyses at the end of 1998 involving 4388 workers with an average follow-up of 12.5 years showed that total cancer registrations for men and women were close to expected figures. However, although the actual number of cancer cases was low, the authors estimated a higher than expected incidence of cancer of the lung, stomach, and breast in women, and a four-fold increase in brain cancers in men. Lack of useful information regarding the latent period between exposure to semiconductor chemicals and tumour diagnosis and work-related patterns of life prohibited definite conclusions. The earlier studies conducted by Sorahan in Birmingham evaluated, firstly, 1807 semiconductor workers and reported three patients with melanoma which could not be explained by work-related patterns of UV light exposure. The results of his follow-up analysis failed to confirm a significant increase in cancers of the skin or other organs and actually showed a lower mortality rate than in the general population of England and Wales. Interestingly, Beall and his colleagues in Birmingham, Alabama, also confirmed a lower than expected mortality among 126,836 workers engaged in two semiconductor facilities and one storage device plant.[7] A small number of cancers of the central nervous system were associated with process equipment maintenance in one facility and 18 cases of prostatic cancer were reported in storage device manufacture, but further evaluation of workplace exposures and occupational groups is necessary to establish a work-related causation. Similarly unconvincing observations were reported in a large cohort of 89,954 men employed in similar occupations in a later study in the Alabama plants.[7] A standardised incidence ratio for all cancers combined was 81, compared to 87 in storage device manufacturing. Analyses of cancer incidence patterns allowing for job specification, duration of employment and lifestyle did not indicate that cancer excesses in any organ were due to exposure to semiconductor materials.

References

1. P. Bhattacharya, *Semiconductors Optoelectronic Devices*, Prentice Hall, Englewood Cliffs, NY, 1994.
2. G. S. May and S. M. Sze, *Fundamentals of Semiconductor Fabrication*, Wiley, New York, 2003.
3. A. Diaz, in *Chemistry: Foundations and applications*, ed. J. J. Lagowsi, Thomson-Gale, New York, 4004, p. 4, 117.
4. W. C. O'Mara, *Handbook of Semiconductor Silicon Technology*, Noyes Publications, Park Ridge, NJ, 1990.

5. D. M. McElvenny, A. J. Darnton, J. T. Hodgeson, S. D. Clarke, R. C. Elliott and J. Osman, *Cancer Among Current and Former Workers at National Semiconductor (UK) Ltd, Greenock*, Health and Safety Executive (HSE) Books, HMSO, London, 2001.
6. T. Sorahan, D. J. Pope and M. J. MacKiernan, *Brit. J. Indust. Med.*, 1992, **49**, 215.
7. C. Beall, T. J. Bender, H. Cheng, R. Herrick, A. Kahn, R. Matthews, N. Sathiakumar, M. Schymura, J. Stewart and E. Delzell, *J. Occup. Environ. Med.*, 2005, **47**, 996.
8. B. A. Fowler, H. Yamouchi, E. A. Connor and M. Akkerman, *Scand. J. Work. Environ. Med.*, 1993, **19**(Suppl. 1), 101.
9. B. A. Fowler, M. Schubaauer-Berigan and C. J. Hines, Citation for IARC Monographs, 2006, **86**, 16.
10. T. Sorahan, J. A. H. Waterhouse, M. J. MacKiernan and R. H. R. Aston, *Brit. J. Indust. Med.*, 1985, **42**, 546.
11. National Toxicology Program, *Natl. Toxicol. Program Tech. Rep. Ser.*, 2000, **492**, 1.
12. B. C. Gottschling, R. R. Maronpot, J. R. Hailey, S. Peddada, C. R. Moomaw, J. E. Klaunig and A. Nyska, *Toxicol. Sci.*, 2001, **64**, 28.
13. K. Nogano, T. Nishizawa, Y. Umeda, T. Kasai, T. Noguchi, K. Gotoh, N. Ikawa, Y. Ikawa, Y. Kuwasumi, T. Yamauchi, H. Arito and S. Fukushima, *J. Occup. Health*, 2011, **53**, 175.
14. H.-W. Chen, *Bull. Environ. Contamination Toxicol.*, 2007, **78**, 123.
15. International Agency for Research on Cancer (IARC), *IARC Monographs on the Evaluation of Carcinogenic Risks to Humans*, 2006, vol. 86.
16. Environmental Protection Agency (EPA), *Guidelines for Carcinogen Assessment*, US EPA, Washington, DC, 2005.
17. US EPA, IRIS, *Toxicological Review of Thallium and Compounds*, US EPA, Washington, DC, 2008, EPA/635/R-08/001.
18. C. R. Chitamber, *Int. J. Environ. Res. Publ. Health*, 2010, **7**, 2337.
19. P. L. Goering, R. R. Maronpot and B. A. Fowler, *Toxicol. Appl. Phamacol.*, 1988, **92**, 179.
20. L. R. Bernstein, J. J. M van der Heoven and R. O. Boer, *Ann. Cancer Agents Med. Chem.*, 2011, **111**, 585.
21. G. A. Kerckaert, R. A. LaBoeuf and R. J. Isfort, *Fundam. Appl. Toxicol.*, 1996, **34**, 67.
22. D. F. Lewis, C. Ioannides and D. V. Parke, *Environ. Health Perspect.*, 1996, **104**(Suppl. 5), 1011.
23. National Toxicology Program (NTP), *Technical Report on the Toxicology and Carcinogenesis Studies of Indium Phosphide in F344/N Rats and B6C3F1 Mice (Inhalation Studies)*, NIH Publication, Research Triangle Park, NC, 2001, Tech. Rep. Ser. No. 499, p. 340.
24. S. M. Becker and K. L. McCoy, *J. Pharmacol. Exp. Therap.*, 2003, **307**, 1045.
25. M. T. Harrison, C. B. Hartmann and K. L. McCoy, *Toxocol. Appl. Phrmacol.*, 2003, **186**, 18.

26. L. Nichols and T. Sorahan, *Occup. Med. (Lond.)*, 2005, **55**, 625.
27. S. Ohyama, N. Ishinishi, A. Hisanaga and A. Yamamoto, *Appl. Organometallic Chem.*, 1988, **2**, 333.
28. A. Tanaka, *Toxicol. Appl. Pharmacol.*, 2004, **198**, 405.
29. American Conference of Governmental Industrial Hygienists (ACGIH), *Gallium Arsenide*, ACGIH, Cinninatti, OH, 2005.
30. Occupational Safety and Health Administration (OHSA), *Arsenic*, OHSA, Washington, DC, 2012.
31. D. G. Tuck, in *Chemistry: Foundations and applications*, ed. J. J. Lagowski, 2004, vol. 2, p. 214.
32. M. Yukawa, K. Amano, M. Suzuki-Yasumoto and M. Terai, *Arch. Environ. Health*, 1980, **35**, 36.
33. K. C. Honeychurch, *World J. Analyt. Chem.*, 2013, **1**, 8.
34. The Government of Canada, *Chemicals Management Plan*, Ottawa, 2001.
35. R. H. Guy, J. J. Hostýnek, R. S. Hinz and C. R. Lorence, *Metals and the Skin: Topical effects and systemic absorption*, Marcel Dekker, New York, 1999.
36. X. D. Liu, P. Van Espen and F. Adams, *Fresnius' Zeitschrift fur analytische Chemie*, 1987, **327**, 659.
37. A. Fisher, *Contact Dermatitis*, Lea and Febiger, Philadelphia, 1987.
38. K. Miyaki, K. Hosoda, M. Hirata, A. Tanaka, Y. Nishiwaki, T. Takabayashi, N. Inoue and K. Omae, *J. Occup. Health*, 2003, **45**, 228.
39. O. Taguchi and T. Chonan, *Nihon Kokyuki Gakkai, Sasshi*, 2006, **44**, 532.
40. M. Nakano, K. Omae, A. Tanaka, M. Hirata, T. Michikawa, Y. Kikuchi, N. Yoshioka, Y. Nishiwaki and T. Chonan, *J. Occup. Health*, 2009, **51**, 513.
41. A. Tanaka, M. Hirata, Y. Kiyohara, M. Nakano, K. Omae, M. Shiratani and K. Koga, *Thin Solid Films*, 2010, **518**, 2934.
42. T. Hamaguchi, K. Omae, T. Takabayashi, Y. Kikuchi, N. Yoshioka, Y. Nishiwaki, A. Tanaka, M. Hirata, O. Taguchi and T. Chonan, *Occup. Environ. Med.*, 200, **65**, 51.
43. US National Library of Medicine, Hazardous Substances, Data Bank, NLM, Bethesda, MD, 2011.
44. H.-W. Chen, *Bull. Environ. Contamination Toxicol.*, 2006, **77**, 289.
45. T. Chonan, O. Tguchi and K. Omae, *Eur. Resp. J.*, 2007, **29**, 317.
46. H. Nogami, T. Shimoda, S. Shoji and S. Nishima, *J. Jpn. Resp. Soc.*, 2008, **46**, 60.
47. Y.-l. Xaio, H. R. Cai, Y.-h. Wang, F.-Q. Meng and D.-P. Zhang, *Chinese Med. J*, 2010, **123**, 1347.
48. K. Omae, M. Nakano, A. Tanaka, M. Hirata, T. Hamaguchi and T. Chonan, *Int. Arch. Occup. Environ. Health*, 2010, **84**, 471.
49. D. Lison and M. Delos, *Am. J. Resp. Crit. Care Med.*, 2010, **181**, 458.
50. (a) S. Homma, A. Miyamoto, S. Sakamoto, K. Kishi, N. Matoi and K. Yoshimura, *Eur. Resp. J.*, 2005, **25**, 200; (b) J. Lalov, I. Crazzari, J. Mullr, V. Rabolli, N. Panin, F. Huaux, I. Fenoglio and B. Fubini, *Toxicol. Sci.*, 2009, **108**, 472.

51. D. Lison, J. Lalov, I. Corazzari, J. Muller, V. Rabolli, N. Panin, F. Huaux, I. Fenoglio and B. Fubini, *Toxicol. Sci.*, 2009, **108**, 472.
52. K. Oda, *Ind. Health*, 1997, **35**, 61.
53. A. Tanaka, A. Hisananga, M. Hirata, M. Omura, Y. Makita, N. Inoue and N. Ishinishi, *Fukuoka Igaku Zasshi*, 1996, **87**, 108.
54. K. Yamazaki, A. Tanaka, M. Hirata, M. Omura, Y. Makita, N. Inoue, K. Sugio and K. Sugimachi, *J. Occup. Health*, 2000, **42**, 169.
55. B. L. Upham and J. G. Wagner, *Toxicol. Sci.*, 2001, **64**, 1.
56. K. Nagana, T. Nishizawa, Y. Umeda, T. Kasai, T. Nochi, K. Gotoh, N. Ikawa, Y. Eitaki, Y. Kawasumi, T. Yamaaaauchi, H. Arito and S. Fukushima, *J. Occup. Health*, 2011, **53**, 175.
57. T. Uemura, K. Oda, K. Omae, T. Takebayashi, T. Nomiyama, C. Ishizuka, K. Hosoda, H. Sakurai, K. Yamazaki and I. Kabe, *J. Occup. Health*, 1997, **39**, 205.
58. G. Repetto and A. del Peso, in *Patty's Toxicology*, ed. E. Bingham and B. Cohrssen, John Wiley, Inc, Hoboken, NJ, 2012, vol. 1, p. 257.
59. Agency for Toxic Substances and Disease Registry (ATSDR), *Thallium*, SU, Department of Health and Human Services, Atlanta, GA, 1995.
60. E. Weinig and P. Zink, *Arch Toxikol.*, 1967, **22**, 255.
61. N. P. Singh, J. D. Bogden and M. M. Joselow, *Arch. Environ. Health*, 1975, **30**, 557.
62. C. Minoia, E. Sabbioni, P. Apostoli, R. Pietra, L. Pozzoli, M. Gallorini, G. Nicolaou, L. Alessio and E. Capodaglio, *Sci. Total Environ.*, 1990, **95**, 89.
63. R. L. Marcus, *J. Soc. Occup. Med.*, 1985, **35**, 4.
64. S. N. Kales and D. C. Christiani, *Hair in Toxicology*, ed. D. J. Tobin, Royal Society of Chemistry, Cambridge, 2005, p. 125.
65. D. E. Rusyniak, R. B. Furbe and M. A. Kirk, *Ann. Emerg. Med.*, 2002, **39**, 307.
66. T. Heyl and R. J. Barlow, *Brit. J. Dermatol.*, 1989, **121**, 787.
67. A. Léonard and G. B. Gerber, *Mutat. Res., Rev. Mutat. Res.*, 1997, **387**, 47.
68. International Program on Chemical Safety (IPCS), *Environmental Health Criteria, Thallium*, National Institutes for Safety and Health, IPCS, 2002, CS0077.
69. M. N. Hughes, W. K. Man and B. C. Waller, *Chem. Biol. Interactions*, 1978, **23**, 85.
70. L. Faveri and M. Mourelle, *J. Appl. Toxicol.*, 2006, **5**, 32.
71. C. E. Hanzel, M. S. Villavarde and S. V. Verstraeten, *Toxicology*, 2005, **207**, 510.
72. S. Galvin- Arzate and A. Santamaria, *Toxicol. Lett.*, 1998, **99**, 1.
73. Environmental Protection Agency (EPA), *Toxicological Review of Thallium Compounds*, EPA, Washington, DC, 2009, EPA/635/R-08/001F.
74. K. H. Schalle, G. Manke, H. J. Raithel, G. Bühlmeyer, M. Schmidt and H. Valentin, *Int. Arch. Occup. Environ. Health*, 1980, **47**, 223.
75. Integrated Risk Information System (IRIS), *Thallium Oxide*, IRIS, US EPA, Washington, DC, 2011.

76. American Conference of Governmental Industrial Hygienists (ACGIH), in: *Documentation of the Threshold Limit Values for Substances in Workroom Air*, ACGIH, Cincinnati, OH, 3rd edn, 1971, p. 256.
77. J. B. Cavanagh, N. H. Fuller, H. R. Johnson and P. Rudge, *Quart. J. Med.*, 1974, **43**, 293.
78. W. L. Downs, K. Scott, L. T. Steadman and E. A. Maynard, *Indust. Hyg.*, 1960, **21**, 399.
79. F. W. Mather and D. G. Low, *J. Am. Vet. Med. Assoc.*, 1960, **137**, 544.
80. S. Galvin-Arzate, A. Martinez, E. Medina, A. Santamaria and C. Rios, *Toxicol. Lett.*, 2000, **116**, 37.
81. T. Kada, K. Hirano and Y. Shirasu, *Chemical Mutagens: Principles and Methods for Their Detection*, Plenum Press, New York, 1980, p. 149.
82. N. Kanematsu, M. Hara and T. Kada, *Mutat. Res.*, 1980, **77**, 109.
83. L. E. Loveless, E. Spoerl and A. Weisman, *J. Bacteriol.*, 1954, **68**, 637.
84. I. Singh, *Mutat. Res.*, 1983, **117**, 149.
85. G. D. Zaskhina, I. M. Vasilyeva, U. I. Kenesariev and P. G. Butenko, *Mutat. Res. Genet. Toxicol.*, 1983, **124**, 163.
86. N. E. Garrett and L. Lewtas, *Environ. Res.*, 1983, **32**, 455.
87. T. J. Bender, C. Beall, H. Cheng, R. F. Herrick, A. R. Kahn, R. Matthews, N. Sathiakumar, M. J. Schymura, J. H. Stewart and E. Delzkll, *Occup. Environ. Health*, 2007, 64.

CHAPTER 16

Thorium and Titanium

16.1 Introduction

Thorium and titanium are classified in Group IV of the Periodic Table but they differ greatly in their chemical and biological behaviour. Whereas titanium is a fairly common stable lightweight resilient metal (named after Titan in Greek mythology) and the ninth most common element in the Earth's crust, thorium is a radioactive element of the actinide series (aptly named after the god Thor, the bringer of thunder and lightning in Norse mythology) and is only slightly more abundant than uranium. Both elements exhibit a tetravalancy, although titanium does form compounds as Ti(II) or Ti(III). The US Department of Health and Human Services (US DHHS) and the International Agency for Research on Cancer (IARC) have identified thorium dioxide, a heavy, crystalline sand-like powder once used in diagnostic medicine, as a known human carcinogen, whereas titanium dioxide, which belongs to a group of sparingly soluble compounds with low toxicity, was classified as a non-carcinogen after much conjecture by successive working parties.[1,2]

16.2 Thorium

Thorium is a radioactive rare earth element of the actinide series. It has six isotopes, of which ^{232}Th comprises almost 100% of natural abundance, others being present at <0.0001%. None of these natural isotopes is stable.^{232} Th is a strong α-emitter and decays in 1.41×10^{10} years, ultimately to produce stable ^{208}Pb (Figure 16.1). The decay process, involving α- and β-emissions, leads to the production of a series of intermediary products including isotopes of radium, actinium, polonium, radon, bismuth and thallium. Much of this pioneering work was conducted by the Nobel Laureate Ernest Rutherford in 1902.[3–5] The radiation emitted in this decay chain poses serious human health risks with ^{232}Th releasing high levels of radioactivity per unit when inhaled or

Issues in Toxicology No. 18
The Carcinogenicity of Metals: Human Risk through Occupational and Environmental Exposure
By Alan B. G. Lansdown

Published by the Royal Society of Chemistry, www.rsc.org

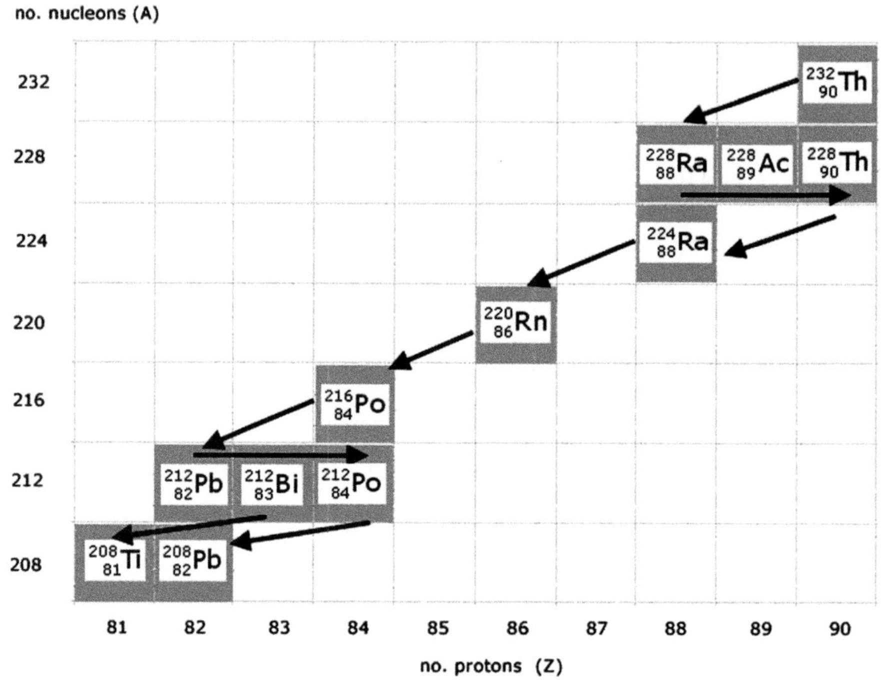

Figure 16.1 Decay chain of ^{232}thorium.

ingested; all radioactive decay products are hazardous and potentially carcinogenic.[6] Thorium dioxide is specifically classified as a human carcinogen in the 12th RoC on the grounds of sufficient human and experimental evidence.[1,2,7] Radiation physics and the health risk associated with radiation is a large subject and outside the scope of the present book. Further reading is available from the Health Physics Society, the National Research Council (NRC) and other authoritative sources.[8–12]

Thorium occurs naturally in complexes with rare earth elements including cerium, uranium and neodymium and is recovered commercially from rare earth ores such as monazite, which contains cerium phosphate and <9% thorium dioxide. More than a dozen other thorium-containing minerals exist including thorite/huttonite ($ThSiO_4$), graylite (Th,PbCa)$PO_4 \cdot H_2O$ and thorianite ($ThO_2 \cdot UO_2$), with thorium content ranging from 2 to 88%. Extraction of thorium from certain ores exposes workers to health risks from lead, radium, uranium, zirconium, silica and chromium in addition to radioactive emissions. Infusible thorium minerals glow with an exaggerated incandescence in the flame test and with Sr, Mg, Zr and Ca they produce a "lime light" effect.[13] Much of the internal heat in the Earth's core is attributable to the decay of thorium and uranium. Pure thorium is an air-stable silvery white lustrous element that is malleable, ductile and can be drawn, swagged or cold rolled. Thorium turnings ignite on heating in air and burn with a brilliant white light (pyrophoric). The principal compound of commercial value, thorium dioxide, has a very high melting point of 1750 °C and boiling point of 4790 °C, the highest of all metal oxides. Thorium and its inorganic compounds are hazardous and are handled with extreme care.

Major use is made of thorium as a nuclear power source and it is well known for its value in incandescent mantles for gas lamps but other major uses of thorium and its compounds include coating for tungsten wire in electronic equipment, alloyed with magnesium to provide high strength, temperature resistant components for electronics, and laboratory glass crucibles with a low dispersion and high refraction index. Thorium dioxide is used as a catalyst in converting ammonia to nitric acid, sulfuric acid production and in petroleum cracking. A second and once important use of thorium dioxide was in radiographic imaging.

Two major uses of thorium in medicine concern the products Thorotrast and Thorium X. Neither is used these days on account of the radiation risks but Thorotrast is still of clinical interest in accounting for the aetiology of tumours arising in workers exposed up to 40 years ago, before Thorotrast was withdrawn on safety grounds.[14,15] Thorotrast, as a 19–20% colloidal solution of thorium dioxide ($^{232}ThO_2$), was originally introduced in Germany in the early 1900s for cerebral angiography and up until 1947 had been used widely as a radio-contrast medium for visualisation of the spleen, liver and reticuloendothelial tissues.[4,14,16] Thorotrast was injected intravenously for diagnostic purposes and up until the 1940s was deemed to be without risk, until a chance observation of endothelial cell sarcoma in the liver of a 70-year-old woman in 1947 revealed its carcinogenicity and led to its abandonment for clinical use.[17] Thorotrast injections had been given to this patient for visualisation of the liver. She died 12 years later with an aggressive epithelial-cell carcinoma. Subsequent studies have established that the liver and bone marrow are prime targets for Thorotrast radiation damage and that angiosarcoma and primary hepatic tumours develop in 16–68% of patients within 15–45 years after initial injection.[18,19]

Thorium X is a natural decay product of thorium and an isotope of radium (^{226}Ra) and emits α-particles. It disintegrates to produce radioactive thoron and ultimately thorium. Thorium X was developed for use in dermatology in the form of alcoholic or aqueous paints and emollients for treating dermatoses, skin cancers, vascular naevi, pigmented birth marks and the so-called "port-wine stain".[20–22] Thorium X penetrates the skin to accumulate mainly in epidermal structures, sweat ducts and hair follicles. Basal cell carcinomas reported in skin treated to remove in port wine stains, many years after Thorium X therapy, led to its withdrawal from dermatological use.[21,23,24] Carcinogenicity was attributed to residual radioactivity, but the minimum Thorium X dosage required to evoke neoplasia is not known owing to wide variations in the reported dosing schedules and investigation times.

The carcinogenicity of thorium is illustrated by several studies conducted in mining and refining.[16,24–29] Chen *et al.* demonstrated that a 20-year follow-up of 638 miners exposed showed lung concentrations of 11.11 Bq radioactive thorium. Several authors cite long-term inhalation of dust as a cause of lung tumours although on occasions inhalation of thorium has been compounded by silica and decay products of thorium such as thoron (^{220}Radon).[30] Jevtić *et al.* discussed the prevalence of thorium and its breakdown products in the atmosphere in the home and in the region of mines, landfill sites and in bedrock, as well as risk incurred through using contaminated building material.[31] From these sources, thorium, radium, radon and other daughter products can be inhaled, leading to deposition of radioactive materials on pulmonary epithelia. Depending on particle size, they computed that the average dose of radiation to the basal epithelium in the lung is about 5- to 8-fold higher than the total dose to the entire pulmonary system.

Epidemiological evidence shows that thorium entering the human body *via* inhalation accumulates initially within in the reticuloendothelial system; it is widely disseminated to liver, lungs, kidney, bone marrow, spleen and bone. In the liver, thorium particles are phagocytosed by mononuclear cells to be disseminated throughout the body.[32–36] Energy dispersive X-ray microanalysis spectrometry has been employed to monitor thorium uptake in the liver tissue, and electron microscopy shows that most of the particles are phagocytosed by the cells of the reticuloendothelial system including macrophages/phagosomes of the portal triad and Kupffer cells in the hepatic sinusoid. Larger aggregates accumulate in the cytoplasm with no visible limiting membrane. Thorium emits α-radiation over long periods and is a cause of inflammatory and cytogenic changes.[36,37] Farid and Conibear conducted an epidemiological study of 275 former workers in a thorium refinery and showed that radon (^{220}Rn) and bismuth (^{212}Bi) exhaled as decay products provided an indication of the body burden of thorium.[28] Hepatotoxic changes were marked by elevated serum aspartate aminotransferase, globulins and bilirubin.

Whereas the risks of primary and metastatic malignancies involving liver, lung and bone, and leukaemia are listed in several studies, in others carcinogenicity seems to be rare. A cohort of 278 former workers in a thorium refinery exposed to thorium and breakdown products (^{212}Bi and ^{220}Ra) showed no evidence of pre-neoplastic change other than increases in liver cell enzymes (alkaline phosphatase and aspartate aminotransferase).[29] Lung cancer in exposed workers was estimated to be about 50 excess cases per million person–years at risk per working-level month.[16] Where carcinogenicity is reported in processing plants, in mining and other occupational surveys, it is rarely possible to distinguish between effects attributable to thorium and those due to breakdown products.[16,25,26] Stehey showed that workers exposed to thorium in mines for 3–24 years accumulated the element in their lungs, bronchial lymph nodes, liver and bone but estimated that the radiation levels emitted were compatible with health studies that found no significance in mortality to the general population.[25,37] When mine workers were exposed to thorium, thoron and uranium for up to 25 years, a group of 592 workers with at least 1 year's experience exhibited an increased mortality from pancreatic cancer as well as liver cancer but smoking was implicated as a possible causative factor for an undefined number of fatalities.[26] Epidemiologists have failed in many studies to distinguish between occupational thorium exposure as a cause of cancer as opposed to the cancer risk presented by natural sources of radon emanating from shoals and igneous and metamorphic rocks.[2,31,38] Other imponderables in analysing such epidemiological studies relate to insufficient information on worker job profiles, durations of exposure and smoking patterns.[25,26,30,37,39]

16.2.1 Thorotrast

The carcinogenicity of Thorotrast has received extensive clinical and experimental attention in the past 40 years and patterns of carcinogenicity seen in humans have been replicated in several species and strains of rat, mouse, rabbit and monkey.[4–43] Information provided by cytogenetic studies illustrates the genetic basis of induced tumours, evidence of point mutations and implications involving *p53* gene mutations.[44] Research has demonstrated that Thorotrast is taken up avidly by the reticuloendothelial system following parenteral injection and is a cause of cirrhosis and liver malignancy, peritoneal tumours, leukaemia, haemangiosarcoma, osteosarcoma, lung cancers and soft tissue sarcomas up to 35 years (mean 26 years) after exposure.[45–49]

Thorotrast injected into the human body is disseminated rapidly to many tissues where it decays to liberate daughter isotopes, which accumulate in bone marrow cells and lead to a high frequency of chromosome aberrations and leukaemia.[37,46,50] Thorotrast accumulating in liver was a direct cause of primary hepatocellular carcinoma, angiosarcoma and cholangiocarcinoma, but also squamous cell carcinomas of the lung and gastrointestinal malignancies.[28,46] Experimental studies indicate that these changes may be modulated by thorium-induced oxidative stress or induced mutations in the *p53* or *Ras* genes.[44,51,52]

Experimental tumours in mice, rats, rabbits and guinea pigs resulting from Thorotrast administration have been classified as:

a. Local tumours at injection sites, the so-called "Thorotrastomas"
b. Leukaemia
c. Tumours arising at sites of selective internal Thorotrast uptake – liver, spleen, lymphnodes and bone marrow.[53]

The incidence is directly proportional to the dose and duration of the radiation emitted rather than to the biochemical, genotoxic or mutagenic properties the element may have.[4,53,54] Whilst α-emissions leading to radiation-damaged cells and chromosomal aberrations progressing to malignancy might be considered a plausible pathogenic mechanism, tumour incidence is shown to be a function of the accumulated α-particle emissions.[48,49] This view seems to be contrary to clinical experience.[44] Cytogenic changes and chromosomal aberrations do occur, but in a study of 36 patients given Thorotrast 11–31 years previously, cytogenic changes did not correlate with either the dose of Thorotrast administered or the period between treatment and laboratory investigation.[55] No non-radioactive isotopes of thorium exist, but in an attempt to establish the role of α-emissions as a principal cause of carcinogenicity, Bensted and Crookall injected mice with Thorotrast or the non-radioactive contrast medium Zirconotrast (a solution of non-toxic zirconium hydroxide).[54] The two groups exhibited a similar number of liver tumours, suggesting that the physic-chemical configuration of the contrast media was more important than the radioactive emissions in causing malignant change.

Clinical experience on the carcinogencity of Thorotrast is based largely on older studies such as that conducted in 50,890 Japanese war veterans (1975–1978), and a comprehensive study by MacMahon published in 1947 which led to the withdrawal of Thorotrast from clinical use.[17,56] Kido *et al.* reported that 150 patients injected with Thorotrast showed a three-fold higher mortality than normal and that rate ratios for liver cancer, liver cirrhosis and leukaemia were 35, 7.5 and 18.2 respectively.[56] A more comprehensive evaluation of 506 Japanese war prisoners who had received Thorotrast confirmed reduced life expectancy with the main causes of mortality being liver malignancy, blood dyscrasias and tumours of the extrahepatic bile duct, peritoneal tumours, osteosarcomas and haemangiosarcoma of the spleen; in each case the histopathology of the tumours was specific.[46] The workers estimated that the carcinogenic dose of thorium necessary to induce cancer in 423 patients was 10^4 Gy per person and that a 10-year latent period should be expected.

16.2.2 Radium

Radium as a decay product of thorium is implicated in cancers seen in young girls involved in painting watch and clock dials in the early 1900s.[57] At the time, people were fascinated by the luminescence produced when radium is mixed with a phosphor, and glow-in-the-dark watch and clock faces became immensely popular. Industries sprang up to manufacture consumer products such as hair tonic, toothpaste, ointments and elixirs containing radium. The majority of these have been withdrawn for health and safety reasons, but luminescent paints were still widely available until at least the end of World War II. The soft glow of radium's luminescence made aircraft dials, gauges and other instruments visible to their operators at night. Many stories exist concerning the so-called "Radium Girls" who painted the dials of watches in large numbers (Figure 16.2). They were instructed to shape their brushes to a fine point with their lips and then to paint the minute numbers and images on the watches; each worker was expected to paint 250 dials daily for five and a half days a week, at low wages. The glow in the dark effect was used in humour and these young girls would paint their lips, faces and fingernails with the radium paint.[58] In 1929, Martland and Humphreys were the first to report the appearance of osteogenic sarcoma in 2 of 15 young girls as a result of exposure to radium in the paint.[59,60]

16.3 Titanium

Titanium is the ninth most abundant metal in the Earth's crust. It is a strong, lightweight silvery white metal and thought to contribute 6.600 ppm of the Earth's crust, or 5.63 g kg^{-1}. Its principal ores are ilmenite ($Fe.TiO_3$), rutle

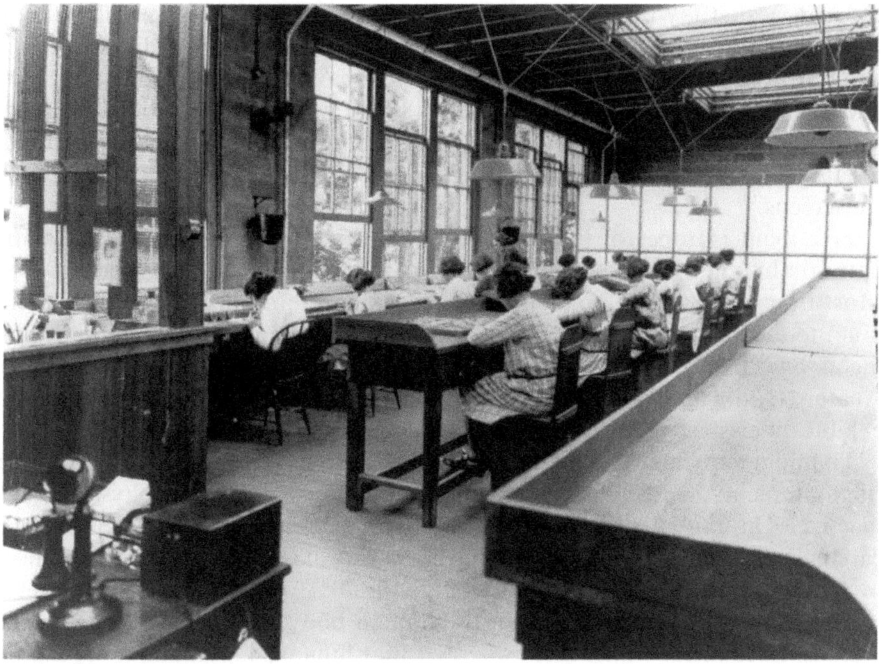

Figure 16.2 Watch dial painters working in the US Radium Corporation (Argonne National Laboratory).

(TiO_2) and sphene ($Ca.TiSiO_5$). Ninety percent of the world's titanium supply is provided by ilmenite deposits in United States, Russia, Canada, Japan, Kazakhstan and Spain. Titanium is a stable metal with a high melting point, tensile strength and resistance to corrosion, and shows appealing capacity to alloy with palladium, ruthenium, nickel, iron, aluminium, manganese and molybdenum. Titanium and its alloys exhibit a unique combination of mechanical and physical properties and corrosion resistance which have made them desirable for aeronautical engineering, industrial, chemical and energy industry service. Additionally, titanium has established value in products demanding resistance to environmental pollution, marine erosion and robustness. Principal uses include rockets and jet engine rotating applications and airframes, electron-beam melting and high quality alloys for aerospace engineering and airframes, power generating plants, where saline, brackish or polluted waters are used as the cooling medium, and in petroleum exploration and production, where its light weight and flexibility make it an excellent material for deep sea production risers. It is as strong as steel but 45% lighter. Titanium has been used increasingly in bone prostheses in recent years on account of its corrosion resistance and robustness. Titanium alloys are used in orthopaedic implants, synthetic joints and cardiac pacemakers, where they remain in contact with body tissue for long periods. Clinical and experimental studies show that titanium is entirely biocompatible with osteoblasts *in vitro* and *in vivo*.[61–63] An experimental study in baboons showed that some titanium is mobilised from orthopaedic implants to the lungs, lymph nodes, spleen and adjacent muscle without ill-effect.[64] Titanium dioxide is employed as a white pigment in paints and plastics. More than 1300 commercial coatings, paints, plastics, cosmetics, sunscreen agents, and products containing titanium dioxide (TDO) nanomaterials are currently marketed.[65] It absorbs UV light and is available in some sunscreens.[66]

Five main isotopes of titanium exist, of which ^{48}Ti comprises 74% of the natural abundance of the metal. Other isotopes each contribute about 5–8%. Most titanium is used in metallic form or as titanium dioxide (TiO_2) but titanium tetrachloride ($TiCl_4$) occurs as an intermediate in the production of pure metal and in the production of titanium dioxide and titanium pigments. Anatase, rutile and brookite have the same chemistry as titanium dioxide but exhibit different crystal structure and stoichiometry.

Titanium tetrachloride is a toxic compound which hydrolyses in water whereas titanium metal is largely inert biologically and considered to be physiologically inactive. On the other hand, titanium dioxide has been considered to be carcinogenic in the production of pigments and has been subject to extensive investigation. Paint manufacture accounts for 70% of the use of the element.

Minute quantities may occur in the human body but the metal has no recognised trace metal value. It is estimated that humans ingest about 0.8 mg per day in food but most is excreted without being adsorbed. The human body can tolerate titanium in large doses without ill-effect. Titanium is abundant in the soil but is poorly absorbed by plants and animals and as such is a minor constituent of human food chains. Some nuts and fruit kernels contain 0.5–6.1 $\mu g\ g^{-1}$ titanium. Titanium concentrations in urban air[67,68] are mainly below 0.1 $\mu g\ m^{-3}$, and river water concentrations in USA range from 2 to 107 $\mu g\ l^{-1}$.[69,70] Titanium in 81% of 42 municipal water supplies in the USA had a mean concentration of 2.1 $\mu g\ l^{-1}$ with a range of 0.5–15 $\mu g\ l^{-1}$.

Metallic titanium is poorly absorbed or retained in the human body following ingestion or dermal contact but its metabolism in the human body is imperfectly understood. Gastrointestinal and percutaneous absorption seems to be negligible. No evidence has been seen that titanium induces or binds metallothioneins, but *in vitro* studies indicate that titanium dioxide particles bind human serum albumins and γ-globulins and that binding is dependent upon particle size (<50 nm) and molecular configuration (*i.e.* anatase, rutile or brookite) and concentration.[71] The true genotoxicity of titanium dioxide is complicated by the existence of the three distinct crystalline forms with differing surface properties and toxicity in cultured cell lines.[72–74]

Principal concerns relate to chronic inhalation of titanium and titanium dioxide dust in industry as a cause of lung cancer. Several large epidemiological studies conducted in Canada, China, Taiwan, the USA and Europe have shown that inhalation of titanium dust is a cause of respiratory distress and pleural disease, that severe pulmonary irritancy and insufficiency occur, and that proteinosis and sarcoid granuloma are exacerbated by silica dust, asbestos and smoking among workers.[75–83] Evidence that titanium dust or inhalation of titanium dioxide may be carcinogenic in humans derives largely from *in vitro* tests and animal trials. Increased cancer-related mortality was not observed in epidemiological studies, and causes of death were as expected for population trends, lifestyle, smoking and exposure to other contaminants.[84,85] Epidemiological studies in workers involved in the chipping and processing of titanium ores, where workers are exposed to aerosols of metal dust, titanium tetrachloride ($TiCl_4$), titanium dioxide, titanium oxychloride ($TiOCl_2$) and unspecified contaminants over more than 6 months, have not revealed any excess lung cancer mortality attributable to titanium, TDO or other inorganic titanium compounds. A pulmonary papillary carcinoma associated with interstitial fibrosis and titanium granules diagnosed in the lung of a titanium dioxide packer after about 13 years' exposure was attributed to a 40-year history of tobacco smoking.[86] Granulomatous lung disease was reported in a 45-year-old man exposed to titanium dust inhalation during 13 years of employment as a furnace-man in aluminium smelting.[87] Laboratory tests revealed a lymphoproliferative response to titanium as evidence of contact sensitisation following inhalation.[88] Lung biopsy of a painter dying of pulmonary alveolar proteinosis showed birefringent titanium dioxide particles ($60–129 \times 10^6$ cm^{-3}) and an overwhelming saturation of alveolar clearance mechanisms involving alveolar macrophages (dust cells).[89] Local cell damage and epigenetic changes were not associated with tumour formation.

In vitro studies in cultured human fibroblasts and astrocytoma U87 cells have shown that micro- and nanoparticles of titanium dioxide are phagocytosed by human polymorphonuclear cells, and release of reactive oxygen metabolites lead to cell death.[90,91] Cultured human lung cells endocytosed titanium dioxide nanoparticles which localised within the cytosol near to but with no observable influence on the nucleus, mitochondria or ribosomes. No evidence of DNA breakage was provided by the Comet assays.[92] On the other hand, *in vitro* studies suggest that titanium dioxide does induce reactive oxygen species and an indirect form of genotoxicity mediated by DNA adduct formation, but this has not been ratified. Titanium dioxide was not clastogenic in cultured rat hepatocytes or Chinese hamster ovary cells.[93] Recent evidence provided by *in vitro* experiments claims that nanoparticulate titanium dioxide is genotoxic in human epidermal cells, lymphocytes and mouse embryo fibroblasts through generation of reactive oxygen species.[94–96] The effect, possibly involving induction of a strongly oxidant peroxynitrite ion, was inhibited by concurrent treatment with Nystatin, an effective inhibitor of titanium dioxide endocytosis. Only relatively high concentrations (100 $\mu g\ ml^{-1}$) of nanoscale titanium dioxide were cytotoxic and inflammation was observed in immortalised human lung epithelial cells.[97] Anatase was 100 times more toxic than equivalent samples of rutile. The more cytotoxic nanoparticle samples were more effective in generating reactive oxygen species.

There is speculation that these ultrafine particles penetrate the epithelial lining of the respiratory tract to cause inflammatory change, oxidative stress, fibrosis and possibly lung cancer, as indicated by experimental studies in rats. Titanium as Ti(IV) combines directly with globulins and albumins and nanoparticles are phagocytotically taken into cells, but at the moment there is no evidence to show direct interaction with DNA. The National Institute for The Occupational Safety and Health Administration (OSHA) permissible exposure limit for TiO_2 is 15 mg m^{-3} based on the airborne mass fraction of total TiO_2 dust.[98] This contrasts with recommendations by Occupational Safety and Health (NIOSH) who recommended safety exposure limits of 2.4 mg m^{-3} for titanium dioxide and 0.3 mg m^{-3} for ultrafine particles for exposures in the workplace of up to 10 hours per day, or 40 hours per week (time weighted average), recognising it as a potential carcinogen.[65]

The IARC have re-classified titanium dioxide as a possible human carcinogen on the basis of "sufficient evidence in laboratory animals".[99] They show clearly that no obvious risk is associated with oral or dermal exposure and that risks lie principally through inhalation of particulate material. However the experiments are contradictory. Early experiments in rodents provided no evidence of carcinogenicity through inhalation, oral/ingestion or dermal exposure at concentrations close to permitted levels.[100,101] Thus, administration of 5% titanium dioxide-coated mica to Fischer F344 rats in their diet for 130 weeks was without obvious effects, possibly reflecting minimal intestinal absorption.[102] Mice dosed orally for two weeks exhibited wide distribution of the material in the body and functional changes in liver and kidney, but no precancerous change.[103] When the National Toxicology Program (NTP) exposed Fischer F344 rats and $B6C3F_1$ mice to massive doses of 25 000 or 50 000 ppm titanium dioxide in the diet for two years, survival rates in female mice were significantly depressed, and female rats exhibited a statistically significant increase in C-cell thyroid tumours.[104] Titanium dioxide is not absorbed percutaneously on account of its insolubility in water or lipids of skin, therefore cancers arising through direct dermal contact are unlikely.[66,105] Experimental studies provide no evidence that nanoparticulate titanium dioxide acts as an initiator or promoter of dermal tumours in a mouse skin carcinogenicity model using 7,12-dimethylbenzanthracene and tetradecanoylphorbol-13-acetate (TPA) as known tumour initiator and promoter respectively.[106]

An overview of experimental data produced from long-term (up to two years) inhalation or intratracheal administration studies in three strains of rat [Fischer F344, Sprague–Dawley (SD) and Wistar] has shown that intra-alveolar accumulations of insoluble ultrafine titanium dioxide particles led to a saturation of intrinsic clearance mechanisms, local inflammation with release of cytokines and fibrogenic mediators, proliferation of type II pneumocytes, proteinosis and keratinisation in alveolar ducts.[107,108] Animals exposed to 0.1–10 mg m^{-3} titanium chloride or 10–250 mg m^{-3} titanium dioxide for up to two years developed a massive dose-related build-up of alveolar macrophages laden with particles, with progressive development of "cystic keratinising squamous pulmonary carcinomas".[109–112] The carcinogenicity of titanium dioxide in the rat lung was claimed to be directly related to the size, surface characteristics, crystal structure and duration of administration, and induction of hydroxyl radicals through inflammatory cytokines, oxidative stress and DNA-adduct formation.[113–116] Uptake of particles (20 nm diameter) in bronchiolar lavages was low: 0.06–0.12% within 24 hours.[117–119] Hext noted that mice inhaling 250 mg m^{-3} ultrafine particles over 90 days developed a similar pulmonary burden of pulmonary titanium dioxide but their clearance rate was greater and lung pathology much lower than seen in rats.[120] Hamsters showed a high particle clearance rate and minimal responses to inhaled titanium dioxide.

In conclusion, the carcinogenic hazard indicated by chronic inhalation and intratracheal administration of insoluble titanium dioxide in rats has to be viewed with great caution. The so-called cystic keratinising squamous pulmonary carcinomas seem to be peculiar to certain strains of rat and are not listed in the histological typing of human lung tumours by the World Health Organization.[121] They should be regarded as a species/strain-specific response to pulmonary overload situations of insoluble inhalates. Available studies provide no evidence that titanium dioxide is catabolised within the human lung, or that it invokes any of the normal detoxification mechanisms or metabolising enzymes expressed by metabolically competent cells for bioactivation of xenobiotics.[122,123]

References

1. International Agency for Research on Cancer (IARC), *Monographs on the Evaluation of the Carcinogenic Risk of Chemicals to Man. Some internally deposited radionuclides*, IARC, World Health Organization, Geneva, 2001, vol. 78.
2. US Department of Health and Human Services (US DHHS), *12th Report on Carcinogens*, National Toxicology Program, Research Triangle Park, NC, 2011.
3. E. Rutherford, *Philosoph. Mag.*, 1900, **49**, 1.
4. J. P. Guimaraes, L. F. Lamerton and W. R. Christensen, *Brit. J. Cancer*, 1955, **9**, 253.
5. P. Radvanyi, *Chem. Internal.*, 2011, **33**, 1.
6. A. Makhijani, *Presentation, Institute for Energy and Environmental Research*, 2004.
7. Agency for Toxic Substances and Disease Registry (ARSDR), *Toxicological Profile for Thorium*, US Enviromental Protection Agency, 1990.
8. E. Meyer, *Chemistry of Hazardous Materials*, Prentice Hall, Englewood Cliffs, NJ, 2nd edn, 1989.
9. J. J. Pickering, *Radiation Safety Training for General Laboratory Workers*, Pickering Enterprises Publishing, Livermore, CA, 2nd edn, 2002.
10. J. Shapiro, *Radiation Protection: A guide for scientists and physicians*, University Press, Cambridge, MA, 1990.
11. National Research Council of the National Academies (NRC), *Health Risks from Exposure to Low Levels of Ionizing Radiation*, Beir VII, Phase 2, NRC, Washington, DC, 2013.
12. Health Physics Society, *Radiation Risk in Perspective*, Health Physics Society, McLean, VA, 1996.

13. J. Emsley, *The Elements*, Clarendon Press, Oxford, 1991, p. 251.
14. A. M. Krasinskas, J. Minda, S. H. Saul, A. Shaked and E. E. Furth, *Modern Pathol.*, 2004, **17**, 117.
15. M. Gatto, M. C. Bragazzi, R. Semararo, C. Napoli, R. Gentile, A. Torrice, E. Gaudio and D. Alvaro, *Digest. Liver Dis.*, 2010, **42**, 2516.
16. H. M. Solli, A. Andersen, E. Stranden and S. Langård, *Scand. J. Work Environ. Health*, 1985, **11**, 7.
17. E. H. MacMahon, A. S. Murphy and M. I. Bates, *Am. J. Pathol.*, 1947, **23**, 585.
18. B. Farid, *J. Am. Med. Assoc.*, 1982, **248**, 550.
19. W. Baker, R. L. Beck, D. P. Shale and B. P. Swinyer, *Clin. Appl. Thromb. Hemostat.*, 1995, **1**, 230.
20. V. H. Witten, M. S. Ross, E. Oshry, A. B. Hyman and V. A. Holmstrom, *J. Invest. Dermatol.*, 1951, **17**, 311.
21. J. Natkunarajah and S. Cliff, *Clin. Exp. Dermatol.*, 2009, **34**, 189.
22. F. S. Airy, *Proc. Roy. Soc. Med.*, 1945, **38**, 142.
23. B. Algermissen, H.-P. Berlien and N. Haas, *Acta Derm. Venereol.*, 2004, **84**, 475.
24. L. Scarri and A. E. Navaratnam, *J. Am. Acad. Dermatol.*, 1994, **31**, 796.
25. A. F. Stehney, *Radiat. Res.*, 1999, **152**, S110.
26. A. P. Polednak, A. F. Stehney and H. F. Lucas, *Health Phys.*, 1983, **44**(Suppl. 1), 239.
27. I.P. Likhachev, *Arkh. Pathol.*, 1976, **38**, 18.
28. I. Farid and S. A. Conibear, *Health Phys.*, 1983, **44**(Suppl. 1), 221.
29. X. Chen, Y. Chen, H. Xaio, G. Feng, Y. Deng, Z. Feng, L. Chen, X. Han, Y. Yan, Z. Dong and R. Zhen, *Chin. Med.*, 2003, **116**, 692.
30. D. B. Chambers, *Radiat. Prot. Dosimetry*, 2010, **141**, 351.
31. M. Jevtić, M. Vestović, M. Mirosavljev, I. Bikit and L. Conkić, *Med. Pregl.*, 2001, **54**, 140.
32. H. Irie and W. Mori, *Acta Pathol. Jpn.*, 1984, **34**, 221.
33. A Ødegaard, E. M. Ophus and A. M. Larsen, *J. Clin. Chem.*, 1978, **31**, 893.
34. W. T. Kabisch, *Ann. N.Y. Acad. Sci.*, 1967, **145**, 585.
35. T. Kluge and T. Hovig, *Am J Pathol.*, 1969, **54**, 355.
36. S. E. Glover, R. J. Traub, C. A. Grimm and R. H. Filby, *Radiat. Prot. Dosimetry*, 2001, **97**, 153.
37. N. P. Singh, *Human Exp. Toxicol.*, 1990, **9**, 79.
38. O. Axelson, *Environ. Perspect.*, 1995, **105**(Suppl. 2), 37.
39. X.-A. Chen, Y.-E. Cheng and Z. Rong, *J. Radiol. Prot.*, 2005, **25**, 451.
40. H. Welsch, G. vanKaick, W. Riedel, A. Kaul, K. Wegener, K. Hasenöhrl and H. Muth, *Health Phys.*, 1983, **44**(Suppl. 1), 317.
41. J. A. Humphreys, N. D. Priest and Y. Ishikawa, *Health Phys.*, 1998, **74**, 332.
42. G. N. Taylor, R. D. Loyd and C. W. Mayes, *Health Phys.*, 1993, **64**, 141.
43. K. Wegener, K. Hasenhohrl and H. Welsch, *Health Phys.*, 1983, **44**(Suppl. 1), 307.

44. K. S. Iwamoto, T. Mizuno, A. Kurata, M. Masuzawa, T. Mori and T. Seyama, *Human Pathol.*, 1998, **29**, 412.
45. C. W. Mayes and H. Spiess, *Environ. Res.*, 1979, **18**, 88.
46. T. Mori, K. Fukutomi, Y. Kato, S. Hatakeyama, R. Machinami, H. Tanooka, Y. Ishikawa and T. Kumatori, *Radiat. Res.*, 1999, **152**(Suppl.6), S72.
47. Y. Ishikawa, T. Moria and R. Machinami, *J. Cancer Res. Clin. Oncol.*, 1995, **121**, 567.
48. J. Visfeldt, G. Jensen and E. Hippe, *Acta Pathol. Microbiol. Scand A.*, 1975, **83**, 373.
49. M. Andersson, M. Vyberg, J. Visfeldt, B. Carstensen and H. H. Storm, *Radiat. Res.*, 1994, **137**, 262.
50. W. Riedel, R. Hirschberg, A. Kaul, H. Schmier and U. Walter, *Environ. Res.*, 1979, **18**, 127.
51. A. Kumar, P. Mishra, S. Ghosh, P. Sharma, M. Ali, B. N. Pandy and K. P. Mishra, *Ind. J. Radiat. Biol.*, 2008, **84**, 337.
52. S. Bolvin-Angèle, L. François, O. Froment, A. Spiethoff, M. S. Bogdananffy, K. Wegener, H. Welsch, A. Barbin, B. Bancel, C. Trépo, H. Bartsch, J. Swenberg and M. J. Marion, *Int. J. Cancer*, 2000, **85**, 223.
53. J. P. M. Bensted, *Ann. N.Y. Acad. Sci.*, 1967, **145**(728).
54. J. P. M. Bensted and J. O. Crookall, *Br. J. Cancer*, **17**, 62.
55. K. E. Buckton, A. O. Langlands and G. E. Woodcock, *Int. J. Radiol.*, 1967, **12**, 565.
56. C. Kido, F Saskaki, Y. Hirota, K. Kiyosawa, S. Hayashi, T. Mori and T. Sobue, *Radiat. Res.*, 1999, **152**(Suppl.6), S81.
57. S. A. Fry, *Radiat. Res.*, 1998, **150**, S2.
58. C. Clark, *Radium Girls: Women and industrial health reform, 1910-1935*, University of North Carolina Press, Chapel Hill, NC, 1997.
59. H. S. Martland and R. E. Humphries, *Arch. Pathol. Lab. Med.*, 1929, **7**, 406.
60. A. B. de Gonzalez, A. Kutsenko and P. Rajarman, *Clin. Sarcoma Res.*, 2012, **2**, 18.
61. G. Kazantzis, *Environ. Health Perspect.*, 1981, **40**, 143.
62. G. Larsson, P. Tomsen, B. O. Arondsson, M. Rodahl, J. Lausmaa, B. Kasemo and L. E. Ericson, *Biomaterials*, 1996, **17**, 605.
63. R. K. Sinha, F. Morris, S. A. Shah and R. S. Tuan, *Clin. Orthop. Relat. Res.*, 1994, **305**, 258.
64. J. L. Woodman, J. J. Jacobs, J. O. Galante and R. M. Urban, *J. Orthop. Res.*, 1984, **1**, 421.
65. Paint and Coatings Industry, *J. Architectural Coatings*, Techology Publ. Co., Pittsburg, PA, 2011.
66. A. B. G. Lansdown and A. Taylor, *Int. J. Cosmet. Sci*, 1997, **19**, 167.
67. International Programme on Chemical Safety (IPCS), *Toxicological Evaluation of some Food Colours, Emulsifiers, Stabilizers, Anticaking Agents and Certain Other Substances*, Food Additives Organisation, WHO Nutrition Report Series, 1969, 46A,WHO/FOOD ADD/70.36.

68. R. D. Giauque, L. Y. Goda and N. E. Brown, *Environ. Sci. Technol.*, 1974, **8**, 436.
69. W. H. Durum and J. Heffty, *U.S. Geol. Survey Circular*, 1961, **445**, 1.
70. C. N. Durfor and E. Becker, *J. Am. Water Works Assoc.*, 1964, **56**, 237.
71. M. S. Zaquout, T. Sumizawa, H. Igisu, T. Higashi and T. Myojo, *J. Occup. Health*, 2011, **53**, 75.
72. Y. Lia, W. Que, Q. Jia, Y. He, J. Zhang and P. Zhong, *J. Mater. Chem.*, 2012, **22**, 7937.
73. C. Jin, Y. Tang, F. G. Yang, X. L. Li, S. Xu, Y. Y. Huang and Y. J. Yang, *Biol. Trace Elem. Res.*, 2011, **141**, 3.
74. J. Ferrin and G. Oberdorster, *Am. Indus. Hyg. Ass. J*, 1985, **46**, 6.
75. J. C. Pimental, *Acta Med. Port.*, 19992, **5**, 307.
76. J. L. Chen and W. E. Fayerweather, *J. Occup. Med.*, 1988, **30**, 937.
77. W. E. Fayerweather, M. E. Karns, P. G. Gilby and J. L. Chen, *J. Occup. Med.*, 1992, **34**, 164.
78. W. R. Parkes, *Proc. Roy. Soc. Med.*, 1977, **70**, 289.
79. D. H. Garabrant, L. J. Fine, C. Oliver, L. Bernstein and J. M. Peters, *Scand. J. Work Environ. Health*, 1987, **13**, 47.
80. P. Boffetta, V. Gaborieau, L. Nardon, M. F. Parent, E. Weiderpass and J. Siemiatycki, *Scand. J. Work Environ. Health.*, 2001, **27**, 227.
81. P. Boffetta, A. Soutar, J. W. Cherrie, F. Granath, A. Andersen, A. Artilla, M. Blettner, V. Gaborieau, S. J. Klug, S. Langard, D. Luce, F. Merletti, B. Miller, D. Mirabelli, E. Pukkala, H. O. Adami and E. Weiderpass, *Cancer Causes Control*, 2004, **15**, 697.
82. C.-M. Liao, Y.-H. Chang and C.-P. Chio, *J. Hazardous Materials*, 2009, **162**, 57.
83. E. D. Ellis, J. Watkins, W. Tankersley, J. Phillips and D. Girardi, *J. Occup. Environ. Health*, 2010, **52**, 303.
84. J. P. Fryzek, B. Chadda, D. Marano, K. White, S. Schweitzer, J. K. McLaughlin and W. J. Blot, *J. Occup. Environ. Med.*, 2003, **45**, 400.
85. A. V. Rammanakumar, M. E. Parent, B. Latreille and J. Siemiatycki, *Int. J. Cancer*, 2008, **122**, 183.
86. I. Yamadori, S. Ohsumi and K. Taguchi, *Acta Pathol. Jap.*, 1986, **36**, 783.
87. S. Redline, B. P. Barna, J. F. Tomashefski and J. L. Abraham, *Br. J. Indust. Med.*, 1986, **43**, 652.
88. A. Fisher, *Contact Dermatitis*, Lea and Febiger, Philadelphia, 1987.
89. C. A. Keller, A. Frost, P. T. Cagle and J. L. Abraham, *Chest*, 1995, **108**, 277.
90. M. Hedenborg, *Int. Arch. Occup. Environ. Health*, 188, **61**, 1.
91. J. C. Lai, M. B. Lai, S. Jandhyam, V. V. Dukhande, A. Bhushan, C. K. Daniels and S. W. Leung, *Int. J. Nanomed.*, 2008, **3**, 533.
92. K. Bhattachayra, M. Devoran, J. Boertz, R. P. Shins, E. Hoffmann and E. Dop, *Part. Fibre Toxicol.*, 2009, **21**, 6.
93. K. Linnainmaa, P. Kivipensas and H. Vainio, *Toxicol. in Vitro*, 1997, **11**, 329.
94. A. Xu, Y. Chai, T. Nohmi and T. K. Hei, *Part. Fibre Toxicol.*, 2009, **6**, 3, doi:10.1186/1743-8977-3.

95. R. K. Shukla, V. Sharma, A. K. Pandey, S. Singh, S. Sultana and A. Dhawan, *Toxicol. in Vitro*, 2011, **25**, 231.
96. M. Ghosh, M. Bandyopadhyay and A. Mukherjee, *Chemosphere*, 2010, **81**, 1253.
97. C. M. Sayes, R. Wahl, P. A. Kurian, Y. Liu, J. L. West, K. D. Ausman, D. B. Warheit and V. L. Colvin, *Toxicol. Sci.*, 2006, **92**, 174.
98. National Institute of Occupational Safety and Health, *Occupational Exposure to Titanium Dioxide*, NIOSH, Current Intelligence Bulletin 63, US, DHHS, 2011.
99. IARC, *IARC Monographs on the Evaluation of Carcinogenic Risks to Humans*, 2010, vol. 93, p. 193.
100. J. Thyssen, G. Kimmerle, S. Dikhaus, E. Emminger and U. Mohr, *J. Environ. Pathol. Toxicol.*, 1978, **1**, 501.
101. J. Ferin, G. Oberdorster and D. P. Penney, *Am. J. Resp. Cell Molec. Biol.*, 1992, **6**, 535.
102. B. K. Bernard, M. R. Osheroff, A. Hoffmann and A. Mennear, *J. Toxicol. Environ. Health*, 1990, **29**, 417.
103. J. Wang, G. Zhou, C. Chen, H. Yu, T. Wan, Y. Ma, G. Jia, Y. Gao, B. Li, J. Sun, Y. Li, F. Jiao, Y. Zhao and Z. Chai, *Toxicol.Lett.*, 2007, **168**, 176.
104. National Cancer Institute, *Bioassay of Titanium Dioxide for Possible Carcinogenicity*, National Cancer Institute, Technical Report Series, 1997, No.97, p. 1.
105. M. H. Tan, C. A. Commens, L. Burnett and P. J. Snitch, *Australas. J. Dermatol.*, 1996, **37**, 185.
106. F. Furukawa, Y. Doi, M. Suguro, O. Morita, H. Kuwahara, T. Masunaga, Y. Hatakema and F. Mori, *Fd. Chem. Toxicol.*, 2011, **49**, 744.
107. K. P. Lee, D. P. Kelly, P. W. Schneider and H. J. Trochimowicz, *Toxicol. Appl. Pharmacol.*, 1986, **83**, 30.
108. K. P. Lee, N. W. Henry, H. J. Throchimowicz and C. F. Reinhardt, *Environ. Res.*, 1986, **41**, 144.
109. H. J. Trochimowicz, K. P. Lee and C. F. Reinhardt, *J. Appl. Toxicol.*, 1988, **8**, 383.
110. H. Muhle, B. Kittel, H. Ernst and R. Mermelstein, *Scand. J. Work Environ. Health*, 1995, **21**(Suppl. 2), 27.
111. D. B. Warheit and S. R. Frame, *J. Occup. Environ. Med.*, 2006, **48**, 1308.
112. D. B. Warheit, T. R. Webb, K. L. Reed, S. Frerichs and C. M. Sayes, *Toxicology*, 2007, **230**, 90.
113. A. Churg, B. Gilks and J. Dai, *Am. J. Physiol.*, 1999, **277**, L975.
114. K. Donaldson, X. Y. Li and W. MacNee, *J. Aerosol Sci*, 1998, **29**, 553.
115. H. Christie, R. J. Mackay and A. M. Fisher, *Am. Industr. Hyg. Ass. J.*, 1963, **24**, 42.
116. R. Baan, K. Straif, Y. Grosse, B. Secretan, F. El Ghissassi and V. Cogliano, *Lancet, Oncol.*, 2006, **7**, 295.

117. U. Mohr, H. Ernst, M. Roller and F. Pott, *Exp. Toxicol. Pathol.*, 2006, **58**, 13.
118. M. Geier, M. Casaulta, B. Kupferschmid, H. Shulz, M. Semmier-Behnke and W. Kreyling, *Am. J. Respir. Cell Mol. Biol.*, 2008, **38**, 371.
119. J. R. Roberts, R. S. Chapman, V. R. Tirumala, A. Kaim, B. T. Chen, D. Schwegler-Berry, A. B. Stefaniak, S. S. Leonard and J. M. Antonini, *J. Toxicol. Environ. Health*, 2011, **74**, 790.
120. P. M. Hext, J. A. Tomlinson and P. Thompson, *Ann. Occup. Hyg.*, 2005, **49**, 461.
121. World Health Organisation, *International Histological Classification of Tumours No.1*, WHO, 2nd edn, 1981, Geneva.
122. J. V. Castell, M. T. Donato and M. J. Gómez-Lechón, *Exp. Toxicol. Pathol.*, 2005, **57**, 189.
123. G. Scheuch, M. J. Kohlheaufl, P. Brand and R. Siekmeier, *Adv. Drug Deliv. Rev.*, 2006, **58**, 996.

CHAPTER 17

Arsenic, Antimony and Bismuth

17.1 General Introduction

Arsenic, antimony and bismuth are chemically similar elements presently classified in Group 15 (formerly Vb) of the Periodic Table. Arsenic and antimony are metalloid elements whereas bismuth is a metal. They are found throughout the world in soil and water courses and exhibit an interesting range of biological properties and uses. They may occur in the human body at low concentrations but do not fulfil the criteria of trace metals.[1,2] The human body has been estimated to contain serum concentrations of $10.6 \pm 1.3\,\mathrm{ng\,ml^{-1}}$ arsenic (or total body concentrations of 10–20 mg),[3] whereas the body content of antimony and bismuth is normally negligible (?<1 ng). All three elements have been known for at least 500 years and have a wide and interesting history, and up until relatively recent times have been used therapeutically in the treatment of bacterial, fungal and protozoal infections.[4,5] Antimony has been used as a treatment for bilhartzia, schistosomiasis and leishmaniasis and drugs such as tartar emetic, Bilharcid EP (piperazine diantominonyl tartrate), Pentostam® (sodium stibogluconate) and Glucantime® (meglumine antimonate) are listed.[6,7] Recent research suggests that antimony and possibly bismuth have a limited value as anticancer agents.[8,9] In contrast, arsenic trioxide has shown promise in clinical trials in the USA as a chemotherapeutic against promyelocytic leukaemia (APL).[10–12] It is able to induce apoptosis in melanoma cells and cytotoxicity in lung carcinoma cells.[13,14]

Although the three elements are similar in their chemical properties, they differ greatly in their toxicity and carcinogenicity. Arsenic is appreciably more toxic than antimony or bismuth. They form trivalent and pentavalent compounds, with trivalent compounds being more toxic in each case. They are methylated by environmental bacteria, fungi and yeasts, and this can have a profound impact on their toxicity and carcinogenicity in the human body.[15,16]

Issues in Toxicology No. 18
The Carcinogenicity of Metals: Human Risk through Occupational and Environmental Exposure
By Alan B. G. Lansdown

Published by the Royal Society of Chemistry, www.rsc.org

Environmental bacteria also have a role in mobilising arsenic in sediments thereby influencing the biogeochemical cycling and reduction of As(V) to As(III) in ground water deposits in countries such as Bangladesh.[17,18]

17.2 Arsenic

Arsenic is the oldest known and most contentious of all human carcinogens and a major international health problem. In the early days, its toxicity and lethality was recognised in several classical cases of suicide and murder including the mysterious deaths of Francesco de Medici and Bianca Capello in 1587, Hannah Russell who used arsenious oxide to murder her husband in 1826, and Napoleon Bonaparte, who is supposed to have died through chronic arsenic poisoning in St. Helena.[19–21] The last issue has been subject to great conjecture but Harwell Nuclear Research Laboratory confirmed that authenticated hair removed from Napoleon did in fact contain arsenic. However, he is believed to have been killed by administration of Calomel (mercuric chloride); it is conceivable that the arsenic in Napoleon's hair may have resulted from the common mortician's practice of adding a pinch of arsenic trioxide as a preservative to hair samples cut as mementoes and/or the use of arsenic-containing hair powder.[22] The lethality of arsenic in rat poison led to the publication in 1851 of the Arsenic Act in Great Britain which was aimed at preventing secret poisoning; until that date, arsenic could be purchased or sold without legal restriction.[23]

Arsenic is a metalloid element found naturally in more than 200 different minerals as inorganic complexes with copper, lead, iron, nickel, cobalt, gold and other metals. The main arsenic-bearing minerals include arsenopyrite (AsFeS), realgar (AsS) and orpiment (As_2S_3). Gold often occurs as tiny deposits within the arsenopyrite. Arsenic concentrations in the Earth's crust may be as high 5 ppm but higher concentrations are found in some igneous and sedimentary rocks. Soils formed by the weathering and breakdown of rock to clays may contain between 0.1 and 40 ppm. Arsenic is released into ground and surface waters by the erosion, dissolution and weathering of rocks; some leaches into rivers and estuaries. Arsenic-rich sediments, peat and ultimately coal may form in the lower reaches of rivers. Combustion of coal is a source of environmental arsenic pollution in urban areas. Additionally, water draining from coal mines and mineral waste sites may have elevated levels of arsenic. Eisler emphasised that arsenic released from gold mining sites becomes concentrated in plants, the consequent risks to wildlife and food animals thereby causing ecological problems.[24] Oral administration of arsenicals at low levels is fatal to 50% of species of birds (17–48 mg kg^{-1}) and mammals (2.5–33 mg kg^{-1}).[24] Gold miners exposed to arsenic occupationally show an increased risk of fatal lung, stomach and lung cancers. Other workers recovering gold from high arsenic-containing ores are exposed to arsenic vapours released from arsenopyrite in the high temperature roasting process.[25] Lung cancers are also an occupational risk of arsenic exposure in copper smelters.[26–28] The general public is exposed to arsenic, antimony and selenium through release of gases and fumes released into the atmosphere from coal-fired power stations and combustion of waste materials.[29–31] Theoretically, even chimney sweeps are at risk from inhaling arsenic dust but the level of health risk has not been quantified. A bizarre source of arsenic poisoning and respiratory disease is reported in the incineration of poultry litter and carcases, when organic arsenicals have been added to feed to eliminate intestinal parasites.[32] Poultry litter is reported to contain 15–35 ppm arsenic!

Arsenic is found the soil and drinking water in many countries as inorganic and organic As(V) or As (III) compounds, particularly in Taiwan, Bangladesh, South America, certain parts of the USA and in the People's Republic of China.[33–38] Blackfoot, an endemic vascular disease prevalent in south west Taiwan, is attributable to consumption of high levels of arsenic in artesian well waters.[35,39–41] Arsenic accumulates also in rice, which forms the staple diet in many South and East Asian countries, as well as tobacco used in cigarettes.[42–45] For many years, arsenic was known to be present in many types of cigarette but the risks have declined in recent times as arsenicals in pesticides have been phased out.

17.2.1 Industrial and Commercial uses of Arsenic and Human Exposures

The principal commercial and industrial applications of arsenic and related compounds are conveniently summarised under:

a. Medical applications: antifungal, antiprotozoal and anticancer agents
b. Non-ferrous low weight, corrosion resistant alloys: solders used in automotive body manufacture, bearings, lead batteries, armour plating, ammunition
c. Electrical applications: semiconductors (such as gallium or indium arsenides), lasers and integrated circuits, televisions, light-emitting diodes and solar cells
d. Pesticides, preservatives: copper arsenate, lead arsenate, roxarsone (3-Nitro®; 1,3-nitro-4-hydroxyphenylarsonic acid), used in crop protection, vineyards, wood preservatives, leather tanning, and farm animal bedding and food
e. Pigments for paints, porcelain, glass, wallpaper.

Arsenic has a long history of use in medicine and legend has it that Hippocrates, Aristotle and Pliny the Elder used *arsenium* in the treatment of infected wounds and ulcers.[33] In the 16th and 17th centuries amulets containing white arsenic were placed around the neck to ward off plague. Later medicinal uses of arsenic included the introduction of Fowler's Solution in the 18th century, and by the early 1900s more than 500 organic arsenic preparations were available for the treatment of syphilis, dysentery and tuberculosis.[47–59] Fowler's solution, also known as *"Liquor Potassii Arenitis"*, *Kali Arsenicosum*, or *Kali arseniatum*, a 1% solution of potassium arsenite (KH_2AsO_3) was introduced by Dr Thomas Fowler, an English physician from Stafford, England in 1786 for a range of ailments including malaria, chorea and syphilis. It was withdrawn on account of its serious side effects including cirrhosis of the liver, idiopathic portal hypertension, cirrhosis and angiosarcoma of the liver, and cancers of the skin and urinary bladder

By the early 1900s pharmacopoeias listed arsenic trioxide, arsenous anhydride, arsenous oxide, white arsenic, arsenous acid and organic As(III) and As(V) compounds as efficacious remedies for conditions ranging from nutritional disturbances, neuralgia, asthma, diabetes, skin diseases and a range of infective conditions including malaria, tuberculosis, syphilis and amoebic dysentery.[5] Other notable arsenic compounds introduced at about this time included tryparsamide, arsanilates, cacodylates, amido-phenylarsonic acids (carbarsone) and the arsphenamines (Salvarsan, neoarsphenamine, sulpharsphenamine). All were claimed to be highly effective antibacterial and antiprotozoal agents, but they were toxic and found to cause gastrointestinal disturbances, haematological dyscrasias, renal degeneration, neurological change and cancer. These arsenicals decompose in the body to release ionic arsenic which is effective against protozoan parasites. Most have now been removed from the international pharmacopoeia for safety reasons. Paul Ehrlich in 1910 claimed that the arsphenamines showed a high efficacy against syphilis and were less toxic than quinines and other agents available at the time (Figure 17.1).[46] Salvarsan and arsphenamines were used with some success against spirochaete and protozoal infections including yaws, relapsing fever, rat-bite fever, trypanosomiasis and amoebic dysentery.

Arsenic trioxide (AsO_3), as transparent crystals or a white powder, was introduced originally as a wood preservative,[57,58] but in recent years has been shown to have commendable anticancer action. Raymond Warrell of the Sloane Kettering Institute in New York and others have demonstrated that patients treated daily with 0.06–2 mg kg^{-1} arsenic trioxide showed a remission

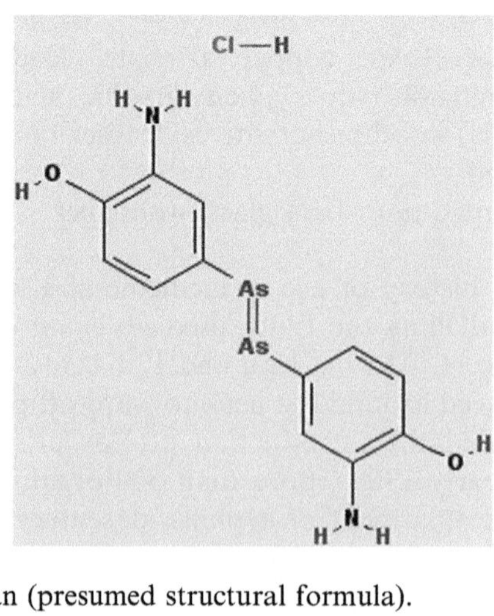

Figure 17.1 Salvarsan (presumed structural formula).

of relapsed promyelocytic leukaemia.[56] Arsenic trioxide is cytotoxic to small cell lung carcinoma cells, uveal and metastatic melanoma, and HeLa human tumour cells.[13,14,60–64] Few adverse effects were recorded with the low doses used.

Arsenic has been used in pesticides and preservatives for 200 years and nowadays this market comprises up to 70% of the global production of the element. Arsenic dips containing white arsenic have been used for treating cattle to ward off flies and tropical parasites, but this usage is now limited on account of the toxicity of the element and its ability to cause liver failure. The arsenic-related pigments Scheele's green [(Schloss green, cupric hydrogen arsenite, $AsCuHO_3$)] and Paris green [Emerald green, Schweinfurt green, $3Cu(AsO_2)_2.Cu(C_2H_3O_2)_2$] have been used in paints and insecticides since Victorian times. Paris green, a brilliant yellowish green double salt of copper arsenite and copper acetate is an extremely poisonous, bright green powder formerly used as a pigment (*e.g.* in wallpaper, glass and ceramics). Arsenic trioxide (As_2O_3, also known as arsenious acid, arsenious oxide, arsenious trioxide, arsenous acid, anhydride, crude arsenic, arsenolite and white arsenic) is produced as a by-product of copper smelting and has been used in the synthesis of other arsenicals and in preservatives, herbicides and insecticides. It has been used in the manufacture of special glasses and ceramics, and as a reductant in dye manufacture.

Lewisite (2-chlorovinylarsonous dichloride) is an arsenical war gas that was developed in the USA for use during the World War I.[65,66] It is a highly toxic blistering agent and powerful lung irritant which may be fatal when inhaled. Lewisite, which acquired the name "dew of death", is absorbed rapidly through the skin and mucous membranes and penetrates ordinary clothing and even rubber. Large quantities were produced in 1918, but there is no evidence that it was ever used. Experimental studies have shown that Lewisite is not mutagenic in bacterial tests and so far no epidemiological evidence has been seen of its carcinogenicity in laboratory animals or humans.[67]

17.2.2 Human Exposure and Arsenic Absorption

Dermal keratoses, lung and bladder cancer are tell-tale signs of chronic arsenic poisoning or excessive environmental or occupational arsenic exposure. Arsenic is absorbed into the body through ingestion of contaminated food and drinking water, inhalation of dust and factory effluents, cigarette smoking and by percutaneous absorption, and levels of exposure are confirmed by increases in blood, urine, hair and toenail concentrations. Cigarette smoke contains up to $70.2\,\mu g\,g^{-1}$ As_2O_3.[68]

Large population studies in Bangladesh and Taiwan demonstrate clearly that absorption and metabolism of bioactive arsenic is dependent upon the age, sex, race, genetic factors, socio-economic status, body mass, health state and nutrition of individuals as well as environmental and geochemical factors. The geochemical form, solubility and concentration of the arsenic present in the environment and the action of environmental bacteria in oxidation–reduction and methylation influence the bioavailability of arsenic.[17,69–71] As(III) ions bind 13-fold more strongly to cytosolic SH-rich proteins than As(V) and are appreciably more toxic.[72,73] Vahter maintained that people eating sea-food consume up to 1000 μg per day of the organo-arsenicals arsenobetaine, arsenocholine and methylarsonic acids but these moieties bind avidly to sulfydryl groups and are rapidly excreted in urine.[74] Evidence for the carcinogenicity of organic arsenic compounds is limited mainly to experimental studies with the metabolites monomethylarsonic (MMA) and dimethylarsinic acids (DMA).

17.2.2.1 Metabolism

The susceptibility of individuals to the toxicity and carcinogenicity of arsenic and its compounds is closely related to their solubility in water or body fluids and capacity to methylate. Once absorbed arsenate and arsenite residues are methylated to MMA and DMA, which are excreted in urine.[37,75–79] On average, 10–30% of As(III) ingested is in the form of arsenites, 10–20% is methylated MMA and 60–80% is DMA. Arsenic poisoning and carcinogenicity are more profound when the amount absorbed into the body exceeds the rate of methylation and where genetic upregulation of key metabolising enzymes such as As(III)-methyl transferase, methylene-tetrahydrofolate reductase and glutathione *S*-transferases is depressed.[80–82] Betaine-remethylation of homocysteine is a prerequisite for the methylation of arsenite and men seem to be more efficient than women in methylating arsenites.[78,83] Arsenite methylation is less efficient with advancing age and is inhibited by alcohol consumption, cigarette smoking and tobacco chewing.[84–89] Children and adolescents are more efficient in methylating arsenic residues than adults, possibly on account of their higher metabolic activity.[90–92] As(III) passes transplacentally and prenatal exposure to arsenic has been shown to increase risks of arsenic-related toxicity in early childhood and vulnerability to malignant and non-malignant pulmonary disease later in life.[93–95]

Population studies in Argentina show that the methylation and excretion of arsenic are determined by genetic polymorphisms and expression of selected genes.[96] MMA is associated more with arsenic toxicity and bladder cancer but DMA has the lowest body retention of all As metabolites. MMA and DMA show a lower tendency to interact with and denature cell membranes and organelles than unmethylated arsenites or arsenates. Earlier views held that MMA and DMA were less toxic that arsenites and arsenates and that methylation was a form of detoxification process but as discussed later, this cytoprotective mechanism has been challenged by experimental results which have shown that both MMA and DMA are carcinogenic in rats.[97–100] The metabolites promote carcinogenesis in genetically altered "cancer-susceptible" strains of mice and can promote carcinogenic changes induced by experimental carcinogens such as nitrosamines.

Mechanisms and optimal conditions for the biomethylation of As(III) are still incompletely understood. Vahter was of the opinion that methylation occurs primarily in the liver by transfer of methyl groups from *S*-adenosylmethionine to As(III) residues.[79] The liver is a target organ for arsenic toxicity and changes attributable to membrane damage, elevated serum aminotransferases and reduction of glutathione and antioxidant enzymes are reported.[101,102] Hepatitis B surface antigen, renal disease and dietary factors are believed to influence the metabolism and elimination of arsenic and its availability to evoke carcinogenesis.[81,103] Dietary factors seem to play a crucial role in the carcinogenesis of arsenic in humans, and administration of folate to individuals with low serum levels has been shown to reduce blood MMA and enhance urinary DMA excretion.Metal-binding proteins also play a cytoprotective role in blocking arsenic bioactivity in exposed people.[10,104,105] People in poorer countries living on low protein diets are more vulnerable to ingested arsenic than those on a high quality protein diet.[76] The former group excrete a higher proportion of ingested arsenic as MMA than DMA whereas those on high protein diets show a higher ratio of DMA to MMA in urine. Clearly, people fed nutritionally balanced diets replete with folate, vitamins, sulfur-rich proteins and essential minerals are appreciably less susceptible to arsenic-related cancers.[36,106] Limited information is available at present that the drug dimercapto-succinate may be an efficacious therapy in patients poisoned with arsenic.[107]

Arsenic is not a cumulative toxin in the human body but has been shown to induce and bind metallothioneins (MT) through transcription factor 1.[105,108,109] Arsenic trioxide induces MT-II, MT-II and MT-III in human glioblastoma cells and this may function as a cytoprotective mechanism against arsenic-induced brain tumours, but a lot more research is still needed here. Variations in the sensitivity of different people to arsenic-related toxicity are in part attributable to genetic polymorphisms influencing the expression of intracellular MT synthesis and arsenic binding or complexing.[96]

17.2.2.2 Absorption

17.2.2.2.1 Drinking Water and Food. Concentrations of arsenic in drinking water worldwide range from <0.5 to 7550 $\mu g\, l^{-1}$. The European standard for arsenic in drinking water was 50 $\mu g\, l^{-1}$, but this level was reduced to 10 $\mu g\, l^{-1}$ in 1993 following guidelines set by the World Health Organization (WHO).[110,111] Inorganic arsenic compounds are readily absorbed gastrointestinally and metabolised throughout the human body, possibly in the form of MT–As complexes. About 140 million people are poisoned annually by arsenic in their drinking water and estimates indicate that populations in Taiwan, Bangladesh and South America account for more than half of the global total.[36,41,112–114] Artesian water in Bangladesh contains 300–4000 $\mu g\, l^{-1}$ and agricultural workers exposed to intense heat are expected to consume up to 15–20 litres daily.[115] Skin lesions, blackfoot and increased cancer mortality are endemic in these areas.[34,39–41] Analyses of arsenic in natural water sources, ground water and water from mine drainage sites has shown that 13 000 $\mu g\, l^{-1}$ is inorganic As(III), 3700 $\mu g\, l^{-1}$ As(V), and less than 100 $\mu g\, l^{-1}$ is

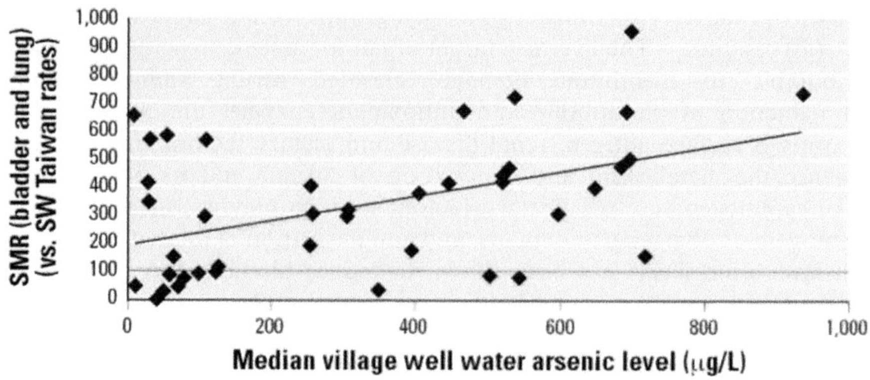

Figure 17.2 Combined bladder and lung cancer mortality rates (standardised mortality ratios) in 42 villages in Southwest Taiwan in relation to median well water arsenic concentration.[118]

in methylated form.[116] Arsenic in soils seems to be highly labile and is released into plants, animal life and into the atmosphere.[117] Samples of food crops in the vicinity of the Kumasi and Obuasi goldfields in Ghana ranged from 0.07 to 7.20 $mg\,kg^{-1}$ dry weight.

A population study of 115,903 people in Bengal showed that chronic exposure to arsenic levels of 10–49 $\mu g\,l^{-1}$ led to increased mortality rates through non-accidental causes, but at higher concentrations (50–149 $\mu g\,l^{-1}$) cancer mortality and cardiovascular diseases (including blackfoot) were greatly increased.[34] Similar dose-related trends are evident in Taiwan and in the Cordoba region of Argentina, both of which show a close association between ingestion of arsenic in village well water and lung, bladder, skin and other cancers (Figure 17.2).[112,118–120] In a Chilean study, children under 20 years of age exposed to arsenic in drinking water in the period from 1950 to 2000 showed an increased incidence of leukaemia, liver and brain cancers.[94] Boys born in this period were 8.9-fold more vulnerable to liver cancer whereas in girls the relative risk was even higher at 14.1.

Although arsenic was mined in areas of southwest England for more than 3000 years and this was at one time the world's largest source of the element, the true incidence of arsenic-related cancer in the UK is not known. People living in former mining areas with average levels of arsenic in garden soil of 365 $\mu g\,g^{-1}$ showed increased levels of arsenic in their urine following ingestion of food contaminated with arsenic residues or inhalation of arsenic dusts in the air.[121,122]

17.2.2.2.2 Air. Inhalation is a primary route of arsenic uptake and cause of lung cancer in occupational circumstances and through smoking cigarettes. Arsenic residues in dusts in gold mining and copper smelting are absorbed readily through alveolar membranes.[27,43,45,53,123,124] According to a report from the California Air Resources Board and the Department of

Health Services, smokers breathe an estimated 0.8–2.4 μg of inorganic arsenic per pack of cigarettes and approximately 40% of this is deposited in the respiratory tract. Arsenic absorbed *via* cigarette smoking or chewing tobacco is incompletely methylated and is metabolised to the skin as a primary site for carcinogenesis.[89]

17.2.2.2.3 Dermal Contact. Dermal keratoses, discolorations and skin tumours are characteristic of arsenic exposure. However, only small amounts of bioactive arsenic are absorbed through the skin since much of the free As(III) ion binds epidermal sulfydryl groups in keratin. Wester, using radiolabelled arsenic, showed that less than 5% of arsenic from wood preservatives penetrated the skin and that urinary levels were only marginally higher than background levels.[125,126] The Consumer Products Safety Commission in USA conducted a safety evaluation of wooden playground equipment treated with arsenic preservative and demonstrated that the quantity of arsenic leaching out was very low (1 in 1 million) and that the risks of skin cancer in children were negligible.[127] The average amount of arsenic absorbed by 66 children exposed to the playground equipment was 0.05 μg, appreciably lower than the expected daily intake in food and drinking water.[128]

17.2.3 Toxicity, Mutagenicity and Carcinogenicity

Arsenic and inorganic arsenic compounds are amongst the most toxic substances known and are currently listed in the 12th RoC and IARC monographs as known human carcinogens, on the basis of sufficient evidence of carcinogenicity in humans.[60,61,106,129] The skin is the most sensitive of human organs to arsenic, but depending upon the duration and magnitude of exposure, lifestyle and demographic factors, virtually any organ in the body can be affected.[33,76,80,81,91,120,130] Large population studies show that long-term exposure to arsenic leads to multi-system carcinogenicity and that tissue sensitivity and patterns of carcinogenicity change with advancing age; arsenic is also carcinogenic through transplacental transfer.[131–134]

Large-scale population studies in Southeast Asia and South America provide substantial knowledge regarding the carcinogenic risk associated with environmental arsenic exposure. Additional information is provided by exposures to arsenic in mining, copper and gold production, wood preservation, pesticides and in medicinal products. Only in recent years have experimental animals been found to be useful surrogates for studying the pathogenicity and mechanisms of carcinogenicity of arsenicals under controlled conditions, and for identifying patterns.[98,99,134,135] *In vitro* studies are available to illustrate the ability of arsenic to generate genetic and epigenetic changes in target tissues and the molecular and genetic basis of arsenic methylation.

True diagnosis of arsenic-related cancers in population studies has been seriously hampered by the long periods of latency between presumed arsenic exposure/consumption and appearance of cancer.[48,136,137] A survey of 115,903 people aged 15 years or older in 1991–2000 in Bangladesh recorded that 9015 (8%)of the individuals had died and 22,488 (19.4%) were lost to follow-up.[34] Even at low levels of arsenic (10–49 ppm), the hazard ratio was estimated to be 1.16, rising to 1.44 at higher levels (50–149 ppm) after 20–30 years of exposure.

17.2.3.1 The Skin and its Appendages

Hyperpigmentation, hyperkeratosis of the palms of the hands and feet, and white and black papules are characteristic of arsenic exposure by any route.[36,101] Buccal mucous membrane melanosis and discoloration of the tongue, lips and gums are additional features of arsenic poisoning. Keratoses and discoloration of the skin have been interpreted as preneoplastic lesions which progress to nodular rough skin with profound melanosis and ultimately squamous carcinoma, basal carcinoma, Bowen's disease and possibly melanoma (Figure 17.3). Blackfoot disease (BFD) is a feature of arsenic toxicity in certain areas of southwest Taiwan.[39–41,112,118] It is a peripheral vascular condition with arterial occlusion in the extremities leading to ulceration and gangrene with spontaneous or surgical amputation. The earliest evidence of arsenic-related skin disease, which concerned palmar and plantar keratoses transforming to Bowen's disease, squamous cell carcinoma and superficial basal cell carcinoma, was reported following treatment with drugs aimed at venereal infections, such as arsenobenzene, acetarsone and the older therapy *liquid arsenicalis*.

Arsenical dermatoses and severe jaundice without malignancy were observed also in 321 cases following injection of the organoarsenic drug arsenobenzol ($C_{12}H_{10}As_2$) for syphilis.[138] Fowler's solution is quoted in many reports,

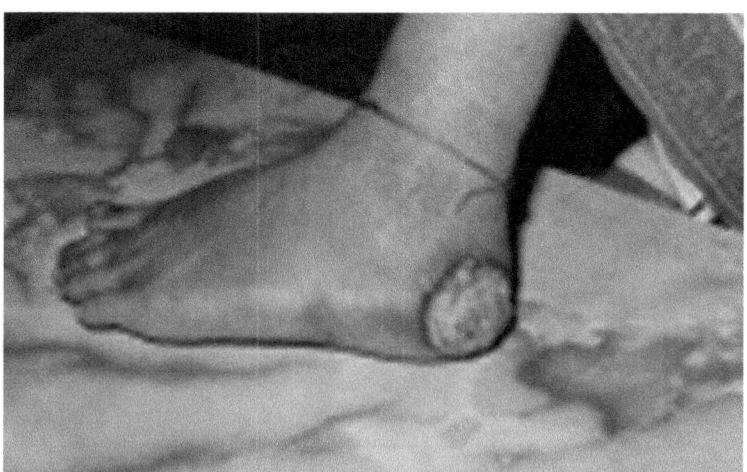

Figure 17.3 Squamous cell carcinoma on the foot of a patient from an area of Bangladesh with arsenic in ground water.[36]

including that of a 34-year-old clerk who developed a dark discoloration, roughness and keratoses of his skin with multiple "corns" on his hands and feet after 14 years.[47] His hands were amputated after 20 years as a result of severe fungating growths and the patient died 18 months later with multiple tumour metastases in his lungs, kidneys, scrotum, adrenals and ribs.

Epidemiological studies in the Reichenstein region of Silesia provided the earliest evidence that arsenic was a cause of keratoses, melanosis, skin tumours, perforated nasal septum and visceral cancers (Reichenstein's disease).[139] Patients were reported to show a 50% mortality rate in areas where drinking water was heavily contaminated with arsenic. More recent studies have confirmed that arsenic absorbed into the body following ingestion or inhalation accumulates specifically in the skin and that keratinocytes, melanocytes and Langerhans cells are particularly vulnerable to arsenic-related carcinogenicity.[36,140] Isolated human keratinocytes exposed to arsenic with or without UV light undergo transformation with release of cytokines and growth factors which, in animal models at least, are shown to participate in the formation of papillomas.[141]

Injection of arsenical agents as a cause of skin or other cancer is unusual these days but two cases are noteworthy. One concerned a patient with Hailey–Hailey disease (a benign familial pemphigus) who developed a squamous cell carcinoma following injection of the drug Salvarsan.[142] In a second man, multi-site cancers developed 40 years after a course of Salvarsan injections for syphilis.[143] At the time, carcinogenic mechanisms for arsenic were poorly understood and the cancers were attributed to local trauma predisposing to arsenic-related neoplastic change. Both tumours resembled Bowen's type intra-epidermal carcinoma but cancers of the vocal cords, stomach and colon were also reported.

The aetiology of Bowen's disease and its relationship to arsenic is equivocal.[136,144–146] Fowler's solution, topical application of arsenicals, ingestion of arsenic in drinking water, and inhalation of arsenic dust are implicated in Bowen's disease and the so-called "arsenical dermatoses" (Figure 17.4). Whereas, at one time, arsenic exposure was thought to be an underlying factor in the causation of both Bowen's disease and related internal neoplasms",[136,147–149] the British Association of Dermatologists now consider that Bowen's disease in most cases develops as a result of chronic solar radiation or that a human papilloma virus may be implicated.[150] Bowen's disease is identified pathologically as an *in-situ* intradermal squamous carcinoma (ISC) which may metastasise to internal organs later in its clinical history. Bowenoid cancers have also been described as squamous cell carcinomas of an intra-epithelial type, with constituent cells showing atypical form and highly invasive patterns with metastases in the form of invasive carcinoma.[119,136,146,151,152] Tumour down-growths show pore invasion with penetration of the epidermal basement membrane. Chronic arsenicism and Bowen's disease are associated with a significant reduction in Langerhans cells and a loss of dendrites in peri-lesional skin areas, consistent with the ability of arsenic to depress immune responsiveness to tumour-specific antigens.[144,153]

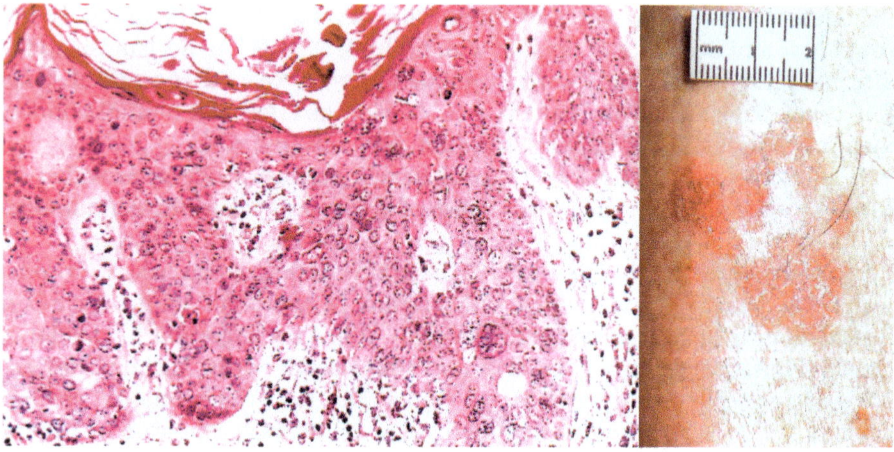

Figure 17.4 (left) typical pathological features of Bowen's disease showing loss of polarity of the epidermis and the presence of atypical mitoses and giant cells; (right) area of Bowen's disease on lower leg. From R. Mackie in Rook's Textbook of Dermatology 5th Edition, eds Champion, Burton and Ebling, 1993, chapter 32, pp. 1482.

Gawkrodger reviewed human skin cancers and concluded that arsenical keratoses are pathognomonic and that "focal or multiple punctuate lesions located on palms of hands and soles of feet are not seen with any other condition".[146] They are usually multiple and may progress to squamous cell carcinoma, Bowen's disease or multiple basal cell carcinomas, mostly but not exclusively on light-exposed areas of the legs, arms and face. Similar observations have been reported in case studies where the onset of Bowenoid lesions followed an average latent period of 6–8 years since the last recorded exposure to arsenic.[136,147] Bowenoid lesions concentrate arsenic more avidly than control tissues from non-arsenic exposed patients. Contradictory evidence for arsenic exposure in the aetiology of Bowen's disease was provided by Callan and Headington who studied 72 patients with characteristic Bowenoid symptoms and internal malignancies.[154] They reported that only 5.7% of their cases showed concurrent internal malignancies and 95% were unaware of having been exposed to arsenic. Just 5% gave a case history of arsenic medication!

A study of 520 patients with skin cancer in Singapore showed that 42.6% had a history of arsenic exposure, and 16.7% exhibited Bowen's disease affecting non-sun exposed areas of the trunk and upper extremities.[145] These authors reported basal cell carcinoma (36.5%), squamous cell carcinoma (24.4%) and malignant melanomas (2.7%) in their patients and remarked that skin tumours are more prevalent in Chinese people (78.1%) and less frequent in the more pigmented races (9.4%). Guo *et al.* were of the opinion that arsenic-induced skin cancers are "cell-type specific".[119] Tumour incidence was highest in the 51–70 years age group; men were appreciably more at risk than women (1.72 : 1). Other studies have inferred that skin cancer incidence in men did not correlate well with arsenic concentrations in drinking water.[112] Enterline *et al.* found no excess in skin tumours in 2808 men working in a copper smelter for more than one year but, in their experience, arsenic-related skin tumours rarely occur before the age of 30 years and are difficult to diagnose.[27] Skin tumour incidence following arsenic exposure increases with age and duration of exposure, possibly reflecting a latent period of 25 or more years between initial exposure and appearance of tumours.

Melanomas are reported in cases of arsenic exposure but the true incidence of arsenic-related malignant melanoma is not known. A survey of 373 cutaneous melanoma cases diagnosed in 1999–2000 in Iowa showed that patients had significantly higher toenail arsenic concentrations.[137] Increased levels of malignant melanoma in areas of southern England may also be attributable to arsenic ingestion in drinking water, but this is unsubstantiated.[155] In the UK, malignant melanoma has an annual prevalence of about 10 in 100 000.[146] Malignant melanomas may result also from long-term solar irradiation. An epidemiological study of 2369 people in 243 townships in Taiwan showed that ingestion of arsenic was associated more with a raised incidence of squamous and basal cell carcinomas and not malignant melanoma.[119]

A rare case of Merkel cell tumour was reported in a 72-year-old factory worker with co-existing Bowen's disease following 4 years of handling inorganic arsenic.[156] The lesion was preceded by arsenical keratosis and a red macule which transformed rapidly to frank tumour with trabecular downgrowths of Merkel cells separated by basement epithelium.

Environmental arsenic concentrations predisposing to skin lesions have mainly been in excess of $100\,\mu g\,l^{-1}$ and where dietary levels of protein, vegetable fibre, folate and essential minerals are low.[157] In Bangladesh, people consume about 4 litres of water contaminated with 100 to $>300\,\mu g\,l^{-1}$ inorganic arsenic (arsenates and arsenites) daily. People fed nutritionally adequate diets were more resistant to the effects of arsenical skin lesions at concentrations of up to $400\,\mu g\,l^{-1}$. Neuropathy with sensory and/or motor nerve damage are additional complications of arsenic-related response toxicity.[36] Blackfoot disease, which seems to be a specific arsenic-related condition endemic in southwest Taiwan, is more common in men subject to cumulative exposure to 671 ± 149 ppb inorganic arsenic (including 21.0 ppb of unidentified insoluble salts in artesian water.[39–41] Blackfoot disease is associated with increased incidence of lung and bladder cancers.

Experimental studies in mice and cultured human keratinocytes have demonstrated that arsenite and UV radiation act as co-carcinogens in the induction of malignant skin cancer and that oxidative DNA damage and genetically regulated apoptotic factors are central to this interaction.[158] Dietary arsenite ($1–25.5\,mg\,l^{-1}$ for 182 days) promoted skin carcinogenesis in mice given UV radiation. Studies in cultured human keratinocytes have demonstrated that induction of oxidative stress and regulation of apoptotic factors are related to arsenite levels and that UV enhances the malignant phenotype of arsenite-transformed cells.[159]

17.2.3.2 Lung

The vast majority of lung cancer cases are attributable to cigarette smoking but in countries with high arsenic contamination of well water, lung and bladder cancers rank among the highest causes of fatality. Allan Smith of the University of California estimated recently that a lung cancer mortality rate ratio estimate of 8.0 (95% CI 3.2–16.5, $p<0.001$) occurs where urine concentrations averaged 1179 $\mu g\, l^{-1}$ following inhalation. He indicated an odds ratio estimate of the lung cancer incidence rate ratio as 7.1 (95% CI 3.4–14.8, $p<0.001$) for an estimated average urine concentration of 825 $\mu g\, l^{-1}$ following ingestion.[130] His research pointed to the lung cancer risk being dependent upon the total amount of arsenic absorbed and not the route of uptake. Population studies in southwest Taiwan with water arsenic contamination at 42–229 $\mu g\, l^{-1}$ showed an estimated threshold at 151 $\mu g\, l^{-1}$ for induction of lung and bladder cancers and a clear dose-related trend.[112,118] Statistical analyses in a copper smelter showed a linear relationship between cumulative arsenic inhalation and respiratory cancer, and that the carcinogenic effect was greater when arsenic was delivered in high concentrations than at low levels over a longer period.[124]

The first indication that arsenic might be a cause of lung cancer was provided by Sir Thomas Legge in 1903, who noted that workers engaged in manufacture of sheep dip from white arsenic developed irritation of the upper bronchioles consistent with hyperkeratosis and pigmentation of the skin, and pulmonary adenocarcinoma.[113,131] Since then many cases of lung cancer have been reported following a variety of occupational and environmental situations including copper and gold smelting and people living close to derelict mines and in areas of arsenic-rich water .[24,26–28,124]

Implication of arsenic as a respiratory tract carcinogen is frequently complicated in occupational situations by co-exposure of workers to dusts containing silica, beryllium, cadmium, asbestos, copper, sulfur dioxide, *etc.*, any of which is a potential cause of the lesions diagnosed.[28,124,160,161] Pneumoconiosis, parenchymal lung disease and related conditions were once widely reported in industrial settings.[162,163] Lung cancer seen in copper smelters is principally of the small cell and epidermoid carcinoma type rather than adenocarcinomas, which were only marginally increased in two studies of workers inhaling arsenic.[26–28] It is conceivable that, as Guo and his colleagues suggest, arsenic may give rise to different pathogenic mechanisms and cell targeting in induction of lung tumours according to the route of exposure.[164] In their study of 37,290 lung cancer patients in Taiwanese townships with endemic arsenic contamination, squamous and small cell carcinomas predominated over adenocarcinomas; squamous cell tumours were more common than basal cell tumours in the skin. Enterline studied records of 2802 men working in a copper smelter for >1 year and concluded that a positive relationship existed between inhalation of arsenic and the incidence of lung, kidney and bone cancer but not other cancers.[27]

Men who smoke have a significantly higher risk of developing squamous cell lung cancer and skin lesions than non-smokers and odds ratios increase with arsenic concentrations.[165] However, in implicating arsenic in cigarette smoke as a putative lung carcinogen, it has to be emphasised that tobacco and mainstream smoke as inhaled contains more than 50 carcinogens acting singly or as co-factors in inducing mutagenic and pre-neoplastic change.[45]

Lewisite is an arsenical war gas suspected of causing squamous intra-epidermal carcinoma on the limbs as well as being a cause of respiratory cancer in workers involved in gas manufacture, but epidemiological evidence is weak and unsubstantiated.[54] Lewisite is not mutagenic. Industrially related lung diseases in metal industries are commonly associated with cardiovascular pathology, haemolymphatic malignancy and liver cirrhosis.[26,166]

17.2.3.3 Urogenital Tract

Bladder tumours are frequently part of the arsenic-related toxicity profile. Bladder and lung cancer mortality in three townships was consistent with epidemiological and toxicological experience with arsenic exposure; this fitted an inverse linear relationship model with an estimated threshold at 151 $\mu g\,l^{-1}$ As (95% CI, 42–229 $\mu g\,l^{-1}$).[112] Bladder cancer rates were slightly higher in men than women, but the incidence was lower in populations fed high protein diets. More than half of the townships showed water contamination levels exceeding the detection limit of 40 $\mu g\,l^{-1}$; the "high" exposure group were presumed to be exposed to 178 $\mu g\,l^{-1}$ of arsenic. (This study indicated also a small trend in arsenic-related liver cancer mortality in low, medium and high arsenic exposure groups, but increased skin cancer mortality was only significant in women in high arsenic areas).

Other population studies conducted in Taiwan and Southeast Asia have confirmed that arsenic exposure is a potential cause of prostatic cancer. Close relationships have been observed between arsenic levels in drinking water (<0.30 ppm, 0.30–0.59 ppm and $\geq$0.60 ppm) and age-adjusted prostatic cancer mortality.[167,168] These studies suggest that androgen-dependent tissues are more vulnerable to arsenic carcinogenicity. This is not supported by standard mortality data produced by Hopenhayn-Rich's 5-year study in the arsenic-rich areas of Argentina with a population of 2,750,000.[120,169]

Studies in Cordoba, Argentina have indicated a close correlation between arsenic in drinking water and renal and bladder cancers.[169] At environmental levels of arsenic ranging from 40 to 176 $\mu g\,l^{-1}$ mortality due to renal and bladder cancer was similar to that of lung cancer, but rates of liver cancer were lower. Interestingly, men exhibited a slightly higher cancer rate than women but were particularly sensitive in the low exposure group ($\leq$40 $\mu g\,l^{-1}$).

17.2.3.4 Gastrointestinal and Other Tissues

Evidence provided by environmental and occupational studies indicates that arsenic is a cause of cardiovascular change (blackfoot disease) and increased mortality from liver cirrhosis and cancer, leukaemia, myeloma and neurological diseases.[28,170] In a study of 8014 copper smelter workers employed for at least one year, analyses showed excess relative risk of lung cancers at 0.21 $mg\,m^{-3}$ associated with airborne arsenic but no other cause of death was related to arsenic exposure.[124] Elsewhere, prolonged exposure to arsenic in a copper smelter at low doses in 1941–1976 was associated with significant increases in cancers of bone and large intestine, and standardised mortality ratios of >150 for cancers of kidney, rectum, buccal cavity and pharynx, but numbers were low.[27]

A case study of a former arsenic miner showed that after 41 years he developed Bowen's disease, gastric cancer, squamous carcinoma of the lung, an adenoid cystic carcinoma originating from the salivary gland, and carcinoma and oncocytoma of the kidney but death was due to respiratory distress and heart failure.[171] A second unusual case concerned hepatic angiosarcoma in a patient 46 years following treatment with Neo-salvarsan.[172]

17.2.3.5 Transplacental Carcinogenesis

Arsenic diffuses across the placental barrier in humans and, as seen in a recent South African study, levels of arsenic in cord blood were as high as 80% of those seen in the maternal circulation.[173] However, there is a lack of clear information on the outcome of such pregnancies. A population study conducted in Bangladesh and Mongolia failed to demonstrate that pre- or postnatal arsenic exposure exerts a detrimental effect on child development up to 18 months of age,[174] although arsenic >50 $\mu g\,l^{-1}$ in drinking water was associated with an increased risk of neonatal death.[175] A trend in reduced birth weight was reported in the children of Bangladeshi women consuming drinking water containing <100 $\mu g\,l^{-1}$ arsenic.[176] Investigations into the effects of arsenic exposure early in life in an arsenic-rich area of Northern Chile have shown that mortality from childhood leukaemia and brain cancer was not increased, but that deaths from liver cancer were higher than expected.[94] Liver cancer mortality in children (0–19 years of age) was higher in females than males (relative risk 14.1 *vs.* 8.9). This preliminary evidence indicating that transplacental arsenic presents a carcinogenic risk should form an incentive to obtain further epidemiological information in an important area.

17.2.4 Experimental Animal Studies

The 12th RoC in 2011 listed experimental studies in which rodents, hamsters, rabbits and dogs had been exposed to metallic arsenic, arsenic trioxide, and arsenates or arsenates, by the oral or dermal routes, inhalation, subcutaneous or intravenous injection, or by intratracheal instillation, provide irrefutable evidence of arsenic-related carcinogenicity.[57] Earlier workers were of the opinion that animal models were of limited value in studying the carcinogenicity of arsenic and its compounds and that many published studies were either flawed and incomplete or they failed adequately to reproduce levels of risk in humans as indicated by epidemiological evidence.[133,135,177–179] In 1973 and 1980 at least, the IARC concluded that there was "a lack of convincing evidence of carcinogenicity of arsenic in animal studies".[129] Since then, competent research using controlled *wild-type* and genetically defined animal models has provided an understanding of:

- Tissue sensitivity
- Patterns in methylation and metabolism of arsenates and arsenites

- Genetic and molecular mechanisms of arsenic related carcinogenesis[97]
- Interspecies variations in As(III) and As(V) metabolism and tissue specificity[80,180]
- Transplacental carcinogenesis.[98,134,179,181,182]

Professor Vahter has contributed much to the understanding of metabolic pathways of inorganic arsenic compounds. She emphasised factors affecting the reduction of As(V) to As(III), patterns of methylation and the formation of MMA and DMA, and pointed out that rabbits and hamsters are suitable for validation on the basis of tissue affinity, absorption and metabolic rate constants. DMA is the principle excretion product of arsenite or arsenates in most animal species.[74,180] Experimental studies in rodents have used *wild-type* or genetically transformed strains to establish that arsenic is a complete carcinogen following administration by oral or dietary means, inhalation, parenteral injection or through transplacental passage.[98,99,179,182,183] Biomethylation products including DMA are capable of inducing urinary bladder tumours in rats and lung tumours in mice.[97,106]

Interspecies physiological differences may render direct extrapolation of studies of arsenic from animals to humans extremely difficult. As Tokar emphasised, it cannot be assumed that the bioactive proportion of a dose of arsenic or any other element administered to rodents will be comparable to that seen in humans and that interspecies and inter-strain variations in tissue sensitivity are inevitable.[98] He summarised 44 animal studies conducted up until 2010, and emphasised that rats and mice given arsenites in drinking water or feed for up to two years developed different patterns of tumorigenicity according to the route and magnitude of doses administered and the species/genotype involved.[98] Thus, tumours in F344 rats given DMA were principally bladder tumours (papilloma or carcinoma), whereas A/J mice showed adenoma or carcinoma of the lungs. F344 rats dosed with trimethyl arsine ($200\,mg\,l^{-1}$) for the same period showed a modest dose-related incidence of liver adenoma.[186,187] Lung tumours (adenoma and carcinoma) were induced in hamsters given repeated 0.025 mg doses of calcium arsenate by intratracheal administration for periods up to lifetime exposure.[188,189] Other studies in hamsters showed that arsenic trisulfide was highly toxic when given intratracheally but non-carcinogenic in surviving animals. Experimental studies in rodents show that arsenite and certain metabolites induce lung and bladder tumours through oral or intratracheal administration, inhalation or parenteral administration.[98,134,182]

Studies with transgenic mice show that the principle metabolite, DMA, is a complete carcinogen and induces bladder cancers in *wild-type* animals more effectively than in transgenic ($p53^{+/-}$ or K6/ODC) animals.[183] Other evidence of direct carcinogenicity is provided by a study in which the pesticide calcium arsenate, administered intratracheally as a single dose of 0.1 ml, induced multifocal bronchogenic adenocarcinoma in a BD-strain of rats.[184] (The dose was much lower than might be expected in workers spraying crops in a vineyard!) Hairless mice fed sodium arsenite ($5\,mg\,l^{-1}$) continuously in drinking water from 21 to 42 days of age and then given UV light for 182 days to promote carcinogenic change showed papillomas in the skin whereas animals given arsenite or UV alone showed epidermal hyperplasia.[185]

Comparative studies have demonstrated that DMA is a lot more toxic than MMA in F344 rats when administered in drinking water, but female animals were more susceptible to urothelial changes in the bladder.[97,99,118] Male animals receiving DMA in drinking water for 104 weeks at 200, 50 or 12.5 mg kg^{-1} showed bladder tumours in 12 of 31, 8 of 31 and 0 of 33 respectively, with carcinogenesis related directly to the rate of urinary excretion. MMA and trimethylarsine oxide were not carcinogenic in this model. Other work has demonstrated that F344 rats dosed orally with $<$200 mg kg^{-1} DMA for two years developed papillary or nodular hyperplasia progressing to bladder tumours, but no tumours in other organs.[97] Genetic analysis of DMA-induced tumours revealed a low rate (10%) of H-*ras* mutations and biochemical evidence of oxidative stress (ROS), and increased cyclo-oxygenase (COX-2) expression, but no obvious alteration in *p*53 expression.

The carcinogenicity of arsenic compounds in laboratory animals is not entirely related to their solubility in water or body fluids. Thus, the sparingly soluble compound gallium arsenide was shown to be carcinogenic in rats inhaling concentrations of up to 50 mg m^{-3} for up to two years.[190] Animals showed a low dose-related incidence of lung adenomas and carcinomas, but an upward trend in tumours (phaeochromocytomas) of the adrenal medulla and mononuclear cell leukaemias. The mechanisms involved are unclear, particularly as tumours were seen only in female animals and not males. Gallium is not a carcinogenic metal, as discussed in more detail in Chapter 15, and it seems unlikely that MMA or DMA metabolites would be released from this compound in body fluids. Gallium arsenide is a very stable compound which decomposes when heated to 480 °C to release arsenic vapour; it reacts with strong acid reducing agents to produce arsine gas.

Desesso assembled and assessed older human and animal investigations on peri- and prenatal exposure to arsenic and concluded that inorganic arsenic compounds pose virtually no danger to developing offspring following maternal exposure that occurs by relevant routes (oral and inhalation) and at concentrations that are likely to be experienced in the environment or in the workplace.[191] However, other experimental work indicates that arsenite may be both teratogenic and carcinogenic in mice. Thus male offspring born to mothers exposed to arsenite at 8–18 days gestation were susceptible to hepatocarcinoma, and female offspring developed an increase in ovarian tumours and preneoplastic lesions of the reproductive tract.[181,192] Neonatal administration of the experimental tumour promoter 12-*O*-tetradecanoylphorbol 13-acetate (TPA) to mice increased the risk of arsenic-induced liver and lung tumours. Mice treated transplacentally with arsenic with or without TPA were not susceptive to dermal carcinogenesis. Whole-life exposures to arsenite from *in utero* though post-natal into adulthood (two years) resulted in increased carcinomas of the lung, liver, gall bladder, adrenal, uterus and ovary.[134] Of these, liver and lung tumours were more aggressive and frequent than those seen following exposure during the prenatal period and were highly enriched with stem cells.

A toxicogenomic analysis of gene expression in neonates exposed to 42.5 or 85 ppm arsenite prenatally showed >2-fold increase in aberrant gene expression in tumour sites.[193] In non-neoplastic liver tissues, 10% of genes examined expressed related to liver tumour formation later in life. Results of experiments in mice prompt further investigation into the potential risks posed to pregnant women in arsenic-rich countries.[93]

17.2.5 *In Vitro* Cytogenicity, Genotoxicity and Mutagenicity

The methylated metabolites of As(III), MMA and DMA, are clastogenic and capable of invoking DNA damage including double strand breaks and cross-linkages.[172] Both metabolites exhibit a capacity to initiate and promote tumour formation in rats and mice (including transgenic and gene knock-out strains), but this has not been confirmed in human studies.[98,182,183,194] Inorganic arsenic is a cytoplasmic poison in mammalian cells.

Bacterial tests, including the Ames bacterial reversion test and *in vitro* mutagenic assays, have failed to confirm the mutagenicity of As(III) or its methylated metabolites.[54,195] However, arsenite, arsenic trioxide and the metabolites MMA and DMA are antimitotic, inhibit DNA repair, induce gene amplification and are mutagenic in mouse lymphoma cells and other assays.[133,183,196,197] As(III) compounds are highly clastogenic *in vitro* and are potent inducers of mitochondrial damage, sister-chromatid exchanges and chromosomal aberrations in human and animal cells through their avid binding to sulfydryl groups.[198] As(III) toxicity is cell specific. Whereas, in human studies, the As(III) ion appears to target epithelial cells in the skin and urinary bladder, hepatocytes and pneumocytes, cellular responses in cultured cells are markedly different.[164,199,200] This is attributed to the greater ability of cytotoxic As(III) cation to traverse cell membranes, methylate MMA and DMA and to accumulate in and bind to cytoplasmic proteins. Human urothelial cells *in vitro* accumulated As(III) more readily than hepatocytes but methylated complexes migrated more readily to ribosomes, nuclei and mitrochondria in hepatocytes.

The As(III)-induced mutagenicity is related to production of reactive oxygen species (ROS).[97,195] The organo-arsenicals arsenobetaine and tetraphenyl arsenium were negative in this assay. Oxygen radical induction was increased in cells showing greater cytotoxic and genotoxic changes. Induction of ROS has also been demonstrated *in vivo* in F344 rats.[97]

17.2.6 Mechanisms of Carcinogenesis and Conclusions

Arsenic is an environmental contaminant found in water, air and soil. It is shown to be a human carcinogen mainly through ingestion of contaminated food and drinking water and through inhalation of dusts in occupational environments. Mechanisms of arsenic-induced carcinogenesis are complex and relate to the intracellular penetration of the As(III) ion, its ability to bind to and denature proteins, induce oxygen stress through induction of ROS, evoke genotoxic change and to impair gene signalling in susceptible cells.[195,201]

Tokar *et al.* view the basis of arsenic related cancer as a disease of stem cells, where arsenic or its bioactive metabolites block the passage of cells into normal pathways of functional differentiation and expose them to neoplastic transformation.[194] All tissues retain a number of stem cells throughout life, but squamous epithelia subject to higher rates of cellular proliferation contain higher levels of stem cells that than liver, endocrine tissues, brain and kidney. Proliferating populations of stem cells in the skin and bladder urothelium are targets in arsenic-related carcinogenesis in human and animal studies. Stem cells as a target for arsenic-related carcinogenesis are well illustrated by experiments in prenatal and neonatal animals where genetic damage and cellular transformation translate to increased neoplasia later in life.

Experiments with cultured mammalian cells show that As(III) and As(V) compounds invoke cytotoxic effects, inhibit RNA/DNA synthesis and inhibit cell growth.[60] Arsenic trioxide, through its action in inducing apoptosis in melanoma cells lines, induced DNA fragmentation and blocked key mitochondrial enzymes leading to necrosis.[13,63] Other keys to the subcellular mechanism of arsenic toxicity include levels of intracellular retention of bioactive ion and its penetration through nuclear membranes and action on ribosomes. Dopp *et al.* showed that inorganic and methylated arsenicals generate reactive oxygen radicals in cultured human hepatocytes cultures and that the action was cell specific.[199,200] In each case, As(III) ion was more effective in inducing cytotoxic change and ROS induction than As(V) ion. In mouse lymphoma cell lines, As(III) compounds such as sodium arsenite and arsenic trioxide invoked a clear dose-related pattern of gene mutation as a basis for carcinogenic change. Clearly, *in vitro* models show that no intracellular compartment in target cells is tolerant of arsenic-related effect and that, whereas MT is likely to be cytoprotective against low levels of arsenic insult, it is inadequate in offsetting the carcinogenic effects of long-term exposure to high concentrations.

Exposure to As(III) (arsenite) is profoundly more carcinogenic than As(V) (arsenate) and levels of risk are a measure of the retention of the methylated metabolites in the body. Experimental studies in rodents emphasise the carcinogenic risks of DMA, particularly in the bladder; it is also shown to be a cause of oxidative damage and production of reactive oxygen radicals. Although there is insufficient evidence that arsenites or arsenates or their metabolites are mutagenic in standard bacterial screens, *in vitro* studies provide evidence that arsenite may impair cell proliferation or otherwise alter genetic signalling and cell transformation

17.3 Antimony

Antimony is a metalloid element which displays certain properties characteristic of metals whilst behaving as a non-metallic element. It has been known since early times and is recorded as being a cosmetic in ancient Egyptian times. Pure antimony is a silvery grey element with metallic appearance but if melted and allowed to recrystallise it displays a distinctive profile known colloquially as the "star of antimony". Antimony rarely occurs naturally in its elemental form but is found mostly in the form of ores such as stibnite (SbS_3) and over 100 different minerals, usually including sulfur. Only stibnite is mined commercially as a source for metallic antimony. Antimony is found in trace amounts in arsenic, silver, copper and lead ores and it is economically possible and environmentally desirable to extract the antimony from these ores when they are smelted. Workers engaged in mining and refining antimony are exposed to these other toxic elements as an occupational risk.[202,203]

Antimony is a Group Vb element with two principle natural isomers, ^{121}Sb and ^{123}Sb, and at least 45 other radioactive isotopes and isomers are now recognised. Metallic antimony is bluish white and an extremely brittle metal of a flaky, crystalline texture. It has a metallic lustre and reacts in air at room temperature. It burns brilliantly when heated with the formation of white fumes of antimony trioxide (Sb_2O_3). It is a poor conductor of heat and electricity. Antimony is ubiquitously present in the environment as a result of natural processes and human activities. It exists mainly as Sb(III) and Sb(V) in environmental, biological and geochemical samples.[204] A Russian study has shown that antimony occurs widely in the environment and is particularly high in the vicinity of factories.[205] Antimony occurs in igneous rock at 0.2 ppm; shales at 1.5 ppm; sandstones at 0.05 ppm; limestone at 0.2 ppm; sea water at 0.00033 ppm; soil at 2–10 ppm; land plants at 0.06 ppm and land animals at 0.006 ppm. Published antimony levels in individual foods or animal feedstuffs are not very extensive, although it has long been known that foods stored in enamel vessels and cans may contain appreciable antimony concentrations.

Antimony has no known nutritional value in the mammalian body but minute quantities may be found in human tissues through ingestion of contaminated food and drink. Nuts are the highest source of dietary antimony ($<200\,ng\,g^{-1}$). Tartar emetic and other antimonial drugs, when injected regularly into the body for treating helminthic worm burdens, lead to a significant rise in concentrations of antimony in most tissues notably erythrocytes. Several workers have demonstrated the affinity of the Sb(III) ion for red blood cells both *in vivo* and *in vitro* and it is suggested that the therapeutic value of the Sb-drugs lies in the capacity of adult worms to digest the erythrocytes. By contrast, human erythrocytes are almost impermeable to Sb(V) compounds. Sb(III) is readily demonstrated in schistosome infections (especially female schistosomes), following experimental injection of tartar emetic in infected mice.[206]

In addition to its early use as a cosmetic chemical, compounds and alloys of antimony have a wide range of commercial and industrial applications, the chief of which are in semiconductors (diodes and infrared detectors), batteries, flame retardants, paints, glass and pottery. Antimonial drugs including tartar emetic (antimony potassium tartrate), sodium stibogluconate, meglumine antimonate and other pentavalent antimonials have proved highly efficacious as antiprotozoal and anthelminthic agents for many years.[207] However, these antimony-containing drugs are toxic and case studies illustrate how patients treated parenterally with antimonials for visceral and cutaneous leishmaniasis, bilhartzia and schistosomiasis in tropical countries develop adverse cardiovascular changes and pathology of the lungs, kidney, pancreas and skin.[208–212] A further disincentive to use antimonial agents is the acquired resistance that builds up in infections, thereby reducing the efficacy of the therapies.

Although antimony is chemically similar to arsenic, its capacity to evoke pathological change is quite different. Cutaneous spots in the form of generalised eruptions and papules in the region of sweat and sebaceous glands in the skin are characteristic of antimony poisoning.[212–214] Antimony spots are accompanied by an intense itching sensation. They are particularly uncomfortable during hot moist weather. Antimony trioxide used as a fire-retardant in cot mattresses was widely publicised in the 1990s as a possible cause of sudden infant death syndrome (SIDS).[215,216] Despite intense investigation and speculation, no association has yet been proven.[217,218] At the time, there was speculation that antimony trioxide (Sb_2O_3) was reduced through the action of the mould *Scopulariosis brevicaulus* or other microorganisms present in cot mattresses with release of toxic stibine gas (SbH_3) in the near vicinity of the young children.[219] This was not substantiated.

No evidence has been seen to show that intravenous injection of tri- or pentavalent antimonial drugs is a cause of human cancer. Toxic risks arising through chronic ingestion of antimony in drinking water or contaminated food are not well documented and the risks of long-term exposure are not known. Antimony is not commonly regarded as a cumulative poison in humans but Sb(III) tends to concentrate in erythrocytes and lung tissue.[220,221] Sb(III) and Sb(V) conjugate with glutathione and are excreted in the urine, and in bile.[222,223] The major metabolic pathway in humans for antimony is oxidation. In contrast to arsenic, methylation of Sb(V) or Sb(III) to monomethyl or dimethyl metabolites represents a minor role in excretion of the element.[224,225] Sb(V) compounds are less toxic generally than Sb(III) compounds, which are less well tolerated by humans and experimental animals.[226]

Chronic inhalation of inorganic antimony compounds in industrial dusts and fumes has been associated with perforation of the nasal septum, pneumoconiosis, respiratory tract irritancy and heart disease, but useful epidemiological studies are few.[227,228] Gerhardsson demonstrated that antimony concentrations accumulating in the respiratory tract were 12-fold higher in exposed smelter workers and that the biological half-life of antimony in lungs was long.[221] No association has been demonstrated between pulmonary antimony concentrations and increased lung cancer mortality.[229] McCallum *et al.* evaluated 1366 men employed in antimony processing and maintenance work in the Howden factory (Newcastle, England) in 1961–1983 and reported that, whereas an excess of number of lung cancer deaths had been recorded up to 1960 in both categories of workers, no excess of lung tumours was evident after December 1960.[203,214] The authors attributed this marked decline in industrially related lung cancer mortality to improvements in working practice after 1960, and introduction of effective means of eliminating arsenic contamination.

Experimental studies have demonstrated that 25–27% of female rats inhaling antimony ore concentrate (TWA 45 $mg\,m^{-3}$) or antimony trioxide (36–40 $mg\,m^{-3}$) for up to 52 weeks developed pulmonary tumours; no lung tumours were seen in male animals and other tumour incidences were unrelated to antimony exposure.[230] A second study claimed that 17 of 29 male rats exposed to nickel antimonide (NiSb) by intramuscular injection developed subcutaneous sarcoma within two years, but in view of the 93% incidence of tumours in nickel oxide-treated animals (positive control), the risk apportioned to antimony *per se* must be low.[231]

Antimony has been listed by the NTP, US DHHS and IARC for investigation as a putative human carcinogen, but expert evaluation of data up to 2011 has failed to identify a carcinogenic risk following occupational exposures to elemental antimony, antimony trioxide as a by-product from stibnite ores, antimony trisulfide, or antimony contamination in lead smelting.[232] The carcinogenicity of antimony in mining and associated industrial processes is difficult to establish in view of the frequent exposure to other environmental contaminants, notably arsenic, lead, nickel, silica and zirconium.[202,203,214] Even so, the American Conference of Governmental Industrial Hygienists still maintained that antimony should be viewed as a "suspected carcinogen", pending further epidemiological evidence.[233] This view is complemented by lack of consistent evidence for genotoxicity and carcinogenicity in non-mammalian systems, including the Ames test.[234–238] Antimony potassium tartrate is not mutagenic in the Ames bacterial reversion test and its carcinogenicity is unproven.[239] The limited studies in rodents suggesting a greater a female sensitivity to antimony-induced lung tumours are inadequate.[230,231]

17.4 Bismuth

Bismuth is a crystalline element with chemical similarities to arsenic and antimony, but it is identified as a metal. Pure bismuth is a brittle whitish metal with a pink tint. Only one relatively stable isotope of bismuth (^{209}Bi) is found naturally and this has an estimated half-life of 1.9×10^{19} years, but at least 60 other isotopes have been identified with half-lives ranging from a few nanoseconds to 3.04×10^{6} years (^{210m}Bi). Bismuth is a comparatively rare element and possibly composes only about 9 ppb of the Earth's crust. Although bismuth resembles arsenic and antimony chemically in many ways, it is a unique element. It is the heaviest metal in the Periodic Table and shows the highest diamagnetism and electrical resistance of all metals. It has a particularly high "Hall effect", which indicates the increase in electrical resistance when placed in a magnetic field.[240] However, the thermal conductivity of bismuth is lower than that of all metals, other than mercury.

Bismuth occurs naturally but is more commonly found in the form of sulfide ores and oxides such as bismuthimite (Bi_2S_3), tetradymite (Bi_2Te_2S), cosalite ($Pb_2Bi_2S_5$), and the oxides bismuthite [$Bi_2CO_3(OH)_4$] and bismite (Bi_2O_3). It occurs also as crystals in the sulfide ores of nickel, cobalt, silver and tin. Bismuth is regarded as an indicator of high temperature mineral deposits, forming in veins with quartz and other metallic elements. Bismuth is mainly produced as a by-product from lead and copper smelting, especially in the USA. The chief areas where it is mined are Bolivia, Peru, Japan, Mexico and Canada. As a

Group Vb element, bismuth forms pentavalent and trivalent salts but most Bi(v) compounds are appreciably less stable that Bi(III).[15,16] Bismuth is stable in oxygen and insoluble in water but dissolves in concentrated nitric acid to form bismuth nitrate; this slowly decomposes to bismuth subnitrate. Bismuth forms a range of inorganic salts, mainly with the Bi(III) valancy, all are insoluble in water and most have a low known toxicity threshold in mammalian systems. Bi(III) is methylated by methanobacteria in the environment and by intestinal bacteria. Trimethyylbismuth (TMB^{III}) is a volatile compound.

17.4.1 Uses of Bismuth

The unusual physico-chemical features of bismuth and its low level of human toxicity make bismuth and its organic compounds suitable for a range of commercial and industrial applications:

- Low melting point metal alloys with lead, tin, indium, iron, antimony, cadmium, *etc.*
- Bismuth-alloy castings can be covered by plastic or other materials in the production of intricate machine parts
- Replacement for lead as a minimally toxic element in plumbing, bullets, birdshot, metal alloys, soldering, ceramic glazes and paints
- Catalysts in chemical syntheses, *i.e.* production of acrylic fibres
- Bismuth oxide (Bi_2O_3) used in production of enamels for cast iron and in the manufacture of glass and ceramic products
- Basic bismuth nitrate is used in porcelain painting to fire-on gold decorations and to impart an iridescent surface to pottery
- Cosmetics
- Pharmaceuticals.

Production of metal alloys represents the largest part of the annual production of bismuth, which in 2009 amounted to 8226 tonnes (US Geological Survey). The low melting point alloys provide for specific applications such as sprinkler systems in buildings. Bismuth is alloyed with iron to create what is known as "malleable irons". Many of these alloys expand slightly when they solidify allowing them to be used in moulding and in soldering, tube bending, special machine parts and die-castings.

The basic chloride (bismuth oxychloride) is used in the polishing of synthetic pearls to give them an iridescent finish. This property is also employed in the cosmetics industry where low toxicity and skin penetration make pearls of monocrystalline bismuth oxychloride suitable for opalescent, pearly white lipsticks, eye-shadows and face powders. Bismuth oxychloride pearls are widely employed in decorative cosmetics where they are claimed to improve the "pressability and moulding" characteristics of the skin and provide a lustrous, smooth effect with good skin adhesion.[241] Apart from rare cases of skin irritancy,[242] bismuth-containing cosmetics are not a cause of skin disease. Bismuth is not listed as a contact allergen in humans although reactions are reported in patients using bismuth salicylate, and where the salicylate anion is implicated. Some bismuth salts used in cosmetics darken on exposure to light. Bismuth oxide and subgallate are used presently in suppositories.

Bismuth subcitrate, subnitrate, subcarbonate and many other bismuth compounds have been in use as antimicrobial agents in treating gastrointestinal infections (*Helicobacter pylori*; *Clostridium difficile*), many common nosocomial infections, venereal diseases, and schistosomiasis for many years.[5,243,244] The therapies are claimed to have minimal effect on normal intestinal flora but are superior to many proprietary antimicrobial agents in eliminating pathogenic infections. Organometallic bismuth compounds are extremely toxic and in the 1970s were investigated as possible biocides, but further work has not been seen.[245] Other bismuth-containing pharmaceuticals include bismuth subsalicylate, which is still prescribed as an antacid for treatment of heartburn, indigestion, upset stomach, nausea and diarrhoea. Bismuth subsalicylate is claimed to be efficacious also in preventing travellers' diarrhoea and as an anti-ulcer treatment. Less than 1% of bismuth used in the treatment of dyspepsia is absorbed in the intestine, but it is believed to form a coating on the mucosa which stimulates bicarbonate secretion as a means of bacteriostasis (*Helicobacter pylori*) whilst reducing gastric ulceration.[245,246] In the case of a man with recurrent malignant melanoma, administration of bismuth subsalicylate (Pepto-Bismol®) analgesic led to a transient black discoloration of his tongue.[247] Other cases are known and most people involved were of dark-skinned races.[248]

Research into the possible use of bismuth salts in the treatment of certain forms of human cancer is at an early stage.[249] Preliminary results suggest that certain Bi complexes are bioactive and provide promising evidence as anticancer therapies, including efficacy against *cis*-platin induced tumours in mice.[250] Other recent work has established also that soluble organo-bismuth analogues of *tris*(pentafluorophenol) bismuth (V) inhibit human mammary carcinoma cell lines in culture, possibly through binding nitrogen in purine bases of the DNA molecule.[251] Little is published concerning the bioactivity of organo-bismuth comounds but the neutral organobismuth(III)bis(thiolates), $CH_3Bi(SCH_3)_2$ and $CH_3Bi(p\text{-}SC_6H_4NH_2)_2$, and the ionic mercaptoanilinium derivative.$CH_3Bi(p\text{-}SC_6H_4NH_2CH_3)_2]^{2+}2I^-$ showed promising activity when tested for antitumor properties in the fluid Ehrlich ascites tumour model in mice.[252]

17.4.2 Human Exposure

Bismuth is a rare metal in the human environment and, with rare exceptions, the general population is not exposed through drinking contaminated water or inhaling bismuth dust in the air. Bismuth is not normally found in the human body but minute amounts may be absorbed through food. Absorption of bismuth through drinking water is expected to be negligible on account of the insolubility of most inorganic bismuth salts.[253] Vanhoe *et al.* used inductively coupled plasma–mass spectrometry to demonstrate minute amounts

(0.007–0.067 μg l^{-1}) of bismuth in the serum of 19 healthy patients.[254] Much higher levels of 242 μg l^{-1} measured in a patient treated with bismuth subcitrate (Pepto-Bismol®) for irritable bowel syndrome show that the element is absorbed gastrointestinally.[255] This patient was hospitalised with bismuth-related dementia. In a second study, Boertz *et al.* administered 215 mg colloidal bismuth subcitrate orally to 20 male volunteers and reported measurable amounts of the methylation product trimethylbismuth in the circulation (2.50 pg ml^{-1}) and in exhaled air samples (0.8–458 ng m^{-3}).[256] In a double trial of the bismuth–protein complex and an antacid preparation (Bicitropeptide®) in 43 patients, blood and urine bismuth levels increased from a baseline of 2–11 ng l^{-1} and 2–29 ng l^{-1} respectively to "saturation levels" of 5–20 ng l^{-1} and 63–780 ng l^{-1} within 6 weeks without obvious evidence of toxicity.[257] No details of the diet are provided in this study and it is not known whether free Bi(III) ion was chelated or otherwise bound by dietary proteins or fibre, as is the case with many other metals.

The biokinetics of bismuth were studied in a healthy man injected intravenously with ^{207}Bi.[258] Highest deposits were seen initially in his liver but 55% of the dose was excreted in the urine within 47 hours. In the longer term, excretion patterns were reduced and 0.8% of the tracer remained after 924 days. Experimental studies in rats have shown that thiol substances, including cysteine, homocysteine and mercaptoproprionic acid, in the diet considerably enhance the absorption and urinary excretion of the metal, thereby mitigating its toxicity.[259] It is conceivable however that if a novel technique using bismuth thiols to clear bacteria and biofilm formation in water systems is adopted, then uptake through treated drinking water is a possibility.[260]

Percutaneous absorption of Bi(III) is negligible on account of the relative insolubility of its salts and the expected strong ionic binding of Bi(III) to sulfydryl moieties in epidermal keratin. Little information has been seen concerning the uptake of bismuth compounds following inhalation of contaminated dusts. Bismuth telluride (Bi_2Te_3), an insoluble grey or black powder or hexagonal crystals with metallic lustre, is used in the semiconductor industry and has been shown to produce respiratory distress on inhalation.[261] The average grain size of the final compacts calculated from electron micrographs was obtained as 90 ± 5 nm. Exposure to bismuth telluride doped with selenium sulfide in construction work is considered by the Occupational Safety and Health Association (OSHA) as a cause of non-malignant respiratory effects but there is no evidence for its absorption or ability to cause lung cancer.[262] Experimental studies in dogs, rabbits and rats exposed to undoped bismuth telluride at the high concentration of 15 g m^{-3} showed the respiratory effects expected as a result of inhalation of inert granules, but failed to show evidence of carcinogenicity.[263]

Bi(III) entering the circulation through ingestion, inhalation or injection is known to induce and bind metallothioneins (MT).[250,264] As a heavy metal, it displaces lighter metals such as zinc and cadmium from MT to form the stable Bi–MT complex. It is expected that this mechanism provides an effective means of cytoprotection in vulnerable tissues. Absorbed bismuth concentrates predominantly in the kidney, with significant amounts also in the lung, spleen, liver, brain and muscle.[265]

17.4.3 Toxicity, Mutagenicity and Carcinogenicity

Bismuth is commonly regarded as being a non-toxic element and without evidence of carcinogenicity in humans or experimental animals though expected routes of exposure. Preliminary evidence suggests that bismuth is mutagenic in the chromosomal aberration test but not in Ames bacterial reversion and sister-chromatid aberration tests in mammalian cells.[266] Gurnani *et al.* compared the clastogenicity of antimony and bismuth trioxides in male mice at doses approaching 20% of the oral LD_{50} dose (1000 mg kg^{-1}) daily for up to 21 days and reported a dose-related incidence of chromosomal aberrations suggesting a cumulative effect.[234,235] Further evidence of clastogenicity was reported in cultured human erythrocytes, hepatocytes and lymphocytes at concentrations of >4μ M, >130 μM and >430 μM after 24 hours' exposure.[267] These workers showed that, in cell cultures, bismuth as Bi(III) locates in the cytosol (erythrocytes > hepatocytes > lymphocytes), and in erythrocytes cytotoxic change was evident at >4 μM whereas in hepatocytes and lymphocytes changes were evidence at 130 and >430 μM respectively. Clastogenic changes were recognised by increased frequency of chromosomal damage, sister-chromatid exchanges and intracellular production of free radicals.

Experimental study on the carcinogenicity of bismuth compounds is limited to one chronic study in rats.[268] Thus, dosing of BD-rats with high oral doses of bismuth oxychloride (1, 2 or 5%) for two years has failed to provide evidence of carcinogenicity or major organ toxicity. This may in part be attributable to the action of MT in irreversibly binding the small amount of bioactive Bi(III) passing into the circulation, or be a reflection on the lack of carcinogenic potential of the Bi(III) ion.[250]

No tangible evidence is seen to show that bismuth or bismuth compounds are a cause of cancer in humans.[269] Encephalopathy is the most frequent clinical effect of high-dose bismuth therapies used for gastrointestinal disturbances, but reports are seen also of bismuth-induced renal dysfunction, osteoarthropathy, stomatitis, colitis and hepatitis.[265,270–272] A rare case of transient patchy or macropapular exanthema, angioedema and erythema in a man occurred two days following use of bismuth subcitrate.[242] Bismuth is eliminated from the circulation *via* the kidney, and degenerative changes in the renal tubular epithelium are reported in some cases of encephalopathy.[255,273–275] No evidence of toxicity was identified in a clinical trial involving 43 patients treated with the bismuth protein complex Bicitropeptide, and where blood and urine bismuth levels ranged from 5 to 20 ng l^{-1} and 63 to 780 ng l^{-1} respectively.[257] Bismuth-related encephalopathy has been reported in many studies since the 1970s and affected patients variously displayed symptoms ranging from acute confusion to myoclonus, dysarthria, ataxia, dementia and even epileptic changes. An "epidemic outbreak" during 1973–1980 in France and Australia was attributed to excessive use of bismuth subnitrate or bismuth subgallate (10 g per day).[276,277]

Cases of encephalitis are reported following application of bismuth iodoform paraffin paste (BIPP), used for packing chronic wound cavities in the head and neck following surgery. Symptoms increased with dosage but regressed following administration of metal chelator therapies and/or discontinuation of bismuth treatment. Blood bismuth concentrations in encephalopathy cases ranged from 10 to 4600 $\mu g\,l^{-1}$ and bismuthaemia may persist for several days after stopping treatment.[277] Ericson was unable to detect bismuth collected from four patients dosed with PeptoBismol (4×56.7 g) over 24 hours.[278] Since 1980, case reports of bismuth encephalopathy have declined greatly.[257,279] In the 40 years of the use of bismuth medications for gastrointestinal disease, no evidence has been presented to show that bismuth, organic or inorganic, is a cause of cancer.

The suspected mechanism for encephalopathy is a bismuth-related impairment in oxidative phosphorylation in the brain, binding of thiol groups on key enzymes, a reduction in cerebral blood flow, and serum protein binding.[280,281] Bi(III) is known to bind serum transferrin, albumins and other proteins, and is known to interact with iron and other trace metals in key intracellular metabolism. Bismuth is taken up by erythrocytes and as such may impair the normal function of the blood in the uptake, utilisation and storage of iron.

Bismuth is not a cumulative poison but in a rare case a man injected with a single dose of bismuth (^{207}Bi) citrate retained 0.6% of the dose in his body for at least 924 days.[258] Organo-bismuth compounds developed in the 1970s as warfare agents are irritant and extremely toxic.[245]

Methylation of bismuth compounds is not well studied.[15,267] Available information shows that Bi(III) is biomethylated by the so-called methanogenic bacteria (*Clostridium*, *Lactobacillus* and *Eubacterium*) in the environment and through microorganisms in the intestine. Stable mono- or dimethyl derivatives of Bi(V) are not known but monomethyl bismuth (MMB^{III}), dimethylbismuth (DMB^{III}) and TMB^{III} have been characterised. MMB^{III} is absorbed more readily than bismuth citrate or bismuth conjugated to glutathione.

It is concluded, from the fragmentary and incomplete information available, that bismuth and inorganic or organic bismuth compounds are not carcinogenic in humans or experimental animals. Further studies are warranted into the detrimental influence of long-term inhalation of bismuth alloy dusts in industry. Whereas methylation products of arsenites and arsenates are now implicated as potential carcinogens on the basis of experimental studies, no such evidence exists that TMB^{III} or other methylated bismuth compounds are similarly active.

References

1. E. J. Underwood, *Trace Elements in Human and Animal Nutrition*, Academic Press, New York, 1977.
2. S. G. Gropper, J. L. Smith and J. L. Groff (ed.), *Advanced Nutrition and Human Medicine*, Wadsworth Pub. Co, 5th edn, 2008.
3. D. R. Mayer, W. Kosmus, H. Pogglitsch, D. Mayer and W. Beyer, *Biol. Trace Elem. Res.*, 1993, **37**, 27.

4. N. C. Norman, *Chemistry of Arsenic, Antimony and Bismuth*, Springer, Blackie Academic and Professional, *London*, 1998.
5. T. Sollemann, *A Manual of Pharmacology*, Saunders, Philadelphia, 1942.
6. K. B. Vu, M. D. Kaminski and L. Nuñez, *Argonne National Laboratory, Chemical Engineering Division,* Report ANL-CMT-03/2, Chicago, IL, 2003.
7. P. Biaocco, G. Colitti, S. Francheschini and A. Ilari, *J. Med. Chem.*, 2003, **52**, 2603.
8. E. R. T. Tiekink, *Crit. Rev. Oncol. Hematol.*, 2002, **42**, 217.
9. Y. Nan and H. Sun, *Prog. Chem.*, 2009, **21**, 856.
10. S. L. Soignet, P. Maslak, Z. G. Wang, S. Jhahwar, E. Calleja, L. J. Dardashti, D. Corso, A. DeBlasio, J. Gabrilove, D. A. Scheinerg, P. P. Pandolfi and R. P. Warrell, *New Engl. J. Med.*, 1998, **339**, 1341.
11. A. Ghavamzadeh, K. Alimoghaddam, S. H. Ghaffari, S. Rostami, M. Jhani, R. Hosseini, A. Mossavi, E. Baybordi, A. Khodabadeh, M. Iravani, B. Bahar, Y. Mortazavi, M. Totonchi and N. Aghdami, *Ann. Oncol.*, 2006, **17**, 131.
12. E. Longfelder, W.-K. Hofmann and D. Nowak, *Leukemia*, 2012, **26**, 433.
13. M. J. Chen, P. Y. Yang, Y. Z. Ye, D. N. Hu and M. F. Chen, *Am. J. Chin. Med.*, 2010, **38**, 1131.
14. H. M. Patterson, A. Pietras, M. M. Persson, J. Karlsson, M. C. Shoshan and S. Påhlman, *Mol. Cancer Tharp.*, 2009, **8**, 160.
15. R. Bentley and T. G. Chasteen, *Microbiol. Mol. Biol. Rev.*, 2002, **66**, 250.
16. R. O. Jenkins, Biological Chemistry of Arsenic, Antimony and Bismuth, 2010, ch. 7.
17. P. R. Dowdle, A. M. Laverman and R. S. Oremland, *Appl. Environ. Microbiol.*, 1996, **62**, 1664.
18. F. S. Islam, A. G. Gault, C. Boothman, D. A. Polya, J. M. Charnock, D. Chatterjee and J. Lloyd, *Nature*, 2004, **430**, 68.
19. F. Mari, A. Polettini, D. Lippi and E. Bertol, *Brit. Med. J.*, 2006, **333**, 1299.
20. R. J. Flanagan and K. D. Watson, *Med. Sci. Law*, 2009, **49**, 153.
21. B. Weider and J. H. Fournier, *Am. J. Forensic Med. Pathol.*, 2000, **20**, 378.
22. J. T. Hindmarsh and J. Savory, *Clin. Chem.*, 2008, **54**, 2092.
23. P. Bartrip, *Med. Hist.*, 1992, **36**, 53.
24. R. Eisler, *Rev. Environ. Contam. Toxicol.*, 2004, **180**, 133.
25. R. R. Fernandez, A. Collins and E. Marczak, Gold Recovery From High Arsenic Containing Ores at Newmont's Roasters, Society for Mining and Exploration Annual Meeting, Denver, 2007, pp. 07–037.
26. M. J. Wicks, V. E. Archer, O. Auerbach and M. Kuschner, *Am. J. Ind. Med.*, 1981, **2**, 25.
27. P. E. Enterline, R. Day and G. M. Marsh, *Occup. Environ. Health*, 1995, **52**, 28.
28. O. Axelson, E. Dahlgren, C.-D. Jnsson and S. O. Rehnlund, *Br. J. Ind. Med.*, 1978, **35**, 8.

29. T. Narukawa, A. Takatsu, K. Chiba, K. W. Riley and D. H. French, *J. Environm. Monitor.*, 2005, **7**, 1342.
30. K. Luo, X. Zhang, C. Chen and Y. Lu, *Chinese Sci. Bull.*, 2004, **49**, 2183.
31. C. Hogstedt, K. Andersson, B. Frenning and A. Gustavsson, *Scand. J. Work Environ. Health*, 1982, **8**, Suppl. 1, 72.
32. M. Ewall, *Energy Justice Network,* www.energyjustice.net/fibrowatch/toxics.html, 2007.
33. W. R. Cullen, *Special Report on Ingested Arsenic*, Royal Society of Chemistry, Cambridge, 2008, ch. 1, p. 1.
34. N. Sohel, L. A. Persson, M. Rahman, P. K. Streatfield, M. Yunus, E. C. Ekström and M. Vahter, *Epidemiology*, 2009, **20**, 824.
35. M. M. Rahman, R. Naidu and P. Bhattachayra, *Environ. Geochem. Health*, 2009, **31**, Suppl. 1, 9.
36. U. K. Chowdhury, B. K. Biswas, T. R. Chowdhury, G. Samanta, B. K. Mandal, G. C. Basu, C. R. Chanda, D. Lodh, K. C. Saha, S. K. Mukherjee, S. Roy, S. Kabir, Q. Quamruzzaman and D. Chakraborti, *Environ. Health Perspect.*, 2000, **108**, 1.
37. C. Steinmaus, Y. Yuan, D. Kalman, O. A. Rey, C. F. Skibola, D. Dauphine, A. Basu, K. E. Porter, A. Hubbard, M. N. Bates, M. T. Smith and A. H. Smith, *Toxicol. Appl. Pharmacol.*, 2010, **247**, 138.
38. E. Smith, R. Naidu and A. M. Alston, Arsenic in the soil environment, a review, in *Advances in Agronomy*, Academic Press, London, 1998, vol. 64, p. 149.
39. C. H. Tseng, J. Environ. Sci. Health C, Environ. Carcinogen. Ecotoxicol Rev., 2005, **23**, 55.
40. C.-H. Tseng, C.-K. Chong, C.-P. Tseng and J. A. Centeno, *Ambio*, 2007, **36**, 82.
41. S. L. Chen, S. J. Yeh, M. H. Yang and T. H. Lin, *Biol. Trace Elem. Res.*, 1995, **48**, 263.
42. L. A. Cox, *Risk Analysis*, 2009, **29**, 3.
43. C.-L. Chen, L. Hsu, H.-Y. Chiou, Y.-M. Hsueh, S.-Y. Chen, M.-M. Wu and C.-J. Chen, *J. Am. Med. Assoc.*, 2004, **292**, 2984.
44. L. E. Moore, A. H. Smith, C. Eng, S. DeVries, D. Kalman, V. Bhargava, K. Chew, O. A. Rey, C. Hopenhayn, M. L. Biggs, M. N. Bates and F. M. Waldman, *Carcinogenesis*, 2003, **24**, 1785.
45. S. S. Hecht, *J. Natl. Cancer Inst.*, 1999, **91**, 1194.
46. F. Bosch and L. Rosich, *Pharmacology*, 2008, **82**, 171.
47. J. Hutchinson, *Brit. Med. J.*, 1887, **2**, 1280.
48. R. Jackson and J. W. Grainge, *Can. Med. Ass. J.*, 1975, **113**, 396.
49. D. M. Joliffe, *J. Roy. Soc. Med.*, 1993, **86**, 287.
50. J. J. Lander, R. J. Stanley, H. W. Summer, D. C. Boswell and R. D. Aach, *Gastroenterol.*, 1975, **68**, 1582.
51. S. Waxman, Y.-K. Jing, Z. Chen and G.-Q. Chen, *J. Natl. Cancer Inst.*, 1999, **91**, 1691.
52. J. W. Sheehy and J. H. Jones, *Am. Ind. Hyg. Ass. J.*, 1993, **54**, 61.
53. M. Shenker, *Tuber. Lung Dis.*, 1992, **73**, 27.

54. S. G. Gilbert, *A Small Dose of Toxicology*, Healthy World Press, Seattle, WA, 2nd edn, 2012.
55. F. Nevens, J. Fevery, J. van Steenbergen, R. Sciuot, V. Desmet and J. De Groot, *J. Hepatol.*, 1990, **11**, 80.
56. J. Cuzick, S. Evans, M. Gillman and D. A. Price Evans, *Br. J. Cancer*, 1982, **45**, 904.
57. US Department of Health and Human Services, *12th Report on Carcinogens,* US DHHS, Research Triangle Park, NC, 2011.
58. Agency for Toxic Substance and Disease Registry (ATSDR), *Draft Toxicological Proposal for Arsenic,* ATSDR, US DHHS, Research Triangle Park, NC, 1998.
59. S. L. Soignet, S. R. Frankel, D. Douer, M. S. Tallman, H. Kantarjian, E. Calleja, R. M. Stone, M. Kalaycio, D. A. Scheinberg, P. Steinherz, E. L. Sievers, S. Coutré, S. Dahlberg, R. Ellison and R. P. Warrell, *J. Clin. Oncol.*, 2001, **15**, 3852.
60. F. I. Abdullaev, R. Rivera-Luna, A. Garcia-Carrancá, F. Ayala-Fierro and J. J. Espinosa-Aguirre, *Mutat. Res.*, 2001, **493**, 31.
61. J. Zhu, Z. Chen, V. Lallemand-Breitenbach and H. de Thé, *Nat. Rev. Center*, 2002, **2**, 705.
62. K. B. Kim, A. Y. Bedikian, L. H. Comacho, N. E. Papadopoulos and C. McCullough, *Cancer*, 2005, **194**, 1687.
63. C. Wang, B. Li, H. Zhang, G. Shi, W. Li and J. B. Jonas, *Ophthalmol. Res.*, 2007, **39**, 302.
64. T. E. Bael, B. L. Peterson and J. A. Gollob, *Melanoma Res.*, 2008, **18**, 147.
65. M. Goldman and J. C. Dacre, *Environ. Geochem. Health*, 1989, **13**, 179.
66. T. C. Marrs, R. L. Maynard and F. Sidell, *Chemical Welfare Agents; Toxicology and Treatment*, Wiley, Chichester, 2007.
67. J. A. Vilensky and K. Redman, *Ann. Emerg. Med.*, 2003, **41**, 378.
68. E. J. Bailey, E. L. Kennaway and M. E. Urquhart, *Brit. J. Cancer*, 1957, **11**, 49.
69. I. R. Rowland and M. J. Davies, *J. Appl. Toxicol.*, 1981, **1**, 278.
70. R. Mukhopadhyay, B. P. Rosen, L. T. Phung and S. Silver, *FEMS Microbiol. Rev.*, 2004, **26**, 311.
71. J.-S. Jean, J. P. Maity, S. Kar, C.-Y. Chen, J.-H. Liu and C. C. Liu, *Geographical Society of America, Annual Meeting Denver, Abstracts with Programs,* 2010, **42,** 216.
72. G. M. Bogden, A. Sampayo-Reyes and H. V Aposhian, *Toxicology*, 1994, **93**, 175.
73. M. E. Hughes, *Toxicol. Lett.*, 2002, **133**, 1.
74. M. Vahter, *Appl. Organomet. Chem.*, 2002, **8**, 175.
75. H. Bartsch and E. Hietanen, *Environ. Health Perspect.*, 1996, **104**, Suppl. 3, 569.
76. C. Steinmaus, K. Carrigan, D. Kalman, R. Atallah, Y. Yuan and A. H. Smith, *Environ. Health Perspect*, 2005, **113**, 1153.

77. M. Vahter and G. Concha, *Pharmacol. Toxicol.*, 2001, **89**, 1.
78. M. Vahter, A. Akesson, C. Lidén, S. Ceccatelli and M. Berglund, *Environ. Res.*, 2007, **104**, 85.
79. M. Vahter, *Sci. Progress.*, 1999, **82**, 69.
80. M. Vahter, *Toxicol. Lett.*, 2000, **112–113**, 209.
81. B. Normell, A. L. Lindberg, M. Rahman, M. Berglund, L. A. Persson, S. El Arifeen and M. Vahter, *Environ. Res.*, 2008, **106**, 212.
82. K. Schläwicke Engström, B. Nermell, G. Concha, U. Stromberg, M. Vahter and K. Broberg, *Mutat. Res.*, 2009, **667**, 4.
83. L. Li, E. C. Ekström, W. Goessler, B. Lönnerdahl, B. Nermell, M. Yunus, A. Rahman, S. El Arifeen, L. A. Persson and M. Vahter, *Environ. Health Perspect.*, 2008, **116**, 315.
84. C. R. Gross and O. A. Nelson, *Am. J. Public Health Nations Health*, 1934, **24**, 36.
85. J. C. Stavrides, *Free Radical Biol. Med.*, 2006, **41**, 1017.
86. Y. C. Chen, H. J. Su, Y. L. Guo, Y. M. Hsueh, T. J. Smith, L. M. Ryan, M. S. Lee and D. C. Christiani, *Cancer Causes Control*, 2003, **14**, 303.
87. M. Rahman, M. Vahter, N. Sohel, M. Yunus, M. A. Wahed, P. K. Streatfield, E. C. Engström and L. A. Persson, *Environ. Health Perspect.*, 2006, **114**, 1847.
88. M. Rahman, M. Vahter, M. A. Wahed, N. Sohel, M. Yunus, P. K. Streatfield, S. El Arifeen, A. Bhula, K. Zaman, A. M. Chowdhury, E. C. Ekström and L. A. Persson, *J. Epidemiol. Community Health*, 2006, **60**, 242.
89. A. L. Lindberg, N. Sohel, M. Rahman, L. A. Persson and M. Vahter, *Environ. Health Perspect.*, 2010, **118**, 533.
90. C. Hopenhayn-Rich, S. R. Browning, I. Hertz-Picciotto, C. Ferreccio, C. Peralta and H. Gibb, *Environ. Health Perspect.*, 2000, **108**, 667.
91. A.-L. Lindberg and M. Vahter, *Arsenic in Groundwater, A World Problem,* NNC-IAH Publication, No.5, Utrecht, Netherlands, 2008, p. 64.
92. M. Vahter, *Basic Clin. Pharmacol. Toxicol.*, 2008, **102**, 204.
93. H. Autrup, *Environ. Health Perspect.*, 1993, **101**, Suppl. 2, 33.
94. J. Liaw, G. Marshall, Y. Yuan, C. Ferreccio, C. Steinmaus and A. H. Smith, *Cancer Epidemiol. Biomarkers Prev.*, 2008, **17**, 1982.
95. M. Kippler, Y. Wagatsuma, A. Rahman, L. A. Persson, R. Raqib and M. Vahtere, *Reprod. Toxicol.*, 2012, **34**, 504.
96. K. Schläwicke-Engström, K. Broberg, G. Concha, B. Nermell, M. Warholm and M. Vahter, *Environ. Health Perspect.*, 2007, **115**, 599.
97. M. Wei, H. Wanibuchi, S. Iwai, K. Yoshda, G. Ndo, D. Nakae and S. Fukushima, *Carcinogenesis*, 2002, **23**, 1387.
98. E. J. Tokar, L. Benbrahim-Tallaa, J. M. Ward, R. Lunn, R. L. Sams and M. P. Waalkes, *Crit. Rev. Toxicol.*, 2010, **40**, 912.
99. J. Shen, H. Wanbuchi, M. P. Waalkes, E. I. Salim, A. Kinoshita, K. Yoshida, G. Endo and S. Fukushima, *Toxicol. Appl. Pharmacol.*, 2006, **210**, 117.

100. E. Dopp, A. D. Kligerman and R. A. Diaz-Bone, *Met. Ions Life Sci.*, 2010, **7**, 231.
101. G. Mazumder, *J. Indian Med. Ass.*, 2001, **311**, 314, 318.
102. J. Liu and M. P. Waalkes, *Toxicol. Sci.*, 2008, **105**, 24.
103. Y. M. Hsueh, G. S. Cheng, M. M. Wu, H. S. Yu, T. L. Kuo and C. J. Chen, *Br. J. Cancer*, 1995, **71**, 109.
104. H. H. Lin, L. Y. Han, H. L. Zhang, C. J. Zheng, B. Xie, Z. W. Cao and Y. Z. Chen, *BMC Bioinformatics*, 2006, **7**, Suppl. 5, S13.
105. M. Toyama, M. Yamashita, N. Hirayama and Y. Murooka, *J. Biochem.*, 2002, **132**, 217.
106. International Agency for Research on Cancer (IARC), *IARC Monographs on the Evaluation of Carcinogenic Risks to Humans, Monograph News,* IARC/WHO, Lyons, France, 2012.
107. H. Q. Dong, K. L. Wang and Y. J. Ma, *Zonghua Nei Ke Za Zhi*, 1993, **32**, 813.
108. X. He and Q. Ma, *J. Biol. Chem.*, 2009, **284**, 12609.
109. I. Falnoga, A. Z. Pevec, Z. Slejkovec, M. Tušek Znidarić, I. Zajac, S. M. Mlakar and J. Marc, *Biol. Trace Elem. Res.*, 2012, **149**, 331.
110. Environmental Protection Agency (EPA), *Arsenic in Drinking Water,* US EPA, water.epa.gov/lawregs/rules regs/sdwa/arsenic/index.cfm, 2010.
111. F. Pearce, *New Scientist,* 1993, 30th October, p. 5.
112. S. H. Lamm, A. Engel, M. Feinlib, C. A. Penn and R. Chen, *Environ. Health Perspect.*, 2007, **115**, A340.
113. B. Gontijo, *Brazilian Dermatol. Ann.*, 2005, **80**, 91.
114. Royal Geographical Society, Media Release, Arsenic in drinking water a global threat to health, RGS, London, 2007, 29th August.
115. J. Beard, *New Scientist,* 1998, 28th March, p. 10.
116. A. J. Bednar, J. R. Garbarino, M. R. Burkhardt, J. F. Ranville and T. R. Wildeman, *Water Res.*, 2004, **38**, 355.
117. E. H. Amonoo-Neizer and E. M. K. Amekor, *Environ. Health Perspect.*, 1993, **101**, 46.
118. S. H. Lamm, A. Wengel, C. A. Penn, R. Chen and M. Feinleib, *Environ. Health Perspect.*, 2006, **114**, 1077.
119. H. R. Guo, H. S. Yu, H. Hu and R. R. Monson, *Cancer Causes Control*, 2001, **12**, 909.
120. C. Hopenhayn-Rich, M. L. Biggs, A. Fuchs, R. Bergoglio, E. E. Tello, H. Nicolli and A. H. Smith, *Epidemiol.*, 1996, **7**, 117.
121. P. Kavanah, M. E. Farago, I. Thornton, P. Elliott, W. Goessler and K. J. Irgolic, *Occup. Environ. Med.*, 1997, **54**, 840.
122. P. Kavanah, M. E. Farago, I. Thornton, W. Goessler, D. Kuehnelt, C. Schlagenhaufen and K. J. Irgolic, *Analyst*, 1998, **123**, 27.
123. W. N. Rom, G. Varley, J. L. Lyon and S. Shopkow, *Br. J. Ind. Med.*, 1982, **39**, 269.
124. J. H. Lubin, L. E. Moore, J. F. Fraumeni and K. Cantor, *Environ. Health Perspect.*, 2008, **118**, 1661.

125. Y. W. Lowney, M. V. Ruby, R. C. Wester, R. A. Schoof, S. E. Holm, X.-Y. Hui, S. Barbadillo and H. I. Maibach, *Toxicol. Ind. Health*, 2005, **21**, 1.
126. R. C. Wester, X. Hui, S. Barbadillo, H. I. Maibach, Y. W. Lowney, R. A. Schoof, S. E. Holm and M. V. Ruby, *Toxicol Sci.*, 2004, **79**, 287.
127. Consumer Products Safety Committee, 1990, www.hickson.com/conprod.html.
128. E. Kwon, H. Zhang, Z. Wang, G. S. Jhangri, X. Lu, N. Fok, S. Gabos, X.-F. Li and X. C. Le, *Environ. Health Perspect.*, 2004, **112**, 1375.
129. IARC, IARC Monographs on the Evaluation of Carcinogenic Risks to Humans, IARC/WHO, Lyons, France, 2004, vol. 84.
130. A. H. Smith, A. Ercumen, Y. Yuan and C. M. Steinmaus, *J. Exposure Sci. Environ. Epidemiol.*, 2009, **19**, 343.
131. A. N. Currie, *Br. Med. Bull.*, 1946, **4**, 402.
132. G. Pershagen, *Environ. Health Perspect.*, 1981, **40**, 93.
133. M. J. Mass, *Environ Geochem. Health*, 1992, **14**, 49.
134. E. J. Tokar, W. Qu and M. P. Waalkes, *Toxicol. Sci.*, 2011, **120**, S192.
135. J. Huff, P. Chan and A. Nyska, *Toxicol. Sci.*, 2000, **66**, 17.
136. J. H. Graham and E. B. Helwig, *Arch. Dermatol.*, 1961, **83**, 738.
137. L. E. Beane-Freeman, L. K. Dennis, C. F. Lynch, P. S. Thome and C. L. Just, *Am. J. Epidemiol.*, 2004, **160**, 679.
138. P. C. P. Ingram, *Brit. J. Venereol. Dis.*, 1934, **10**, 274.
139. L. Geyer, *Arch. Dermatol. Syphilol.*, 1898, **43**, 221.
140. H. Yamouchi, K. Takahashi, M. Mashiko and Y. Yamamura, *Am. Ind. Hyg. Assoc. J.*, 1989, **50**, 606.
141. M. I. Luster, J. L. Wilmer, D. R. Germolec, J. Spalding, T. Yoshida, K. Gaido, P. P. Sineonova, F. G. Burleson and A. Bruccoleri, *Toxicol. Lett.*, 1995, **82-83**, 471.
142. S. I. Chun, K. C. Whang and W. P. Su, *J. Cutan. Pathol.*, 1988, **15**, 234.
143. K. Murata, T. Iwazawa, T. Takayama, K. Yamashita and K. Okagawa, *Surg. Today*, 1994, **24**, 1115.
144. B. J. Wang, Y. Y. Lee, C. P. Mak, H. F. Kao, M. L. Hsu and J. R. Hsien, *J. Formosa Med. Assoc.*, 1991, **90**, 1093.
145. S. H. Tan, S. N. Tham and C. L. Goh, *Int. J. Dermatol.*, 1995, **34**, 770.
146. D. J. Gawkrodger, *Occupational Med*, 2004, **54**, 458.
147. J. H. Graham, G. M. Mazzanti and E. B. Helwig, *J. Invest. Dermatol.*, 1961, **37**, 317.
148. I. B. Sneddon, *Br. J. Dermatol.*, 1963, **2**, 405.
149. S. Elmariah, R. Anolik, R. F. Walters, K. Rosenman, M. K. Pomeranz and M. R. Sanchez, *Dermatol. Online J.*, 2008, **14**, 242008.
150. British Association of Dermatologists (BAD), *Bowen's Disease, Information Sheet*, BAD, London, 2010.
151. R. M. MacKie, in *Textbook of Dermatology*, ed. R. H. Champion, J. L. Burton, and F. J. G. Ebling, Blackwell Scientific, Oxford, 1993, ch. 32, p. 1459.
152. S. B. Ball and R. P. Dawber, *Australas. J. Dermatol.*, 1998, **39**, 63.
153. H. S. Yu, W. T. Liao and C. Y. Chai, *J. Biomed Sci.*, 2006, **13**, 657.

154. J. P. Callan and J. Headingon, *Arch. Dermatol.*, 1980, **116**, 422.
155. R. Phillip, A. O. Hughes, C. Roberson and T. F. Mitchell, *Bristol Med. Chir. J.*, 1983, **98**, 165.
156. D. Tsurata, T. Hamada, K. Mochida, K. Nakagawa, H. Kobayashi and M. Ishi, *Brit. J. Dermatol.*, 1998, **139**, 291.
157. S. R. Mitra, D. N. G. Mazumder, A. Basu, G. Block, R. Haque, S. Samanta, N. Ghosh, M. M. H. Smith, O. S. von Ehrenstein and A. H. Smith, *Environ. Health Perspect.*, 2004, **112**, 1104.
158. F. J. Burns, A. N. Uddin, F. Wu, A. Nádas and T. G. Rosssman, *Environ. Health Perspect.*, 2004, **112**, 599.
159. Y. Sun, C. Kojima, C. Chignell, R. Mason and M. P. Waalkes, *Toxicol. Appl. Pharmacol.*, 2011, **255**, 242.
160. R. K. Wolff, G. Obminski and M. T. Newhouse, *Br. J. Ind. Med.*, 1984, **41**, 499.
161. P. Kelleher, K. Pacheco and L. S. Newman, *Environ. Health Perspect.*, 2000, **108**, 685.
162. A. I. G. McLaughlin, *Industrial Diseases of the Iron and Steel Industry*, Her Majesty's Stationery Office, London, 1950.
163. P. S. Burge, *Eur. Respir. J.*, 1994, **7**, 1032.
164. H. R. Guo, N. S. Wang, H. Hu and R. R. Monson, *Cancer Epidemiol. Biomarkers Prevent.*, 2004, **13**, 638.
165. M. G. Mostafa, J. C. McDonald and N. M. Cherry, *Occup. Environ. Med.*, 2008, **65**, 765.
166. J. M. Peters, D. Thomas, H. Falk, G. Oberdörster and T. J. Smith, 1986, **70**, 71.
167. W. Chen, J. L. Martindale, N. J. Holbrook and Y. Liu, *Mol. Cell. Biol.*, 1998, **18**, 5178.
168. L. Benbrahim-Tallaa and M. P. Waalkes, *Environ. Health Perspect.*, 2008, **116**, 158.
169. C. Hopenhayn-Rich, M. L. Biggs and A. H. Smith, *Int. J. Epidemiol.*, 1998, **27**, 561.
170. A. M. Lee and J. F. Fraumani, *J. Natl. Cancer Inst.*, 1969, **42**, 1045.
171. A. Tanimoto, T. Hamada, H. Kanisaki, K. Matsuno and O. Koide, *J. Univers. Occup. Environ. Health*, 1990, **12**, 89.
172. M. Salgardo, M. Sans, X. Forns, M. Bruguera, A. Castalls, M. Navasa and J. Rodés, *Gastroenterol. Hepatol.*, 1995, **18**, 132.
173. C. V. Rudge, H. B. Röllin, C. M. Nogueira, Y. Thomassen, M. C. Rudge and J. Ø. Odland, *J. Environ. Monit.*, 2009, **11**, 1322.
174. J. D. Hamadani, S. M. Grantham-McGregor, F. Tofail, B. Nermell, B. Fängström, S. N. Huda, S. Yesmin, M. Rahman, M. Vera-Hernández, S. E. Arifeen and M. Vahter, *Int. J. Epidemiol.*, 2010, **39**, 1206.
175. S. L. Myers, D. T. Lobdell, Z. Liu, Y. Xia, H. Ren, Y. Li, R. K. Kwok, J. L. Mumford and P. Mendola, *J. Epidemiol. Community Health*, 2010, **64**, 325.
176. M. R. Karagas, *Lancet*, 2010, **376**, 213.
177. A. Leitch and E. L. Kennaway, *Brit. Med. J.*, 1922, **2**, 1107.

178. P. L. Goering, H. V. Aposhian, M. J. Mass, M. Cebrián, B. D. Beck and M. P. Waalkes, *Toxicol. Sci.*, 1999, **49**, 5.
179. H. Wanibuchi, E. I. Salim, A. Kinoshita, J. Shen, M. Wei, K. Morimura, K. Yoshida, K. Kuroda, G. Endo and S. Fukushima, *Toxicol. Appl. Pharmacol.*, 2004, **198**, 366.
180. S. Mann, P. O. Droz and M. Vahter, *Toxicol. Appl. Pharmacol.*, 1996, **137**, 8.
181. M. P. Waalkes, J. Liu and B. A. Diwan, *Toxicol. Appl. Pharmacol.*, 2007, **222**, 271.
182. J. P. Wang, L. Qi, M. R. Moore and J. C. Ng, *Toxicol. Lett.*, 2002, **133**, 17.
183. K. T. Kitchen, *Toxicol. Appl. Pharmacol.*, 2001, **172**, 249.
184. S. Ivankovic, G. Eisenbrand and R. Preussmann, *Int. J. Cancer*, 1979, **24**, 786.
185. H. Wanibuchi, E. Salim, J. Kinoshita, J. Shen, M. Wei, K. Morimura, K. Yoshida, K. Kuroda, G. Endo and S. Fukushima, *Toxicol. Appl. Pharmacol.*, 2004, **196**, 366.
186. M. Wei, H. Wanibuchi, S. Yamamoto, W. Li and S. Fukushima, *Carcinogenesis*, 1999, **20**, 1873.
187. J. Shen, H. Wanibuchi, M. P. Waalkes, E. I Salim, K. Yoshida, G. Edo and S. Fukushima, *Toxicol. Appl. Pharmacol.*, 2006, **210**, 171.
188. G. Pershagen and N. E. Bjorklund, *Cancer Lett.*, 1985, **27**, 99.
189. A. Yamamoto, A. Hisanaga and N. Ishinishi, *Int. J. Cancer*, 1987, **40**, 220.
190. National Toxicology Program, *NTP Tech. Rep. Ser.*, 2000, **492**, 1.
191. J. M. DeSesso, *Teratology*, 2001, **63**, 170.
192. M. P. Waalkes, J. Wad and B. A. Diwan, *Carcinogenesis*, 2004, **25**, 133.
193. S.-X. Liu, M. M. Davidson, X. Tang, W. F. Walker, M. Ather, V. Ivanov and T. K. Hei, *Cancer Res.*, 2005, **65**, 3236.
194. E. J. Tokar, B. A. Diwan, J. M. Ward, D. A. Delker and M. P. Waalkes, *Toxicol. Sci.*, 2011, **119**, 73.
195. T. K. Hei, S. X. Liu and C. Waldren, *Proc. Natl. Acad. Sci.*, 1998, **95**, 8103.
196. T.-C. Lee, S. Wang-Wuu, R. Y. Huang, K. C. C. Lee and K. Y. Jan, *Cancer Res.*, 1986, **46**, 1854.
197. C. Soriano, A. Creus and R. Marcos, *Mutat. Res.*, 2007, **634**, 40.
198. A. D. Kligerman, C. L. Doerr, A. H. Tennant, K. Harrington-Brock, J. W. Allen, E. Winkfield, P. Poorman-Allen, B. Kundu, K. Funasaka, B. C. Roop, M. J. Mass and D. M. DeMarini, *Environ. Mol. Mutagen*, 2003, **42**, 192.
199. E. Dopp, L. M. Hartmann, A. M. Florea, U. von Recklinghausen, R. Pieper, B. Shokouhi, A. V. Hirner and G. Obe, *Toxicol. Appl. Pharmacol.*, 2004, **201**, 156.
200. E. Dopp, U. von Recklinhausen, L. M. Hartmann, I. Stueckradt, I. Pollok, S. Rabieh, L. Hao, A. Nussler, C. Katier, A. V. Hirner and A. W. Rettenmeier, *Drug Metab. Disposition*, 2008, **36**, 971.

201. J. E. Klaunig, L. M. Kanendulis and B. A. Hocevar, *Toxicolog. Pathol.*, 2010, **38**, 96.
202. R. I. McCallum, *J. Roy. Coll. Phys. Lond.*, 1989, **23**, 28.
203. R. I. McCallum, *J. Environ. Monit.*, 2005, **7**, 1245.
204. M. Fillela, N. Belziel and Y.-W. Chen, *Earth Sci. Rev.*, 2002, **57**, 125.
205. I. B. Rafel, I. Popov and R. M. Zakususilova, *Vopr. Pitan.*, 1985, **5**, 65.
206. M. M. Molokhia and H. Smith, *Bull. WHO*, 1969, **40**, 123.
207. A. Davis, *J. Toxicol. Envir. Health*, 1975, **1**, 191.
208. A. Henderson and D. Joliffe, *Br. J. Clin. Pharmacol.*, 1985, **19**, 73.
209. N. C. Hepburn, I. Siddique, G. F. Howie and P. C. H. Beckett, *Lancet*, 1993, **242**, 238.
210. J. P. R. Veiga, E. R. Wolff, R. N. Sampio and P. D. Marsden., *Lancet*, 1983, **ii**, 569.
211. K. A. Winship, *Adverse Drug React. Acute Poisoning Rev*, 1987, **6**, 67.
212. F. A. De Wolff, *Brit. Med. J.*, 1995, **310**, 1216.
213. C. J. Stevenson, *Trans. St. John's Hosp. Dermat. Soc.*, 1965, **51**, 40.
214. R. I. McCallum, *Proc. Roy. Soc. Med.*, 1977, **70**, 756.
215. B. A. Richardson, *J. Forensic Sci. Soc.*, 1994, **34**, 199.
216. S. Chantler, *Pediatric Update*, 1996, Feb, 115.
217. C. White, *Br. Med. J.*, 1998, **316**, 1553.
218. P. J. Fleming, M. Cooke, S. M. Chantler and J. Golding, *Br. Med J.*, 1994, **309**, 1594.
219. P. N. Gates, J. B. Pridham and J. A. Webber, *Br. Med. J.*, 1995, **345**, 386.
220. C. G. Elinder and L. Friberg, in *Handbook on the Toxicology of Metals*, ed. L. Friberg, G. F. Nordberg and V. Voulk, Elsevier, Amsterdam, 2nd edn, 1986, vol. 11, pp. 26–42.
221. L. Gerhardsson, D. Brune, G. F. Nordberg and P. O. Wester, *Scand. J. Work Environ. Health*, 1982, **8**, 201.
222. R. Bailly, R. Lauwerys, J. P. Buchet, P. Mahieu and J. Konings, *Br. J. Indust. Med.*, 1991, **48**, 93.
223. R. Lüdersdorf, A. Fuchs, P. Mayer, G. Skulsuksai and G. Schäcke, *Int. Arch. Occup. Environ. Health*, 1987, **59**, 469.
224. R. Bentley and T. G. Chasteen, *Microbiol. Mol. Biol. Rev.*, 2002, **66**, 250.
225. Y. Ogra, *Analyt. Sci*, 2009, **25**, 1189.
226. K. L. Stemmer, *Pharmacol. Ther.*, 1976, **1**, 157.
227. V. Pokonjak and M. Pavlovich, *Int. Arch. Occup. Environ. Health*, 1983, **51**, 19; *Ibid.* 1983, **51**, 299.
228. A. H. Smith, *Arch. Environ. Health*, 1988, **43**, 124.
229. IARC, *Monographs on the Evaluation of Carcinogenic Risks to Humans*, IARC, Lyon,1989, vol. 47, p. 129.
230. D. H. Groth, L. E. Stettler, J. R. Burg, W. M. Busey, G. C. Grant and L. Wong, *J. Toxicol. Environ. Health*, 1986, **18**, 607.
231. F. W. Sunderman and K. S. McCully, *Canc. Invest.*, 1983, **1**, 469.
232. European Safety and Health at Work, *Antimony and Antimony Compounds: Evaluation of the carcinogenicity and genotoxicity,* Health Council of the Netherlands, The Hague, 2011.

233. American Conference of Governmental Industrial Hygienists (ACGIH), *Documentation of threshold limit values and biological exposure indices*, ACGIH, Cinncinnati, OH, 5th edn, 1986.
234. N. Gurmani, A. Sharma and G. Talukder, *Biometals*, 1992, **5**, 47.
235. N. Gurmani, A. Sharma and G. Talukder, *Cytobios*, 1992, **70**, 131.
236. S. El Nahas, S. A. Temtamy and H. A. de Hondt, *Environ. Mutagen.*, 1982, **4**, 83.
237. B. M. Elliott, P. MacKay, P. Clay and J. Ashby, *Mutat. Res. Genet. Toxicol. Environ. Mutag.*, 1998, **415**, 109.
238. G. Grosskopf, T. Schwerdtle, L. H. F. Mullenders and A. Hartwig, *Chem. Res. Toxicol.*, 2010, **23**, 1175.
239. Carcinogenic Potency Project, *Antimony Potassium Tartrate*, *Database*, *CAS 28300-74-5,* 2007.
240. P. H. Craig, *Phys. Rev.*, 1926, **27**, 772.
241. P. A. Riley, *Pouchers Perfumes, Cosmetics and Soaps*, ed. H. Butler, Kluwer Acad. Publ., Dordrecht, Netherlands, 2000, 10th edn, pp. 161–165.
242. J. P. Ottervanger and B. H. Stricker, *Ned. Tijdschr. Geneeskd.*, 1994, **138**, 152.
243. D. E. Mahoney, S. Lim-Morrison and H. Maguire, *Antimicrobial. Ag. Action*, 1999, **43**, 582.
244. B. J. Marshall, J. A. Armstrong, G. J. Francis, N. T. Nokes and S. H. Wee, *Digestion*, 1987, **37**, Suppl. 2, 16.
245. T. Klapötke, *Biol. Metals*, 1988, **1**, 69.
246. G. Qvigstad and H. L. Waldum, *Tidsskr. Nor. Laegeforen.*, 1995, **115**, 1353.
247. P. R. Cohen, *J. Drugs Dermatol.*, 2009, **8**, 1132.
248. L. Zala, T. Hunziker and L. R. Braathen, *Dermatology*, 1993, **187**, 288.
249. E. R. Tiekink, *Crit. Rev. Oncol. Hematol.*, 2002, **42**, 217.
250. M. Satoh, Y. Kondo, M. Mita, J. Nakagawa, A. Naganuma and N. Imura, *Cancer Res.*, 1993, **53**, 4767.
251. V. K. Tiwari, S. K. Shukla, V. S. Chauhan, R. Kant, S. Rani and I. C. Tiwari, *Int. J. Pharm. Anal.*, 2010, **2**, 9.
252. P. Köpf-Maier, *Inorganica Chem. Acta*, 1988, **152**, 49.
253. M. Filella, *J. Environ. Med.*, 2010, **12**, 90.
254. H. Vanhoe, J. Versieck, L. Vanballenberge and R. Dams, *Clin. Chem. Acta*, 1993, **219**, 79.
255. W. K. Summers, *J. Alzheimer's Dis.*, 1998, **1**, 57.
256. J. Boertz, L. M. Hartmann, M. Sulkowski, J. Hippler, F. Mosel, R. A. Diaz-Bone, K. Michalke, A. W. Rettenmeier and A. V. Hirner, *Drug Metab. Depos.*, 2009, **17**, 352.
257. W. J. Serfontein, R. Mekel, S. Bank, G. Barbezat and B. Novis, *Res. Commun. Chem. Pathol. Pharmacol.*, 1979, **26**, 383; *Ibid.* 1979, **26**, 391.
258. D. Newton, R. J. Talbot and N. D. Priest, *Hum. Exp. Toxicol.*, 2001, **20**, 601.
259. D. Chalell, F. Lefevre, P. Allain and G. J. Martin, *J. Inorg. Biochem.*, 1981, **15**, 213.

260. F. Codony, P. Domenico and J. Mas, *J. Appl. Microbiol.*, 2003, **95**, 288.
261. M. Saleeemi, M. S. Toprak, S. Li, M. Johnsson and M. Muhammed, *J. Mater. Chem.*, 2012, **22**, 725.
262. Occupational Safety and Health Administration (OSHA), *Bismuth Telluride (Undoped)*, OHSA, CAS: 1304-82-2, US OHSA, Final Rule on Air Contaminants Project, 1988, ref. 54FR2324.
263. American Conference for Governmental Industrial Hygienists (ACGIH), *Bismuth telluride (undoped)*, ACGIH, Cincinnati, OH, 1986, ex. 1–3, 59.
264. H. Sun, H. Li, I. Harvey and P. Sadler, *J. Biol. Chem.*, 1999, **274**, 29094.
265. A. Slikkerveer and F. A. de Wolff, *Bismuth: Biokinetics, toxicity and experimental therapy of overdosage*, ed. A. Slikkerveer, Leiden University, Netherlands, 1992.
266. K. Asakura, H. Satoh, M. Chiba, K. Serizawa, M. Nakano and K. Omae, *J. Occup. Health*, 2009, **51**, 498.
267. U. von Recklinghausen, L. M. Hartmann, S. Rabieh, J. Hippler, A. V. Hirner, A. W. Rettenmeier and E. Dopp, *Chem. Res. Toxicol.*, 2008, **21**, 1219.
268. R. Preussmann and S. Ivankovic, *Food Cosmet. Toxicol.*, 1975, **13**, 543.
269. B. A. Fowler, E. F. Madden and S. Chou, *Patty's Toxicology*, ed. E. Bingham and B. Cohrrsen, John Wiley, Chichester, 2012, pp. 1, 475.
270. A. Slikkerveer and F. A. de Wolff, *Med. Toxicol. Adverse Drug Exp.*, 1989, **4**, 303.
271. G. Monseu, M. Struelens and M. Roland, *Acta Neurol. Belg.*, 1976, **76**, 301.
272. M. Teepker, H. M. Hamer, S. Knake, O. Bandmann, W. H. Oertel and F. Rosenow, *Epileptic Disord.*, 2002, **4**, 229.
273. R. J. Playford, C. H. Matthews, M. J. Campbell, H. T. Delves, K. K. Hla, H. J. Hodgson and J. Calam, *Gut*, 1990, **31**, 359.
274. J. A. Szymańska, J. Chmielnicka, A. Kaluzński and W. Pierz, *Biomed. Environ. Sci*, 1993, **6**, 134.
275. M. Sarikaya, A. Sevinc, R. Ulu, F. Ates and F. Ari, *Nephron*, 2002, **90**, 501.
276. J. D. Lowe, *Med. J. Austr.*, 1974, **2**, 664.
277. G. Martin-Bouyer, C. Barin, A. Beuget, J. Cordier and H. Guerbois, *Gastroenterol. Clin. Biol.*, 1978, **2**, 349.
278. C. D. Ericsson, H. L. Dupont and L. K. Pickering, *J. Am. Med. Assoc.*, 1980, **244**, 1435.
279. A. Bruinink, P. Reiser, M. Müller, B. H. Gähwiler and G. Zbinden, *Toxicol. in Vitro*, 1992, **6**, 285.
280. H. Sun, H. Li, A. B. Mason, R. C. Woodworth and P. J. Sadler, *J. Biol. Chem.*, 2001, **23**, 276.
281. H. Sun and K. Y. Szeto, *J. Inorg. Biochem.*, 2003, **94**, 114.

Part 3
Metals and Metalloid Elements as Carcinogens

CHAPTER 18

Discussion and Conclusions

18.1 Introduction

The human environment contains a vast array of metals and metallioid elements and related compounds. In the course of a lifetime, people are exposed to most of these elements in their work, recreation and home life or through medical therapies. Radioactive isotopes such as strontium-90, radon and breakdown products of uranium occur at low levels in the atmosphere, particularly in the vicinity of mining sites. As industrial development advances, so workers are exposed to different environments and products. At the same time, the World Health Organization and other regulatory authorities are providing realistic guidelines for worker safety and drinking water quality based on epidemiological experience and laboratory research gained over many years. The chemical environment in which we live is continually changing and previously unrecognised human health problems are emerging. Several, including the use of silicone implants in mammoplasty, antimony in child cot mattresses and aluminium in cosmetics as a cause of breast cancer, have been resolved but questions still remain as to the safety of metal-on-metal prostheses and the risks associated with manufacture of semiconductors.

Metal and metalloid elements have specific actions in the human body. Over many millennia, the human body has evolved to depend upon at least 15 metal or metalloid elements as electrolytes, components or cofactors of essential enzymes, structural proteins or physiological systems.[1–5] Elements of macro- and micronutrient value interact with each other and with xenobiotic metals but many questions remain regarding regulatory mechanisms of uptake, tissue accumulation and excretion, ionic interaction and protective mechanisms. Most elements absorbed into the human body are biologically inactive in their elemental form and require extracellular or intracellular oxidation, methylation or other forms of transformation to convert them to biologically active forms. Micro-nutrients such as chromium, nickel, cobalt, strontium, manganese and

Issues in Toxicology No. 18
The Carcinogenicity of Metals: Human Risk through Occupational and Environmental Exposure
By Alan B. G. Lansdown

Published by the Royal Society of Chemistry, www.rsc.org

vanadium are potentially toxic and carcinogenic if present to excess, or where blood levels are sufficiently high to overwhelm inherent protective mechanisms afforded by metal-binding proteins including metallothioneins, albumins and macroglobulin, or other intrinsic factors. The threshold between the maximum tolerated dose and minimal carcinogenic level in blood is not clearly defined for these elements.

The position of *cis*-platin and other platinum compounds (oxaloplatin, carboplatin) as human carcinogens has been equivocal for many years. They are efficacious intravenous anticancer therapies with specific recommendations for testicular, ovarian, lung, head and neck and bladder cancers, but they are shown to be mutagenic *in vitro* and induce tumours in two species of rodents at different sites following intraperitoneal injection.[24] *Cis*-platin at least has also been shown to induce preneoplastic change in mice and rats following intraperitoneal dosing of females late in pregnancy. In the last situation preneoplastic or initiated carcinogenic changes are promoted postnatally by sodium barbital or other non-carcinogenic agents. Experimental studies show also how metallothioneins protect the body or otherwise mitigate carcinogenic change.

The World Health Organization, International Agency for Research on Cancer (IARC), Environmental Protection Agency in the USA, the Council for Europe and others have reviewed health risks encountered by human populations exposed to all metals in work and recreation and have made far-reaching recommendations. However, environmental control over contamination and mining of arsenic in southeast Asia and South America, and lead contamination in urban communities, present ongoing difficulties. Lead residues derived from lead piping persist in many inner city areas; car exhaust emissions remain in pasture lands adjacent to major roads and will contaminate plants and food animals for many years after removal of lead from petrol. Arsenic, lead, mercury and cadmium are seen in human food chains, although they can be eliminated in modern food processing and preparation. Research in the UK and USA is now being pioneered to remove arsenic from drinking water, but this is a mammoth task bearing in mind the millions of people in Bangladesh, Taiwan, southeast China and south America who are exposed.[5] Radon emissions from uranium decay in rock formations deep in the ground represent a third continuing issue in environmental carcinogenicity and control, but new building techniques are being developed to limit infiltration of radon gas into homes built in high radon areas.[6]

The present review has discussed safety issues relating to a large number of metal and metalloid elements for which some evidence of carcinogenicity is provided by human studies and laboratory evaluations using animal models and *in vitro* systems. Opinions, predictions and regulatory guidelines are based upon the adequacy of information available at the time. It is clear that data derived from many studies conducted 50 years or more ago is less reliable than expected these days. Even in 1989, Ian Purchase pointed out that the research environment of the 1950s was substantially different from that expected today, at earlier times, protocols required for animal experiments were not as rigidly defined by regulator edict as presently.[7] In the 1980s mutagenicity studies were seen more as a series of competing theories, whereas nowadays they form an integral part of carcinogencity evaluations.[8–10] The scientific community faces major changes in toxicological practice and evaluators and regulators experience difficulties of relativism, over-regulation and indifferent science.[11] Protocols for epidemiological and experimental appraisal currently expect stringent ethical appraisal, elimination of bias and humanitarian considerations. In animal studies, the Animals (Scientific Procedures) Act in the UK and the US Centers for Disease Control and Prevention (CDC) emphasise that experimenters have a legal and moral responsibility to ensure that animals are treated in accordance with highest standards of animal welfare and care.[12,13]

18.2 Diagnostic Methods

18.2.1 Experimental Animal Models

Regulatory-style studies in experimental animals form an integral part of the identification and classification of carcinogenic agents.[10] Most metals and metalloid elements and related compounds are capable of inducing one or more forms of cancer in laboratory animals by oral/in-diet administration, inhalation, dermal application, intraperitoneal, intramuscular or subcutaneous injection. To achieve a dose-related effect, the doses used are frequently greatly in excess of expected human daily and lifetime exposures, thereby exceeding physiological or inherent protective mechanisms. Administration of test materials is by routes not always in keeping with human exposure expected from clinical practice. This is evident in evaluation of titanium dioxide where intratracheal instillation of ultrafine dispersions led to a massive pulmonary overload, inflammatory change and saturation of clearance systems progressing to carcinoma.[14–16] In the case of anticancer platinum drugs, clinical guidelines specify intravenous administration, whereas intraperitoneal doses have been given to rodents in order to induce tumours.

Animal tests, notably lifetime oral or in-diet dosing studies in rodents, have been conducted for many years as a principal means of identifying carcinogenic materials. A major criticism of published animal tests concerns the narrow range of strains used and their susceptibility to spontaneous lesions. Thus, the National Toxicology Program (NTP) of the USA has used inbred Fischer F344 rats and hybrid B6C3F1 mice as choice species for regulatory-style experiments, with outbred Sprague–Dawley and Wistar rats, and ICR and Swiss mice being selected on occasions for carcinogenicity studies.[17–19] In 2010, the NTP opted to retain the B6C3F1 mouse as a primary test strain in view of earlier experience but to extend the number of rat strains to include the hybrid FBNF1 (F344 crossed with Brown Norway), or Wistar Han rats, but so far no studies using these additional strains and relating to metals or metalloid elements (especially cobalt–tungsten hard metal complexes) have been published.[20,21] A further criticism concerns the statistical power of assays conducted in inbred strains of rodent with high spontaneous tumour incidence, and their predictive value in detecting human carcinogens. As a geneticist, Dr Michael Festing was of the opinion that use of a multi-strain assay could improve the NTP carcinogenesis bioassay.[22]

Animal models for human carcinogens show considerable inconsistency according to route of dosage and compounds administered. This is well illustrated with reference to lead, which in human experience is a cause of increased incidence of lung, stomach, and bladder cancer mortality following occupational exposure.[23,24] Metallic lead powder was a weak carcinogen when administered intragastrically to F344 rats, although when injected intramuscularly it induced a low incidence of leukaemia and injection site sarcomas.[25] Lead chromate was identified as a potent carcinogen on the basis that it induced a 64% incidence of injection site sarcomas in rats following intramuscular injection, and a small number of renal carcinomas. The compound was toxic to NIH Swiss mice at doses tolerated by rats, but failed to induce a significantly significant increase in tumour incidence at lower concentrations.[25] Elsewhere, rats fed basic lead acetate, lead subacetate or lead phosphate developed a 94% incidence of renal tumours.[26–28] Coogan *et al.* treated over 1000 Sprague–Dawley rats with lead subacetate by the oral, intraperitoneal or subcutaneous route and also reported a high incidence of renal cortical carcinomas.[29] Poirier *et al.* showed that lead subacetate injected intraperitoneally induced lung adenomas in strain A mice, but that simultaneous administration of equimolar calcium or magnesium acetate inhibited tumorigenic activity.[30] Human lung tumours attributable to chronic lead exposure were mostly carcinomas.

The relevance and diagnosic value of the repeated subcutaneous injection model is seriously questioned. Iron–dextran complex, for example, is categorised as an "anticipated human carcinogen" on the basis of local sarcomas induced in rats, mice, hamsters and rabbits,[24,31–33] yet there is minimal evidence that this therapy is carcinogenic to the many thousands of patients treated annually for iron-deficiency anaemia. The significance of spindle cell carcinomas, fibrosarcomas, rhabdomyosarcomas, and osteosarcomas, as have been observed in rats, mice and hamsters following repeated subcutaneous or intramuscular injection of compounds of iron, lead, cadmium, mercury, bismuth or other elements, is questioned in the light of experiments with water-soluble food additives with no known carcinogenic properties in humans.[34–36] Grasso and Golberg showed clearly that local tissue reactions to injected materials predisposing to neoplastic change were largely determined by physicochemical factors, concentration, surface activity, lipophilia, *etc.*, rather than genotoxic damage and mutagenicity.[35] Early experiments demonstrating that repeated injections of hot or cold distilled water, hypertonic glucose, fructose or saline were a cause of sarcoma in rats and mice must invalidate the use of subcutaneous sarcoma assays in predicting the carcinogenicity of metals, metalloids or any other materials.[37–39]

Transgenically derived mouse and rat strains are increasingly used in experimental research but there is limited evidence so far as to their value in predicting human carcinogens.[40] In their predictive carcinogenicity study,

Table 18.1 Animal models in experimental carcinogenesis.

Species	*Longevity (years)*	*Spontaneous tumour incidence*
Rat	1.5–3	High incidence of tumours of endocrine tissues; tumours in other tissues according to strain
Mouse	1–2	Tumours in all tissues according to sex and inbred/outbred strains; lymphocytic and thymic leukaemias
Hamster (Chinese and Syrian)	1–2	Low incidence
Rabbit (New Zealand White)	5–8	Very low, occasional tumours of uterus, intestine and liver
Dog	>5 years	Ostosarcomas, mammary carcinoma, oral melanoma, oral squamous cell carcinoma, nasal tumours, lung carcinoma, soft tissue sarcomas, and malignant non-Hodgkin's lymphoma
Cat	>5 years	Similar to dog but some strains are subject to leukaemias (viral)

Pritchard *et al.* focused on three extensively used strains of transgenic mice, Trp53 +/−, Tg/AC, and RasH2, in comparing the carcinogenicity of 99 chemically dissimilar substances in the traditional two-year rodent bioassay.[41] The individual transgenic models made the "correct" determinations (positive for carcinogens; negative for noncarcinogens) for 74–81% of the chemicals, with an increase to as much as 83% using combined strategies (*e.g.* Trp53 +/− for genotoxic chemicals and RasH2 for all chemicals). In contrast, traditional two-year rodent studies yielded a 69% correct determination. Greater harmonisation of experimental protocols, conduct of experiments in approved laboratories, and appropriately qualified operators improve the quality of the studies but will rarely, if ever, provide a surrogate for human investigations.[19] Extrapolation of the results of animal tests in terms of human hazard depends much on experience of the observers, knowledge of age-related patterns of spontaneous tumours in test species, chemical and behavioural similarity of the test chemical to known human carcinogens, and levels of statistical significance of increased tumour incidence (Table 18.1).[42,43]

18.2.2 *In Vitro* Tests for Mutagenicity and Clastogenicity

The majority of metal and metalloid compounds have been tested for mutagenicity in the Ames reverse transcription assay developed in the 1970s,[44–46] and in the battery of *in vitro* screens documented by the IARC.[47] Purchase evaluated mutagenicity tests in mammalian cell lines and in bacterial assays and concluded that the Ames test, together with cell transformation tests, was the most accurate means of predicting carcinogenicity with greater than 90% accuracy.[7] Ashby and Tennant confirmed a 90% correlation between mutagenicity in the Ames test and rodent carcinogenicity data for a total of 523 synthetic organic compounds.[48,49] Of these, they identified two groups of carcinogens within their database, those acting by a genotoxic pathway and those categorised as non-genotoxic. However, Strauss pointed out that in his view, point mutations were central to carcinogenic mechanisms, as supposed by older research but not supported by more recent genetic analyses, and that "the class of non-mutagenic carcinogens is considerable".[50] It is notable that, in these assays, very few of the compounds evaluated were metals or metalloid compounds, most being synthetic organic compounds such as methylnitrosourea.

Since the introduction of *in vitro* mutagenicity tests the International Programme on Chemical Safety (IPCS) Harmonized Scheme for Mutagenicity Testing has issued strategies for testing and compilation of numerous databases.[51] The World Health Organization decided to update this harmonisation in a workshop in San Fransisco in 2005. They redefined a "mutation" resulting from chemical action as "a permanent change in the structure and/or amount of the genetic material of an organism that can lead to heritable changes in its function, and it includes gene mutations as well as structural and numerical chromosome alterations." Genotoxicity is taken more widely to include changes in the cell which can give rise to mutations. They defined genotoxicity as "the capability of substances to damage DNA and/or cellular components regulating the fidelity of the genome—e.g. spindle apparatus, topo-isomerases, DNA repair systems, DNA polymerases and all adverse effects on genetic information". The revised approach for the identification of mutagenic and genotoxic agents is as applicable to metal and metalloid elements and related compounds as it is to the wide range of unrelated compounds in the human domain. The comprehensive strategy incorporating bacterial assays and mammalian cells cultures is designed to identify the ability of substances to evoke point mutations, clastogenicity (structural chromosome aberrations) and aneuploidy (numerical chromosome aberrations).[51]

The IPCS and WHO harmonised system envisages a second phase of *in vivo* tests in *wild-type* and transgenic animal models to examine chromosome aberrations, gene mutation, covalent DNA adducts, oxidative lesions, DNA strand breaks, alkali-labile lesions, and alterations in the cell cycle in proliferating tissues.

18.3 Mechanisms of Carcinogenesis

Toxic and carcinogenic risks of metallic and metalloid elements are specific and relate closely to biochemical properties, valency and concentration of bioactive ions, and the duration and mode of exposure.[52] In each case, bioactivity, mutagenicity and neoplastic change are determined by the bioavailabily of the free ion and its capacity to circumvent or otherwise overcome inherent protective systems within the body and target cells. Mechanisms of carcinogenic action by metals and metalloid elements are complex and can be expected from published research to involve direct action on the nucleus, DNA/RNA synthesis and gene expression, or to act by an epigenetic pathway, or a combination of both. Chromium [Cr(VI)] and cadmium are examples of agents acting directly on DNA that evoke chromosomal damage leading to mutagenicity, whereas arsenic, nickel and beryllium act mainly through non-genotoxic mechanisms.[21] Epigenetic pathways are defined as "heritable changes in gene function that occur without any direct effect on DNA sequence". Such changes compromise mitotic and meiotic aspects of gene expression and may involve covalent modifications in amino acid residues in histones, changes in methylation status of dinucleotides and gene silencing. Induction of oxidative stress through production of reactive oxygen species (ROS) is contributory to the pre-neoplastic process and can influence the multi-step carcinogenic process proposed by Berenblum and Shubik in 1947.[53,54] ROS-induced damage is held to signal intracellular transduction pathways and lead to activation of transcription factors and altered gene expression. Nickel [Ni(II)] exerts its carcinogenic action principally through non-genotxic epigenetic mechanisms and is dependent upon the intracellular concentration and its capacity to evoke damage to cell organelles.[55]

Arsenic [As(III) compounds], nickel, chromium [Cr(VI], beryllium, lead and probably cadmium compounds are examples of complete carcinogens that induce irreversible mutagenic changes in target cells with or without metabolic transformation, followed by transformation/promotion of stem cells to tumour formation.[54–56] Close correlations exist between the genotoxic capacity of these elements and their potency as inducers in carcinogenic change. Keodrith and Seo argued that "interference in metal homeostasis may result in oxidative stress which represents an imbalance between production of free radicals and the system's ability to readily detoxify reactive intermediates".[57] This leads to DNA damage, deregulation in DNA repair, lipid peroxidation, protein modification and changes in cell signal transduction pathways influencing cell growth and development. Physico-chemical properties govern uptake, intracellular distribution and binding of metal compounds and interactions with proteins (zinc finger structures *etc.*). In the opinion of some, these appear to be more relevant for metal carcinogenicity than DNA binding alone.[58] Costa and his colleagues studied mechanisms of carcinogencity of nickel and its compounds specifically and demonstrated that water-soluble compounds are highly carcinogenic relative to water-insoluble compounds.[59–61] Water insoluble Ni(II) compounds are readily absorbed phagocytically and intracellular dissolution yields highly bioactive Ni^{2+} which binds strongly to proteins and the oxygen moiety of the DNA–phosphate backbone. Carcinogenic Ni(II) compounds potentiate DNA methylation, possibly by substitution of trace metals such as magnesium, or evoke gene-silencing as part of their carcinogenic mechanism. Nickel binds DNA bases directly by co-ordinate covalent bonding.

The carcinogencity of elements with nutritional value in the human body creates some difficulty in diagnosis and classification. The 15 elements discussed above show defined uptake and metabolic pathways yet show mutagenic, genotoxic and clastogenic effects in target cells, exhibit competitive receptor-binding to intra- and extracellular metal binding proteins with possible effects on cell membrane function and critical ion-ion balances.[62–64]

The carcinogencity of minor trace nutrient ions such as chromium, nickel and possibly cobalt is difficult to evaluate and questions arise such as:

- What is the maximum tolerable concentration for the element as a nutrient?
- What is the minimal level liable to induce cancer in any tissue in the most vulnerable individuals?

Recommended daily dietary intakes are documented but, at the moment, no threshold blood or tissue concentrations have been identified for their carcinogenicity in any tissue. Xenobiotic elements including lead, cadmium and mercury exert their toxic and pre-carcinogenic influences through interaction with nutrients such as zinc, iron, calcium, *etc*., displacing them from metal-binding proteins, key enzymes or other essential pathways.

The carcinogenic risk of several metals including lead, chromium and nickel rests heavily on lung cancers seen following occupational inhalation of industrial dusts and fumes. Alveolar macrophages or dust cells represent a first-line protective mechanism against these materials, but epidemiological studies have shown that carcinogenic responses arise through a complex sequence of massive inflammatory changes, foreign body reactions and alveolar cell damage associated with release of cytokines, growth factors, *etc.* which modulate gene expression, immunological factors and tissue responsiveness.[65,66] In these situations, the protective function of macrophages is saturated and foreign body reactions, inflammatory cytokines and release of reactive oxygen radicals constitute epigenetic changes predisposing to carcinogenesis.[21,67–69] Few epidemiological studies reveal the true nature of the initiating process or the contribution played by lifestyle elements to the carcinogenic changes seen. Excess cancers of the respiratory tract result from prolonged exposure to atmospheres contaminated with dispersions of silica, arsenic, chromium, nickel, hydrocarbons and toxic gases (radon, hydrogen sulfide, *etc*.).[70,71] Diagnosis of the true aetiology of lung cancers is complicated by latent periods of 20 years or more between initial exposure to potential carcinogens and detection of cancers.

The essential nutrients calcium, copper, magnesium, manganese and zinc are not carcinogenic in humans but experimental studies in rats implicate them as initiators or promoters in long-term dietary studies or through repeated subcutaneous injection.[24,63,64,72–74] At best, they function as tumour promoters through the induction of key enzymes involved in the cell cycle. The present publication discusses experimental studies showing their capacity to advance the growth of tumours induced by non-metallic carcinogens including 9,12-dimethyl benz[1,2]anthracene.[75–79]

18.4 Radioactivity

Radioisotopes of certain metal and non-metal elements have an important role in diagnostic and palliative medicine, but all forms of ionising radiation are potentially carcinogenic.[24] They include X-radiation, neutron emissions and α-, β- and γ-rays emitted by certain radioactive elements. All tissues are potentially at risk of radiation-induced genetic damage, mutations, chromosomal aberrations and DNA strand breaks. Inhalation of radon gas (^{222}Rn, half-life 3.824 days) is a known cause of human lung tumours with cytogenetic change and specific chromosomal damage closely allied to radiation dose.[80] The present review has discussed the carcinogenicity relating to radioactive gold isotopes and the decay products of thorium and Thorium-X as used therapeutically. Fuller discussion on the carcinogenicity of radioactive isotopes in humans and experimental animals is not within the remit of the present book, but it is noteworthy that radioisotopes of several non-carcinogenic elements are a prominent cause of bone tumours.[81] External, local or whole-body radiation with α-, β- or γ-emitters has been shown to cause osteogenic or osteoblastic sarcomas in rats. Animals exposed orally, intraparenterally or by intraosseous injection with phosphorus (^{32}P), calcium (^{45}Ca), strontium (^{89,90}Sr) and barium (^{149}Ba) developed osteogenic tumours.[82,83]

Radionuclides of phosphorus, calcium and strontium are metabolised systemically and induce carcinogenic changes throughout the skeletal system, particulary osteogenic areas of the metaphysis and diaphysis of growing immature bone and in sites of endochondrial ossification.[80] Spongy bone and metabolically active regions of the endosteum, periosteum and Haversian canal systems were shown to be particularly vulnerable to β-emissions. This is illustrated by a study in which Wistar rats were exposed to strontium (^{90}S), where osteogenic sarcoma developed within 8–10 months following intramuscular injection.[84] This and other studies illustrate that carcinogenesis related to radioactive emissions is directly related to the type of emission, route or exposure and tissue penetration, duration of exposure and the age and susceptibility of the tissues. At high doses, β-emitting isotopes are potentially fatal in laboratory animals before tumours develop. Luz and Gössner reported that the majority of osteosarcomas induced in mice by α-particle radiation were of vertebral origin but,[88] unusually, in the case of the β-emitter ^{90}Sr, the tumours induced with high doses occurred in spinal regions whereas low-dose radiation led to increased carcinogenesis in diaphyseal regions.[85–87] On occasions, ^{90}Sr was shown to induce chondro-osteoblastic tumours

18.5 Interactions between Elements in Carcinogenesis

Massive growth in manufacturing industries in major industrialised societies in the 21st century has led to increased exposure to non-ferrous elements including chromium, arsenic, nickel, cobalt, vanadium, titanium, tungsten and hard metal alloys. In an industrial environment with an atmosphere containing a variety of toxic and potentially carcinogenic material and where increases in lung, skin and other cancers are recorded, it can be extremely difficult to determine:

a. Which element or compound is the principal carcinogen
b. To what extent other contaminants in the environment influence the carcinogenic process, or act as co-carcinogens

c. To what extent is a low-grade carcinogenicity of one element promoted or otherwise modified by others.

This problem is illustrated in the study of arsenic exposure and mortality in a Swedish copper smelter, where workers were exposed to inorganic arsenic, sulfur dioxide, copper and other unspecified agents. Of the patients supposedly dying from arsenic-related lung cancer, 83% were held to be smokers.[87] Copper dust and sulfur dioxide are pulmonary irritants and copper-related fume fever is recognised.[70,71] Tobacco smoke and its contaminants are associated with and can exacerbate lung cancers, possibly through a direct mutagenic mechanism or through an epigenetic route.[89–91]

18.6 Future Considerations

Case studies and epidemiological studies will continue to be the mainstay providing information relating to the carcinogenicity of metal and metalloid elements (Table 18.2). Epidemiological studies supported by laboratory research have identified which metal and metalloid elements should be classified as carcinogens, those which may be carcinogenic but require further

Table 18.2 Metal and metaloid elements as carcinogens.

Carcinogenic agent	*IARC classification*	*Human exposures*	*Commercial use/ exposures*	*Cancer sites (human)*
Arsenic and inorganic arsenicals	1	Food, drinking water, minining contamination, tobacco	Pesticides, preservatives, metal alloys Anticancer therapies	Lung, skin, prostate, liver, vascular, bladder, brain, *etc.* Transplacental
Beryllium	1	Industrial dusts and fumes, food and drink	Metal alloys, ceramics, electronic equipment	Lung, bone
Cadmium and related compounds	1	Food and drink, inhalation, tobacco	Metal industries, paint pigments, batteries, fungicides	Lung, prostate, kidney, bladder
Chromium [Cr(VI)] compounds	1	Air, water soil, food, inhalation dermal contact	Paint pigments, metal alloys, welding	Lung, leukaemia, sinus, bone
Cis-platin	2A	Drug industry	Anticancer drug	Lung, skin Transplacental[b]
Cobalt (sulfate)	2B	Food, drinking water	Electroplating, metal industries, alloys, animal feeds, paint pigments, mining	Lung

Table 18.2 (*Continued*)

Carcinogenic agent	*IARC classification*	*Human exposures*	*Commercial use/ exposures*	*Cancer sites (human)*
Iron–dextran	2B	Iron supplement	None	Intramuscular injection sites[b]
Haematite mining	3	Mining operations	Mining operations	Lung[a]
Lead and related compounds	2A	Food, drink, water pipes, *etc.*, lead fumes, tobacco, petrol	Metal industries, batteries, pigments, mining, plumbing, welding, printing	Lung, stomach, bladder, kidney, brain
Nickel and related compounds	1, 2B	Air, food, drink, dermal contact, Ni-metal implants	Metal alloys, mining smelting, electro-plating, welding casting, prostheses, jewellery	Lung, nasal sinus
Selenium (sulfide)	3	Antidandruff shampoos	Hair treatment production and usage, packaging	Lung, liver[b]
Crystalline silica	1	Environmental and industrial dusts and fumes	Construction and building, abrasives, fillers and plastics	Lung
Thorium (dioxide)	1	Injection of Thorotrast radio-opaque media	Thorotrast use and production	Liver, leukaemia, bone

[a]Lung cancers attributed to radon emissions.
[b]Experimental studies.

epidemiological evidence for confirmation, and those elements that are a cause of ill-health but, on the basis of available information, cannot be implicated as carcinogens.[24,33,92–95] Human responses to environmental carcinogens vary greatly according to the age, sex and health state of exposed people; children and occasionally unborn infants are at special risk. Transplacental carcinogenesis is a relatively new area of study in carcinogenicity testing and epidemiological study but has been demonstrated with arsenic and nickel. The conduct and interpretation of epidemiological studies have come under increasing criticism and scrutiny in recent years.[9,11,66] To be acceptable they should be unbiased and scientifically and ethically sound. Professor Sir Colin Berry maintained that a more vigorous assertion of the particular values of the scientific viewpoint and a willingness to incorporate new knowledge into regulatory decision making is necessary in gaining optimal appraisal and predictions of health risk.[11] This is particularly true of the numerous occupational health studies where excesses of lung and respiratory tract cancers have been observed through long-term inhalation of substances in factory dust.

References

1. A. T. Miles, G. M. Hawksworth, J. H. Beattie and V. Rodilla, *Crit. Rev. Biochem. Mol. Biol.*, 2000, **35**, 35.
2. I. Bremner, *Prog. Food Nutr. Sci.*, 1987, **11**, 1.
3. W. Mertz, *Proc. Nutr. Soc.*, 1974, **33**, 307.
4. N. S. Scrimshaw and V. R. Young, *Sci. Am.*, 1976, **235**, 51.
5. Building Research Establishment, *Radon: Guidance on protective measures for new dwellings*, CRC Ltd, London, 1999, BR 211, ISBN 1 86081 328 3.
6. J. E. Underwood, *Trace Elements in Human and Animal Nutrition*. Academic Press, Inc., London, 1977.
7. I. F. H. Purchase, *J. Roy. Soc. Med.*, 1989, **82**, 254.
8. W. C. Hueper, *Cancer Res.*, 1961, **21**, 842.
9. H.-O. Adami, C. L. Berry, C. B. Breckenridge, L. L. Smith, J. A. Swanberg, D. Trichopoulos, N. S. Weiss and T. P. Pastoor, *Toxicol. Sci.*, 2011, **122**, 223.
10. R. D. Storer, *Toxicol. Lett.*, 2000, **112-113**, 557.
11. C. R. Berry, *Toxicology*, 2010, **267**, 7.
12. The Animals (Scientific Procedures) Act (Commencement) Order 1986, HMSO, London, 1986.
13. Centers for Disease Control and Prevention (CDC), *Animal Care and Use in CDC Laboratory Research*, CDC, USA Government, Atlanta, GA, 2013.
14. K. P. Lee, D. P. Kelly, P. W. Schneider and H. J. Trochomowicz, *Toxicol. Appl. Pharmacol.*, 1986, **83**, 30.
15. H. J. Trochimowicz, K. P. Lee and C. F. Reinhardt, *J. Appl. Toxicol.*, 1988, **8**, 383.
16. D. B. Warheit, T. R. Webb, K. L. Reed, S. Frerichs and C. M. Sayes, *Toxicology*, 2007, **230**, 90.
17. J. K. Haseman, J. R. Hailey and R. W. Morris, *Toxicol. Pathol.*, 1998, **26**, 428.
18. A. P. King-Herbert, R. C. Sills and J. R. Bucher, *Toxicologic Pathol.*, 2010, **38**, 180.
19. A. B. G. Lansdown and D. M. Conning, *Experimental Toxiclogy*, Royal Society of Chemistry, Cambridge, UK, 1988.
20. National Toxicology Program (NTP), US Department of Health and Human Services, *Report on Carcinogens: Listing status for cobalt-tungsten carbide powders and hard metals in the 12th RoC and the scientific justification for the recommendation, Fed. Reg.*, **74**, 27, 2009.
21. E. J. Tokar, B. A. Diwan, J. M. Ward, D. A. Delker and M. P. Waalkes, *J. Toxicol. Sci.*, 2011, **119**, 73.
22. M. F. Festing, *Environ. Health Perspect.*, 1995, **103**, 44.
23. International Agency for Research on Cancer (IARC), *Monographs on the Evaluation of Carcinogenic Risks to Humans, Inorganic and Organic Lead Compounds*, IARC, Lyon, 2007, France, vol. 87.

24. US Department of Heath and Human Services (US DHHS), *12th Report on Carcinogens*, US DHHS, National Toxicology Program, Research Triangle Park, NC, 2011.
25. A. Furst, M. Sclauder and D. P. Sassmore, *Cancer Res.*, 1976, **36**, 1779.
26. H. V. Zollinger, *Virchows Arch. Path. Anat. Physiol.*, 1953, **323**, 694.
27. E. Boyland, C. E. Dukes, P. L. Grover and B. C. Mitchley, *Br. J. Cancer*, 1962, **16**, 283.
28. G. J. Van Esch, H. van Genderen and H. H. Vink, *Br. J. Cancer*, 1962, **16**, 289.
29. P. S. Coogan, L. Stein, G. Hau and G. Hass, *Lab. Invest.*, 1972, **26**, 473.
30. L. A. Poirier, J. C. Theiss, L. J. Arnold and M. B. Shimkin, *Cancer Res.*, 1984, **44**, 1520.
31. H. G. Richmond, *Br. Med. J.*, 1959, **1**, 947.
32. F. J. C. Roe and R. L. Carter, *Int. J. Cancer*, 1967, **2**, 370.
33. R. L. Carter, *Br. J. Cancer*, 1969, **23**, 559.
34. S. D. Gangolli, P. Grasso and L. Golberg, *Food Cosmet. Toxicol.*, 1967, **5**, 601.
35. P. Grasso and L. Golberg, *Food Cosmet. Toxicol.*, 1966, **4**, 297.
36. P. Grasso, S. D. Gangolli, L. Golberg and J. Hooson, *Food Cosmet. Toxicol.*, 1971, **9**, 463.
37. K. Warabioka, *Gann*, 1959, **49**, 147.
38. Y. Nishiyama, *Gann*, 1938, **32**, 85.
39. W. C. Hueper, *Cancer Res.*, 1965, **25**, 440.
40. D. Gulezian, D. Jacobson-Kram, C. B. McCullough, H. Olson, L. Recio, D Robinson, R. Storer, R. Tennant, J. M. Ward and D. A. Neumann, *Toxicol. Pathol.*, 2000, **28**, 482.
41. J. B. Pritchard, J. E. French, B. J. Davis and J. K. Haseman, *Environ Health Perspect.*, 2003, **111**, 444.
42. E. Faber, *Cancer Res.*, 1964, **44**, 4217.
43. P. Bannasch and W. Gösner, *Pathology of Neoplasia and Preneoplasia in Rodents, EULEP Color Atlas*, Schattauer, Stuttgart, 1994.
44. B. N. Ames, W. E. Durston, E. Yamasaki and F. D. Lee, *Proc. Natl. Acad. Sci. USA.*, 1973, **70**, 2281.
45. H. Greim, W. Göggelmann, K. H. Summer and T. Wolff, *Arch. Toxicol.*, 1980, **46**, 31.
46. K. Mortelmans and E. Zeiger, *Mutat. Res.*, 2000, **455**, 29.
47. International Agency for Research on Cancer (IARC), *IARC Monographs on the Evaluation of Carcinogenic Risk of Chemical to Humans, Supplement 2*, IARC, Lyons, France, 1980, vol. 85.
48. J. Ashby and R. W. Tennant, *Mutat. Res.*, 1991, **257**, 229.
49. J. Ashby, R. W. Tennant, E. Zeiger and S. Stasiewicz, *Mutat. Res.*, 1989, **223**, 73.
50. B. S. Strauss, *Cancer Res.*, 1992, **52**, 249.
51. D. A. Eastmond, A. Hartwig, D. Anderson, W. A. Anwar, M. C. Cimino, I. Dobrev, G. R. Douglas, T. Nohmi, D. H. Phillips and C. Vicker, *Mutagenesis*, 2009, **24**, 341.

52. K. T. Kitchen, J. L. Brown and A. P. Kulkarni, *Mutat. Res.*, 1993, **290**, 155.
53. J. E. Klaunig, L. M. Kamendulis and B. A. Hocevar, *Toxicol. Pathol.*, 2010, **38**, 96.
54. I. Berenblum, *Carcinogens: Identification and mechanisms of action*, A. C. Griffin and C. R. Shaw, Raven Press, New York, 1979, p. 25.
55. K. Salnikow and A. Zhitkowicz, *Chehm. Res. Toxicol.*, 2007, **21**, 28.
56. H. Sun, *Biological Chemistry of Arsenic, Antimony and Bismuth*, J. Wiley, New York, 2010.
57. P. Keodrith and Y. R. Seo, *Int. J. Mol. Sci.*, 2011, **12**, 9576.
58. D. Beyersmann and A. Hartwig, *Arch. Toxicol.*, 2008, **82**, 493.
59. M. Costa, *Toxicology of Metal*, CRC Press, Boca Raton, FL, 1996, 245.
60. M. Costa and C. B. Klein, *Environ. Health Perspect.*, 1996, **107**, A438.
61. M. Costa, *Environ. Health Perspect.*, 1999, **107**, A438.
62. A. B. G. Lansdown, B. Sampson and A. Rowe, *J. Anat.*, 1999, **195**, 375.
63. A. B. G. Lansdown, *Wound Rep. Regen.*, 2002, **10**, 271.
64. A. B. G. Lansdown, U. Miratschijski, N. Stubbs, E. Scanlon and M. S. Ågren, *Wound Rep.Regen.*, 2007, **15**, 2.
65. B. L. Stauffer and C. A. De Souza, *J. Appl.Physiol.*, 2010, **109**, 230.
66. D. L. Weed, *J. Med. Philosoph.*, 2004, **29**, 313.
67. K. Grønbaek, C. Other and P. A. Jones, *APMIS*, 2007, **115**, 1039.
68. E. J. Tokar, W. Qu and M. P. Waalkes, *Toxicol. Sci.*, 2011, **120**(Suppl.), S192.
69. E. J. Tokar, L. Benbrahim-Tallaa and M. P. Waalkes, *Met. Ions Life Sci.*, A.Sigel, H. Sigel and R. K. O. Sigel, *Royal Society of Chemistry*, 2011, **8**, 375.
70. P. Kelleher, K. Pacheco and L. S. Newman, *Environ. Health Perspect.*, 2000, **108**(Suppl.), 685.
71. R. K. Wolff, G. Ominski and M. T. Newhouse, *Br. J. Ind. Med.*, 1984, **41**, 499.
72. International Agency for Research on Cancer (IARC), *Monographs on the Evaluation of Carcinogenic Risks to Humans*, WHO, Geneva, 2012, vol. 100C.
73. J. Kaltenberg, L. M. Plum, J. L. Ober-Blobaum, A Hörnscheid, L Rink and H. Haase, *Eur. J. Immunol.*, 2010, **40**, 1496.
74. L. Tomatis, *Environ. Health Perspect.*, 1993, **101**(Suppl. 5), 149.
75. A. Léonard, G. B. Gerber and F. Léonard, *Mutat. Res.*, 1986, **168**, 343.
76. M. Walters and F. J. C. Roe, *Fd. Cosmet. Toxicol.*, 1965, **3**, 271.
77. P. Collery, L. J. Anghileri, P. Coudoux and J. Durlach, *Magnes. Bull.*, 1981, **1**, 11.
78. K. S. Kasprzak and M. P. Waalkes, *Adv. Exp. Med .Biol.*, 1986, **206**, 497.
79. B. Bobrowska-Korczak, D. Skrajnowska and A. J. Tokarz, *J. Biomed. Sci.*, 2012, **16**, 19.
80. L. Dano, M.-N Guilly and M. Muleris, *Genes, Chrom. Cancer*, 2000, **29**, 1.
81. A. B. G. Lansdown, in *General and Applied Toxicology*, B. Ballantyne, T. C. Marrs and T. Syversen, Wiley, Chichester, UK, 3rd edn, 2009, vol. 3, p. 1491.

82. N. N. Litvinov, Radiation Injuries to the Skeletal System, *Medgiz, Moscow*, 1964, 163–303.
83. N. N. Litvinov and J. N. Soloviev, in *Tumours of the Rat*, V. Turasov and U. Mohr, *IARC Scientific Publications*, Lyon, 1990, vol. 1, pp. 659–676.
84. B. Graf, J. Lafuma, C. Parmentier and N. Parmentier, *Bull. Cancer*, 1970, **57**, 381.
85. A. Nilsson, *Strontium-induced Bone and Bone Marrow Changes*, Almquist and Wikell, Uppsala, 1962.
86. A. Nilsson, *Acta Radiol:Therap., Phys., Biol.*, 1970, **1**, 107.
87. A. Luz and W. Gossner, Neoplastic bone lesions in the mouse, in *Pathology of Neoplasia and Pre-neoplasia in Rodents*, ed. P.Bannasch and W. Gossner, Schattauer, Stuttgart, 1994.
88. O. Axelson, E. Dahlgren, C.-D. Jansson and S. O. Rehnlund, *Br. J.Ind. Med.*, 1978, **35**, 8.
89. S. S. Hecht, *J. Natl. Cancer Inst.*, 1999, **91**, 1194.
90. J. C. Stavrides, *Free Radical Biol. Med.*, 2006, **41**, 1017.
91. G. P. Pfeifer, M. Denissenko, M. Olivier, N. Tretakova, S. S Hecht and P. Hainaut, *Oncogene*, 200, **21**, 7435.
92. IARC Working Group on the Evaluation of Carcinogenic Risks to Humans, *IARC Monograph on the Evaluation of Carcinogenic Risks to Humans*, IARC, Lyons, France, 1987, Suppl. 0250-9555, p. 7.
93. WHO/IARC, *Monographs on the Evaluation of Carcinogenic Risks to Humans*, WHO/IARC, Lyons, 1999, vol. 74.
94. European Union, Scientific Panel on Food Additives, Flavourings, Processing Aids and Materials in Contact with Food, *EFSA J.*, 2006, **391**a,b,c,d, 1.
95. World Health Organization, in *Guideline in Drinking Water Quality*, WHO, Geneva, 2nd edn, 1998, Addendum to Volume 2.

Acknowledgements

I should like to acknowledge, with grateful thanks, the help and advice provided by the following, who have also provided photographs and illustrations to support the text and to provide further interest:

Mr David Carlisle and staff of the Yorkshire Dales Mining Museum

Dr Superb Misra, Department of Mineralogy, Natural History Museum, London

Professor José Rodríguez, Fundación Universitaria Iberoamericana, Mexico

Dr Michael P. Waalkes, National Toxicology Program Laboratory, Division of the National Toxicology Program, Inorganic Carcinogenesis Program, USA

The National Trust of England and Wales who are dedicated to preserving memories of Britain's rich heritage of metal production

Dr Holly Hedgeland, Cambridge

Issues in Toxicology No. 18
The Carcinogenicity of Metals: Human Risk through Occupational and Environmental Exposure
By Alan B. G. Lansdown

Published by the Royal Society of Chemistry, www.rsc.org

Subject Index

References to figures are given in *italic* type. References to tables are given in **bold** type.